The
Structure
of
Humanity

AKH Taylor

Paperrata

Paperrata

ISBN: 978-0-987-49451-1

I thank Pam and Colin for their assistance, particularly proofreading the text.

The
Structure
of
Humanity

Arranged marriage systems *explain* human evolution

AKH Taylor

Paperrata

Contents

Chapter 1

Introduction

This is an argument that humans evolved using a complex form of marriage – the bilateral cross-cousin marriage system. The cross-cousin marriage system is a form of cooperation and I argue that it played the major role in the evolution of many human traits. The cross-cousin marriage system has been well studied in anthropology, where it has long been recognised that humans exchange reproduction for cooperation. Rather than *Homo sapiens*, the 'wise man', we should instead define humans as 'the apes that exchange cooperation for reproduction'.

Currently, almost all our understanding of human behaviour and morphology reflects a Western perspective of the world. This is both unsurprising and highly misleading. For instance, most people are aware that arranged marriages are common in Africa, the Middle-East, and Asia. They may also be aware that several centuries ago in western Europe arranged marriages were common, and perhaps universal. Despite this, arranged marriages are almost entirely ignored in our understanding of human traits.

Arranged marriages are a form of both reproduction and cooperation. Until the existence of central governments, people organised their cooperative groups through marriage. They did this by exchanging daughters to marry their sons. By doing so, they could increase the number of cooperative males. The girls are betrothed at a young age, often soon after they are born, and the future husband's family is expected to cooperate with the family betrothing the girl. This is a contract, rather like a mortgage, and binds the two families for more than a decade.

The implication of this should be immediately obvious to anyone who understands evolutionary biology: reproduction equals cooperation. Reproduction equals long-term cooperation. The parties need to be able to see into the future to be able to form the contract, and, after many years of cooperation, the designated male in the cooperative family receives the girl.

By using such a system, without trustworthy central governments, people were able to create large groups of cooperative people. This enabled cooperation in a single community or cooperation between communities.

With our Western perspectives we wrongly believe that human cooperation is dependent upon the market economy. In reality, human cooperation is based on arranged marriage systems, as has been studied in anthropology for more than a century. Only recently has the market economy become so successful that arranged marriage systems disintegrated to leave the mirage of the market economy as an imperative. This transition appears to have occurred in western Europe between 1500 and 1850 (or perhaps even later, for the royal families). It is occurring in India and China as this book was written. The person with the Western perspective who looks for the all-encompassing market economy in the Middle Ages in Europe will not find one; instead, what will be found is a web of arranged marriages with a market economy in its infancy.

It is generally assumed that people who lived in the 'stone-age' enjoyed lives not very different to our own. A man and a woman met: he chose her because she was attractive, and she chose him because he was a good provider. No doubt they might have met at the 'stone-age' equivalent of the discotheque or bar, perhaps over a fruit-based alcoholic drink. Evolutionary psychologists, primatologists, paleoanthropologists, archaeologists, and biologists assumed that all people have the same pattern of life. Yet nobody seems to be able to explain even the most basic question in human evolution. A classic example is: why do females hide their ovulation from their long-term partners, whom they chose. There are at least five different answers to this question, yet none contain even the smallest amount of logic, since we wouldn't expect a female who chose her long-term partner to hide her ovulation from him. Female gorillas certainly don't.

I came up with the ideas in this book after reading Napoleon Chagnon's *Yanomamö: The End of Eden*, a popular version of his remarkable monograph on the Yąnomamö. Chagnon's research concerned an Amazonian people, on the border between Venezuela and Brazil, numbering in the region of 20,000 individuals, who were largely unaffected by Western or other cultures when Chagnon studied them.[1] Remarkably, Chagnon had performed statistical research on people that had never previously seen a non-Yąnomamö.

The people Chagnon described were very different to the assumptions made about human beings in the 'stone-age'. The people engaged in arranged marriages; shelter, food and protection from potential predators were relatively simple concerns; warfare was the most serious concern rather than provision; non-lethal conflict within the community was common; and the interaction between people within the community and with surrounding communities was political and complex.

Anybody who reads Chagnon's work will be struck with the similarity between arranged marriage systems and the legal framework that governs our own lives. The arranged marriage system is an alternative method of government, cooperation and dispute resolution. The central government, for instance, is replaced by alliance formation. Alliance formation is a personal affair, something that people create themselves, out of their own flesh and blood through marriage. Men married their daughters to other men to gain those men's support. Rather like getting the numbers in parliament, people used arranged marriages to create a large number of supporters when it came to a dispute. Winning a 'legal' dispute meant having the numbers in a political sense, rather than supporting statutes or precedents. Having the numbers meant having an overwhelming number of people willing to use force to enforce their wishes. Following on from Claude Lévi-Strauss, Chagnon described the arranged marriage system as a method of cooperation.

To anyone with an evolutionary perspective, this should ring bells as loud as any imaginable. Cooperation and reproduction were linked. If a man didn't cooperate, he couldn't get an arranged marriage, and he couldn't have children. If a woman didn't cooperate, she had fewer men to protect herself and her children. This was not merely how people cooperated without money and central government, it was how people reproduced.

Cooperation equalled reproduction. The passing on of genes down the generations was entirely dependent upon the existence of cooperation: this is the implication of the arranged marriage.

The arranged marriage system was not merely a different cooperative system, but a different mating system. The obvious difference is that people don't select each other. In these communities, a woman can have children without ever choosing a male, and without a male ever choosing her. People reproduced because of a larger system at work. People gave up their right to choose their sexual partner in exchange for being part of a larger cooperative network. Non-choice of a long-term partner must be the single most unusual method of reproduction amongst mammals.

I thus applied the various questions concerning human traits to this different mating and cooperative system. It felt as if a long row of dominoes had toppled over.

Explaining all this is not easy, since an understanding of the workings of arranged marriage systems is necessary. In his introduction to *African Systems of Kinship and Marriage*, A.R. Radcliffe-Brown wrote:

> For the understanding of any aspect of the social life of an African people – economic, political, or religious – it is essential to have a thorough knowledge of their system of kinship and marriage. This is so obvious to any field anthropologist that it hardly needs to be stated. But it is often ignored by those who concern themselves with problems relating to economics, health, nutrition, law, or administration amongst the peoples of Africa...[2]

When one considers that the arranged marriage systems are the equivalent of Western legal systems, the importance of Radcliffe-Brown's words makes sense. Ignoring the kinship and marriage systems of African peoples is the equivalent of entering a Western nation and ignoring all government and legal systems. Even where African people have a central government, this might actually be dominated and controlled by kinship and marriage systems. People in Western nations simply consider marriage as a pact between a single man and a single woman and cannot understand how marriage could be of any greater importance. Most men simply think of it as something for sex, children, and romance novels. Yet, for the sixty percent of people that continue to use arranged marriages in the world, these arranged marriage systems may be the most dominant force in their life.[3] Many of these people will live in societies without either a government or a trustworthy government: the only manner in which they can protect themselves and exist is through the cooperative structures that exist through kinship and marriage.

In order to understand the marriage and kinship systems, we might follow the practice of Claude Lévi-Strauss and concentrate on structures in perfect models. While this might give us a good understanding of how the systems are meant to operate, it would also give us only a simplistic understanding of how they actually do operate.

The systems used by people in hunter-gatherer or hunter-horticultural societies have specific rules of who is meant to marry whom. However, some people within the community break the rules to gain an advantage or to maintain the system. Rather than a perfect legal system, what exists is an imperfect legal system that people within the society frequently attempt to manipulate to their own advantage. Cooperation through arranged marriage is a highly malleable imperfect system. How people manipulate the arranged marriage system is linked to the birth of the *majority* of children in these societies and thus there are enormous implications for human behaviour and morphology.

The importance of arranged marriages is generally ignored within Western culture. For instance, a few centuries ago arranged marriages were a normal part of life in western Europe, yet the dramatic change from arranged to selected marriages is viewed to be of little importance compared to the industrial revolution. This makes the biggest single change in human sexual reproduction less important than the steam engine! People assume that central government was the consistent cooperative mechanism before and after the industrial revolution, and ignore the disappearance of the main method of human cooperation throughout human history.

A further reason why arranged marriages are ignored is that there tends to be a fairly patronising view of hunter-gatherers and hunter-horticulturists. The assumption is: 'How hard could it be? Find enough food, fight a few predators, feed the children, and that's life'. With this in mind, is it any wonder that modern science struggles to explain the most basic questions of human existence? If life for these people is so simple, then why are we so intelligent? Why do we need such a big brain in order to run away from predators and find enough to eat, when a mouse can do this? It is only once people begin to understand the arranged marriage systems and their complexity that the need for our complexity begins to make sense. We are complicated, and a complicated explanation is needed to explain our existence.

These marriage systems create societies, which, in their own way, are as complex as our own. In these societies, people are the lawyers, soldiers, doctors, nurses, builders, and so on, performing, personally, every major aspect of life. The same person that one day may be a doctor, will, the next day, personally have to decide whether to engage in a war. By contrast, while Western societies are more complex in their entirety, people work in specialist areas. The observations of Bob Connelly, a documentary film-maker who observed the conflict amongst Papua New Guinean communities that were in an awkward interval between a modern and 'stone-age' life, reflect the differences in lifestyle:

> I thought of the Lutheran pastor in *Joe Leahy's Neighbours* who preached Christian forgiveness as the Ganiga sharpened their spears and readied for war with the Gaimelka; his crushed expression as old Dubai – that quintessentially moral man – stalked away with his fellow warriors, leaving the pastor standing there with his prayer book, humiliated; the gut-wrenching pathos of the pastor's "turn the other cheek" plea in a culture shaped, defined and delineated by revenge. What a place, what a crazy, chaotic, *big* place this was, where so much that happened was beyond any writer's power of invention. People lived big lives here, lives of raw immediacy that continually tested them.[4]

We can only imagine what it would be like to go home from work and have to go to a community meeting to decide whether to attack a neighbouring suburb with the intent of lethal violence. We can only imagine the personal pressures involved in such decision making, particularly since we would be the ones required to implement the decision, and deal with the consequences, whether injury or retaliation. People in Western societies are generally free of such levels of responsibility, and can leave such decisions to professional soldiers and politicians. It is worth attempting to imagine what it would be like to live in a society where serious levels of conflict are a normal part of life. Where you, personally, need to enforce your legal rights or your right to exist. A society with no police force, no army, and if you have a problem with another person or a group of people it is up to you to create your own police force or army to deal with the problem or do it personally. Nobody to help you, nobody to step in to prevent fights, and nobody to keep the peace. A society where you, basically, are the law. Once a person with some imagination can place himself or herself in such an environment, they can quickly imagine that being a hunter-gatherer could be a lot more difficult than their own life. We have the luxury of going to horror movies or going on show rides in an attempt to experience being scared: for some people being scared is a daily reality.

In most texts, hunter-gatherers are presented as simple people living simple lives. We are given stories of the difficulty of finding a tuber or succeeding in a hunt. This book will take a completely different approach and argue that the most difficult aspect of people's lives is succeeding in a competitive environment where the greatest threat is other people. It will argue that cooperation and maintaining this cooperation is the most complex issue that faced our ancestors. It will argue that the method of cooperation has tremendous implications, and is the major evolutionary force behind our remarkable traits, such as brain size, language, and unusual sexuality.

This book will present a comprehensive explanation of human evolution. All major aspects of human behaviour and morphology will be considered and a step-by-step guide to how they evolved will be given. All of these behaviours and morphologies will be a response to the method of cooperation. It is the method of cooperation that links every argument.

I won't go through every alternative theoretical argument, which would extend the book's length unnecessarily, and which can easily be found elsewhere.[5] Instead I will focus on what is the first complete argument to consider the implications of the bilateral cross-cousin marriage system for human behaviour and morphology from an evolutionary perspective.

Below, I will discuss the materials I have used and why, and also why I have created certain definitions. This is not essential; however, it is important to explain why I used some materials rather than others.

Materials used

This book will assume that the best evidence of how people evolved comes from anthropological studies, that is, the studies of people who were actually living in the type of environment into which we evolved. Other studies, such as those of university students, will be treated as additional, imperfect material. The reason for this is that people in Western societies live in a very different environment; the difference in marriage patterns (selected as against arranged) is a good example of why such studies cannot be considered as the best evidence.

There are a vast number of studies of different peoples in anthropology and so a major issue is which studies to use. The two main studies I use are Napoleon Chagnon's *Yąnomamö* and Lloyd Warner's study of the Murngin people in northern Australia, *A Black Civilization*. The reasons I decided to use these two studies are that: (1) both peoples had little contact with Western societies (unacculturated); (2) the peoples were still engaged in warfare; (3) both studies are extensive and highly detailed; and (4) both studies are also extremely similar in what they reveal about the people studied. To support these texts I also use various studies of Aborigines and Helena Valero's statement of her time living with the Yąnomamö, and, less commonly, studies of Papuans. Valero was a Brazilian girl kidnapped by a Yąnomamö community; she lived with several communities over three decades and was married to a headman. The details of her life closely reflect Napoleon Chagnon's study and there could not be a better, more objective viewpoint to support his work. She also gives a woman's perspective, which counterbalances Chagnon's work, which mainly concerns the role of men. I also use a similar, but less substantial personal account of life within Australian Aboriginal societies, *The Adventures of William Buckley*, written by someone who actually lived in such a society before Western intrusion. William Buckley, an escaped convict, was incorporated into an Australian Aboriginal community for more than three decades.

I preferred to use studies of the Yąnomamö and Aborigines because these studies were often very detailed and extremely consistent. By contrast, my readings of studies of other peoples, such as Papuans, revealed a lack of consistency in the studies. Aborigines are probably the best studied people living in a hunter-gatherer existence, while Chagnon's study of the Yąnomamö, which lasted five years over three decades, is the best single study of a hunter-horticultural people.

Many studies are imperfect since the people are often affected by Western influences. For instance, in Mervyn Meggitt's text, *Desert People*, the Aborigines have even taken on Western names (though, evidently, without a thorough understanding, since one man's given name is Hitler). I used such texts if the arranged marriage systems were still being used. A major benefit of Mervyn Meggitt's text is that he gives a very detailed account of disputes in the community, and his wife assisted him, thus giving him access to the female perspective.

Some studies are too problematic to use. One example is the various studies of the Aché of Paraguay, which are commonly cited as authorities. In his summary of the historical and social contexts of various hunter-gatherer societies, Daniel Stiles comments on the Aché: 'Extensive arguments are presented of how today's foraging by the Aché is the same as precontact times, even though they live on a Catholic mission and grow crops.'[6] Stiles goes on to explain how various researchers have used the Aché as an example of a pre-contact people, despite the fact that some use shotguns, bring food such as maize on their hunting trips, and earn money from selling crafts.[7] From my own reading, perhaps the most obvious example of how these people now demonstrate unusual patterns of behaviour is in marriage, since the Aché males marry in their late teens, and even get to choose their own partner, which even heavily acculturated Australian Aborigines do not do. Added to this is that the number of Aché is barely more than 100 people. The greatest problem of this work was that the emphasis of the research was on food collection and its implications. For instance, the ability of men to obtain food has been linked to both marriage and extra-marital affairs, yet such links seem dubious as the people live on a Catholic mission and receive nutritional support.[8]

Texts of people living on missions are of questionable benefit in terms of analysing food acquisition, but they can be of interest if the people use arranged marriage systems, even if the cooperative system is heavily affected by external influences, since they can provide some useful information regarding the disputes arising from marriage arrangements. They can also be essential in obtaining a female perspective, since no female anthropologist has ever produced a high quality work of unacculturated peoples still involved in warfare. The closest is possibly Phyllis Kaberry's *Aboriginal Woman: Sacred and Profane*, which, rather sadly, has been generally ignored.

What is most vital about the studies of the Yąnomamö and Aborigines is that the details of the arranged marriage systems they use are very thorough. In these studies, we are not merely presented with systematic diagrams, but we are often given details of how individual marriages occurred and the disputes that arose.

The benefit of using well-studied peoples is that the flaws in individual studies can be removed. There are many studies where people engage in unusual behaviour for humans, but it is almost never confirmed in further studies, or studies of neighbouring peoples. A high degree of skepticism needs to be applied to anthropological studies, where it might be revealed that the men perform most of the childcare, or that women have sexual relations with any man they like. In his text, *Studying the Yąnomamö*, Napoleon Chagnon's first rule for collecting Yąnomamö genealogies was: 'Never assume that a new informant will tell you the truth', or 'Always assume that he *is* lying'[9] (Chagnon's italics). Amongst the Yąnomamö, genealogies were carefully guarded secrets, possibly because knowledge of genealogies allowed Yąnomamö men to manipulate the arranged marriage system and genealogies were thus the equivalent of trade secrets. One might equally apply such a rule to anthropological monographs. It may not be that the anthropologist has lied, but that the anthropologist's informants lied. Some anthropologists looking for noteworthy material may not have been very skeptical of the information they collected and many studies were short in duration. Anyone using anthropological monographs must be skeptical of everything that is written, and look for confirmation from alternative sources.

Even when there is a consistent picture of life in these communities, one of the tremendous difficulties is trying to understand exactly what is going on. Often the people themselves do not know. This is particularly true where there is a divide between what the women know and what the men know, or in the knowledge between competing communities. This is true for our own lives. Unless we frequent a particular social environment ourselves, opportunities to get an insight into what occurs 'behind closed doors' are rare; in some cases, glimpses can be high profile events.

This book is concerned with studies that present the cut and thrust of what actually occurs in 'stone-age' societies: which men get married and which don't, which men get more wives, how people instigate extra-marital affairs, how people cooperate, what undermines or enhances their cooperation, and so on. This book will argue that it is these details that are vital to understanding human evolution, and unless these details are understood, we will always be attempting to understand human evolution through a different social system – our own.

For instance, evolutionary psychologists argue that women want men with resources. The problem with this argument is that while a woman in a modern society might prefer a man likely to buy an expensive car and house, people in traditional societies live in a subsistence lifestyle. There is simply no correlation between an expensive car and food, since people in both societies are well fed. In fact, not merely are there tremendous differences in resource acquisition, there are also tremendous differences in marriage patterns, the enforcement or non-enforcement of monogamy, and the systems of government. The social systems are so completely different that every assumption of modern life must be questioned. This is not to question the existence of human instincts and behaviours, but rather the context of the behaviours.

Definitions

I have used the terms 'husband-by-arrangement' or 'wife-by-arrangement' instead of 'husband' and 'wife' to differentiate arranged marriages from Western marriages. I have used the term 'cooperatively-simple' to label 'stone-age' societies. By this I mean a society where the people live in a subsistence manner which uses marriage as a form of exchange, and uses no other important method of exchange. To that extent, the people of Papua, when they were studied, would not fit into this category since they have pigs to exchange as well as marriage. By using these terms I want to emphasise that it is cooperation and the method of cooperation that defines a society, rather than other factors. By contrast, the term 'hunter-gatherer' is the equivalent of 'butcher-supermarketer' for Western societies. What is unusual about humans is not that they hunt and gather, (bears do this), but that they are highly cooperative, and this should be the focus in any definition of their societies. Western societies would be labelled 'cooperatively-complex' since the people use money rather than marriage. Societies that use both arranged marriages and other forms of valuable currency might be labelled 'cooperatively-moderate'. This reflects older terminology, which is now unusable due to the terms being viewed as offensive rather than as defining terminology (savagery, barbarism and civilisation were once terms used to define people's social systems).

Concluding Remarks

This book is intended as a series of theoretical arguments. Sometimes there is a tremendous body of evidence to support an argument; sometimes an argument is entirely original and there is little relevant evidence, since no-one has ever considered the argument and looked for the evidence. The main argument remains that arranged marriages are normal and human traits can only be understood through this reproductive method.

Chapter 2

Competitive Cooperation

Human beings are the most cooperative animal in existence apart from insects. Unlike insects, human beings are able to cooperate with members of their community who are not closely genetically related. This book will focus on the method by which humans use to cooperate and its implications. Before considering specifically how humans cooperate, I will consider the underlying theoretical basis for why such cooperation evolved.

Charles Darwin has provided us with two theories with which to explain animal behaviour and morphology: natural selection and sexual selection. One hundred and thirty years later, these remain the major theories of selection. He explains the two theories as follows:

> Sexual selection depends on the success of certain individuals over others of the same sex, in relation to the propagation of the species; whilst natural selection depends on the success of both sexes, at all ages, in relation to the general conditions of life.[1]

Natural selection concerns the ability of an animal to survive in its natural environment, while sexual selection concerns the ability of an animal to reproduce successfully in competition with members of its own species. Darwin's theories suggest that every animal is the result of selection, with some animals surviving and others not, and, of the survivors, some animals reproducing while others not. Over time, animals evolved according to these pressures, with the unsuccessful traits disappearing and the successful traits defining the animal.

Human cooperation could then reflect either of these two selective pressures, natural selection or sexual selection. There does not appear to be any third element that could cause our cooperation other than survival within the environment and procreation. All of our ancestors, stretching back millions of years, must have successfully found the food they needed, avoided predators, and survived diseases and parasites. Then they must have procreated successfully and the successful infant was equally as successful in finding a sexual partner, finding enough sustenance, and overcoming predators and parasites. It is always astonishing to consider how our own personal existence is dependent upon this occurring time after time, without a single dead end, stretching back millions of years, much in the same way as it is astonishing to consider the seemingly endless stars in a rural night-time sky. Both concepts push the boundaries of our comprehension.

Natural Selection as a Basis for Human Cooperation

Firstly, it must be considered whether human cooperation could have evolved due to natural selection.

I will argue that the possibility that we evolved to cooperate in order to find enough food, to avoid predators and survive parasites is impossible.

As to parasites, grooming is something in common amongst primates, but has never yielded anything like complex, cooperative behaviour. Our lack of body hair would also suggest that grooming is less essential amongst humans than other primates. Since the major area of hair left on our bodies is small compared to other primates, it would be remarkable if the cooperation that led to Western Civilisation was based on the search for lice in two areas each measuring not much more than 15 cm x 15 cm. Other parasites, such as viruses, cannot be linked to our cooperation. No doubt our level of cooperation assisted ill community members to survive, but the immune system evolves to respond to viruses, cooperation does not. This is self-evident when one regards the lack of cooperation in other species.

Avoiding predators is important, but it has never yielded complex, cooperative behaviour in animals generally. The idea that predators were an issue for our ancestors is dubious. If predators were a serious problem for our ancestors, it seems doubtful that their response would be to use two feet rather than four. We seem to have done our best to evolve to make ourselves perfectly incapable of escaping from a predator. Compared to four-legged primates, we are slow, less able to climb trees, and have reduced canine teeth that cannot be used as weapons. The only benefit in terms of movement in using two legs appears to be over long distances, which is inappropriate for escaping predators, which invariably use ambush as a technique.

The discovery in Germany of 400,000 year old spears, fashioned in a modern-style, gives an indication of how long our pre-human ancestors have been able to protect themselves with complex implements.[2] Robin Dennell commented on the spears:

> The spears have other exciting implications. First, the time and skill needed to make them: each is made from the trunk of a 30-year-old spruce tree; in each, the end with the tip come from the base of the trunk, where the wood is hardest; and each has the same proportions, with the center of gravity a third of the way from the sharp end, as in a modern javelin. These represent considerable investment of time and skill in selecting an appropriate tree, in roughing out the design and in the final stages of shaping. In other words, these hominids were not living within a spontaneous 'five-minute culture', acting opportunistically in response to immediate situations. Rather, we see considerable depth of planning, sophistication of design, and patience in carving the wood, all of which have been attributed only to modern humans.[3]

This suggests that for at least 400,000 years our ancestors have been able to make proper javelins and use them for defence. That these pre-human ancestors were able to fashion modern javelins gives some idea of the intelligence of our pre-human ancestors. This discovery certainly suggests that the difference in intelligence between ourselves and those hominids living 400,000 years ago is much less than we assume. It also begs the question, if our ancestors of 400,000 years ago were able to fashion modern javelins, why did any further evolution occur at all? This question will be answered in this book.

Before javelins it is easy to imagine that our ancestors were capable of using other implements as projectiles, such as rocks. Though we have lost much of our physical strength, which is probably due to improved penetrative technology as the German spears implies, our earliest ancestors would have possessed levels of strength closer to that of chimpanzees. Many have argued that we used projectiles for protection against predators.[4]

There is no doubt that large predators can be a problem for human beings. Napoleon Chagnon, in his study of the Yąnomamö, observed:

> Indeed, the Yąnomamö are terrified at the prospect of having to spend the night in the jungle without fire, for they know that Jaguar hunts at night and they know they are no match for him in this realm.[5]

> Jaguars are also occasional daylight hunters: in August 1990 a large male attacked a party of armed Yąnomamö men in the village I was then studying and killed three of them without a single arrow being fired at him. He dragged one of them off and ate most of him before the others found his body and cremated what was left. Several years earlier, a crazed jaguar walked into a large Yąnomamö village and began attacking residents in broad daylight. The men managed to spear him to death with bowstaves before he killed anyone, but he left a lot of ugly claw scars on many of them.[6]

Unsurprisingly, the jaguar features in many of the Yąnomamö myths:

> A theme that repetitiously appears in Yąnomamö myths is about Man's relationship to Jaguar. In mortal form the jaguar is an awesome and much-feared beast, for he can and does kill and eat men. He is as good a hunter as the Yąnomamö are and is one of the few animals in the forest that hunts and kills men – as the Yąnomamö themselves do. He is in that sense like Man, but unlike Man, he is part of Nature, not of Culture. This distinction is fundamental in Yąnomamö conceptions, for the Yąnomamö separate themselves invidiously from the animals, and point out, as proudly as we do, that humans have Culture and the animals do not. Thus, they bifurcate the world into *urihi tä rimö* ('things of the forest') and *yahi tä rimö* ('things of the village'). The former is Nature; the latter is Culture. It is tempting to speculate about the nearly universal distribution of this opposition and whence it comes and what it means.[7]

While the jaguar is feared, in the myths, he is portrayed as the lesser species:

> But in their stories about him, he is consistently portrayed as a stupid brute, constantly being outwitted by Man, and constantly subjected to the most scathing, ridiculous, and offensive treatments at the hands of Man.[8]

Despite the danger of large predators, it seems implausible that our levels of cooperation could be a response to predation. Firstly, no other animal has human-like cooperation. Other primates have remained quadrupeds and possess simple levels of intelligence and cooperation. Much is made of the concept of preyed-upon hominids walking the African savannah, yet Hamadryas baboons, which also utilise the savannah in Africa, protect themselves by sleeping on cliff faces and retaining long, razor sharp canines.[9] If predation led to human-like cooperation, we should expect to observe it in many animals.

Secondly, humans evolved physical characteristics that limit our capacity to flee from predators. It remains implausible, to say the least, that our ancestors, fearing the lions of Africa, responded by decreasing their short-distance speed and manoeuvrability, reducing their ability to climb trees, and losing their canines.

Thirdly, pre-human ancestors were able to fashion weapons that could be used for protection from predators, and have possessed these weapons for more than two hundred thousand years before humans evolved. If these pre-human ancestors possessed weapons such as spears, why did any further human evolution occur if predation is the cause of increased cooperation?

Fourthly, high levels of predation have not been observed in anthropological studies. While Napoleon Chagnon recorded several deaths due to predation amongst the Yąnomamö, he recorded at least 282 deaths due to human violence,[10] and that approximately twenty-five percent of all adult male deaths were due to human violence.[11]

Certainly, it is plausible that basic levels of cooperation – such as the tolerance to competitors sleeping on the same cliff face – could evolve due to the issue of predation, but it is impossible to think that our complex levels of cooperation could have evolved due to predation. If predation were the initiating factor causing our evolution, as soon as these first bipeds were capable of dealing with the large predators of Africa, the evolution of cooperation would have ceased to evolve any further. We would never have evolved past *Homo australopithecine*. Of course, if *Homo australopithecine* could not have survived the large predators of Africa, it would have gone extinct, and we would not exist. Our ancestors must have been able to deal with predators for millions of years.

As to natural selection, a third argument concerns the need for food. Nutrition is generally believed to be a major reason for our cooperation, with man the hunter or woman the gatherer as good examples of these arguments. Superficially, these arguments seem to be plausible. Considering the evidence of the spears above, it appears that our ancestors did hunt and this would require cooperation.

Yet, the fact remains that the need for nutrition will never create a selection pressure that leads to increased complexity through evolution.

We are often presented with arguments that climate change reduced nutritional levels and this led to evolutionary change in our ancestors; however, starving animals do not reproduce. In fact, they have a tendency to die rather than to reproduce. The only animals that reproduce are healthy animals. The need for increased nutrition can immediately be discounted as a factor in increased complexity in our ancestors. This applies to humans and all other animals. I suspect that such a statement will require further elucidation.

Is nutrition a factor in evolution? Of course it is. If an animal lives in an environment where it continually faces nutritional shortages then this will have the following outcomes: shorter lifespan and/or smaller body size and/or smaller brain size. In a year when it can reproduce, it will do so, and the young produced will attain a body size and level of complexity appropriate for the amount of nutrition available. If they are a long lived animal, a shortage of nutrition will lead to a stunting of the juvenile animal so that it attains a smaller adult body size. What we should thus expect is that a species of animal that faces nutritional shortages will become smaller until its new body size and level of complexity reflects the amount of nutrition available annually. Large, complex, unhealthy animals will evolve into smaller, simpler, healthier animals capable of successful reproduction. The large version will become extinct, and the smaller version will become common.

The change towards increased size, complexity and population density occurs because of an increase in levels of annual nutrition, that is, a lack of hunger. If small, simple, healthy animals move into an environment where there are greater levels of nutrition, then they will evolve into larger, more complex, healthy animals. Our ancestors were smaller, simpler, healthy animals and they evolved into larger, more complex, healthy animals. Starvation and food shortages were never a factor.

The reader should notice that this argument allows for the factor that hunting and gathering played a role in human evolution. But such evolution has nothing to do with natural selection.

I have no doubt that this argument will cause some confusion. Surely, people will say, that the cheetah is capable of such speed because without such speed it would have starved and become extinct. This is incorrect. Only healthy cheetahs are capable of reproducing; starving cheetahs will not reproduce. At each stage of increasing speed, the predecessor of cheetahs must have been able to reproduce. The fact remains that every predecessor of the cheetah must have been born of a healthy mother that was not starving, otherwise they would have become stunted or died.

This stunting of animals due to nutritional shortages has been observed in humans and other animals; food variations for humans can be found throughout the world, and does not reflect life or extinction, but height and weight. Malnutrition has been observed in human beings, and it has led to both impaired growth and stunting:

> An adequate supply of food is essential for individuals to achieve a normal growth pattern and maximize their potential as adults. The intake of food supplies a growing child with its nutritional requirements and there are many studies which have shown the close relationship between this supply and growth. The evidence for this received considerable support in analysis of data collected during the 1914-18 and 1939-45 wars where famines which occurred during the hostilities retarded the growth of many children. For example, in the Netherlands, the winter famine of 1944-45 severely impaired children's growth. In post-war Germany, under similar near famine conditions, children lagged 10-20 months behind normal growth performance, and in Japan the mean height of adults fell between 1945 and 1949 and only returned its previous levels in 1956.[12]

Studies of rats and mice have found that delays in maturity can equal the entire lifetime of a normal animal.

Human malnutrition involves not only a deficiency of calories, but a lack of specific foodstuffs, and it is impossible to separate the two factors. Several experiments on supplementation of inadequate diets have been carried out under conditions of widespread malnutrition and famine and these have shown that the preadolescent growth period is most severely affected. There are ethical problems involved in studies in this field so most of our knowledge on the relation of nutrition to growth derives from animal work.

When rats and mice are fed on a diet deficient in calories but otherwise satisfactory, they cease to grow but can resume growth when an adequate supply of calories is restored. Rats have been kept immature for up to 1000 days on such a diet, and were still juvenile while their fellows on a normal diet were senile. On starting normal feeding, the rats rapidly grew to adult size, although those which had been kept on the deficient diet for more than 3 years could not always begin growing again. Subsequent work showed that the diet which results in the maximal final size of experimental animals is one which, as judged by appetite and by growth rate, represents a moderate degree of caloric under-feeding.[13]

Babies and adolescents require high levels of nutrition; a period of malnutrition can be overcome through 'catch-up growth' but can also lead to stunting:

The human need for calories varies with the phase of development. In the first year of life a baby requires about twice as many calories per unit weight as a male adult doing moderately heavy work, and a schoolboy going through his adolescent spurt may require as much as 3000kcal a day. He also needs about 50% more calcium, nitrogen, and vitamin D than he did before the spurt began.

Nutrition also affects the rate of maturity. Malnourished children exhibit delayed growth and also a delayed puberty. Body proportions at birth such as weight related to length are also a good indicator of a time of restraint of growth during pregnancy. After a period of human or animal malnutrition has ended, growth accelerates in an attempt to compensate for the loss in height and weight. This 'catch-up growth' can be successful in restoring normality if the period of malnutrition has been short. Sometimes complete compensation proves impossible as, even if puberty is delayed, the longer growth period available cannot compensate for the lost growth in preadolescence due to the dietary deficiencies.[14]

Cooperation requires genes for cooperation to exist and to become transmitted amongst members of the community or species. This would take many generations. By contrast, stunting requires merely one generation to occur. Climate change is currently being used as a cause for hominid evolution. The argument suggests that there was a change in climate, that the hominid faced reduced nutrition, and responded by evolving complex methods of obtaining food, such as hunting. Yet, this would never happen, because it would only take one generation for the individuals to become stunted or to die. Assuming that reduced nutrition was not catastrophic, the generation experiencing nutritional deficiencies would grow to a body size that reflected the lower levels of available nutrition. The selection pressure for novel methods of obtaining food would disappear in a single generation, as soon as the progeny took on a smaller body size.

This leads to a rather interesting question. If our ancestors evolved to hunt, why would they do so, since they were already healthy? In fact, why did our ancestors evolve many complex methods of increasing their caloric intake, such as cooking, when, in order to reproduce, they were already healthy? Many academics make statements as to the importance of food without identifying the selection pressure underpinning these changes. They need to explain why an animal would begin to hunt or use fire for cooking, when only healthy animals reproduce and that if there are food shortages, then either extinction or stunting will quickly ensue. This is particularly true where these academics argue that the change they are describing is fundamental and would involve the need for genetic changes in the animal. It is simply impossible to argue that a species maintains a state of near starvation generation after generation until the genetic changes have occurred in order to explain the evolution of novel food gathering methods. The nutrition argument contradicts research on the impact of decreases in caloric intake in humans and rodents.

My argument is that depressed levels of available nutrition will lead to smaller, less complex progeny or the dispersal of the group, or both. Where dispersal occurs we would still expect that the progeny would be smaller, unless they were able to find new territories with similar levels of nutrition to exploit that could maintain historical adult size. Such stunting will occur before any genes that may benefit the animal to exploit new sources of nutrition can take hold. Stunting only takes a single generation: gene dispersal takes many. Depressed levels of nutrition thus leads to smaller, less complex animals, not a more complex, larger brained, more cooperative, physically larger species, as in human beings.

No doubt some will argue that our ancestors, faced with food shortages, may have sought out novel ways of finding food. Such an outcome may be expected in any animal. Any animal, facing starvation, may attempt to eat almost anything. But we are attempting to explain significant genetic changes that require many generations. Let's suggest that a hominid, facing starvation, began to eat a novel food source, such as a carcass. Let's suggest that this is a successful manner of avoiding starvation. So, exactly the same animal, with no genetic changes, no alteration of physical or behavioural characteristics, now knows it can, if necessary, eat the meat on a carcass. So, where is the evolution? The animal has not evolved.

Starvation or depressed levels of nutrition are literally, in terms of attempting to explain increased complexity, a dead end. Starvation and depressed nutrition are indeed selection pressures, but they are selection pressures for reduced body size and complexity. Such selection pressures can take effect in only a single generation. The notion that our ancestors maintained levels of starvation for many generations until they evolved more complex behaviours and morphologies is a fantasy that goes against medical knowledge.

This fantasy applies not merely to evolution but also to assumptions concerning life in a natural environment. There tends to be a view that any animal living in 'the wild' must be in a constant state of desperation. However, there is certainly no evidence that people in cooperatively-simple societies live in a state of need. This reality of the true levels of health of hunter-gatherer and hunter-horticulturist peoples was recognised by Marshall Sahlins, who described such peoples as the 'original affluent society':

The anthropological disposition to exaggerate the economic inefficiency of hunters appears notably by way of invidious comparison with neolithic economies. Hunters, as Lowie put it blankly, "must work much harder in order to live than tillers and breeders." On this point evolutionary anthropology in particular found it congenial, even necessary theoretically, to adopt the usual tone of reproach. Ethnologists and archaeologists had become neolithic revolutionaries, and in their enthusiasm for the Revolution spared nothing in denouncing the Old (Stone Age) Regime. It was not the first time philosophers would relegate the earliest stage of humanity rather to nature than to culture. 'A man who spends his whole life following animals just to kill them to eat, or moving from one berry patch to another, is really living just like an animal himself.'[15]

Marginal as the Australian or Kalahari desert is to agriculture, or to everyday European experience, it is a source of wonder to the untutored observer 'how anybody could live in a place like this.' The inference that the natives manage only to eke out a bare existence is apt to be reinforced by their marvelously varied diets. Ordinarily including objects deemed repulsive and inedible by Europeans, the local cuisine lends itself to the supposition that the people are starving to death.[16]

Sahlins quotes Sir George Grey, writing in 1845, as to his observations of Australian Aborigines:

To render palpable 'the ignorance that has prevailed with regard to the habits and customs of this people when in their wild state,' Grey provides one remarkable example, a citation from his fellow explorer, Captain Stuart, who, upon encountering a group of Aboriginals engaged in gathering large quantities of mimosa gum, deduced that the 'unfortunate creatures were reduced to the last extremity, and, being unable to procure any other nourishment, had been obliged to collect this mucilaginous.' But, Sir George observes, the gum in question is a favourite article of food in the area, and when in season it affords the opportunity for large numbers of people to assemble and camp together, which otherwise they are unable to do. He concludes:

'Generally speaking, the natives live well; in some districts there may be at particular seasons of the year a deficiency of food, but if such is the case, these tracts are, at those times, deserted. It is, however, utterly impossible for a traveller or even for a strange native to judge whether a district affords an abundance of food, or the contrary ... But in his own district a native is very differently situated; he knows exactly what it produces, the proper time at which the several articles are in season, and the readiest means of procuring them. According to these circumstances he regulates his visits to different portions of his hunting ground; and I can only say that I have always found the greatest abundance in their huts.'[17]

Sahlins observed of the general perception of hunter-gatherers and the reality:

When Herskovits was writing his Economic Anthropology (1958), it was common anthropological practice to take the Bushmen or the native Australians as "a classic illustration of a people whose economic resources are of the scantiest," so precariously situated that "only the most intense application makes survival possible". Today the "classic" understanding can be fairly reversed - on evidence largely from these two groups. A good case can be made that hunters and gatherers work less than we do; and, rather than a continuous travail, the food quest is intermittent, leisure abundant, and there is a greater amount of sleep in the daytime per capita per year than in any other condition of society.[18]

Of the Bushmen, Sahlins summarises the work of Richard Lee:

Although food collecting is the primary productive activity, Lee writes, "the majority of the people's time (four to five days per week) is spent in other pursuits, such as resting in camp or visiting other camps":

"A woman gathers on one day enough food to feed her family for three days, and spends the rest of her time resting in camp, doing embroidery, visiting other camps, or entertaining visitors from other camps. For each day at home, kitchen routines, such as cooking, nut cracking, collecting firewood, and fetching water, occupy one to three hours of her time. This rhythm of steady work and steady leisure is maintained throughout the year. The hunters tend to work more frequently than the women, but their schedule is uneven. It is 'not unusual' for a man to hunt avidly for a week and then do no hunting at all for two or three weeks. Since hunting is an unpredictable business and subject to magical control, hunters sometimes experience a run of bad luck and stop hunting for a month or longer. During these periods, visiting, entertaining, and especially dancing are the primary activities of men."

The daily per-capita subsistence yield for the Dobe Bushmen was 2,140 calories. However, taking into account body weight, normal activities, and the age-sex composition of the Dobe population, Lee estimates the people require only 1,975 calories per capita. Some of the surplus food probably went to the dogs, who ate what the people left over. "The conclusion can be drawn that the Bushmen do not lead a substandard existence on the edge of starvation as has been commonly supposed."[19]

Dogs, specifically dingoes, are a common domesticated animal amongst the Australian Aborigines, and one might wonder how such animals could be maintained if there weren't plentiful levels of food.

In the 1970s another reason for the invocation for desperation was invoked: warfare and cultural anthropology. Napoleon Chagnon, in his studies of the Ýanomamö, produced data that the people were engaged in warfare, which was completely at odds with the assumption of the 'noble savage' – that people living in a subsistence manner must be living in perfect harmony. Since materialism causes warfare, those living without materialism or personal property must be harmonious. The response from various cultural anthropologists to Chagnon's findings was that the Yąnomamö must be desperate for food and that they were fighting over protein, rather than women, as Chagnon and the Yąnomamö themselves argued.[20] In response, various studies were performed in South America, where the amount of daily meat was weighed. There were 8 different peoples studied (Jivaro, Yąnomamö, Wayana, Boni, Maimande, Bari, Ye'kwana, and Siona-Secoya) in 11 different studies, by 8 independent researchers or groups of researchers.[21] It was discovered that the average daily intake was 64.6 grams per capita and 88.8 grams per adult (7 of the 11 studies showed daily adult consumption of between 95 and 119 grams, two were of 77 or 75 grams, and two were of only 36 grams). The average daily intake for the highest meat-eating societies amongst industrialised societies was 70 grams (Australia and New Zealand) and 66 grams (U.S.A. and Canada). Thus it was found that the average South American indigenous person ate as much or more meat on a daily basis than the average person living in industrialised societies with the highest levels of meat consumption. (Unsurprisingly, upon receipt of this information, cultural anthropologists simply changed their argument, this time arguing that it was the occasional gifts of steel tools by the anthropologists that the indigenous peoples were fighting over.)

As to Australian Aborigines, in his 1914 publication, Herbert Spencer found that:

> Physically the finest natives whom I met with were those on Melville and Bathurst Islands, that is, the men, because there was often a marked contrast in size between the men and women. The latter were seldom more than four feet six or eight, the former were often five feet ten or even six feet.[22]

In his earlier 1899 publication on the people of central Australia, Spencer took measurements:

> In the matter of height, the average of twenty adult males measured by us, was 166.3 cm. The tallest was 178.2 cm., and the shortest 158.2 cm. The average of ten adult females was 156.8 cm.; the tallest was 163 cm., and the shortest was 151.5 cm. The average chest measurement of the same twenty men was 90.33 cm.; the greatest being 97 cm., and the least 83 cm.[23]

These people were unaffected by Western culture. Anybody who has been to central Australia – which must be one of the harsher environments in which any human has thrived, being mostly red soiled desert – would be impressed that in such an environment any man could reach the height of 178 centimetres and that the average was over five feet six. That in the slightly better environments of the Melville and Bathurst Islands Aboriginal men could be six feet in height is truly remarkable. One might thus suggest that Australian Aborigines possessed higher levels of meat consumption than their South American counterparts. Anybody who is five feet eight, let alone six feet, must possess a high level of nutrition on a regular basis. Yet, if one were to read the general assumptions concerning hunter-gatherers, we would believe that these were desperate people, living a hand-to-mouth existence, and only surviving by their thrifty genes that allow for greater food intake during difficult times.

Edward Eyre, in his 1845 summary of his expeditions into central Australia from Adelaide, also gave a summary of the 'Manners and Customs of the Aborigines'. He noted:

> The male is well built and muscular, averaging from five to six feet in height with proportionate upper and lower extremities. ... The men have fine broad and deep chests, indicating great bodily strength, and are remarkably erect and upright in their carriage, with much natural grace and dignity of demeanour.[24]

He wrote at length on the diet of the Aborigines:

> Throughout the greater portion of New Holland, where there do not happen to be European settlers, and invariably where fresh water can be permanently procured upon the surface, the native experiences no difficulty whatever in procuring food in abundance all the year round. It is true that the character of his diet varies with the changing seasons, and the formation of the country he inhabits; but it rarely happens that any season of the year, or any description of country does not yield him both animal and vegetable food. Amongst the almost unlimited catalogue of edible articles used by the natives of Australia, the following may be classed as the chief:- all salt and fresh-water fish and shell-fish, of which, in the larger rivers, there are vast numbers and many species; fresh water turtle; frogs of different kinds; rats and mice; lizards, and most kinds of snakes and reptiles; grubs of all kinds; moths of several varieties; fungi, and many sorts of roots; the leaves and tops of a variety of plants; the leaf and fruit of the mesembryanthemum; various kinds of fruit and berries; the bark from the roots of many trees and shrubs; the seeds of leguminous plants; gum from several species of acacia; different sorts of manna; honey from the native bee, and also from the flowers of the banksia, by soaking them in water; the tender leaves of the grass-tree; the larvæ of insects; white ants; eggs of birds; turtles or lizards; many kinds of kangaroo; opossums; squirrels, sloths, and wallabies; ducks; geese; teal; cockatoos; parrots; wild dogs and wombats; the native companion; the wild turkey; the swan; the pelican; the leipoa, and an endless variety of water-fowl, and other descriptions of birds.
>
> Of these articles, many are not only procurable in abundance, but in such vast quantities at the proper season, as to afford for a considerable length of time an ample means of subsistence to many hundreds of natives congregated in one place; and these are generally the kinds of food of which the natives are particularly fond.[25]

It should be noted that the areas from which this food was obtained were viewed by Europeans as an extremely parched and difficult environment.

While many people's ideas of hunter-gatherers procuring food involves men chasing large animals over the land, the reality is entirely different. Edward Eyre noted that 'Kangaroos are speared, netted, or caught in pit falls.'[26] Eyre noted four main methods of capture: (1) sneaking up on the kangaroo, (2) following for two days until it tires, (3) creating a hide and ambushing the kangaroo near a water hole, or (4) the women using fire to drive the kangaroos towards waiting men with spears. Other methods include: driving the kangaroos over a precipice or into a river; placing a net along a path the kangaroos regularly use and alarming the kangaroo so that it takes flight on its usual path only to become entangled in the net; or digging pitfalls near water holes (though this is rarely practised).[27] The emu is also caught using the methods (1), (3) and (4) as well as the following manner:

> The emu is frequently netted by night through a peculiarity in the habits of the bird, that is well-known to the natives, and which is, that it generally comes back every night to sleep on one spot for a long time together. Having ascertained where the sleeping place is, that natives set the net at some little distance away, and then supplying themselves with fire-sticks, form a line from each end of the net diverging in the distance. ... The emu finding only one course free from fire-sticks, viz. that towards the net or apex of the triangle, takes that direction, and becomes ensnared.[28]

Even in December, one of the hottest periods in southern Australia when temperatures regularly reach over 35°C, food was plentiful:

> An unlimited supply of fish is also procurable at the Murray about the beginning of December, when the floods, having attained their greatest height, begin again to recede; and when the waters, which had been thrown by the back water channels of the river into the flats behind its banks, begin again to reflow through them into the river as it falls in height. At this time the natives repair to these channels, and making a weir across them with stakes and grass interwoven, leave only one or two small openings for the stream to pass through. To these they attach bag nets, which receive all the fish that attempt to re-enter the river. The number procured in this way in a few hours is incredible. Large bodies of natives depend upon these weirs for their sole subsistence, for some time after the waters have commenced to recede.[29]

> The tops, leaves, and stalks of a kind of cress, gathered at the proper season of the year, tied up in bunches, and afterwards steamed in an oven, furnish a favourite, and inexhaustible supply of food for an unlimited number of natives. When prepared, this food has a savoury and an agreeable smell, and in taste is not unlike a boiled cabbage. In some of its varieties it is in season for a great length of time, and is procured in the flats of rivers, on the borders of lagoons, at the Murray, and in many other parts of New Holland.
> There are many other articles of food among the natives, equally abundant and valuable as those I have enumerated: such as various kinds of berries, or fruits, the bulbous roots of a reed called the belillah, certain kinds of fungi dug out of the ground, fresh-water muscles, and roots of several kinds, &c. Indeed, were I to go through the list of articles seriatim, and enter upon the varieties and subdivisions of each class, with the seasons of the year at which they were procurable, it would at once be apparent that the natives of Australia, in their natural state, are not subject to much inconvenience for want of the necessities of life. In almost every part of the continent which I have visited, where the presence of Europeans, or their stock, has not limited, or destroyed

their original means of subsistence, I have found that the natives could usually in three or four hours, procure as much food as would last for the day, and that without fatigue or labour.[30]

The reader must be reminded that these environments depicted are some of the most difficult environments inhabited by human beings. Australia is well known as being the driest continent. One would expect that most environments that humans inhabited offered far fewer difficulties in obtaining food. By contrast to Australia and the deserts of the Kalahari, most of Africa would have provided an abundance of food far greater than Australian Aborigines experienced.

The reality is that these people are entirely healthy. This was also the conclusion of James Neel, a geneticist who led a team investigating the health of Amazonian peoples. In the wake of the nuclear bomb dropped in Japan in World War II, the peoples of the Amazon were selected as the control group for comparison, that is, as people entirely untouched by Western Civilisation, as part of a long-term study to investigate the long-term impact of radiation. Rather than starving, desperate people, he found healthy, robust people, who effectively controlled their population.[31] He observed from his studies:

> There are two commonly held beliefs about the health of tribally organised unacculturated populations which I wish to challenge in this brief presentation. The first is that the mortality pattern of such populations requires that they reproduce at near capacity simply to hold their numbers even.[32]

> Instead, I will argue that the health of at least some newly contacted groups, when viewed on a scale that encompasses the health of a wide variety of human populations over the past several hundred years, is really pretty good. I will further suggest that the high morbidity and mortality of at least some of these groups when confronted with the infectious diseases of civilization is due much more to certain epidemiological characteristics of these populations than to innate, genetically determined susceptibilities.[33]

> The definition of 'health' remains elusive, and so then does the evaluation of just how healthy a given population is. Our group has now conducted complete physical examinations of the inhabitants of entire villages, among three Amerindian tribes (the Yanomama, Xavante and Makiritare) who, well past the stage of first contacts, may nevertheless be regarded as minimally acculturated at the time of our studies. By our usual medical standards, these groups have appeared relatively healthy, the average young man, in particular, exhibiting the physique of a trained athlete...[34]

> However, these are cross-sectional studies, and leave open the possibility that these apparently healthy persons represent the relatively few survivors of a much larger birth cohort. We have therefore devoted a considerable effort to developing life tables for one of these groups, the Yanomama, not easy for a preliterate group where all ages must be estimated and it is considered dangerous to mention the names of one's

departed ancestors, lest their spirits hear and be displeased. To this end, a sex-estimated age pyramid has been developed based on a census of 29 Yanomama villages.[35]

Neel went on to show that the Yąnomamö possess a life expectancy that is superior to that of Indians in 1900, a society based on agriculture.[36] Previously, in 1962, James Neel was the person who suggested the 'thrifty-gene' hypothesis to explain high levels of obesity and diabetes, particularly in indigenous peoples.[37] The paper was entitled: 'Diabetes mellitus: A 'thrifty' genotype rendered detrimental by 'progress'?' – note the final word that encapsulates the assumptions. This paper suggests that 'stone-age' peoples have a genetic predisposition to a feast and famine lifestyle, and thus, when confronted with a Western lifestyle, tend to gorge themselves. It suggests that our 'stone-age' bodies simply are not compatible with a modern environment, and thus the tendency for problems concerning obesity and diabetes. In a 1998 paper, he and collaborators observed:

> Finally, it is now clear that the original thrifty genotype hypothesis, with its emphasis on feast or famine, presented a grossly overly simplistic view of the physiological adjustments involved in the transition from the lifestyle of our ancestors to life in the high-tech fast lane. Eaton et al., have emphasized how different in composition the Stone Age body that received that intermittent alimentation was. Although trained athletes retain the relative muscle mass of early man – at least until the competitions are over – modern man is characterized by a striking sarcopenia, with the interstices within and between muscles well padded with fat. Because fat and skeletal muscle cells have strikingly different insulin sensitivities, the overall sensitivity to insulin of skeletal muscle is profoundly altered. This differential is emphasized by the high conditioning of the muscle of early man, with the corresponding greater efficiency of insulin utilization. Whereas the total daily energy expenditure in adult members of hunter-gatherers and traditional agricultural societies was in the neighborhood of 3,000 kcal/day, now in industrialized societies it is of the order of 2,000 or less. Finally, the composition of the diet – and more specifically the use of highly refined carbohydrates, with the resulting almost instantaneous "sugar highs" – has of course altered dramatically with civilization, as has the mix of dietary components. Exactly how this might influence the development of insulin resistance is a matter of active debate.[38]

These conclusions, by James Neel are in stark contrast to the assumption of the half-starved, bedraggled hunter-gatherer. He compares the male muscle mass of 'early man' to modern athletes, and he accepts that these people have a 'daily energy expenditure... in the neighborhood of 3,000 kcal/day' – 1,000 greater than the needs of those in industrialized societies. He compares the muscle tone of modern man, whose muscles are well padded with fat, to 'the high conditioning of the muscle of early man, with the corresponding greater efficiency of insulin utilization'. If the thrifty-gene hypothesis was correct, we should expect the opposite, that 'early man's' body would be packed with fat, in preparation for the feast and famine lifestyle. If people living in a traditional lifestyle possessed a feast and famine lifestyle, one would expect that their bodies would have receptacles of fat. Instead, the contrary was found. Despite this, Neel and colleagues conclude:

> Despite all these advances in our understanding of NIDDM, the basic genetic maladjustments that result in the disease remain obscure. Given the intensity of the current effort to localize and characterize the genes whose functioning seems to be compromised in NIDDM, speculation at this time concerning their nature seems of little value. *However, the concept of a thrifty genotype remains as viable as when first advanced,* and it now seems desirable, in view of the direction this essay is taking, to begin to put the concept of a compromised thrifty genotype into the broader context of the "diseases of civilization."[39] (My italics.)
> [Non-insulin dependent diabetes mellitus = NIDDM.]

It is one thing to suggest that the human body was originally well adapted to its normal environment, and that the people ate high energy diets and exercised regularly, but this is not the argument. The argument is that people in their natural environment live an *unhealthy* lifestyle, that their 'thrifty-genes' enable them to survive their feast and famine 3,000 calorie per day existence, and then when these people finally get enough to eat on their 2,000 calorie per day requirements in a Western society, their body malfunctions because it is engineered for a deficient diet. But where is the evidence of the deficient diet, where's the famine? Neel's conclusions suggest the exact opposite of the hypothesis. The evidence produced found that the people were very healthy, that their high-energy dietary needs were met, and that the men possessed very little storage of fat. The 'thrifty-gene' argument is even applied to Polynesians, such as Maoris and Tongans, whose heavy physiques are apparently underpinned by a 'thrifty-gene' lifestyle.[40]

If equivalent studies could be made of various hominids, such as *Homo neanderthalensis*, *Homo australopithecine*, or *Homo erectus*, would we expect conclusions concerning their health to be any different? No, we should naturally expect that the same conclusions as to the health of these hominids. We should expect this because if these hominids were not healthy, they would not be able to reproduce! If these hominids were at all in need of nutrition for long periods of time, surviving infants and adolescents would become stunted, and their new physical size would take on a size appropriate to the nutrients available. Or they would have become extinct.

None of this should be surprising. Humans are very long-lived, and this would only occur if humans had a consistent source of nutrition. The idea that any woman could be pregnant and breast-feed a child for at least two years on a feast and famine nutritional source is impossible. The idea that an infant could take fifteen years to sexually mature based on a feast and famine nutritional source is equally ridiculous. Any animal that possesses a feast and famine nutritional source would be short-lived. All long-lived animals will possess predictable levels of nutrition to sustain them regularly into adulthood, and if they didn't, they would be short-lived. No animal would risk requiring fifteen years to mature and then suddenly be faced with starvation. The very fact of humanity's existence would suggest that all humans, and their ancestors, have had a consistent supply of nutrition for millions of years. If this were not the case, how could our remarkable brains – which slowly increased in size to unprecedented levels for an animal of its body size – ever have evolved, considering the vast nutritional resources needed to sustain them? Why would an animal risk such an oversized brain if it lived a feast and famine existence? Plainly, it would not.

Furthermore, where is the evidence of starvation or famine? The assumption is regularly made, but biologists, medical researchers and other scientists never feel any compulsion to provide even a scrap of evidence. Yet, as anyone who understands science would appreciate, incorrect assumptions that are used as the basis for research can entirely invalidate research and render it obsolete. Every assumption has to be proven otherwise it is mere conjecture. Anyone who wishes to argue that feast and famine exists must trawl through the medical, anthropological or other literature to support their argument, otherwise they cannot make the assumption. Non-human animal studies are barely applicable since these animals are not human beings, with the specific human traits that we are attempting to explain. The preferred animals for mammalian experiments, short-lived rodents, which can maximise reproduction in a very short time period, are a poor basis for long-lived, slow reproducing human beings. These animals usually live for less time than it takes for a human infant to reach the state of being weaned! When such animals have been used for research, low levels of nutrition lead to stunting and delayed sexual maturity.

The argument that a lack of nutrition led to human evolution is impossible on both evidentiary and logical grounds. There is no evidence that humans in their natural environment lack nutrition, even in the most desperate environments such as central Australia or the Kalahari Desert. Research into protein levels has found that people living in their natural environment eat more meat than those living in Western societies. Physical examinations by medical researchers of unacculturated peoples in South America have found the people to have a high level of health and for the males to be comparable to modern athletes. Furthermore, the very notion that a lack of nutrition can lead to increased complexity is illogical: starving animals do not reproduce, and starving infants or adolescents grow up stunted. There never will be a medical experiment in which an animal responds to lowered caloric intake by selecting genes for greater brain size. The idea that starvation leads to preferential selection of increased brain size, body size and group size is a contradiction in terms – it should lead to the opposite.

The idea of a lack of nutrition leading to increased human cooperation is thus implausible. Indeed, this should be so obvious that a simple explanation – food restriction leads to starvation and stunting, not complexity – should be all that is required; however, such is the popularity of the argument, I felt compelled to give it a lengthy discussion. Indeed, it is troubling that medical researchers continue to search for the 'thrifty-genes' of our ancestors, and believe that our ancestors lived a feast and famine lifestyle, when their own research found the opposite, and the entire argument is illogical. No animal would evolve with a human-sized brain while living with a feast and famine lifestyle. Indeed, one could argue that the entire argument is simply based on Western arrogance, as Marshall Sahlins suggests – that the 'neolithic revolution' does not make sense without the assumption that prior to agriculture people were in a state of starvation. The narrative is linked with the idea of Western progress: that our ancestors were starving until the neolithic revolution, which led, in turn to our own plenitude. Without such a nutritional pressure, the 'neolithic revolution' lacks a logical explanation. Indeed, every aspect of human evolution, whether cultural or genetic, appears to be based on natural selection and desperation. Unfortunately, there does not appear to be any evidence to support such a narrative: it is a myth. If medical researchers and others wish to continue to pursue this argument, they should be forced to find supporting material of humans in their natural environment. The fact that many indigenous people with Western diets are overweight and have diabetes isn't evidence that all humans evolved in a state of feast and famine any more than the fact that many people in Western environments wear glasses is evidence that humans evolved with poor eyesight.

The issue in evolution is one of selection pressures. In order for a trait to evolve, we need a 'must' argument: 'the trait evolved because it *must* have evolved'. Neither a lack of nutrition, nor the need to remove parasites, nor the need to avoid predators create the selection pressures (the *must* arguments) we require to explain complex traits. This is particularly true for such remarkable, unique traits that humans possess.

Sexual Selection

Charles Darwin recognised that natural selection simply could not explain unusual animal traits, and responded with the theory of sexual selection. In *The Descent of Man, and Selection in Relation to Sex* he argued that humans evolved through sexual selection. This fact has been ignored by almost everyone involved in the question of human evolution for more than 130 years. Geoffrey Miller's *The Mating Mind* is a rare and recent exception.

Sexual selection is defined by Darwin as:

> The sexual struggle is of two kinds; in the one it is between individuals of the same sex, generally males, in order to drive away or kill their rivals, the females remaining passive; whilst in the other, the struggle is likewise between the individuals of the same sex, in order to excite or charm those of the opposite sex, generally the females, which no longer remain passive, but select the more agreeable partners.[41]

Darwin's theory of sexual selection involves the competition between healthy members of a species as to who reproduces and who does not. It argues that survival alone is not enough for sexual animals, and that they must then compete to reproduce.

Part of the basis for this is parental investment. Robert Trivers has argued that females generally have a greater parental investment and are therefore the choosier of the two sexes.[42] The female is choosier as they often place large amounts of energy and nutrients into the eggs and the upbringing of young, while the male may invest little more than sperm towards the infant. The male, by contrast, can fertilise many females with only a moderate investment. Occasionally males have the greater parental investment, usually involving tending the female's eggs, and on those occasions the male is the choosier. Females, since they produce eggs, have a greater confidence in their opportunity to reproduce. Males, by contrast, must inseminate a female in order to pass on his genes, and thus he is more limited.

Sexual selection is Darwin's greatest and most controversial theory. The reasons for this are numerous. Firstly, while people can easily imagine why prey might evolve greater speed to escape from predators, they find it more difficult to understand why selection of one reproductive partner over another might lead to evolution. Avoiding death appears to be a greater selection pressure than mere sex to the human brain. Secondly, sexual selection makes many uncomfortable.

Many people involved in human evolution do not simply want any answer, they want an answer that they can live with. Sexual selection makes many academics uncomfortable because it is an emphatic statement of competition, with both winners and losers. The idea that humans evolved to be cooperative for the purposes of survival, by contrast, is quite appealing. This idea suggests that humanity's remarkable abilities of cooperation exist to succeed in a difficult and dangerous environment, and that this evolution was a cooperative, peaceful struggle to survive.

Such an image, however, does not fit with texts that humans have historically used to define their existence. For millennia women were seen as the source of all evil, as is depicted in the stories of Adam and Eve in the Garden of Eden and Pandora's box – both virtually the same story. In one, Adam and Eve eat fruit from the tree of knowledge and become wise; God responds by creating for Adam: 'enmity between thee and the woman'; and for Eve: 'in sorrow thou shalt bring forth children; and thy desire shall be to thy husband, and he shall rule over thee'. In the earlier tale of Pandora, man steals fire from the gods, and in response, Zeus orders Hephaestus to create woman for man:

> So said the father of men and gods, and laughed aloud. And he bade famous Hephaestus make haste and mix earth with water and to put in it the voice and strength of human kind, and fashion a sweet, lovely maiden-shape, like to the immortal goddesses in face; and Athene to teach her needlework and the weaving of the varied web; and golden Aphrodite to shed grace upon her head and cruel longing and cares that weary the limbs. And he charged Hermes the guide, the Slayer of Argus, to put in her a shameless mind and a deceitful nature.[43]

> For ere this the tribes of men lived on earth remote and free from ills and hard toil and heavy sickness which bring the Fates upon men; for in misery men grow old quickly. But the woman took off the great lid of the jar with her hands and scattered all these and her thought caused sorrow and mischief to men.[44]

The implications of this are perhaps best embodied in Helen, daughter of Zeus and the most beautiful woman in the world, whose abduction by Paris leads to a ten year war (Paris is promised Helen by Aphrodite, god of adultery, in exchange for Paris choosing Aphrodite as the fairest of the gods, and thus the recipient of a golden apple). On both occasions man stole knowledge, and in return received reproduction *as a penalty*. It has been suggested that Pandora and Eve are variations of Inanna, 'the Sumerian goddess of love who significantly enough goes down to the underworld and challenges the queen of that realm, Ereshkigal. She eventually causes the death of the male god associated with her, a figure which may be lover, husband or son.'[45] The Sumerians practised agriculture and developed writing, the first language for which there is written evidence, and existed between 7,000 to 4,000 years ago. 'She stirs confusion and chaos against those who are disobedient to her, speeding carnage and inciting the devastating flood, clothed in terrifying radiance. It is her game to speed conflict and battle, untiring, strapping on her sandals.'[46] In early myths, beginning with the Sumerians, women are viewed as the source of evil and violence. They are the mythical embodiment of Darwin's theory of sexual selection.

Needless to say, violence is something that has affected people for millennia. Where historical documents exist allowing an accurate depiction of life at the time, warfare is endemic. European history is often depicted as little more than one war after another. Today, we live in a world where there are more than ten thousand nuclear weapons.

While such facts may make us uncomfortable, human evolution should be a matter of science, not philosophy. Unfortunately, when it comes to human evolution, many academics appear to cross this line. Indeed, this appears to be the reason why the idea that people in their natural environment are generally hungry and always in a dire search for food has held sway, not because it is based on good evidence, for there is no evidence to support this assumption, but because it supports desirable assumptions – that we evolved for the need to overcome the lack of nutrition, rather than to compete over reproduction. Mythical stories of human origins suggest that people knew very well that it was reproduction and not nutrition that led to human conflict.[47]

I will argue that sexual selection is the basis for human cooperation, and that sex, violence and inequality are integral parts of human evolution.

Sexual Selection and Cooperation

In 1971 Robert Trivers considered the evolution of reciprocal altruism – if you scratch my back, I'll scratch yours.[48] Two animals thus cooperate, give each other a benefit, but expend energy in doing so. Examples in the animal kingdom might be the removal of parasites from a member of the same species in return for a reciprocal removal of parasites, or for the parasites to be a source of food, as is the case for more than forty-five species of fish.[49] However, there is a difficulty with reciprocal altruism in explaining cooperation: what if the second animal doesn't reciprocate? Robert Trivers considered the evolutionary arms race between cooperation and cheating, and that this may have played a role in the evolution of the human brain. He used the economists' game Prisoner's Dilemma, as an example of a complex arms race between cooperation and cheating. Prisoner's Dilemma was a common method to explain how cooperative behaviour occurs in economics and is now widely used in biology. Though unnecessary to make the explanation, an explanation that should be self-evident, Prisoner's Dilemma is a simple tool to explain cooperation, and I will use it to demonstrate my argument.

Prisoner's Dilemma is a simple game in which two participants can either cooperate or cheat. The basic idea consists of two people being charged with two crimes: a lesser crime and a serious crime. The police hold them in separate interrogation rooms. They are offered clemency if they accuse the other person of the serious crime. If one accuses the other of the crime (deceit) and the other says nothing (cooperation), then the first is freed and the second receives a very heavy sentence. If they both accuse each other of the serious crime then they are punished, but given a lesser sentence for giving evidence. If neither say anything they are charged with the lesser crime. There are thus four outcomes:

Choice of behaviour		Outcome
A	B	
Cooperate	Cooperate	Both receive a small punishment (X)
Cooperate	Deceit	A receives a heavy punishment, B receives no punishment (Z/W)
Deceit	Cooperate	A receives no punishment, B receives a heavy punishment (W/Z)
Deceit	Deceit	A and B receive a moderate punishment (Y)

The loss is represented as: $W<X<Y<Z$.

The game can be played as a single event or played repeatedly. Points can be assigned and weighted to present different circumstances. Thus instead of jail terms, we might have a game where two animals agree to remove parasites from the other. An animal can have its parasites removed and then refuse to remove the parasites off the other animal (cooperation/deceit); or both can remove parasites (cooperation/cooperation); or both can refuse to remove parasites (deceit/deceit). The only change here is that the differences are counted as benefits rather than losses.

Ten years after Robert Trivers' paper, Robert Axelrod and William Hamilton put the ideas of reciprocal altruism and cheating to the test and held a competition to see which tactic best prevents cheating.[50] The game was played repeatedly on a computer to discover the best tactic over a large number of repeated games. Tit-for-Tat, designed by Anatol Rapaport, in which the tactic was simply to copy the other animal's last move, was the most successful. So if the animal cooperated in the previous encounter by removing the ticks, Tit-for-Tat would cooperate. If the animal was deceptive, then Tit-for-Tat would be deceptive.

This idea that animals cooperate by using a tit-for-tat tactic in reciprocal altruism has become generally accepted as a basis for animal reciprocal cooperation, including human beings.

The problem, however, is that we know humans simply do not cooperate in this manner. Certainly the idea of reciprocal altruism and tit-for-tat might explain the capacity to remove parasites from each other, but it cannot explain New York. Our capacity to cooperate goes far beyond this basic form of cooperation, and it is this unusual cooperation that we are attempting to explain.

Firstly, consider a situation with two male bucks and a female doe in oestrous. Do the males play tit-for-tat and copulate with her one after the other? No. They fight as hard as they can to defeat the other male so that the successful male monopolises the female.

If we interpret this in accordance with Prisoner's Dilemma, the loser 'cooperates' by agreeing not to copulate with the female; the winner 'deceives' his opponent by copulating with her. The weaker party 'cooperates' and the stronger 'deceives' the other. This has been called 'coercive cooperation'.[51]

The game of Prisoner's Dilemma assumes that the two participants are equal. Yet, as I have already stated, sexual selection is based upon the idea that sexual animals are not equal. If the animals were equal, sexual selection would have no effect, since females would not bother to select between equal males. The peacock's tail is an attempt by various peacocks to present an inequality in comparison to other peacocks. Sexual selection concerns the difference in the individual's ability to propagate their genes, and thus it concerns differences in quality between individuals. The males fight to prove their inequality, and the victor is selected by the female.

Let's return to the situation with the two men in the original Prisoner's Dilemma. Let's suggest their names are Craig and Anthony. Both are charged with possession of a large quantity of drugs and carrying an illegal weapon. Craig washes the dishes at the restaurant named 'Anthony's' and occasionally does 'jobs' for Anthony. Anthony owns the restaurant. He is also the leader of a major organised crime group. He has people working for him in extortion and drugs, and is known to have organised the murder of several people. He has men who have worked for him in various jails and he takes care of these men's families while the men are in prison. His networks are extensive. Craig's only important connection is to Anthony himself. In this situation, we know who will cooperate and who will not. This is why we have witness protection programs.

In this case, Anthony and Craig are unequal. Craig is forced to cooperate because of the inequality between himself and Anthony. Status is the label we give to describe the inequality between Craig and Anthony.

An example of this is when a police force enters a 'stone-age' society. The following comes from the first interactions between police and Papua New Guineans. This is the remarkable situation where a man living in a completely 'stone-age', traditional lifestyle is actually being arrested by the police for murder (even more perverse than the Monty Python sketch where the medieval knight Sir Lancelot is arrested by modern British police officers); it is also a perfect example of how the game Prisoner's Dilemma operates in reality:

> Sir Hubert Murray was a lawyer, and held the position of Chief Judicial Officer as well as being the Lieutenant-Governor. It was natural for him to regard the British system of law, transplanted to Papua in the form of the Queensland Criminal Code (Adopted), as one of the highest manifestations of social morality. He may well have been correct in this belief, but since the Papuans did not understand these principles of legality their enforcement merely produced bewilderment and demoralization that have persisted to this day.
>
> The clearest example of this opposition between native custom and British legal precepts is in their respective ideas of individual responsibility. The Tauade realize that acts are the work of individuals but are not concerned, when exacting vengeance, that it should fall on the head of the guilty party. They are satisfied if anyone from the group concerned is made to suffer. ... [T]he following description by a cadet patrol officer of the arrest of a murderer at Zhamora, in the Pilitu Census District, illustrate this:

They [the people of the village] realised that some form of restitution for the alleged killing would be pressed upon them, but could not understand that the person responsible for the crime was the one on whom the punishment should fall. Three successive times I was presented with 'substitutes' (all insignificant gaol fodder in the tribal hierarchy) and when these were sent away after an explanatory talk, I was presented *with* two men, said now to be the joint killers. Later, a sum of £10 was added as an extra inducement. (A.N. Flowers, C.P.O., Pilitu, 23/11/61-29/11/61.)

The government attitude, on the other hand, was that the guilty person, and only he, was to be arrested. This was of course quite incomprehensible to the people.[52]

As would be expected from the manner in which people cooperate, when it came to who admitted guilt for the crime, it was not the killer, but the 'gaol fodder' that was offered. People do not operate as individuals, but as cooperative groups, and, evidently, the killer in this example possessed more cooperative support than the 'gaol fodder'. The killer possessed higher status. A perfect example of how the game Prisoner's Dilemma operates in reality.

This is how humans cooperate. We cooperate competitively. We combine our cooperative forces and then force others to cooperate with us. We combine our status with other people's status, and thus increase our personal status in comparison to the status of others.

We are not the only animal that acts in this manner. Consider the following example from Jane Goodall's work with chimpanzees in Gombe, Tanzania:

Firm friendships, like that between Goliath and David Greybeard, seem to be particularly prevalent among male chimpanzees. Mike and the irascible, testy old J.B. traveled about in the same group very frequently. When I first knew them, J.B. was the higher-ranking of the two, but Mike's strategies with the kerosene cans served to subordinate J.B. along with all the other males. However, once things had settled down, with Mike secure in the top-ranking position, it became apparent that J.B. had also risen in the social ladder. When he was in a group with Mike, J.B. was able to dominate Goliath as well as other males who had held a higher rank than he before Mike's rise. These other males quickly accepted J.B. as second only to Mike, but Goliath asserted his old superiority over J.B. on many occasions when Mike was not part of the group. I well remember one day when Goliath threatened J.B., who had approached his box of bananas. J.B. at once moved away but began to scream loudly looking across the valley in the direction that Mike had taken earlier. Mike must have been quite close, because within a few minutes he appeared, his hair on end, looking around to see what had upset his friend. Then J.B. ran toward the box where Goliath sat, and Goliath, with submissive pant-grunts, hastened to vacate his place – even though Mike took no further active part in the dispute.[53]

In this example we have the following scenario:

Bananas
J.B.   Goliath

Goliath's status is greater than J.B.'s. When Mike reappears, we have the following:

Bananas
J.B.   Goliath
Mike

Mike has joined forces with J.B. and together they can defeat Goliath. Mike is actually higher in status than Goliath. Goliath has greater status than J.B.; however, when Mike is present, J.B. becomes higher in status than Goliath. Mike did not attack Goliath; his mere presence allowed J.B. to attack Goliath and present greater status, despite the fact that individually he was weaker.

This is what I will call 'competitive cooperation', where two or more animals combine their status to defeat one or more other animals of the same species. The most obvious form of competitive cooperation is between parents, where a male and a female combine their forces to obtain a better territory or to defend their young against others of the same species. This is a two-on-two situation, where two mating pairs are competing for a territory. This could be expanded to include a pride of lions, where males are combining their forces to protect the females and young from other males. One should be able to calculate cooperation between parents in accordance with parental investment, so that the loss from cooperation is actually smaller than the benefit in young created. (One can see here how there is a distinct overlap between Trivers' concept of parental investment, Hamilton's concept of inclusive fitness (see page 80), and the Axelrod-Hamilton concept of cooperation. The male cooperates with the female (against others of the same species) where the male's inclusive fitness is increased as against the cost of parental investment and the cooperation. If the inclusive fitness benefit is minimal (e.g. the peacock), then the cooperative input may be minimal. On the other hand, if the benefit is great or essential (e.g. the gorilla), then the cooperative input may be great.) Note that this argument concerns cooperation against others of the same species, and relies upon sexual selection, that is, competition between members of the same species.

As Goodall stated above, such cooperation is common amongst chimpanzees, often between half-brothers. They form an alliance, combining their status to defeat others. The benefits are reproductive:

Faben's privileged position was very apparent at such times for Figan usually shared his sexual possessions with his brother much as he shared precious food items, such as meat. And Figan received a payoff for his generosity: Faben helped to keep an eye on the current lady friend when Figan was momentarily busy elsewhere. However, even Figan and Faben between them could not prevent their female from enjoying occasional clandestine intercourse with one or other of the frustrated lower-ranking males. Such opportunities arose when the attention of the alpha male and his brother was temporarily diverted. Once, for example, when Figan and Faben were intently watching a troop of colobus monkeys with an eye to acquiring monkey meat, three other males copulated with their female in quick succession: neither of the brothers even noticed![54]

Faben was affected by polio and possessed a paralysed arm. It might be expected that he would, individually, have been very low in status compared to other male chimpanzees due to the loss of use in one arm; however, when he and his brother combined their status, together they could defeat all other opponents. The benefit is reproductive: both have greater access to females and the ability to control females.[55]

Chimpanzees take such cooperation further than the simple occurrence of two males cooperating. Chimpanzee communities are known to compete against one another, in what can only be described as warfare. Jane Goodall's researchers at Gombe observed the splitting of a single community into two distinct communities.[56] This involved the following situation:

Kasakela	Kahama
(Including territory and females)	(Territory and females)
Mike	Godi
Rodolf	Dé
Humphrey	Goliath
Figan	Charlie
Faben	Sniff
Jomero	Willy Wally
Evered	Hugh
Satan	
Sherry	

The Kahama community split from the Kasakela community. The Kasakela community proceeded to eliminate each of the Kahama males one by one during, what Goodall described as, 'the four year war'.[57] The Kasakela community took the Kahama community's territory and only the adolescent female, Honey Bee, survived.

The above is an example of competitive cooperation: the Kasakela males were cooperating to compete against the Kahama males and attacked lone Kahama males in groups of three or more.

Such cooperation as the chimpanzees demonstrate is rare. Most mammals compete individually or as a male-female pair; it's rare to observe mammals competing against other mammals in larger numbers, particularly males.

Humans and Competitive Cooperation

Human beings cooperate for the same reason as chimpanzees: we cooperate to compete against members of our own species. The evolution of cooperation in some animals concerns survival, but in humans it concerns competition.

For some people, the idea that humans cooperate to compete is probably completely obvious. One need only look at companies, sporting teams, warfare, boys in the playground, or throughout the anthropological literature, to see men or boys forming groups to compete against their rivals. Women and girls also compete in groups, using both violence and information (gossip) to attack other people, usually other females; however, competition between females tends to be at a much lower level of competition than that between males.

Despite the obvious nature of our cooperation being competitive, this has been ignored by theorists. It is currently accepted in science that humans use reciprocal altruism to cooperate to survive. Yet any consideration of Western Civilisation makes such an idea laughable; are we seriously suggesting that Coca-Cola and Pepsi exist in order for humans to survive, that these two companies are engaged in extensive reciprocal altruism with its consumers, employees and manufacturers? Are we really to suggest that the U.S.A., currently armed with innumerable nuclear missiles, is really an example of extensive reciprocal altruism to ensure the survival of human beings? What of the various examples of sporting teams? Are we seriously to suggest that this is simply some form of extensive 'you scratch my back and I'll scratch yours'? These are examples of pure competition, outright competition, above the needs of survival, and nothing less. They are examples of sexual selection, not natural selection. Every nation is an example of people forming a collective body in order to compete against other nations. A sporting team is an example of a group of people competing against other teams of people. A tribal society is an example of a group of people in competition, whether as nomads or between villages.

Darwin observed the fundamental nature of human competition in *The Descent of Man, and Selection in Relation to Sex*:

> Extinction follows chiefly from the competition of tribe with tribe, and race with race. Various checks are always in action, serving to keep down the numbers of each savage tribe, - such as periodical famines, nomadic habits and the consequent death of infants, prolonged suckling, wars, accidents, sickness, licentiousness, the stealing of women, infanticide, and especially lessened fertility. If any one of these checks increases in power, even slightly, the tribe thus affected tends to decrease; and when of two adjoining tribes one becomes less numerous and less powerful than the other, the contest is soon settled by war, slaughter, cannibalism, slavery, and absorption. Even when a weaker tribe is not thus abruptly swept away, if it once begins to decrease, it generally goes on decreasing until it becomes extinct.[58]

Rather than cooperating in order to survive, Darwin had no doubt as to the ability of humans to overcome all difficulties involved in natural selection:

> Man can long resist conditions which appear extremely unfavourable for his existence. He has long lived in the extreme regions of the North with no wood for his canoes or implements, and with only blubber as fuel and melted snow as drink. In the southern extremity of America the Fuegians survive without the protection of

clothes, or of any building worthy to be called a hovel. In South Africa the aborigines wander over arid plains, where dangerous beasts abound. Man can withstand the deadly influence of the Terai at the foot of the Himalaya, and the pestilential shores of tropical Africa.[59]

Current evolutionary thought depicts humans as miserable animals, pushed to cooperate merely to obtain enough food to survive. Remarkably, these arguments are put forward by people who live in cities of millions of people, who drive cars, fly in airplanes, and live in plenty.

It is worth considering 'reciprocal altruism' in Western society. Consider the purchase of a chocolate bar. I go into a shop that sells chocolate bars; I hand over money, the owner of the chocolate bar hands over the chocolate bar, and thus reciprocal altruism is reflected. Is this simply a matter of 'you scratch my back and I'll scratch yours'? In reality, the following situation occurs:

Chocolate Bar

Purchaser	Shop Owner	Manufacturer
	Other employees.	
Health Commission	Health Commission	
Police	Police	Police
Lawyers	Lawyers	Lawyers
The courts	The courts	The courts
Jail	Jail	Jail

When I go into the shop to buy a chocolate bar, the shop owner has already bought the bar off the manufacturer. The manufacturer and the shop owner are engaged in a game of Prisoner's Dilemma. Depending on who goes first, the manufacturer might deliver the chocolate to the shop owner and the latter might refuse to pay. Alternatively, the shop owner might pay the manufacturer, and the manufacturer might refuse to deliver the chocolate. If the shop owner is the corner shop, and the manufacturer is a gigantic, multi-national, then we should expect that the multi-national, being higher in status, might occasionally or regularly deceive the shop owner.

Even if the shop owner receives the chocolate, the shop owner is then in a second game of Prisoner's Dilemma with the customer. The customer might attempt to deceive the shop owner (steal the chocolate) or the shop owner might attempt to deceive the customer (refuse to hand over the chocolate or sell chocolate that has gone off). The opportunities for deception are numerous, and the status of each group is different, with the manufacturer probably possessing greater status than the shop, which, in turn, probably has greater status than the customer.

The reason the system works so easily, with little manipulation, is because of the government, otherwise known as 'the law'. If the chocolate is bad, the customer or the shop owner can go to the health commission. If any of the parties attempt to steal from one another, they can go to the police. If any of the parties attempt to refuse to pay, they can also go to the courts, whose decisions are enforced by the police. The courts can use fines or jail to punish offenders. The health commission can invariably use fines, jail, or the removal of the product from every shelf in an entire country to punish the offender or prevent others being injured.

The government is thus larger than any of the parties. The government prevents deception by punishing deception. Parties may not even need to punish deception personally, because the government has numerous organisations aimed at preventing deception. The government even has the power to break-up companies into smaller companies if the company becomes too powerful, or to prevent monopolies.

We cannot be fooled for one second into believing that this is simple, basic reciprocal altruism. The purchasing of a chocolate bar involves regulated competition. Manufacturers compete to produce a product that shop owners and customers want to buy; shop owners compete to sell the product to customers; and customers sell their labour to a third party in order to buy the product. This competition is then regulated by the government.

This is the basis of the wealth of Western nations. People produce something to sell, in exchange they receive money, which they use to purchase another product, and all these transactions have been regulated by governments for centuries. Those caught in deception, such as theft, faced punishment. None of this occurs for the reasons of survival, but competition. People using a hunter-gatherer or -horticultural method of living are completely self-sufficient, they produce everything they need and do not need to trade to survive.

The basis of why people in Western Civilisation compete in order to obtain money and products is status. Money buys lawyers, doctors, guards, and weapons. Money buys the cooperation of other people: in exchange for money, they give us their time, expertise and assistance. Most importantly, money buys time in the arena of modern competition: the courts. A person with no money is at the mercy of any person who wishes to take them to court. A bankrupt can be brought to court for virtually any reason by those to whom they owe money. A person does not even have to win a case in order to drain another person's status (money), all a person needs to do is force their opponent to hire a lawyer. Particularly in Anglo nations, a legal case is enough to bankrupt most people.

Cars are an excellent example of the intimidating impact of status. The following is taken from Cialdini's excellent book *Influence*:

> According to the findings of a study done in the San Francisco Bay area, owners of prestige autos receive a special kind of deference from us. The experimenters discovered that motorists would wait significantly longer before honking their horns at a new, luxury car stopped in front of a green traffic light than at an older, economy model. The motorists had little patience with the economy-car driver: Nearly all sounded their horns, and the majority of these did so more than once; two simply rammed into his rear bumper. So intimidating was the aura of the prestige automobile, however, that 50 percent of the motorists waited respectfully behind it, never touching their horns until it drove on.[60]

A car is a symbol of status. It informs another person how much money they have and how many people they can pay to assist them in bankrupting or hurting you. When Marilyn Monroe sang *Diamonds are a Girl's Best Friend* she wasn't singing of feeding her children, she was singing about protection. Her jewels symbolised how many people she, or her partner, could employ to defeat anyone who attempted to harass, harm, or rape her. This is why women, paradoxically, like to wear exactly what some one might want to steal, because the potential thief or rapist knows that the woman may be able to employ numerous assistants. The diamond engagement ring, the necklace of pearls or diamonds, the diamond earrings demonstrate how competitive she will be in being able to buy other people's assistance. This is why expensive engagement rings are associated with the level of love and commitment of the male partner.

For cars the situation is thus:

Conflict

Driver with $50,000 car	Driver with $5000 car
Approx. $50,000 for lawyers	Approx. $5000 for lawyers
Approx. $50,000 for security	Approx. $5000 for security
Approx. $50,000 for debt collectors	Approx. $5000 for debt collectors

Whether the issue is defamation, assault, injury, or negligence, as soon as a person wants to attack someone else the only legal way to do so is in the courts and this costs money. In the above situation, one person has more money available to attack the other person. The car can be replaced by houses, jewellery, clothes, mobile phones or stereos and the argument remains the same. We amass money to compete against other people and we seek symbols that reflect our level of competitiveness. The more money we have to spend, the more money we have to take someone to court, initiate debt collection, protect ourselves with security, and employ private detectives to observe our opponent. In order to make this money, we make or sell chocolate bars or some other product. The less money we possess, the more easily we are intimidated and controlled by others.

This cooperative environment is regulated by 'the law': the courts, the legislature and the executive. Our ability to compete is regulated by the law. Violence is thus outlawed. A poor person and a wealthy person can both buy a gun and shoot their opponent, but such behaviour is heavily regulated. Western nations spend vast amounts of money on courts, judges, jails, the police and prison guards to prevent such behaviour. If such behaviour was allowed, then money would become worth much less, and perhaps worthless.

Money buys cooperation and we respond to money instinctively with fear, the fear of the cooperation that the money can buy.

Status is also relative. A soldier in the army may come from a family with more money than another soldier, but if the second soldier has greater rank, then the latter soldier has greater status. A soldier with more status can obtain the cooperation of more men (other soldiers) to defeat the lower ranked soldier. Status is thus relative to the situation. If a person enters a courtroom, no matter how wealthy the person might be, the court will have greater status (it has the support of the government).[61]

We instinctively feel this; when we enter someone else's house, shop, or office, we enter an environment where we may have limited status. The classic study of this is the Milgram study, where Stanley Milgram attempted to discover how the Jewish holocaust occurred, and how people could be so subordinate as to actually kill millions of people because they were under orders. To do this, Milgram set up an experiment, where an actor – the 'learner' – was strapped into a chair that delivered shocks, and a member of the public – the 'teacher' – was asked to read the learner a question, and, if the answer was incorrect, deliver an electric shock to the learner. The shocks were increased with each incorrect answer until 450 volts were delivered. The teacher was, of course, unaware that the learner was an actor, and the 'learner' acted as if the shocks were real. Leading up to the experiment, members of the psychology fraternity – students, psychology majors, colleagues, and psychiatrists – predicted that one or two percent of the general public would go all the way to the 450 volt measurement. Milgram found that 65 percent of participants delivered the fatal shock, despite the fact that they were simply directed to by the researcher, and there was no penalty for stopping the test. Men and women delivered shocks with equal willingness, and 100 percent of the subjects delivered 300 volts of shock to the learner. Cialdini summarises Milgram's conclusions:

> It has to do, he says, with a deep-seated sense of duty to authority within us all. According to Milgram, the real culprit in the experiments was his subject's inability to defy the wishes of the boss of the study – the lab-coated researcher who urged and, if need be, directed the subjects to perform their duties, despite the emotional and physical mayhem they were causing.[62]

One observer commented:

> I observed a mature and initially poised businessman enter the laboratory smiling and confident. Within twenty minutes he was reduced to a twitching, stuttering wreck who was rapidly approaching a point of nervous collapse. He constantly pulled at his earlobe and twisted his hands. At one point he pushed his fist into his forehead and muttered: "Oh, God, let's stop it." And yet he continued to respond to every word of the experimenter and obeyed to the end.[63]

From Milgram's study we can see that all it takes is a white lab-coat, an assertive pattern of behaviour, and a position of authority to make a person harm or kill another person, without exception. The study has been repeated in Holland, Germany, Spain, Italy, Australia, and Jordan, with similar results.[64] Similar experiments concerning the effect of authority have been performed using business suits, uniforms, titles, and qualifications and similar results have been found.[65]

Milgram's work was aimed at understanding how the Jewish holocaust occurred, with the belief that ordinary people were responsible; however, this does not appear to be the case. In reality, those that performed the most secret exterminations were specifically selected for the position. The Einsatzkommando 1005 "the most secret of all the 'Special Commands' which were active all over Eastern Europe and the Balkans"[66] and responsible for millions of killings were specifically selected. In the words of Prosecutor Kurt Tegge, they were selected by practising on Germans:

> 'The high SS officers who headed the sections of the 1005 were almost all originally trained in the euthanasia programme,' he said. 'That was in the very beginning – in 1938-9. We know now how it worked: they would be ordered to kill all patients in certain institutions and hospitals for the insane, incurably ill or severely retarded.' They would be told that "this is a difficult job, but, for the sake of Germany, and the patients themselves, it has to be done". If it turned out that they were sickened by the assignment – and many were – they would be asked on another occasion to try again, and then once more. By the third time, they had either overcome their objections and revulsion and would then be considered capable of "hard" assignments – or else they had shown themselves incapable of this type of work and were transferred to other duties. It is certain that this was quite deliberately used as the training ground for the SS. It was considered that if these men turned out to be capable of killing sick German children and old people, they'd be capable of anything.'[67]

The idea that anyone is capable of astonishing acts of violence appears not to be true. While it is doubtful that these selected individuals were entirely responsible for the 14 million civilians (6 million Jews, 5 million Russians, 2 million Poles and 1 million gypsies, German free-thinkers, and German mentally or incurably ill[68]) that were murdered by Germans and their allies, they may have been responsible for the majority. Staatsanwaltz prosecutor Richard Dietz found '132 SS men in Byelorussia managed to kill 138,000 men, women and children in seven months by shooting each in the back of the neck as they knelt awaiting the shot, and carefully kept book while doing so.'[69] Astonishingly, these men killed approximately 650 people each day and noted down each death. Milgram's work appears to show that we do obey authority; however, the notion that every person – even those that volunteer for the SS – is capable of killing civilians appears not to be correct, only certain people are capable of this. I am unaware of whether the Soviet NKVD used similar techniques to select exterminators.

We obey money and we obey authority. We instinctively observe how much status an authority figure has and obey that person if their status appears to be greater than ours. When we enter a courtroom, every person instinctively recognises how much power the court has and obeys the court's decision. When we see a police officer, we obey the police officer. Evidently, when we enter a laboratory and are instructed by a man wearing a lab-coat to electrocute another person, to a large degree *every* person obeys the man in the laboratory coat, who appears to possess authority, because in that environment the man in the lab-coat does possess authority. If we did not do this, then Western Civilisation would not exist since chaos would ensue. People can enter entirely strange environments and yet appreciate another person's status very quickly. If we enter a strange environment and do not recognise another person's status then we might apologise 'sorry, I thought you were the...' or be dismayed when if treated a person with far more reverence than we needed to. Many people will underestimate their capacity to obey – particularly young males. This was reflected in the study on how long people take to respond to a stationary vehicle at traffic lights:

Later on, the researchers asked college students what they would have done in such situations. Compared to the true findings of the experiment, the students consistently underestimated the time it would take them to honk at the luxury car. The male students were especially inaccurate, feeling that they would honk faster at the prestige- than the economy-car driver; of course, the study itself showed just the opposite. Note the similarity of this pattern to much other research on authority pressures. As in Milgram's research, the midwestern hospital-nurses' study, and the security guard-uniform experiment, people were unable to predict correctly how they or others would react to authority influence. In each instance, the effect of such influence was grossly underestimated. This property of authority status may account for much of its success as a compliance device. Not only does it work forcefully on us, but it also does so unexpectedly.[70]

(Studies comparing how people predict their actions and their actual actions pose serious questions as to the reliability of self-report questionnaires.)

It is rare to find incidents where individuals acting alone (rather than with supporters) specifically do not cooperate and place their lives at risk by not cooperating. One remarkable example is that of Hugh Thompson Jnr., a U.S. military helicopter pilot, who saw the massacre of hundreds of Vietnamese civilians by U.S. soldiers at My Lai village, on 16th March, 1968. Thompson aimed his guns at his own soldiers, ordered them to stop shooting, and then called in additional air support to assist injured civilians.[71] Yet, one must expect that this was in part due to the U.S. military being one where diverse opinions were not heavily punished. This might be contrasted to the BBC's *World at War* documentary, in its *Making of* episode, where a woman retells the story of her meeting an SS officer who was so disgusted at German behaviour that he placed himself continuously in risky situations in the hope of being killed, yet could not defy his own army.

The reason why people obey others in this manner is because of cooperative competition: when faced with someone who appears to have more allies than ourselves, we obey or defer to them. A young man may thus obey an older frail man, if the latter has greater status, despite the obvious fact that the younger man could easily defeat the older man in a fight between only themselves.

If you don't believe this, and you believe that humans use tit-for-tat or reciprocal altruism, go into your local supermarket and play 'deceit'. We all know what happens if your deceit is observed: the supermarket calls in the government – police, courts and jail – and the government will punish you for playing 'deceit'. We don't have the choice in whether we play cooperate or deceit, we are forced to cooperate by the government, by 'the law'. People are well aware of when they will be punished by the government for behaving in certain ways. One might suggest that we have a code book in our brains that learns the details of what we can be punished for, and constantly updates the details as the rules change.

This does not mean that humans are equal in 'the eyes of the law' either. A person who can afford a lawyer can obtain a ruling against a person who cannot afford a lawyer, and get all their fees paid for by the losing, poorer person. This is especially true in countries using the common law where barristers – rather than the judges in Roman law nations – can practically run or control the proceedings. The legal fees alone may prevent a large number of people from being able to gain damages against another person.

The tendency to obey is also true for a person in a cooperatively-simple society; a person entering a community in which they have no allies will instinctively recognise that a group of allied men, such as several brothers, will possess more status than him or herself. Equally, a person in a hunter-gatherer or hunter-horticultural society who has no allies will recognise that other people *within* the community with more allies (brothers, half-brothers, friends or political allies) will have more status than they possess. Those communities using the hunter-gatherer or -horticultural method are no different to the courtroom. If those people with extensive alliances in the community make a decision, then the man without any alliances has little choice but to accept their decision.

The following situations are thus identical:

Man in white lab-coat	> status 'Teacher'
Judge	> status Plaintiff or
Government	Respondent
John Smith	> status John Robertson
David Smith (brother)	Robert Robertson (brother)
Craig Smith (brother)	Allan Hughes (ally)
Robert Smith (half-brother)	
Ian Smith (half-brother)	
William Jones (ally)	
Joshua Jones (ally)	
Peter Jones (ally)	

In each of the above situations one person has more allies than the other, and the latter has little choice but to obey the decisions of the former (or possibly face dire consequences). In each case, the first person or group of people has greater status than the latter. In each game of Prisoner's Dilemma, the latter feels compelled to play 'cooperate' while the former can decide whether to play 'deceit' or 'cooperate'. A judge can decide to play 'cooperate' or 'deceit' with either plaintiff or respondent. John Smith can decide to assist or harm John Robertson. The 'teacher' cooperates with the man in the lab-coat who occasionally plays 'deceit' and tells the teacher to harm the learner.

In order to compete, in order to prevent ourselves from being manipulated and forced to cooperate, we must increase our status by obtaining more allies. The fewer allies we possess, the more we can be manipulated by other people; the more allies we possess, the less we are manipulated and the more we can manipulate others to our own advantage.

The situation and its implications are beautifully summarised in the following passage by Cicero, in his book on friendship, *Laelius*. Cicero, a Roman statesman, observed that human cooperation is far more complicated and remarkable than a simple exchange of services. Cicero elevates friendship to the point of love, the desire of two individuals to increase each other's position out of mutual affection:

I have thought a lot about friendship. And the more I think about it, the stronger my feeling is that the important question one has to decide is this. Does the fact that people need friendship mean there is some weakness and deficiency in themselves? If so, this would mean that their main purpose in acquiring friends is to exchange services with another person, that is to say to give and receive benefits which it would be beyond their individual powers to grant or obtain on their own account. But surely this giving and receiving constitutes merely *one* feature and consequence of friendship. As for its origins, do these not, rather, lie in something altogether more primeval and noble, something emanating more directly from the actual processes of nature?

For goodwill is established by love, quite independently of any calculation of profit: and it is from love, *amor*, that the word for friendship, *amicitia*, is derived. It is true that you do often find people extracting advantages from someone by pretending they are his friends – the reason they are courting him so deferentially is because they are after some immediate benefit. But when there is real friendship, no element of falsity or pretence can possibly enter into the matter. Friendship simply cannot help being genuine and sincere all through. And that is why one is obliged to conclude that it must be a product of nature rather than of any deficiency. It cannot under any circumstances be derived from any calculation of potential profit. It comes from a feeling of affection, an inclination of the heart.[72]

In *The Roman World*, Martin Goodman further defines this idea of 'friendship' in a dangerous environment where politics and friendship are evidently inseparable:

But, more importantly, advancement to the next rung in this clearly stratified hierarchy was usually possible only with the active support and friendship (*amicitia*) of other politicians, so that the development of reliable political supporters (*amici*), through marriage links or assiduous attendance, was a major concern of all politicians who had any ambition.[73]

Cicero outlines that human cooperation is not due to weakness or the exchange of services for survival, but rather out of a common bond of mutual assistance. Cicero continues of his own friendship with Africanus:

Consider Africanus, for example. Did he *need* me? Of course he did not. Nor, for that matter, did I need him. I was attached to him because I admired his fine qualities; and he returned my feeling because he also, on his side, appeared not to have formed too bad an opinion of my own character. And as we got to know one another better, our mutual affection grew. Certainly the association did bring practical advantages – and on a considerable scale. But in the formation of our friendship no expectations of such a kind played even the slightest part."[74]

The friendship between them was not based on weakness, but rather, the exact opposite – strength. By forming a friendship they increased their mutual status. If chimpanzees could verbally communicate, I have no doubt that in dealing with the fight over the box of bananas with Goliath (who was stronger than J.B., but not Mike) that J.B. or Mike would have said exactly the same thing as Cicero. Did Mike need J.B. or did J.B. need Mike? No, even if J.B. had lost the bananas to Goliath, there was plenty of food available in the forest. But as Cicero profited from Africanus's assistance, so J.B. profited from Mike's. And, no doubt, particularly in mating with females, Mike would have profited from J.B.'s assistance. Their mutual assistance was in producing strength, not out of a need due to weakness.

Thus we have an important instinct: humans have a natural desire for friendship or alliance with other humans in order to increase their personal status. We do not form friendships in order to survive, but to succeed. We do not form friendships and alliances with other people as part of a mutual exchange of goods and services, but out of a desire to assist each other to succeed, a desire to increase each other's status. Without friendship, we are individuals, at best exchanging to survive.

The competitiveness of cooperation is captured in *The Gift* by Marcel Mauss, in which Mauss presented the importance of reciprocal giving in otherwise subsistence societies. This reciprocal giving is not a method of assisting others to survive or even simply forming alliances, but a form of competition. Rival communities or families attempt to present larger gifts in order to show their status relative to the community or family receiving the gift. The capacity for human beings to create excess produce is thus a form of competition in its own right. Mauss's work is summarised by Claude Lévi-Strauss:

> It is well known that in many primitive societies, particularly those of the Pacific Islands and the North-west Pacific coast of Canada and Alaska, every ceremony celebrating an important event is accompanied by a distribution of wealth. Thus in New Zealand the ceremonial offering of clothes, jewellery, arms, food and various goods was a common feature of Maori social life. These gifts were made on the occasions of births, marriages, deaths, exhumations, peace treaties, crimes and misdemeanours, 'and many other things too numerous to mention'.[75]

> These gifts are either exchanged immediately for equivalent gifts or are received by the beneficiaries on condition that at a later date they will give counter-gifts often exceeding the original goods in value, but which in their turn bring about a subsequent right to receive new gifts surpassing the original ones in sumptuousness. The most characteristic of these institutions is the *potlatch* of the Indians of Alaska and the Vancouver region. During the *potlatch* considerable valuables are transferred in this way, sometimes amounting to several tens of thousands of rugs handed over in kind, or in the symbolical form of copper plaques whose face value increases in terms of the importance of the transactions in which they have figured. These ceremonies have a triple purpose, viz., to return gifts previously received, together with an appropriate amount of interest, sometimes as much as 100 per cent; to establish publicly the claim of a family or social group to a title or prerogative, or to announce officially a change of status; finally, to surpass a rival in generosity, to crush him if possible with future obligations which it is hoped he cannot meet, so as to take from him his prerogatives,

titles, rank, authority and prestige.[76]

This might be compared to the attempt of Roman politicians to increase their status by holding festivities (such as gladiatorial contests) or the manner in which modern societies hold sporting events, such as the Olympics, to increase their status.

> Doubtless the system of reciprocal gifts only reaches such vast proportions among the Indians of the North-west Pacific coast, virtuosi who display a genius and an exceptional aptitude for the treatment of the fundamental themes of primitive culture. But Mauss has been able to establish the existence of analogous institutions in Melanesia and Polynesia. For example, it is certain that the main purpose of the feasts of several New Guinea tribes is to obtain recognition of a new *pangua* by an assembly of witnesses, that is, the same function which, according to Barnett, is the fundamental basis of the Alaskan *potlatch*. The same author sees the desire to go one better than anyone else as a characteristic peculiar to Kwakiutl ceremonies, and regards the interest-bearing loan as a preliminary transaction to the *potlatch*, rather than as one of its modalities. Doubtless there are local variations, but the various aspects of the institution form a whole found in a more or less systematized way in North and South America, Asia and Africa. It is a question of a universal mode of culture, although not everywhere equally developed.[77]

This reflects the fact that, in environments similar to early humans, people are far from starving, but instead use their excess produce to compete in exchanges of status. For the Yąnomamö, feasts were also competitive events:

> The hosts, too, have an opportunity to display themselves and strut before their guests. Moreover, the very fact that they have given the feast is in itself a display of affluence and surfeit apparently calculated to challenge the guests to reciprocate with an equally grandiose feast at a later date. Indeed, in some areas so much food is provided that the participants competitively drink enormous quantities of banana soup or peach palm gruel, vomit it up, and return for more.[78]

This is the more positive side to human cooperation; with competitive cooperation we can rise above the basic need for survival. We can multiply our personal strength many times over due to the support of other human beings. We can compete with our surplus. We can also gather our status together (taxes) to fund hospitals, education, and infrastructure.

Yet this cooperation can also lead to competition that is destructive. Keeley, in his book *War Before Civilisation*, came to the following conclusion about warfare in prestate societies:

> One author has very liberally estimated that more than 100 million people have died from all war-related causes (including famine and disease) on our planet during this century. These deaths could be regarded as the price modern humanity has paid for being divided into nation-states. Yet this appalling figure is *twenty times smaller* than the losses that might have resulted if the world's population were still organized

into bands, tribes, and chiefdoms. A typical tribal society lost about 0.5 percent of its population in combat each year. Applying this casualty rate to the earth's twentieth-century populations predicts more than 2 *billion* war deaths since 1900.[79] (Author's emphasis.)

In studies of the South American Yąnomamö and Jivaro; the Papuan Mae Enga, Dugum Dani and Huli; and in the Australian Murngin, death rates from warfare are more than twenty percent of male deaths.[80] Archaeological excavations reveal similar levels of violence. Keeley notes:

> My own first excavation training was on a prehistoric village with obsidian arrowheads embedded in the bones, missing heads, and other signs of violent death were so common that our excavation crew referred to burials as "bad sights." As a matter of fact, one distinctive characteristic of this period in central Californian prehistory is that about 5 percent of all human skeletons contain embedded arrowheads – which, of course, represent only the most obvious evidence of death in warfare. Indeed, several of these prehistoric cases seriously underestimate the number of violent deaths because only individuals with projectiles in their bones are counted as war deaths. ... Contrary to arguments that tribal violence increased after contact with Europeans, the percentage of burials in coastal British Columbia bearing evidence of violent traumas was actually *lower* after European contact (13 percent from 1774 to 1874) than the very high levels (20 to 32 percent) evidence in the prehistoric periods.[81]

Humans in cooperatively-simple societies are very competitive. Warfare is endemic. If a typical tribal society possessed 100 people, an annual death rate of 0.5 percent means that one person would die due to warfare in that community every two years. Since half the population would be children, each of the twenty-five men would have slightly less than a 2 percent chance of being killed every year (assuming that occasionally women or children died, thus the 1 in 50 chance is reduced). For a larger tribal society of 200 people, one person – probably a man – could die due to warfare every year.

Keeley explains the reason for this:

> But how can such high losses be reconciled with the low casualty rates generally observed in primitive battles, where action is often broken off when both sides have suffered a few dead? Part of the answer lies in the higher sortie rate of primitive warriors. As was noted earlier, warfare occurs much more frequently in most primitive societies than in civilized ones. Thus a relatively low loss rate per war, battle, or raid can cumulate very rapidly to catastrophic levels.[82]

To put this into perspective, if a city with one million people suffered a casualty rate of 0.5 percent a year, this would mean that five thousand people would die each year due to warfare, or one hundred thousand over twenty years. A nation that maintained a population of twenty million would thus suffer two million war deaths over a twenty year period. This is, no doubt, a major reason why the human population was at a very low level for tens of thousands of years, until relatively recently. In other words, the formation of larger societies has led to lower levels of death due to warfare, allowing for an increase in the number of humans currently living.

The basis for such violence is sexual selection. In cooperatively-simple societies, most wars begin as fights over women, which can escalate into long-term wars. Men can be polygynous, so they seek to increase the number of wives they have. If one man has twenty allies and another man has two allies and a desirable daughter or wife then we should expect that the former man would consider taking that girl or woman as his wife. Genetically, any man who did so and gained a wife would produce more children and leave more genes in the next generation. Such an attempt to remove a female will obviously initiate a response, which can escalate into long-term conflict. (Indeed, as we shall see in later chapters, women are so central to these societies that the men are not merely fighting over reproduction, but cooperation itself, since people use arranged marriages to create cooperation. The fact that such warfare is not based on need has already been considered earlier in this chapter – people in these societies are healthy and vigorous.)

Human beings cooperate in order to improve their capacity to exist beyond mere survival. However, our cooperation comes at a price: we improve our general welfare, but we increase our level of competition, particularly violent competition. Through friendship men increase their ability to compete and obtain more wives, but they also increase the level of competition to remarkable levels.

In Western societies this system has achieved remarkable levels, with large governments that enforce cooperation. At one end of the spectrum is communism where the government controls every aspect of life, and at the other end is smaller government that enforces fewer rules. The utopian society is simply a variation on the theme of third party enforcement of cooperation.

People in hunter-gatherer societies have no material wealth, no taxes, and no equivalent of a Western government, so the question must be asked: how do they cooperate without a government to force genetically diverse people to be cooperative? The rest of this book will argue that we use the arranged marriage system to form alliances between men for competitive purposes. It is the fact that these people use their own bodies for exchange which makes them 'cooperatively-simple'. It is not the fact that we cooperate to compete that defines us; this is something we share with chimpanzees, yet we are unlike them. It is the method we utilise to cooperate to compete that defines us.

In conclusion, I argue that sexual selection can only operate because sexual animals are unequal. This difference is called status. Sexual animals, particularly males, compete by using their higher status to force other animals to cooperate (coercive cooperation). Humans compete by combining their levels of status to force other humans to cooperate (competitive cooperation). Humans are capable of combining the status of two individuals or millions of individuals.

Chapter 3
Arranged Marriages

History of the Study of Arranged Marriages

The arranged marriage systems of cooperatively-simple societies must be amongst the most carefully studied of human behaviours. In 1851, Lewis Henry Morgan, a lawyer, studied the Iroquois in North America 'a confederation of five tribes formed c.1400 AD by the legendary Hiawatha: Mohawk, Onondaga, Seneca, Oneida and Cayuga. Through their civil and military organization, they dominated neighbouring peoples in the pre-colonial era and successfully resisted and negotiated European encroachment for nearly two centuries afterwards.'[1]

Morgan produced the work *League of the Ho-de-no-sau-nee or Iroquois*, in which he described the method in which the Iroquois people organised their social structure. Morgan's chief interest was how the Iroquois named people in a different manner to Western societies. He argued that the successful confederation of the Iroquois was based on 'an elaborate concept of consanguinity'.[2]

In both Iroquois and Western societies, a person calls his actual mother 'mother'. However, an Iroquois would call his mother's sister 'mother', while a Caucasian would call that woman 'aunt'. An Iroquois would call his father's brother 'father', while a Caucasian would call that man 'uncle'.

In anthropology, what in Western culture would be described as immediate family are described as 'lineal' kin; that is they are part of one's lineal descent, from parents, to children, to grandchildren, and so on. Those outside of this lineal descent are described as 'collateral' kin. In Western society we carefully separate lineal and collateral kin, and would never call a cousin 'brother' or 'sister'. The Iroquois, however, specifically describe cousins as 'brothers' and 'sisters', thus mixing lineal and collateral kin. This description of collateral kin as lineal kin is known as classificatory terminology; in other words, these are not real brothers and sisters but are classified as brothers and sisters – they are 'classificatory' brothers and sisters.

Morgan wondered whether this was unique to the Iroquois, or whether it was common amongst similar peoples. He wrote to numerous missionaries and other people that worked with what were termed 'primitive' people, and asked whether those people used systems similar to the Iroquois. The responses were positive and in 1871 he presented the information given to him in a larger work *Systems of Consanguinity and Affinity of the Human Family*. In this work he argued that 'classificatory' terminology is a universal feature of people living in 'primitive' societies. As part of his argument, he produced 200 pages of tables as evidence of this universality. In other words, it was a common feature of people living in societies without extensive market economies to mix lineal and collateral kin. Thus it was common for people to call their mother's sister 'mother' instead of 'aunt'. This work established anthropology as a discipline in its own right and its importance was based on the possibility of understanding human evolution and the evolution of civilisation. People believed that the study of such people could be a window into the pre-historical past.

Some confusion was created by another aspect of these societies, that as a man had many 'sisters' he also had many 'wives', and as a woman had many 'brothers' she also had many 'husbands'. Faced with this absurdity, in Western terms, Morgan suggested the concept of group marriage: that originally many men were married to many women. In his letters with Morgan, Charles Darwin dismissed this concept as a possibility. Even so, it held sway until further evidence showed that the 'wives' and 'husbands' were simply in that classification. As men labelled 'brother' were not necessarily brothers, so men labelled 'husband' were not necessarily husbands. These were classificatory terms. The issue of who marries whom and what this meant, however, continued to be a major subject of contention. At the centre of this argument is the incest taboo. In 'primitive' societies, people could only marry those in the correct terminological group, 'husbands' with 'wives', and all other marriage was 'incest', despite the fact that the people may not be closely related. Incest fascinated intellectuals, such as Sigmund Freud, in the late-nineteenth and early-twentieth centuries; most people are so ignorant of anthropology that they do not understand Freud's concepts, such as the Oedipus complex, stem from anthropology rather than psychology. Indeed, Freud's work could be interpreted as the greatest moment of influence for anthropology, rather than psychology. (Many people may wonder why Freud was so concerned with men being attracted to their mothers, but in 'primitive' societies, men cannot marry their 'classificatory' mothers or mothers-in-law. These 'primitive' societies, for early psychologists, were a window into the human psyche, stripped of Western influence.)

The implication of Morgan's work has been a source of tremendous debate in anthropology for more than a century. What does it mean if people describe lineal and collateral kin in similar terms? Are these simply different methods of describing people or does it have a larger meaning? Why are there marriage restrictions? This created two schools within anthropology: those who thought that these were simply cultural manifestations, and those who thought this affected the structure of the society. This was further complicated by the fact that there was no universal method of classification. For instance, some Australian Aborigines used four sections (husband/wife, sister/brother, father/mother and father-in-law/mother-in-law) while others used eight. While the systems people used were similar, they were rarely identical.

In 1949 Claude Lévi-Strauss published *The Elementary Structures of Kinship*, it was his attempt to create a universal theory. Previously, another French anthropologist, Marcel Mauss, had advocated the importance of exchange, and applied this to material objects, such as food or blankets, as was described in the previous chapter. Lévi-Strauss extended this idea into marriage; he argued that what was most important was the idea that people were exchanging marriage with one another, and this is what bound them in alliance. Any differences between peoples in kinship terminology or social structure were simply variations on the general theme of exchange. Lévi-Strauss viewed kinship in terms of politics or affinal alliance:

Moreover, for Lévi-Strauss it was marriage, or more precisely affinal alliance –
namely the repetition of marriages over time – that was the significant element in
social organization not descent, as it had been for Radcliffe-Brown. Indeed, although
recognizing the importance of terminologies in principle, Lévi-Strauss virtually
ignored them in practice, mostly preferring to concentrate on the data of marriage rules
in establishing the characteristics of particular kinship systems. In addition, he took
from Mauss ... the notion of exchange and, in combination with exogamy, applied it to
marriage to create a whole theory of affinal alliance. According to this theory, the incest
taboo was the foundation of society because although its boundaries were culturally
variable, it was essentially the negative counterpart to a principle of reciprocity: one's
own group was required to forego its own women in marriage for the sake of giving
them to another group and obtaining theirs in return.[3]

Lévi-Strauss's work thus identified marriage, alliance and exogamy as the important features
of the 'primitive' society. Lévi-Strauss's emphasis on the biological rather than the cultural
aspects of kinship has made *The Elementary Structures of Kinship* his most controversial text.
It is interesting that after writing this highly biological text, he later moved on to concentrate
on more cultural issues – perhaps suggesting that the book was poorly received in a discipline
that tends to view biological issues with disdain. Anthropologists such as Rodney Needham,
continued to attack Lévi-Strauss's work even 25 years later, emphasising terminology's
linguistic character and abandoning any link between terminology and alliance, with its
heavily biological overtones.[4] Though this may seem an abstract debate, one only needs to
consider the words of Edward Tylor, who first suggested the vital importance of exchange and
alliance, to appreciate the potential competitive implications of alliance:

> Exogamy, enabling a growing tribe to keep itself compact by constant unions between
> its spreading clans, enables it to overmatch any number of small intermarrying groups,
> isolated and helpless. Again and again in the world's history, savage tribes must have
> had plainly in their minds the simple practical alternative between marrying-out and
> be killed out.[5]

Considering these words and the tendency for the social sciences to be fairly left-wing, it is easy
to understand why many in anthropology wanted to destroy any link between terminology
and alliance. Edmund Leach, while lauding Lévi-Strauss as the greatest anthropologist of his
generation, was scathing of *The Elementary Structures of Kinship*, and specifically linked Lévi-
Strauss's work with Tylor's:

> The big book starts off with a very old-fashioned review of 'the incest problem'
> which brushes aside the substantial evidence that there have been numerous historical
> societies in which 'normal' incest taboos did not prevail. This allows Lévi-Strauss to
> follow Freud in declaring that the incest taboo is the corner-stone of human society. His
> own explanation of this allegedly universal natural law depends upon a theory of social
> Darwinism similar to that favoured by the English 19th-century anthropologist Edward

Tylor. The latter maintained that, in the course of evolution, human societies had the choice of giving their womenfolk away to create political alliances or of keeping their womenfolk to themselves and getting killed off by their numerically superior enemies. In such circumstances, natural selection would operate in favour of societies enforcing rules of exogamy which Tylor equated with the converse of the incest taboo. So also does Lévi-Strauss.[6]

Indeed, the emphasis of terminology being purely linguistic and cultural has dominated in anthropological departments in Anglo nations. For anybody with a biological perspective, however, *The Elementary Structures of Kinship* is a monumental text and argument.

In this work, Lévi-Strauss argued that these 'primitive' societies are dual societies. They are dual because one half of the society are given terms of blood-relations (mother, father, sister and brother), while the other half are given terms of marriage-relations (mother-in-law, father-in-law, husband, wife).

The society is thus separated into two halves (called moieties). One half marries the other half, which is repeated over generations and this binds the society. If a person married someone in their own half, this is known as 'incest' because those people are breaking the rules that bind the society – not because the people getting married are necessarily closely related.

It is easiest to imagine this by separating a single society into two groups with different surnames, let's say 'Robertson' and 'Miller'. The Miller women marry the Robertson men, and then their daughters, now with the Robertson surname, marry the Miller men. Thus in one generation Miller women are married to the Robertson men, and then, in the next generation, the women are 'returned' when the Robertson women marry Miller men. And the process can then be repeated in subsequent generations.

The following depicts the structure of marriages amongst Kamilaroi people of northern New South Wales:

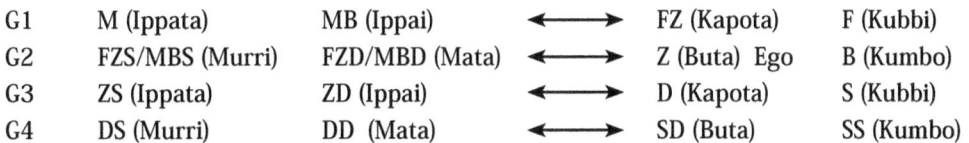

G1	M (Ippata)	MB (Ippai)	⟵⟶	FZ (Kapota)	F (Kubbi)
G2	FZS/MBS (Murri)	FZD/MBD (Mata)	⟵⟶	Z (Buta) Ego	B (Kumbo)
G3	ZS (Ippata)	ZD (Ippai)	⟵⟶	D (Kapota)	S (Kubbi)
G4	DS (Murri)	DD (Mata)	⟵⟶	SD (Buta)	SS (Kumbo)

G = Generation. Marriage = ⟵⟶

Figure 1: Kamilaroi structure of marriage. Note the term 'ego', the reference person to whom all others are relative. People's relationships to 'ego' are described in terms of being either actual kin or as classificatory kin. Arrows indicate marriage, with D or Z marrying into the alternate lineage. The symbols infer the following relationships: B = brother; D = daughter; F = father; M = mother; S = son; Z = sister. Adapted from L.R. Hiatt *Arguments about Aborigines*.[7]

G1	Ippata ♂	Ippai ♀		Kapota ♀	Kubbi ♂

G2	Murri ♂	Mata ♀		Buta ♀	Kumbo ♂

G3	Ippata ♂	Ippai ♀		Kapota ♀	Kubbi ♂

G4	Murri ♂	Mata ♀		Buta ♀	Kumbo ♂

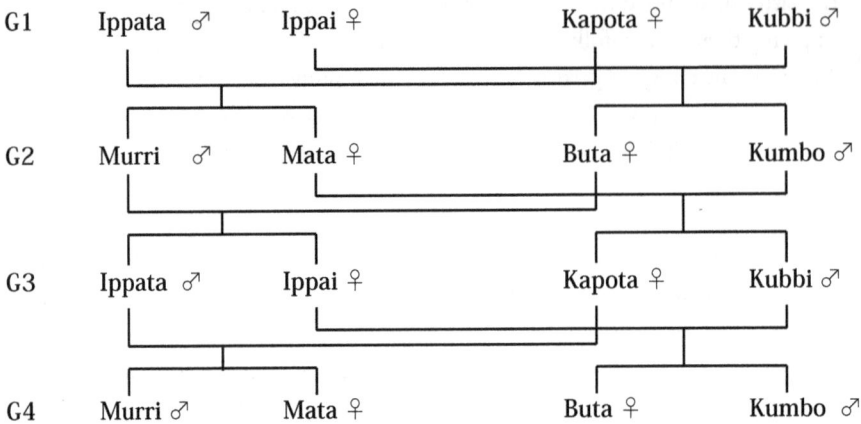

Figure 2: Kamilaroi structure of marriage, simplified view.

In Figure 1, one can see how the classificatory names are repeated in the first and third generations, and then the second and fourth generations. This could then be repeated infinitely. One can see how the brother/sister, husband/wife, brother-in-law/sister-in-law and mother-in-law/father-in-law dualities are created by the system.

The importance of this is that a man will relate to four categories of women. They will be described as 'mother-in-law', 'mother', 'sister' and 'wife'. As expected, he can only marry one of the four groups of women. He calls certain women 'wife' not because they are his wives, but because that is their position in relation to him.

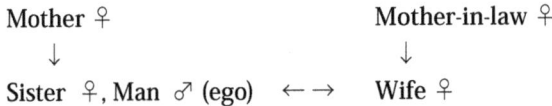

Mother ♀   Mother-in-law ♀
↓   ↓
Sister ♀, Man ♂ (ego)   ← →   Wife ♀

Thus in the middle line of the Kamilaroi example, a man classified 'Kumbo' can only marry a woman classified 'Mata'. He cannot marry the women classified as Kapota (mother/daughter), Ippata (mother/daughter-in-law), or Buta (sister) – this would be 'incest'.

The system classifies according to rules rather than reality. Thus a man may want to marry his 'sister' or 'mother' because these women might be his age and third or fourth cousins. The system regulates marriage according to laws or rules rather than to personal needs or desires. Marrying someone in the incorrect classification is incest because it is breaking the rules of reciprocity that bind the society. The individual's desires are secondary to the important flow of marriage between the two exogamic groups (moieties). It is a system that reduces competition for potential partners.[8]

There are two restrictions: by descent and by generation. Thus a man cannot marry a woman in his mother's or daughter's generation – this would be taking a woman from other generations and 'incest'. In his own generation he can only marry the women in the other moiety and not his own, which would be taking a woman from the other moiety and 'incest'. One can see how the system attempts to regulate and share marriage, and prevent individual biological greed.

The same situation is true from the woman's point of view: men fall into four categories, of which she can marry only one:

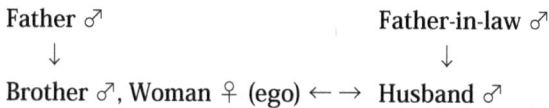

Father ♂   Father-in-law ♂
↓   ↓
Brother ♂, Woman ♀ (ego) ← → Husband ♂

In anthropology these four categories are known as sections. The importance of it is that it regulates quite systematically who can marry whom. Thus a man or woman can only marry one quarter of the available women or men. These societies are polygynous – a man can marry more than one woman – so the limitation of available partners is very important. A man may live in a society where there are various young girls in the community, but he will only be able to marry those whom fall into the correct position relative to himself. If he marries a different woman this is incest. If a man attempts to marry such a woman, he is not merely doing something 'wrong' he is removing a possible wife from those who were expecting to marry that woman.

Such events do occur and often create tremendous turmoil within the society. For instance, a polygynous man may father his first child when he is twenty and his last when he is fifty-five. This means that all these children will be in the same generation, in terms of their classification, despite the fact that there are thirty-five years between them. In these situations, men may attempt to bend the rules. Since there is only a small number of women in the community – let alone in each generation – people attempt to bend the rules in order to ensure that they marry, rather than to stick to the rules. But, of course, any attempt to 'gain' a wife, removes a potential wife from other men, and thus the turmoil.

Lévi-Strauss explains how common these dual societies are:

> The distribution of dual organizations has features which make this type of organization particularly remarkable. These features are not apparent among all peoples but are encountered in all parts of the world, and generally at the most primitive levels of culture. This distribution therefore suggests a functional character peculiar to archaic cultures rather than a single unique origin. Thus in Indonesia traces of dual organization may be found among the Sakai of Sumatra, in the Macassar region, and in central and southern Celebes, on Sumba, Flores, Timor, and in the Moluccas. There is evidence and some suggestions that they existed, and still exist, in the Carolines and Palau of Micronesia. They are found in New Guinea, in the Torres Straits and the Murray Islands. In Melanesia, Codrington, Rivers, Fox and Deacon express their agreement in almost the same words, that dual organizations are the most archaic social structure. Finally, traces or embryonic forms have been observed in the Banks Islands, the New Hebrides, Fiji (by Hocart), Somoa, Tahiti, and perhaps even on Easter Island... It is unnecessary to dwell on Australia, for the division into two exogamous moieties is known to be a frequent feature of Australian aboriginal culture and has nowhere else been subject to such refinements.

Writers of the sixteenth century had already pointed out forms of dualism in Central America and Mexico, and during the same period similar indications were forthcoming from Peru. In North America, moieties extend widely throughout the whole eastern zone, notably among the Creek, Chickasaw, Natchez, Yuchi, Iroquois and Algonquins. They are found in the cultures of the Plains, distinctly or as survivals, among the Sauk-Fox, Menomini, Omaha, Kansas, Osage and Winnebago, and as less and less clear vestiges among the western groups. They are lacking in particular among the Arapaho and the Cheyenne, but reappear in the primitive cultures of central California. Finally, it has been only in the last ten years or so, but with a richness which has all the force of a proof, that dual organization has been observed in the most primitive cultures of South America. If dual organization, which is present at least in principle among the Nuer, the tribes of the Lobi branch, and the Bemba of Northern Rhodesia, nevertheless seems less common in Africa than elsewhere, it could be shown that even where there is no dual organization certain mechanisms of reciprocity persist which are functionally equivalent to this type or organization.[9]

In other words, this concept of the dual society, of two groups allied by marriage exchange, has been found extensively in 'primitive' societies throughout the world – Australia and Oceania, South and North America, Southern Asia and Africa.

Evolutionary Consequences of Arranged Marriages

The logical significance of an arranged marriage is immense. An arranged marriage is the exchange of the ability to reproduce for cooperation. This should attract the mind of anyone interested in human evolution and particularly human cooperation. One might ask how do people cooperate when they do not possess a market and possess no central government? When people have no possessions of any significant value to exchange to facilitate cooperation, they exchange the only items of any significant value – themselves.

This book will argue that the method by which humans cooperate is through the arranged marriage. The arranged marriage is a system by which people exchange the ability to reproduce for cooperation. The benefit of the cooperation has nothing to do with survival; the benefit is that people can increase their status.

The arranged marriage is a contract between two parties. The anthropologist Napoleon Chagnon describes the basic process of the arranged marriage between two men:

> They strike a long-term deal: I'll give you my sisters in exchange for yours, and when I have children I'll give you my daughters to your sons if you give your daughters to my sons. So they carry out the deal and in the next generation all the grandchildren are the grandchildren of both men.[10]

This is also found in Genesis in the Old Testament or the Torah:

> Then will we give our daughters unto you, and we will take your daughters to us, and we will dwell with you, and we will become one people.[11]

Girls are usually betrothed at a young age, sometimes before they are even born. One man promises his daughter, who might be three or younger, on the agreement that the other man cooperates; when the girl reaches puberty the marriage is consummated. The betrothal system is a long-term contract, similar to a mortgage: a man will only receive his betrothed wife if he agrees to cooperate with the other man. This means that one man will need to cooperate with another for at least a decade in order to receive a wife. The contracts can be extended over generations, and bind two groups of males together for long periods of time.

This idea serves the needs of the evolution of cooperation perfectly: the ability to pass on genes is exchanged for cooperation. If, and only if, a man cooperates with another man, can he pass on his genes. If he does not cooperate, then he loses that opportunity. Cooperation and reproduction are thus entwined: no cooperation, no reproduction.

The contract does not merely concern the two men and the women exchanged, but their lineal kin. Thus rather than a single man giving a daughter to a particular man, he is actually forming a contract between his brothers, his half-brothers and himself with another man, his brothers and his half-brothers. The contract may extend to less related kin, but to a lesser extent.

This is made obvious by the operation of the levirate and the sororate. The levirate states that when a man dies his wife is passed on to her dead husband's brother. The sororate states that if a woman dies, another sister will be given by the dead sister's family to replace her. (It also leads to sisters marrying the same man or set of brothers.) Either way, the system is based on a contract. If any of the parties die then the contract is not voided, but other men or women become the physical embodiment of the contract. A man that dies is replaced by his brother: a woman that dies is replaced by her sister.

A further implication is that if two families exchange girls over generations, then their two families will become increasingly genetically related. Thus the ongoing contract of cooperation is not the only aspect binding the two families, but the two groups will become increasingly genetically similar. The two groups are not exchanging unrelated individuals, but nieces and nephews. The ideal arranged marriage between two kin groups is one where there is a continual exchange of girls between the men:

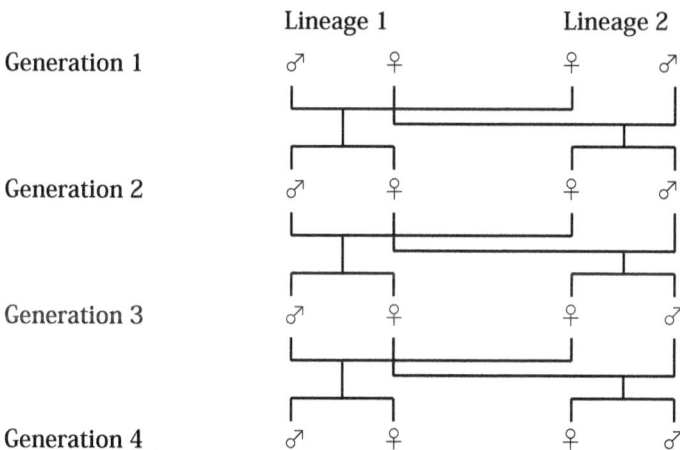

Reasons for the Need to Cooperate

The importance of cooperating is that humans, like chimpanzees, compete in groups. This has profound implications for men and women: men become the objects of violence, while women become the objects of abduction. Even if men are not killed, if their wives are taken from them then their capacity to reproduce is made non-existent.

The threats to males are remarkable. As explained in the previous chapter, Keeley estimated that one in two hundred people will die each year due to warfare in cooperatively-simple societies. For a cooperatively-simple society with perhaps, at the most, two hundred men, women and children (children make up half the population) then this may mean almost one adult male death every year – a remarkable number for such a small community.

Despite women rarely being killed, the threats to their capacity to reproduce are also significant. Some of the starkest examples of the threats a woman and her children can experience come from Helena Valero's experiences. Valero, a young Brazilian girl, was canoeing with her family in the Amazon when attacked by a group of Yąnomamö men. She was shot with an arrow, and her family was forced to leave her behind. Before she had even learned the language properly, she was already not merely the observer of a war but the subject of the war, a war with tremendous implications for the women and children of the Kohoroshiwetari, the people she was living with. It is an example of what can happen to people in these environments.

> One morning two Kohoroshiwetari men had gone to get roots and bark to prepare the curare. From a distance they saw some Karawetari men who were crossing a large bridge suspended over an *igarapé*. The Karawetari saw them too, and shouted: 'You think we are only passing by. No, we shall return to our *shapuno* only when we have carried off all your women.'[12]

The Kohoroshiwetari men and women ran from the much more powerful Karawetari:

> Two men came running; one had a child in his arms. 'Run away again,' they said, 'we can already hear the enemy shouting, "Miserable women, miserable women! Why do they run and fall among these boulders? Do they think we shall leave them alone?"'[13]

The women were finally captured:

> Meanwhile from all sides the women continued to arrive with their children, whom the other Karawetari had captured. They all joined us. Then the men began to kill the children; little ones, bigger ones, they killed many of them. They tried to run away, but they caught them, and threw them on the ground, and stuck them with bows, which went through their bodies and rooted them to the ground. Taking the smallest by the feet, they beat them against the trees and the rocks. The children's eyes trembled. Then the men took the dead bodies and threw them among the rocks, saying: 'Stay there, so that your fathers can find you and eat you.' They killed so many. I was weeping for fear

and for pity, but there was nothing I could do. They snatched the children from their mothers to kill them, while the others held the mothers tightly by the arms and wrists as they stood up in a line. All the women wept. [14]

['Eat you' refers to endocannibalism, where the ashes of the dead are eaten.]

The women were then made to be wives of the Karawetari. It should be noted that a possible reason for the killing of children is that male children are expected to avenge their fathers' deaths when they are adults.

> One woman had a baby girl in her arms. The men seized the little child and asked: 'Is it a boy or a girl?' and they wanted to kill it. The mother wept: 'It's a little girl, you mustn't kill her.' Then one of the men said: 'Leave her; it's a girl; we won't kill the females. Let's take the women away with us and them give us sons. Let's kill the males instead.'[15]

Later, when she was the mother of three boys, Valero's husband, the headman of her village, was killed and the village was dispersed. She was forced to take refuge with a community that contained enemies, but with whom her husband had historically possessed allies:

So we entered the *shapuno*. Rashawe's uncle, brother of his father, was squatting down eating. As we passed, he stared at us, recognized us, and shouted: 'Who's coming? You who are young, take these children, cut off their heads and throw their bodies and their heads onto the same path by which they came. Thus may their uncle learn that we have cut off their heads with machetes, which we have.'[16]

[*Shapuno* = shelter with plaza.]

> After some time Rashawe came up holding his bow and arrows in his hand. I was trembling with fear, because he was the most *waiteri* of all. I did not even dare to look at him and thought of the time when he was so friendly with my sons' father, when they used to take their *epená* together. The *tushaua* loved him very much; whatever he ate, he would send part of it always to Rashawe and Rashawe would send some to him. All this I remembered at that moment; I lowered my head and started to cry. He came close, squatted down and said: 'You have come here?' 'Yes, I have come.' 'Did you remember this *shapuno*?' 'I did not remember; it was my companion that brought me, or I should not have come.' 'Don't be afraid; don't be afraid that anyone will kill your sons, don't be afraid of anything. It is I who ought to kill your sons. It is true, their father has killed my brother... But I feel so much pity for you and I have no anger against your children.' When he said 'The father of these children has killed my brother,' I was afraid. He went on: 'Don't be afraid, not one of these men has the courage to kill the children. They are waiting for me to kill them. They say: "It was not we who killed his brother; he is the one who must avenge his brother; we are other men"; that is what they are saying, that is what I have heard them saying here and there. ... I wept when I saw you coming in and when I heard the old man wanting to kill the children. I come now to talk with you; now that I have ceased to weep. Do not be afraid. Only I could kill them, no one

else would have sufficient courage; but I feel too much pity for you. Sleep here in peace, no one will molest you; I will not allow anyone to worry you. I was a great friend of your sons' father. If you had been another woman, I would have killed you all as soon as I saw you enter the *shapuno*.' He spoke even more, he spoke to the old woman and finally he said: 'Sleep in peace; not even my elder brother will say anything: he is like a woman'; and he returned to his hammock.[17]

[*Waiteri* = fierce or courageous; *epená* = a hallucinogenic drug; *shapuno* = shelter with plaza; *tushaua* = headman.]

One can find similar events in the Old Testament; in Numbers:

NUM:031:006 And Moses sent them to the war, a thousand of every tribe, them and Phinehas the son of Eleazar the priest, to the war, with the holy instruments, and the trumpets to blow in his hand.

NUM:031:007 And they warred against the Midianites, as the LORD commanded Moses; and they slew all the males.

NUM:031:008 And they slew the kings of Midian, beside the rest of them that were slain; namely, Evi, and Rekem, and Zur, and Hur, and Reba, five kings of Midian: Balaam also the son of Beor they slew with the sword.

NUM:031:009 And the children of Israel took all the women of Midian captives, and their little ones, and took the spoil of all their cattle, and all their flocks, and all their goods.

NUM:031:010 And they burnt all their cities wherein they dwelt, and all their goodly castles, with fire.

NUM:031:011 And they took all the spoil, and all the prey, both of men and of beasts.

NUM:031:012 And they brought the captives, and the prey, and the spoil, unto Moses, and Eleazar the priest, and unto the congregation of the children of Israel, unto the camp at the plains of Moab, which are by Jordan near Jericho.

NUM:031:013 And Moses, and Eleazar the priest, and all the princes of the congregation, went forth to meet them without the camp.

NUM:031:014 And Moses was wroth with the officers of the host, with the captains over thousands, and captains over hundreds, which came from the battle.

NUM:031:015 And Moses said unto them, Have ye saved all the women alive?

NUM:031:016 Behold, these caused the children of Israel, through the counsel of Balaam, to commit trespass against the LORD in the matter of Peor, and there was a plague among the congregation of the LORD.

NUM:031:017 Now therefore kill every male among the little ones, and kill every woman that hath known man by lying with him.

NUM:031:018 But all the women children, that have not known a man by lying with him, keep alive for yourselves.

And in Judges:

JUD:021:009 For the people were numbered, and, behold, there were none of the inhabitants of Jabeshgilead there.

JUD:021:010 And the congregation sent thither twelve thousand men of the valiantest, and commanded them, saying, Go and smite the inhabitants of Jabeshgilead with the edge of the sword, with the women and the children.

JUD:021:011 And this is the thing that ye shall do, Ye shall utterly destroy every male, and every woman that hath lain by man.

JUD:021:012 And they found among the inhabitants of Jabeshgilead four hundred young virgins, that had known no man by lying with any male: and they brought them unto the camp to Shiloh, which is in the land of Canaan.

The killing of children is rare, but the abduction of women is a common feature throughout the anthropological literature. Napoleon Chagnon, studying the Yąnomamö, found:

> The fear of abduction gives the women a special concern with the political behavior of their men, and they occasionally try to goad the men into action against an enemy by caustic accusations of cowardice. The men, of course, cannot stand being belittled in this fashion, and they may be badgered into acting.[18]

Helena Valero personally witnessed such badgering, when the men of another community entered her village:

> The women too egged them on: 'Those about whom everybody talks have come, those who think themselves most *waiteri*; take advantage of it, men, to give them enough blows to make their bones all soft. We shall see then whether they really are men.' Then the other Hasubueteri shouted: 'Women, keep quiet! Women always talk when they shouldn't: with an evil mouth against our friends!'[19]
>
> [*Waiteri* = fierce or courageous.]

Warner commented in his study of northern Australian Aborigines (whom he collectively called the 'Murngin') that competition over women was the main reason for continuing violence:

> The causes leading to warfare are the killing of a clansman by a member of another clan, and interclan rivalry for women. The latter is the usual cause of a killing. Clans within a moiety are more likely to be fighting, since clans of the opposite exogamic moieties are not in competition for women. An analysis of battles and killings shows this to be true in most cases, but there are a number of instances in which groups belonging to opposite moieties have been in conflict.[20]
>
> If a camp has been ambushed and the men all killed in a general tribal war, the women and children are taken by the conquerors and given as wives to the men who are in the right relationship to them.[21]

(Note that even in abductions of women, classification may continue to be important in terms of who may marry the abducted woman. Thus classification may not merely reduce competition over resident women, but even abducted women.)

A.W. Howitt noted that abductions also occurred amongst the Australian Aboriginal peoples: the Ta-tathi, Keramin, Kamilaroi, Geawe-gal, Kuinmurbura, Wakelbura, Wotjobaluk, Wurunjerri, Narrang-ga, Narrinyeri, Gringai and Chepara.[22] In his book *War Before Civilisation*, Keeley noted:

> The capture of women was one of the spoils of victory – and occasionally one of the primary aims of warfare – for many tribal warriors. In many societies, if the men lost a fight, the women were subject to capture and forced incorporation into the captor's society. Most Indian tribes in western North America at least occasionally conducted raids to capture women.[23]

The experiences of women in modern warfare have been ignored until relatively recently, but one only has to read Joanna Bourke's *Rape* or Antony Beevor's *Berlin: The Downfall 1945* to find extensive accounts of women in modern warfare being raped, let alone the bombing, killings, and organised genocide of civilians that occurred in World War Two. One might wonder what would have happened to German or Soviet women if either the invading German or Soviet Union armies had allowed polygyny, but, at least in Western nations, the enforcement of monogamy obviously prevents the capture of women in warfare, at least in the long-term. Joanna Bourke concluded, on the issue of rape in warfare, that normal men are potential rapists:

> What the experience of rape in twentieth-century wartime did reveal, however, is the fact that patently 'normal' men – conscripts, even – were capable of raping, given certain circumstances. War is one occasion when rape scripts diverge from the individualist canon of legal and psychiatric practice and enter fully into everyday mass culture. The prevalence of rape in military institutions and during war challenges generally accepted conceptions of 'the rapist'. Ideas about pathological acts or behaviour must be jettisoned. Male sexual aggression became part of a wider social malaise.[24]

Bourke emphasises that rapes occurred to both women of the occupied, defeated nation and women of allied, victorious nations to the occupying nation. Antony Beevor documented that Eastern European women freed from labour or concentration camps were subjected to rape by Soviet soldiers. Rape in warfare appears to be a modern occurrence where male behaviour is limited to prevent abduction but not rape – again, reflecting the importance of human behaviour being limited by social or government control, rather than simply instinctive patterns. Even in the chaos of war, the men are well aware of what they can get away with, and the restriction of monogamy. In recent wars in Africa, women have become forced wives of soldiers, reflecting differing governmental restrictions.

None of the above should be surprising; it is in accordance with Darwin's theory of sexual selection, that males should compete to copulate with females.

A male cannot succeed in such a competitive environment without allies. Through cooperation via marriage alliance, a single male can greatly increase his capacity to compete with other males. If he cooperates with his brothers, and he and his brothers can exchange their sisters and daughters with another group of men, a single man can greatly increase his ability to compete. Those men that successfully increase the number of men cooperating with them, and take women from other communities, will increase the number of their genes in the following generation. Men that did not create such alliances would have been attacked by the larger groups of allied men and had their daughters, sisters, and wives taken from them. The genes of those that were only able to create small groups of cooperative males would quickly lessen in the gene pool, while those that were capable of forming large groups to protect their wives, daughters and sisters from other men, and increase the number of women by abducting women from other communities, would have increased their genetic presence in the gene pool.

Generally, however, the abduction of women tends to be an indirect event of general conflict. This was the conclusion of Napoleon Chagnon:

> Let me point out some things that are definitely not true about Yanomamö warfare, but have crept into the literature as alleged 'interpretations' of what I have published about their violence and warfare. First, intervillage raids are almost never conducted with the explicit intent of capturing women. Their 'wars' are not initiated or intended to capture females from enemies, although if they find a woman at a safe distance from the enemy they may 'capture' her. Although abductions of females is common among the Yanomamö, most abductions take place right at home – when visitors from distant villages come in small numbers and bring their wives and daughters with them. If the risks are small the benefits high, they might retain the females and send the men packing. If the victimized village is small and therefore militarily weak, the risks are small. Another, less frequent way to abduct females is to feign friendship with a group, invite them to a feast, and then treacherously turn on them, often with the help of third-party allies from other villages who have long-term grievances with the victims. This is called a *nomohori* (a treacherous trick). Visiting men are often killed in *nomohoris*, and sometimes large numbers of women and girls are captured. This is a vile and despicable form of violence in Yanomamö eyes, but if they hate an enemy intensely enough, they will do it. The invariable result is a long-term period of raid and counter-raid, sometimes lasting decades.[25]

Warner also noted that the acquiring of women is often an indirect benefit of warfare amongst the Murngin rather than the intent itself:

> Still another method of securing women – by ambush – differs from woman-stealing only in that the husband is first killed; capture is not the object of the fight except as retaliation for an earlier theft of women. The writer has recorded instances of a group of clans acquiring five or six women by this means.[26]

Chagnon emphasises that warfare amongst the Yąnomamö concerns competition between the villages, with men seeking to avenge deaths, but the issue that causes the desire to avenge a death is usually to do with women:

> Most wars are merely a prolongation of earlier hostilities, stimulated by revenge motives. The first causes of hostilities are usually sorcery, killings, or club fights over women in which someone is badly injured or killed. Occasionally, food theft involving related villages also precipitates fighting that, if it leads to a death, precipitates raiding. ... Food theft is often provoked by the intention of intimidating, not by hunger.
> The Yąnomamö themselves regard fights over women as the primary causes of the killings that lead to their wars.[27]

The cooperative method not only allows males and females to compete vigorously, it will also lead to other benefits: the ability to provide protection for the sick or elderly, the ability to protect members of the group from predators (though, as I have discussed, this is a minor issue), and general group assistance. Thus individuals that might, in other animal social systems, have perished, would possess an increased chance of survival. Yet, even this issue of assistance to members within the community would still increase the competitiveness of the community. An adult male that is allowed to recuperate from an injury or disease and able to regain fitness, would be able to prevent the status of the community from being diminished by his death. Every fertile female that survives due to assistance from others will be able to give birth to males and females in the future to increase the status of the community and, through the exchange of the daughters, to increase the ability of her husband's family to bond with other families. Every death is a loss of status (male) or potential future status (female) and thus the prevention of the death of members of the community will stabilise or increase the relative status of the community.

Possible Alliances

Above I explained the idea of the dual society, and gave the simplistic example of only two lineages involved in a cooperative agreement, such as in the Kamilaroi example. While this gives us a simple model, the model can be built upon to create even more extensive networks.

If two lineages can exchange girls in order to reproduce and cooperate, then there is the possibility for further contracts. There is no reason why three or four lineages might not exchange females in arranged marriages:

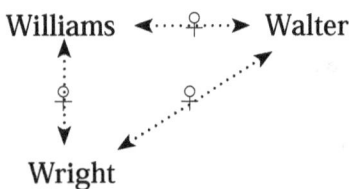

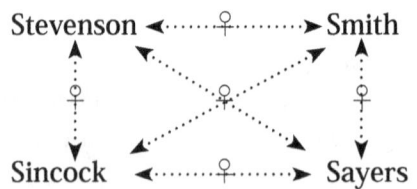

The more assistance men and women obtain, the greater their status, and the greater their ability to compete against other human beings. Instead of having a competitive group of six males (Smiths), they might be able to form a cooperative group with 24 males (Stevenson, Smith, Sincock and Sayers, if each lineage has six males). How this actually occurs is somewhat more complicated and these examples are purposely simplistic. It should be noted that this still works in terms of a dual society: one's own lineage forms one moeity and the other lineages form the other moeity. Thus the duality still exists, as does all the terminology, since one's own lineage will possess brothers, sisters, fathers and mothers, while the other lineages will contain wives, husbands and in-laws. Thus the idea of the dual society is relative to the individual rather than absolute boundaries of simply two groups.

Furthermore, marriage alliances can also occur between communities. A man may marry his daughters to men in his own community, and men in other communities:

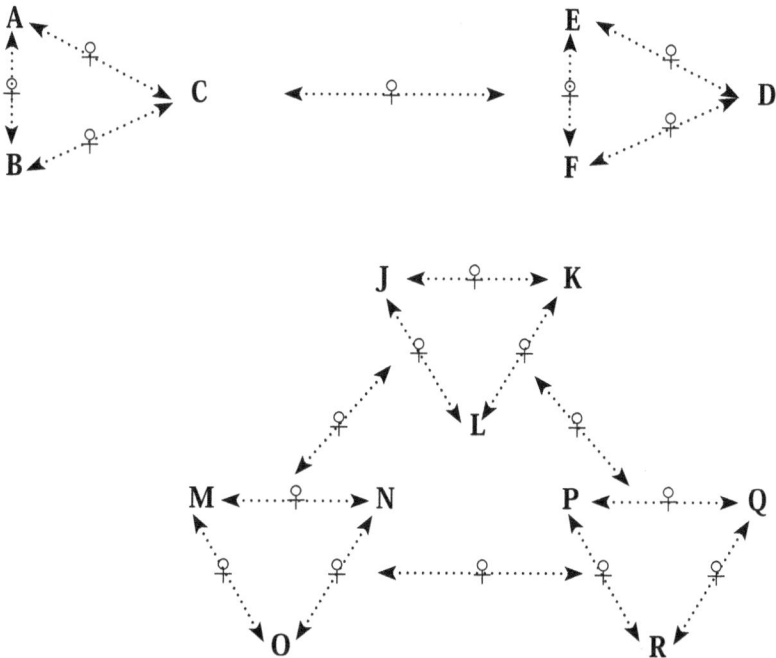

Through the exchange of females, a single lineage can form a cooperative union with several other lineages to form a community, and then link their community to other communities. A community of two or three lineages may create an alliance with another community of a similar size based on the betrothal of young girls who will be exchanged on the basis of a cooperative relationship.

Instead of a single male acting alone, a male can form alliances with other males in his own community and in other communities. He may increase his status from one to twenty or thirty cooperative adult males or more. Once a community has such alliances, other communities have no alternative but to increase their number of alliances or face potential attacks by a much stronger force, the outcome of which has been described above.

Cross-cousin and Parallel-Cousins

The main system of marriage is called the cross-cousin marriage, and it is invariably the method used by cooperatively-simple peoples to facilitate cooperation.

People live in small communities that are inter-married with surrounding communities, which means that almost everyone is related to everyone else, to some degree. Since people must marry in the same generation, they must marry a cousin. Cousins are given two labels: cross-cousin and parallel-cousin. Parallel-cousins are within the same lineage, while cross-cousins exist in two different lineages. Thus, a Smith female and a Williams male, if they are cousins, could marry since they are cross-cousins, while a Smith female and a cousin Smith male could not marry, since they are parallel-cousins. This would be the case whether they were first cousins or fifth cousins. The importance is to bind the two groups, so exogamy is absolutely vital. If people began to marry within their own lineage then this undermines the capacity of the lineage to form alliances with other lineages. The vital factor is cooperation, not biological restraints. [28]

People fall into eight categories: brother, sister, father, mother, mother-in-law, father-in-law, husband or wife. People can only marry those that fall into the category of husband or wife, from their own perspective. These are cross-cousins and are from a different lineage. Those that fall into the category of brother or sister are parallel-cousins.

It is essential that the woman copulates solely with the man to whom she is given. The idea of the arranged marriage is that she is given to the man in order to produce his children. The contracted parties benefit by virtue of one receiving the other's assistance and the other being able to reproduce. If the woman produces an entirely different man's children then this obviously dissolves the reason for the contract – genetic benefits in exchange for cooperation. In their work on Australian Aborigines, Ronald and Catherine Berndt summarise the problems caused by elopements:

> Marriage reciprocity must be seen in terms not just of exchange of men and women, but of gifts, of rights and privileges, obligations and responsibilities. An elopement, where elopements are not institutionalized, represents a threat to this. It upsets the delicately balanced relationship existing between the person, or units, involved in that particular cycle of marriage arrangements.[29]

The woman who elopes with another man undermines the glue holding her community together. If all the women eloped in a cooperatively-simple society, then the community would be reduced to a social system similar to that of gorillas – the single male harem – or to groups of brothers. If this occurred, then these people would be at the mercy of any group of people that could maintain the cooperative alliance system. This creates the negative view towards divorce, adultery and female sexuality since it undermines group cohesion in societies bound by arranged marriages. Interestingly, these concepts continue to exist in Western societies, despite the fact that arranged marriages are no longer used.

The problem created by elopements underlines the importance of the arranged marriage. Mammals do not simply cooperate – they need a reason to cooperate. Arranged marriages bind people together through long-term obligations of cooperation. If women chose their own long-term partner, then, in societies without governments, there is no glue to hold the people together since the people live in a subsistence manner and have no possessions to create a market or to produce taxes. The arranged marriage creates the obligation that forces people to cooperate. The obligation is created by the betrothal. The girl is betrothed at a young age to another man and his close kin; that man and his close kin are expected to cooperate until the girl goes through puberty and the marriage is consummated. If the man and his close kin do not cooperate, then the betrothal is rescinded and the girl betrothed to another man and his close kin.

Concluding Remarks

Before there was money, or any other form of currency, people exchanged the only item of any value – themselves. Arranged marriages held communities together as the basic, most fundamental cooperative contract. It is a long-term contract, organised when a girl is very young, only a baby or child, and concluded when she has entered puberty.

How this system of exchange could have evolved is considered in detail in the penultimate chapter; however, for the moment it is more important to understand the features of the arranged marriage, its importance, and its implications.

In terms of evolutionary significance, it must be understood that the arranged marriage is entirely linked to the passing of genes. Cooperation is bound to reproduction. The complexity of the system obviously has tremendous implications in terms of social organisation, cooperative systems, sexuality, intelligence, and so on. For instance, this system is the product of intellectual thought. People are aware that if they or others marry within their own lineage then this undermines their capacity to form cooperative bonds. Thus they describe such marriage as illegal or incestuous. The only acceptable marriage is one between cross-cousins, people of the same generation that are from different lineages. This maximises people's capacity to form cooperative links, and illegalises selfish acts of people attempting to marry those within their own lineage.

This is a basic legal system. People need to be able to think into the future, to imagine themselves years in advance when the girl reaches puberty. If the girl is a year old, the man betrothing the girl needs to consider the situation of the potential husband-by-arrangement fourteen years into the future. This may mean he needs to understand the concept of that man's mortality, and even his own mortality. He needs to ask himself: 'Will this man be alive at that time? Will I be alive? What will be the status of this man at this future time? Would it be better to betroth her to a younger man or to a young boy who will be at his peak when she passes through puberty? Is this boy dependable and will he cooperate with my sons and sons of my brothers?' People need to understand complex politics and to be able to plan politically years in advance. Today, people continue to possess an instinctive understanding of 'marriage' and of the father 'giving away' his daughter.

Individuals benefit, whether male or female, because it increases their capacity to compete against other groups of individuals (status). Increased complexity of cooperation is rewarded by individuals being able to form larger cooperative groups that have greater capacity to defend and increase their own capacity to reproduce and pass on genes.

The similarities between a Western society and the society I have described are obvious. People engage in long-term contracts, either arranged marriages or mortgages. People engage in a market system of cooperation, based on either themselves or alternative denominations, such as money. People try to form their own police forces, armies, defences, and entire society by organising these cooperative contracts.

Chapter 4

Male Alliances

In the previous chapter I asserted two important ideas: firstly, that males cooperate by forming alliances; and, secondly, that these alliances are formed using arranged marriages. This chapter will concern how exactly men form these alliances.

The fact that men cooperate at all is something we simply take for granted yet most males in the animal kingdom do not cooperate with other males. If all the people in a city were turned into gorillas, instantaneously, with the exception of some father-son relationships, male cooperation in the presence of females would end. The presence of organisations that enforce cooperation – such as the army, police, courts and jails – further enhances the illusion of easy cooperation between males. If these organisations were disbanded, then the idea of forming a company, in a modern sense, would become implausible.

The idea of a society without an army, police force, courts or jails, is a situation that many people can't imagine. Let's try to imagine such a society. You are a man and you have a lovely wife and two attractive daughters, both in their teens. There's a man who likes your daughters so much that he decides he's going to take your daughters away from you and make them his wives. You think this man is the most detestable man you have ever met in your life. Maybe you wonder why this person is not in jail, except that there are no jails, no courts, and no police force. The type of person you may read about in the newspaper as having committed appalling acts is free to do as he pleases. What are you going to do about it? People often think that in a society without the police and an army that the main issue will be stopping people stealing their car. Actually, the issues are far more serious. So you've been informed that the man is going to knock on your door, with his brothers and other allies in tow, and take your daughters by force. As soon as he gets them into his home, he'll have sexual intercourse with them. There is no police force or army to call upon or anybody else who will enforce your daughters' rights. What are you going to do?

Evidence that this actually happens in societies without police forces was produced in the previous chapter. In fact, in the previous chapter not merely were women and girls removed from other communities, but children's lives were threatened. Furthermore, we should expect that humans will do whatever it takes to leave the most descendants. If there are two types of man, one that will take women by force to be his wives and another type of man who only defends women from men, then we should expect, slowly but surely, that the men with the genes to take women by force will slowly dominate the gene pool.

This chapter explains how people respond to this situation.

I use the term 'alliance' rather than 'partnership' or 'friendship' because the term incorporates the ideas of politics and partnership. Invariably the modern view of friendship tends to be one of a partner in fun or common interests, rather than a trusted political partner. The view of male cooperation that this chapter considers may include friendship, but is

best considered as a political alliance in the same way that people may work together successfully but not consider each other as friends.

Books about men's alliances or friendships are rare. It no longer appears to be something worthy of comment. In earlier, less stable times, the issue of men's alliances was deemed
to be something of tremendous importance. The Roman statesman Cicero wrote:

> Even someone who shuns company and spends his life out of town still has great need
> to possess a reputation for goodness and justice, because even he has to have associates
> of some kind. For without them, isolated as he is, he will have no one to support him –
> in which case he may easily become the victim of every sort of persecution.[1]

Hallpike documented killings of strangers in Papua, which appeared possible because they
had been arrested in the enforcement of Queensland laws, had escaped and then had to make
their way back to their own community, facing serious dangers in doing so:

> A Ku— man was arrested by the *kiap*. His name was Ko—. He was taken to Tapini to
> work [in the prison] but he did not like the work, so he escaped, and came to Tanini,
> where he was given food by Kaua Komaepe [the son of Kepe Maia, eldest brother of
> Aimo Kamo]. Aimo Kamo said 'You should kill him' so Kaua took an axe and killed
> him. Beto Aima [of Kataipa] also hit him with another axe. Kumo Kotsi also axed him,
> and they took him to the Iguam stream, and pushed him in the hole where the water
> comes out. A Kataipa man came and took him out and cut him up and fed him to pigs.[2]

> Two of the women whom he had arrested escaped, and came through Goilala on their
> way home. Kamo Beto and Orou Keruvu offered to show the way, and took them down
> to Kataipa, where they copulated with them. Then they killed the two women there.
> E— was killed by Kamo Beto, and Orou killed U—. They buried them in the bank by
> the river. The river washed the soil away, and the pigs came and ate the bodies. The pigs
> were seen carrying away the bones.[3]

In societies where men compete vigorously against other men, a man's allies are
most important to his success. To discover the truth of this we need to consider the books
of ancient writers, where the love between men is firmly stated, and sometimes misinterpreted as homosexuality. Previously I have quoted Cicero's observation that the Latin word
for friendship, *amicitia*, is derived from the word for love, *amor*. Cicero goes on to consider
friendship to be the greatest quality humans possess:

> Real friendship is even more potent than kinship; for the latter may exist without
> goodwill, whereas friendship can do no such thing. You can see its unique power when
> you consider this point. The bonds which nature has established to link one member
> of the human race with another are innumerable; but friendship not only surpasses
> them all but is something so choice and selective that its manifestations are normally
> restricted to two persons and two persons only – or at most extremely few.
> Friendship may be defined as a complete identity of feeling about all things in

heaven and earth: an identity which is strengthened by mutual goodwill and affection. With the single exception of wisdom, I am inclined to regard it as the greatest of all the gifts the gods have bestowed upon mankind. Some people, I know, give preference to riches, or good health, or power, or public honours. And many rank sensuous pleasures highest of all. But feelings of that kind are something which any animal can experience; and the other items in that list, too, are thoroughly transient and uncertain. They do not hang on our own decisions at all, but are entirely at the mercy of fickle chance. Another school believes that the supreme blessing is moral goodness; and this is the right view. Moreover, this is the quality to which friendship owes its entire origin and character. Without goodness, it cannot even exist.[4]

> Take away the bond of kindly feeling from the world, and no house or city can stand. Even the fields will no longer be cultivated. If that sounds an exaggeration, consider the converse situation: note the disasters that come from dissension and enmity. When there is internal hatred and division, no home or country in the world is strong enough to avoid destruction. That shows the value of the opposite situation – friendship.[5]

It is unsurprising that Cicero, living in an age where the state was still an unstable entity, and slavery or an early death was commonplace, would possess a strong view of friendship. In his day, a friend did not exist for mere entertainment, but was a supporter who may risk his life to assist another friend. The human world revolves around alliances between males and without those alliances all recognisable human environments, whether complex or simple, would disintegrate.

The importance of male cooperation is accurately defined by Cicero – without it, no city, village or community would exist. Women would have no choice in this matter; if men gave up cooperating with one another and attacked other males on sight – as gorillas do – then no matter what the female preference, they would live in small, single male harems as do gorillas. If women were larger and more aggressive than males, as female hyenas are, then arguably females could define the group, but women are neither larger nor more aggressive than men.

This chapter will outline how men in cooperatively-simple societies form alliances and it will follow the following Bedouin proverb:

> I against my brother
> I and my brother against our cousin
> I, my brother and our cousin against the neighbours,
> All of us against the foreigner.[6]

This proverb is considered in genetic terms in Diagram 1, below.

```
        .0625                        .125                        .0625
        FFFB ——————————————— FFF ——————————————— FFFZ
          /                                                    \
   .0313  .125                        .25                        .125  .0313
   FFFBS  FFB ——————————————— FF ——————————————— FFZ  FFFZS
       /      /                                              \      \
 .0156      .0625  .25                .5                .25  .0625      .0156
 FFFBSS     FFBS   FB ———————————— F ———————————— FZ   FFZS       FFFZSS
    /         /      /                                      \      \        \
.0078     .0313  .125 .25      .5      1.0   .5      .25  .125   .0313    .0078
FFFBSSS FFBSS  FBS   Half-  Brother Ego  Sister  Half- FZS   FFZSS  FFFZSSS
3rd       2nd    1st  brother                       sister 1st    2nd      3rd
Parallel-cousins                                         Cross-cousins
```

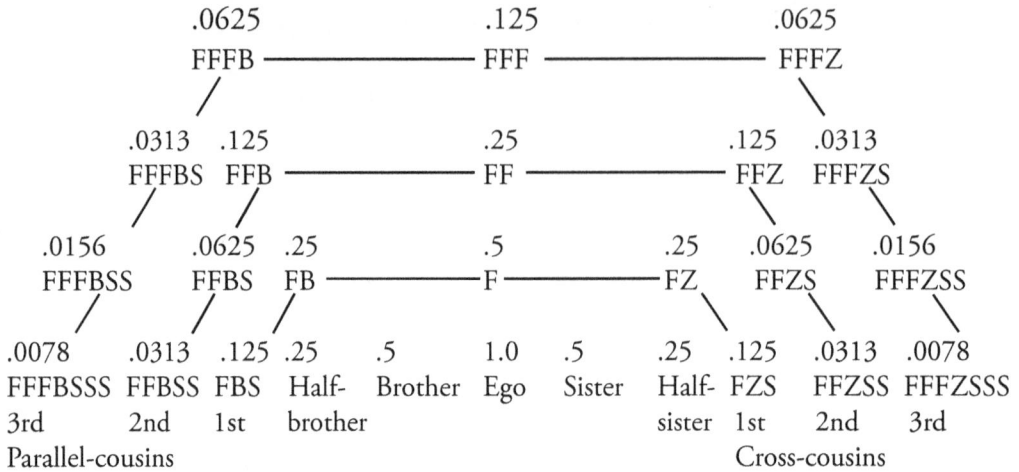

F = father, B = brother, S = son, Z = sister

Diagram 1. Ego, and his genetic relationship to his close kin. Marriage to outside parties is assumed. Parallel-cousins refer to a man's paternal cousin's, and cross-cousins refer to a man's maternal cousins. A man has one set of genes (1.0). His sister or brother (excluding the chance of cuckoldry) possesses half of his genes (0.5). And so on. The emphasis on genes, represents the concept of inclusive fitness as outlined by George Williams and by William Hamilton.[7]

The Individual

I against my brother

Though we expect highly genetically related individuals to cooperate, they can also be competitors. Since a male and his brother will share only fifty percent of their genes, if a male allows his brother to reproduce with a female rather than himself, then the infant will possess only 25% of his genes rather than 50% if it were his own offspring. For half-brothers, it is 12.5% rather than 50%. Therefore, even brothers have good reason to compete.

The relationship between cooperation and genetic relationship is encapsulated in what is known as Hamilton's Rule. Following the ideas of J.B.S. Haldane and R.A. Fisher, William Hamilton showed, mathematically, that an individual should assist another animal where the cost to the giver was lower than the benefit to the receiver with regard to their genetic relationship ($b > c/r$ where b is the benefit to the recipient of the behaviour, c is the cost to the performer of the behaviour, and r is the co-efficient of relationship between the two individuals).[8]

Thus an animal might risk its life in order to save the life of three brothers or sisters ($3 \times 0.5 > 1.0$ where $1 = 100\%$ and $0.5 = 50\%$ of an individual's genes); or five half-brothers or half–sisters ($5 \times 0.25 > 1.0$). But the opposite is also true. A male might fight or even kill a brother if it meant it increased his capacity to leave descendants and the injury or death to his brother had less impact on his genetic presence. His own infants will possess 0.5 or 50% of his own genes while an infant of his brother will only possess 0.25 or 25% of his own genes. Thus, in simplistic reasoning, one might argue that a male might kill his brother if it meant he could sire three infants instead of his brother siring the infants (3 x 0.5 > 0.5 + 3 x 0.25) or sire only one infant instead of his half-brother (1 x 0.5 > 0.25 + 0.125).

When it comes to gaining wives, men should be in conflict with their brothers and all others. If a man has the opportunity to compete with a brother to gain a wife, then we should expect that he would, up until that point where the competition becomes more costly than the extra benefit he might receive. In more complex societies, the equivalent is the competition over property. Killings of brothers do occur: two studies from India of tribal societies found that 6 in 100 and 8 in 90 were brother-brother killings that involved property. In these societies birth order played a significant role in inheritance.[9]

In order to increase his reproductive success, we should expect that men would compete against all others, including his closest genetic male relatives. By contrast, because women can only have their own children, we should expect that the benefits of altruism should be much higher between sisters than they are between brothers. A woman may share her husband with a sister with negligible loss to herself (sororal polygyny), but a man can suffer a serious genetic loss by allowing his brother to copulate with his wife and polyandry is less common. Polyandry has been generally only found in the most extreme environments (such as the northern regions of North America or the Himalayas), where female infanticide may have been heavily practised as a form of population control.

However, a man cannot exist alone; in a very competitive male environment, he needs male supporters to be able to reproduce. His first allies will be his closest genetic relatives – as noted by Cicero, who took for granted that closely related males will support each other, and thus friendship between non-related males was exceptional.

The Brother-Alliance

Ego	Brother

I against my brother
I and my brother against our cousin

A man's most important allies will be his brothers and half-brothers: the brother-alliance. Brothers cooperate to increase their ability to compete as a cooperative unit against other men. This is due to sexual selection (competition), not natural selection (survival).

In the previous chapter, I argued that humans cooperate by exchanging cooperation for the capacity to reproduce. Arranged marriages are the bonds that hold communities together that have no market economy. Groups of brothers exchange sisters and daughters to obtain each other's cooperation.

However, as brothers can cooperate with another group of brothers, so they can also be deceptive. Brothers can organise a contract for cooperation (a betrothal for marriage), but they can also lose this contract since one group can renege on a former promise. In order to prevent such deception, a brother-alliance must present itself as possessing comparable status to others in the community, that it will respond to any deceit with force.

Brothers combine their forces to respond to attacks on their status, such as insults or the death of one of their brothers. The following example comes from Warner's work with the Murngin:

> One of Warlumbopo's older brothers (actual) had been ambushed and mortally wounded by a tribal yukiyuko.
>
> Shortly after this Warlumbopo walked into the camp of the dying man.
>
> Warlumbopo wailed for his dead brother. He refused to allow the wives of the dead man to bury him until he had killed the slayer of his wawa, when he returned to his dead brother and buried him.
>
> When Daurlung, an older brother to Warlumbopo and younger to the dead man, heard the news he said, "We'll get all our Ritarngo people and go fight them." "No," said Warlumbopo. He had not told Daurlung of his killing their brother's slayer since the latter had been a good friend of Daurlung's. Daurlung said, "Yes, we must go kill our brother's slayer." "You can't kill him now." "Why? "He's dead." "Who killed him?" "I did." "My own brother! Good! I'll give you two of my wives."[10]
> [Yukiyuko or Wawa = younger or older classificatory or real brother.]

The attack on a brother is an attack on the status of the group of brothers. The status of the brothers is reconfirmed when one of the actual ('my own') brothers obtains revenge.

Reflecting the greater importance of status over survival, Warner comments that the use of obscenities by one man against another often led to violence, while disputes over food rarely led to fighting:

> The use of obscenity and profanity against a man always results in a camp fight. Profanity usually reflects not only on the man but on his clan and carries an incestuous connotation. Occasionally quarrels result from an unfair division of the game killed in the hunt, but not always open warfare, since those who feel they had a right to the game are near relatives whose solidarity is too strong to allow fighting.[11]

Insults question a man's (or woman's) status. If a man can insult another man without a response, then he proves his greater status. One study of U.S. homicide found 37% were due to 'insult, curse, jostling', while a study of Detroit closed-case homicides found 124 of 512 involved showing off or verbal abuse.[12] Since men compete not as individuals but in groups of close genetic relatives, a man that insults another man not merely insults that man, but also the status of the man's competitive group. In the following example from Warner's study, opossum strings are used as an invitation to extra-marital sexual relations:

> Narnarngo began cursing Bengaliwe and accusing him of an affair with his wife, but not in his presence. A near relative of Bengaliwe heard Narnarngo's comments and rushed up to the men's camp, where he told Bengaliwe what the old man had been saying. The

accused did not take the charges seriously, and went unarmed to Narnarngo's camp. When he was shown the opossum string he admitted it was his, but said that he had made it to wear around his head (opossum strings are usual ornaments). Narnarngo said that Bengaliwe was lying and had tried to fornicate with his wife.

Meanwhile another quarrel had developed in the camp because a father thought one of the black magicians was trying to kill his daughter. This second dispute developed into a general camp brawl in which Balliman, a near relative of Bengaliwe's, took part. He loved a fight for the sheer joy of exercising proficiency in battle. When his own fight had more or less quieted down, although all the participants were still excited, Balliman heard Narnarngo once again call Bengaliwe various profane terms, but not within the hearing of the latter. He went to the accused and told him. "Let us go back and fight him," he said. The two took their spears and went. "Why do you keep on saying I try to lie with your wife? I didn't." Narnarngo called Bengaliwe "deindumeiu" (big testicles). Bengaliwe replied by calling Narnarngo "golitchirtommeru" (big kidneyed). Both terms designate a man who would break all laws of sexual behavior.

When Benaitjimaloi heard the quarrel he came over with his four brothers and stood back of Narnarngo to help him. He is true galle to Narnarngo. Other near male relatives joined each side and a new camp brawl was in progress. No one was killed or very much hurt in the fight that followed. It was later ascertained that a small boy, unaware of the string's significance, had taken it out of Bengaliwe's and placed it in Djowa's wife's basket.[13] [Galle = cross-cousin.]

Note that the suggestion that a man is incestuous is an insult, since it means the man is unco-operative – he places his sexual (genetic) desires before the solidarity of the group.

Closely related people support each other in fights and defend each other's reputation. The film *The Ax Fight* records a visual document of a similar fight amongst the Yąnomamö to the one above. It shows how closely related people are more likely to support each other in a fight than less related people.

Brothers cooperate to increase their status. They cooperate with other groups of brothers and receive wives in exchange. They can also compete with one another for wives. There is thus a conflict: how do brothers resolve their competition with each other for wives?

Warner explains how this is resolved amongst the Murngin; the junior levirate – the passing of wives from an older brother upon his death to a younger brother – occurs while the brothers are living:

The junior levirate functions even before the death of the older brother. A gawel (father-in-law) is supposed to give his waku (son-in-law) all the female offspring from his own wives, but there is a strong feeling that a man who has acquired two wives or all the issue of his gawel's first wife should allow his younger brother to have his share of the women from the gawel and mokul. The older brother, however, has the first right to them, and his gawel will ask his permission before giving the third daughter to the younger man. He will say: "It is better that your yukiyuko have this woman, for you have two already, and he has none. It is better that I have more waku to help me." Unless the older brother is an extraordinarily selfish person, he immediately replies, "Yes, you tell him to feed her" (give her to him for a wife): and he helps his younger brother procure the presents he must give gawel and mokul for this wife.[14]

[Mokul = father's sister; Yukiyuko = younger brother; Waku = nephew.]

Thus younger brothers do not simply assist their older brother to reproduce; they too have the opportunity to reproduce and to gain wives.

Warner also explains that the older brother's reproductive seniority is not dependent upon age; if a younger brother has greater forcefulness, the younger brother may move to the head of the family and gain the greater reproductive advantage:

Wawa always is at the head of a family of brothers and has the strongest voice in their affairs if the father is very old or dead. However, if wawa is not of an assertive type, whereas one of the younger brothers is, the latter could hold equal authority with him and frequently would have the final decision on any problem that confronted them.[15]

[Wawa = older brother]

There is also the possibility of brothers sharing the same wife, whether in a polyandrous marriage, or by an older brother allowing his younger brother to copulate with his second wife as Napoleon Chagnon observed:

There was speculation that Kąobawä was planning to give Koamashima to one of his younger brothers who had no wife; he occasionally allows his younger brother to have sex with Koamashima, but only if he asks in advance.[16]

It appears that the women have little say in the movements of wives between brothers or as to whom they have sexual intercourse with; however, since they have no say in whom they marry to begin with, then this may be seen as very much a part of the arranged marriage system.

Thus the conflict of competition amongst brothers is resolved. There is a definite order of benefit in which the eldest brother will benefit from any marriages bestowed upon he and his brothers from another lineage. The only exception to this might be if a younger brother was more forthright or an elder brother more selfish. The elder brother will be expected, however, after taking the first two or three wives, to allow younger brothers to marry subsequent wives. The eldest brother benefits to the greatest degree, but he is expected to allow younger brothers to also benefit and may allow a younger brother to copulate with one of his wives-by-arrangement. If the elder brother dies, all his wives are given to a younger brother. As people in Western nations inherit financial wealth, in these societies men inherit reproductive wealth.

This has led Chagnon and Warner to an important conclusion: since brothers can be expected to support each other as an unquestioned cooperative group and as a reproductive group, then we cannot accept the idea of the nuclear family as being the basic human reproductive unit (husband, wife, grandparents, children). How can we accept that the nuclear family is the basic reproductive unit if the woman may be expected to copulate with the man's brother and marry him if her husband-by-arrangement dies or decrees? Rather than the nuclear family we need to appreciate that the alliance between brothers is, in cooperatively-simple societies, the fundamental cooperative and reproductive unit.

We might then consider the following as the 'family':

Eldest Brother	Middle Brother	Youngest Brother	Half-brother	Younger Half-brother
2 Wives	Wife		2 Wives	Wife
Children	Children		Children	Children

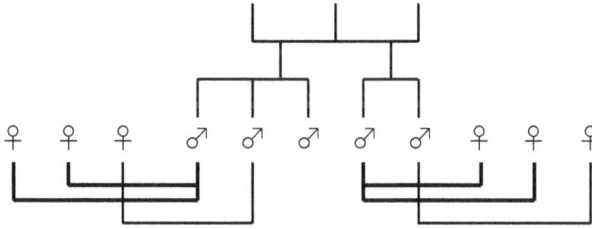

Diagram 2: Example of the family unit.

Above are three brothers and two half-brothers with six wives. If the eldest brother should die, his wives, as part of the levirate, will move to his younger brothers. If their two half-brothers (on the right) should die, then their wives will move to the three brothers (on the left). It is a cooperative, reproductive unit, which fights to retain and keep the wives and provide a safe environment for the children.

This becomes important when the death rates amongst adult males is considered. Invariably, twenty or twenty-five percent of adult males will die due to warfare. Some will die due to other causes, such as disease. Napoleon Chagnon has researched the impact this has on the Yąnomamö and found that if the mere presence of a biological mother and father were considered:

> What Figure 4.16 indicates is that by the time a Yanomamö reaches the ripe old age of about 10 years, only about one out of three live in a family containing his yet-married mother and father, and by the age of 20 years only about one in ten comes from such a family! There are good reasons to believe that this has been true for most of our history. The so-called 'fundamental building block' of human society appears to have a very short half-life when it is viewed statistically.[17]

So, ninety percent of boys and girls will enter adulthood without a biological mother or father as a parent. If we define a nuclear family as possessing a *monogamous* family and a living grandparent, then Chagnon found that the likelihood that a Yąnomamö child will live in a nuclear family by the age of ten is virtually zero:

> When they are between the ages of birth and four years, only about 25% of individuals in this population fit the description of 'most cohesive unit' in band society, i.e., live in a household containing both living married parents (monogamously) to each other,

and one grandparent. The fraction drops to about 12% in the age category five to nine years, and approaches zero by the time the age category 10-14 years is reached! As we shall see in a moment, the rarity of the 'nuclear family' defined in this fashion is largely due to the restriction of including an 'aged dependent', a grandparent in this case, and partly due to the restriction that the parents can only be married monogamously. It is caused also by the divorce of parents after the birth of children, and by the relatively high incidence of mortality among adults, especially males.[18]

In other words, at the age of ten, virtually no Yąnomamö children lived in what we might consider to be the nuclear family – two monogamously married biological parents, and at least one grandparent.[19] The nuclear family is evidently a feature of Western societies, but not of cooperatively-simple societies. In cooperatively-simple societies, the nuclear family exists for a small minority of children. It should be noted that this study is the only dependable study of people living in an unacculturated social environment including life expectancy and death rates.

If a father dies, this does not mean that children grow up with genetically foreign step-fathers; the children will, in accordance with the levirate, be cared for by their father's brother or half-brother. The larger unit absorbs the changes. Equally, if a woman dies, then her children remain in the unit. Those children will be cared for by their father and his brothers. Due to the sorority, it is also highly likely that one of her sisters or half-sisters will be within the group. If one sister marries her male cross-cousin, then it is likely that another sister or half-sister will marry the same man or one of his brothers. Thus if a mother dies, there may be a reasonable chance that a sister or half-sister will be able to care for her children. Warner summarises these findings amongst the Murngin:

> The elementary family tends to be much more stable in Murngin society than it does in our own because of the operation of the sororate and levirate, since if a man obtains two or more sisters for wives and they are looked upon as mothers, the death of any one of them does not disrupt the equilibrium of the family structure; if the husband dies, his brothers inherit the wives and children and there is still no break within the general structure, since the children have, as the classificatory terminology attests, regarded their father's brothers as fathers.
>
> ... This is of extreme importance in the formation of the kinship pattern and partially explains it. To understand this properly, the kinship terms as separate entities must be forgotten and the total kinship system looked upon as a group of interlocking elementary familial units, in each of which there are certain social elements which have an intercellular unity and system of relations of their own.[20]

Most people would understand this in a business sense. In Western nations we understand the idea of a business or a farm being passed on to a relative, often an elder son. However, before the existence of businesses or farms, this idea concerns reproduction and not money or property. Wives were passed on to younger brothers, and families operated not as monetary businesses, but reproductive businesses. The form of exchange was reproduction (women).

Brothers operated as a unit. Women married a single male only in a technical sense but would be passed on to another brother upon his death. Upon her death, her children remained under her husband's care. This is the basis for the consistent notion in fairy-tales that the evil threat to a child was not a step-father but a step-mother, because the step-fathers were the father's brothers and therefore close genetic relatives of his children. The only genetic foreigners in the group were the wives of the father or his brothers. The wives compete on behalf of their children, as with the story of Cinderella, where motherless Cinderella is treated as a servant and her step-sisters live opulently with their mother's preferential treatment. The dangerous step-father is a modern event.[21] However, Napoleon Chagnon notes that there may be other issues beyond health, in the benefit of arranged marriages:

> Given the above data on nuclear family decay it seems clear that few men will have their genetic fathers (or mothers) around to arrange their marriage, and fewer still will have their genetic grandfathers around.[22]

> We cannot continue to ignore the possibility that adult male (or female) marriage arrangers will 'favor' their own sons or grandsons over the less closely related offspring of deceased brothers in marriage arranging. Evidence is accumulating that closeness of kinship introduces a bias in the social interactions of members of many different species, including humans, and points more and more to the necessity of reconsidering some of our assumptions regarding indiscriminate nepotism.[23]

Though there appears to be no data on the issue, brother-brother competition could extend beyond individual competition to competition in regard to their sons, particularly if one brother dies and leaves his sons – and their marriage prospects – in the hands of his close male relatives.

The cooperative and reproductive unit is thus the group of brothers, and not individuals. The first and foremost alliance formed by any male is with his brother. I have named this group the 'brother-alliance', which is the most basic family unit in cooperatively-simple societies, replacing the nuclear family, which rarely exists.

One can see how our understanding of human behaviour is dominated by a Western perspective, since, once we begin to look at more simple societies, then the idea of marriage and the family must be entirely rethought.

Coalitions between Brother-Alliances

Ego	Brother	Cross-Cousins

I against my brother
I and my brother against our cousin
I, my brother and our cousin against the neighbours,

While brothers and half-brothers possess a high degree of genetic relatedness to facilitate cooperation, groups of less related males have less ability to cooperate. This difficulty is overcome by the arranged marriage, a long-term contractual agreement between the men where the capacity to reproduce is exchanged for cooperation. The benefit of these contracts is status: the more men cooperating as a group, then the greater their status (where status equals the capacity to compete against others). A group of four men that can immediately expect the cooperation of another dozen will possess the status of sixteen in disputes, and any group of men that sought to attack the group of four men could expect that they would be fighting sixteen and not four.

This situation is summarised in the following extract concerning a Yąnomamö headman, Kąobawä:

> He leads more by example then by coercion. He can afford to do so, at his age, for he established a reputation as a young man for being forthright and as fierce as the situation demanded, and the others respect him. Also, he has five mature brothers or half-brothers in the village whom he can count on for support, and several other mature "brothers" (parallel-cousins, but called "brothers" in the kinship system) in the village who frequently come to his aid, though not as often as his real brothers. In addition, Kąobawä has given a number of his sisters to other men in the village and has promised his eight year old daughter in marriage to a young man who, as required in marriage alliances, is obliged to help Kąobawä, as are the other members of the young man's family. In short, Kąobawä's natural or kinship following is large, and partially because of this support he does not have to keep displaying his aggressiveness to maintain his position.[24]

At the beginning of the chapter I proposed the idea of what a man might do if another male decides to take all his daughters from him by force. The response by males is to form close bonds with their brothers and half-brothers, and to form contracts of cooperation with other less related males. Thus if a male must defend his own personal needs, rights or desires or those of people closely related to himself, he has bonded with a large group of males that will support him in his defence. He has, in effect, created his own police force and army. His wife or wives, sons and daughters will gain this protection.

Forming this cooperative group is actually very complicated, and this is a major selective pressure for higher intelligence.

A male and his brothers may form a contract with another group or several groups of males. In the previous chapter I gave a simple outline of what this might look like:

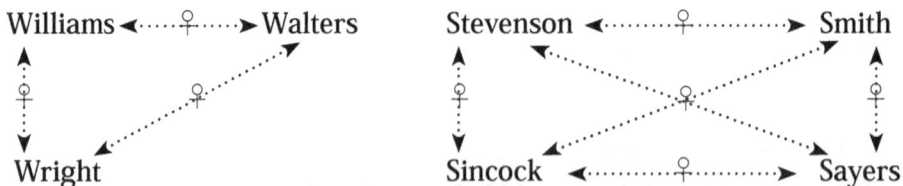

However, life is not so simple. Life in cooperatively-simple societies can be fast and short. In a population of 1466 Yąnomamö, Napoleon Chagnon found that while 23 percent of the community were males between the age of 0 and 10, only 13 percent were males between the ages of 10 and 20, and 8 percent were males between the age of 20 and 30.[25] If males had Western standards of life expectancy, we might expect a slight decline in the percentage of males in each 10 year age group due to death. Instead, we find a remarkable decline. In a generation, deaths can change a robust group of males into a weak group of males, particularly when we are considering such small numbers of people.

A second factor is competition for allies and wives. In the above example of Williams, Walter and Wright, there are three lineages exchanging daughters for cooperation. In a perfect world, this would occur in an easy, organised fashion. Yet, if we consider that a daughter is born in the Wright lineage, then two lineages will be competing for the girl to be betrothed to a man in their lineage. Furthermore, within each lineage there may be several groups of brothers – parallel-cousins (cousins with the same surname) – that are competing with each other to marry the girl. So there can be competition between and within lineages for the future marriage to the girl. Since there will be very few girls born in any given year and that males can be polygynous, there are many males that will be competing for themselves or on behalf of their sons to gain a wife, or for a younger brother to gain a second wife, and so on.

This creates an obvious difference between cousins. Cross-cousins (different surname) are potentially natural allies since they can exchange daughters between them and form cooperative contracts. Parallel-cousins (same surname) are natural competitors. Any marriages between parallel-cousins are viewed as incest or wrong marriages. Such contracts reduce the ability of lineages to form contracts with other lineages and are seen as illegal. Since the arranged marriages are what hold the community together, a lineage that marries within its own members is breaking the rules that hold the society together.

Thus cross-cousins are potential allies while parallel-cousins are potential enemies. Examples come from the Yąnomamö:

> In a biological sense, the inclusive fitness interest of brothers-in-law (who are *also* cognative kin) overlaps strongly, and the warmth of the ties between them is consistent with expectations derived from kin selection theory. They support each other in fights, share food and goods, hunt together, and, as they have children of their own, may arrange the marriages of these children to each other: the children of brothers-in-law are marriageable partners.[26] (Chagnon's emphasis.)

Another example comes from the Murngin concerning dué diramo and galle diramo (respectively, one's father's sister's son and one's mother's brother's son – cross-cousins):

> This is one of the strongest relationships between men. The actual brothers of ego's wife (who is the daughter of mother's brother) and the consanguine father's sister's son (who is the husband of ego's sister) have the strongest bond of any of the male dué and galle. Each always comes to the other for assistance when he gets into trouble. When a man is giving presents to gawel's daughter, he also offers them to galle diramo

because he is brother to the wife.

When Narnarngo got himself into trouble with Bengaliwe his dué immediately came to his assistance... This is only one instance of scores recorded by the writer.

Dué and galle engage in many enterprises together, such as canoe building. The bond between them finds a full expression in all the everyday activities of life. When a man is in trouble and needs a good supply of weapons it is always his brothers, and his dué and galle, who lend him extra spears and clubs, and who come to his assistance.

Male dué and galle have added strength given to their relationship through indirect ties; dué's son and galle's daughter will later be husband and wife; and galle will later be, or is, gawel to dué's son, just as dué is waku to galle's father.[27]

So cross-cousins are strong allies: the men are more likely to assist each other, and they organise for their children to marry. Yet, a man may have many cross-cousins to whom he can promise a daughter. To that extent parallel-cousins are in competition with one another for the daughter of their cross-cousin. As the system provides for cooperation, it also provides for disharmony.

The system is also open to fraud: men can promise a daughter to one man, accept that man's cooperation in fights and the assistance of his family, increase his personal status, and then give his daughter to someone else from whom he has extracted the same promise and services. A man can renege on his promise and refuse to give his daughter to a cooperative man. Alternatively, a man may accept a promise to cooperate in exchange for a man's daughter, and then rarely assist that man, but demand the daughter.

Since many men may only have one wife in their entire life, if another man reneges on a promise of marriage for his daughter, the expectant man might lose his only opportunity to have children. Since the contracts involve supporting other men in fights, then not to gain a benefit after risking one's life or health would cause further anger.

These issues are further complicated by the fact that alliance formations are made between communities. Thus a man may promise a young girl to a man in his own community and to a man in another community, as part of an alliance between communities.

Chagnon comments that the Yąnomamö use specific methods to prevent deceit concerning betrothal:

A significant number of Yanomamö arguments and fights occur when wife-givers renege on a promise. In some cases, the wife-receivers will appropriate the woman by force, dragging her away from her family. In terms of bookkeeping analogy, the wife-givers have overexpended by promising the same girl to several men; they cannot deliver her to more than one, yet they have taken advantage of bride-service from one or more males or have enjoyed the security, if the marriage promise involved two villages, of an alliance with neighbours. Given the possibility of reneging on marriage promises, it would be to the advantage of a father with several daughters to give them *all* to men in those descent groups that are linked by several generations of reciprocal exchange even though at that time he might receive fewer in return for his sons. A nubile female is a kind of resource that cannot be stored and hoarded until a more propitious "investment situation" emerges. If a father must cede her in marriage, then it is to his advantage to give her to those people who would seem more likely to reciprocate in the future and less likely to cheat.[28]

We could thus deduce two major factors that might affect a man's decision as to the betrothal of a girl: trust and reputation. To quote Cicero again:

> ... I feel we should do well to follow the advice of Themistocles. Someone asked him whether he ought to give his daughter in marriage to a man who was honest but badly off or to one who was disreputable but rich. 'Personally,' he replied, 'I like a man without money better than money without a man.'[29]

A man will be most likely to enter into long-term contractual arrangements with those that have shown themselves to be trustworthy and reputable. This is not so much a question of what occurred last week, or last year, but in that man's lifetime and in previous generations. As a modern bank might establish a reputation over decades or a century, so the generations of males will gain a reputation for trustworthiness or dishonesty. The family name is thus important, and we should expect that each generation would place great importance on ensuring that they maintain an excellent reputation. The consequences of gaining a reputation for dishonesty and untrustworthiness might be to lose a long-standing arrangement of marriage exchanges with another group of men, ending a cooperative agreement that may have lasted generations. This does occur, as where one man kills a man in another lineage, which may lead to hostilities, and the fissioning of the community into two groups of enemies.

A second factor will be equality. If a brother-alliance agrees to exchange girls with another group of men, then there will be the opportunity for one of the groups to cheat. If one group is much larger than the other, then there arises the opportunity to accept a girl in marriage but refuse to give one in return, or to promise a girl and then refuse to give her. This is a serious risk if one group possesses higher status than the other. Since the smaller group of men may depend upon the larger group for their ability to compete against other communities, they may have no alternative but to allow the deception to occur.

The answer to this is to cooperate with groups of similar size, to ensure that the chance for deception is limited. Chagnon again:

> Security and the ability to attract and keep wives in such a milieu are difficult, but can be optimized or maximized by entering into long-term reciprocal marriage ties with individuals in other, similar-sized lineages and agreeing to exchange women back and forth over time. Even when there are demographic imbalances, such as a situation where a man has two sons and four daughters, it would be to the benefit of his lineal descendents (sons especially) for him to give all four daughters in marriage and accept just two in return for his sons. He might, for example, give all four daughters to the single son of his brother-in-law in exchange for his brother-in-law's two daughters. This binds the two groups intimately together and "obligates" them to continue in the next generation, even though there is a temporary imbalance. A man with four wives is

certain to have many children, and the female children are subsequently "returned" as marriage partners to the sons of the polygynous man's brother-in-law.[30]

...over several generations imbalances even out ... and each group receives about as many women as it has given to the other. This tends to be true only provided that the two exchanging groups are of approximately the same size. Otherwise, a numerically-dominant group attempts to coerce smaller groups into ceding more women than they themselves are willing to give.[31]

Equality will be extremely important to men, and when two men enter into a cooperative agreement, we should expect that weaker parties would be less willing to enter into cooperative agreements with stronger parties, since this could lead to deception. However, since adult men are particularly at risk of being killed in warfare, it is possible that two equal parties that enter into an agreement to exchange girls for cooperation might suffer casualties and become less equal. Thus a group of men in entering into a cooperative agreement can never be sure that both parties will remain equal and refrain from attempts at deception.

Historically, and perhaps today, this is also observable in modern societies in the manner in which men prefer to remain within their social group. Nowhere is this more obvious than between unions and employers or executives. Men tend to prefer to congregate with people of the same level of status, and tend to be distrustful of those of a higher level of status (measured in money). By contrast, women are more mobile in terms of status: they can marry into higher status.

A man's most important allies outside his brother-alliance are the cross-cousins, with whom he, his sisters, his sons and daughters will marry. These alliances are formed with long-term contractual agreements of exchange, which may overlap over generations. By entering into these agreements men will be able to exchange status for reproduction: cooperation for exchanged marriages. Men will be able to increase their personal status, and the status of their brother-alliance. Cross-cousins that are deemed to be of a similar level of status will be preferred as allies. Stronger groups of cross-cousins will be deemed a risky cooperative partner, while weaker groups of cross-cousins will be deemed to be less desirable but offer opportunities for exploitation.

Thus men are cooperative. However, as men can expect to compete with brothers for the marriage of any girls betrothed to their brother-alliance, so they can expect any cooperative contract with their cross-cousins to be liable to exploitation. At every instance there is the possibility of manipulation and the possibility of an increase or decrease in one's progeny or one's brothers' progeny.

Competition, Further Community Alliances and Marriage

Ego	Brother	Cross-Cousins	Parallel-Cousins

While men can form alliances with their cross-cousins by exchanging marriages for status, they are unable to do so with parallel-cousins. This means parallel-cousins will be competitors for marriages with cross-cousins, and not necessarily allies. This is most obvious when a community fissions into two smaller communities due to tensions in the group. Napoleon Chagnon found that when a Yąnomamö village fissions, men separate from parallel-cousins and retain cross-cousins:

> In addition, adult males separate from other adult males who are primary social competitors for females – usually distant parallel cousins and remote relatives of various categories. Conversely, they surround themselves with primary social allies – adult males who are siblings, half-siblings and cross-cousins.[32]

It makes logical sense that men would prefer to remain with those of either high genetic similarity or with whom they can make long-term contracts of cooperation, rather than those with whom contracts are less possible (marriages are illegal) and are of lesser genetic similarity. Amongst the Murngin, Warner observed that fights among parallel-cousins were more common than warfare between communities (a moiety is a group from which women are exogamic):

> Intra-moiety fights are common; in fact, more common than warfare with clans of the opposing moiety. No clans would join together to fight another people simply because they belong to the same moiety; nor, were this reason advanced, would it prevent fighting within the moiety. The struggle for women among members of the same moiety is too keen to be overcome by an appeal to moiety solidarity.[33]

The genetic similarity of parallel-cousins should also be expected to affect the level of competition between them: first parallel-cousins may be more likely to cooperate than second parallel-cousins, and so on. The sons of those born of the same brother-alliance will be expected to be allies to a reasonable extent. Their father's brother may have assisted them in gaining a wife, fought on their own father's behalf, and increased their own capacity to be successful. But for second or third parallel-cousins, such close kinship may have not existed to the same extent, and the ties may be weaker.[34] One implication of this is that the more men in the community, the more potential conflict, and the greater the difficulty in forming a tightly bound, cooperative society.

This creates, from the Western perspective, the paradoxical situation where to be called 'brother-in-law' is actually a sign of friendliness, while to be called 'brother' is actually to suggest that one is a competitor. Napoleon Chagnon experienced this personally. It is normal for the anthropologist to be incorporated into the kinship system of the people he or she is studying, and the anthropologist may quickly learn who is friendly and who is not. Möawä was a violent and selfish headman:

> My relationships to the village at large were largely predicated on my specific relationships to Möawä and his father-in-law Dedeheiwä. Möawä called me "older brother" (*abawä*) and Dedeheiwä called me "brother-in-law" (*shoriwä*), and in uncanny fashion, my relationships to these men as implied in our mutual kinship usage was very

characteristic. My relationships with Möawä were fairly cold and strained, but with Dedeheiwä, friendly, relaxed and easy-going.[35]

However, real cooperation does exist, and this is demonstrated in that if a man dies, and he has no younger actual brother, then his wives may be given to a parallel-cousin. In other words, the levirate extends beyond brothers and half-brothers to include parallel-cousins. Warner observed this amongst the Murngin:

> If wawa has four or five wives, he may say to a single yukiyuko "You see that one – you may take her and feed her." Yukiyuko usually says, if wawa is an old man, "No, you are an old man. I'll wait until you die, then I'll have them all." Wawa replies, "No, you take her now, yukiyuko, I have many wives and you have none." Yukiyuko then takes her, and she belongs to him permanently. Wawa loses all right to her sexually and economically, and she cannot be returned to him. If yukiyuko died, his new wife would go to his younger brother, even though he were not consanguine but only tribal.[36]
> [Wawa = older brother; yukiyuko = younger brother.]

The relationship can thus be highly cooperative, as well as competitive.

This leads to the question of how this competition between male parallel-cousins is resolved. What affects the success of the brother-alliance, not merely initially, but over generations? What disputes arise? How do men manipulate the system to their advantage?

Since there is competition over women, men need to be able to prove that they can win any competition, and the method used is force. Amongst the Murngin, Warner found that the manner in which a man obtained many wives was through coercion:

> A man wants as many wives as he can get and still keep his peace with the other males of his clan. The average is about three and a half for middle-aged men; but there is a recorded case of one native with seventeen wives, the majority having been obtained by stealing them or killing their husbands. He also married "wrong." There is a close correlation of having many wives with fighting strength, or with being the son of a man powerful in war who had thereby acquired a large number of women whose brothers would be gawels to the son. A dué likes many wives because he tires of cohabiting with one and because such multiplicity creates more sons and daughters, so that he will have more waku, dué, and galle, with stronger bonds between them.[37]
> [Waku = niece and nephew; dué = cross-cousin, husband; galle = cross-cousin, wife; gawels = mother's brother.]

Helena Valero, who was kidnapped as a young girl by the Yąnomamö, and became a headman's wife, observed:

> He who has killed is held in great esteem. If he returns to the fight and kills again, he is considered even more highly. He who has killed one who has killed is called by others *waiteri*, valiant. When he dies, the women sing, weeping: 'He was *waiteri*; *waiteri* in the battle; he was *waiteri*, and he never stayed behind.'[38]

The 'staying behind' might be placed in the following context; from Chagnon's work with the Yąnomamö:

> In all but the most determined raiding parties, a few men drop out for reasons such as being "sick" or "stepping on a thorn." Some of these dropouts privately admitted to me that they were simply frightened. Chronic dropouts acquire a reputation for cowardice (*têhe*) and often become the subject of frequent insult and ridicule, and their wives become targets of increased sexual attention from other men.[39]

The status of any competitive male organisation will thus be determined by the capacity for violence of its members, and the greatest determiner will be whether they are capable of lethal violence.

In his study of several Yąnomamö villages, Napoleon Chagnon found that 137 of 380 men were *unokais*. This does not mean that each man killed an individual man, but he may have participated in the killing with one or two other men (more than one may have shot the same man with arrows). Chagnon found that *unokais* were more reproductively successful than *non-unokais*. For those men aged between 20 and 40, 62 were *unokais* and they had 66 wives, while of the 197 men in this age group who were *non-unokais*, only 100 had wives. Thus *unokais* men were about twice as likely to be married as *non-unokais* men. For those over the age of forty, the 75 men who were *unokais* had 157 wives, while the 46 men who were *non-unokais* had only 53 wives, again nearly twice as many wives. Overall, *unokais* men averaged 1.63 wives each while *non-unokais* men averaged 0.63 wives. While men do not need to kill another man to become married, or even polygynously married, men that have killed another man had twice as many wives in each age group, and more than twice as many wives overall.[40]

In cooperatively-simple societies it is fighting strength that partly determines both the status of the individual and of his brother-alliance. A powerful group of brothers will be:

(1) more desirable as allies because of their capacity to influence disputes;
(2) more capable of enforcing any marriage alliances and reducing deceit;
(3) more capable of forcing lower status groups to cede women in marriage without receiving women in return; and,
(4) more capable of enforcing 'wrong' marriages, such as a marriage between parallel-cousins.

Ultimately, stronger groups are more capable of manipulating weaker groups in the community.[41] This is best considered in terms of modern politics; if you are a strong political group, you are more capable of being an attractive partner for an alliance, and better able at manipulating minor parties. In cooperatively-simple societies, this means more wives and more children.

A most obvious example of the need to attract good fighters comes from Meggitt's study of the Walbiri in Australia:

While they were at Yuendumu, Wally djabaldjari and his sister's husband, Jack djagamara, were allies in a series of disputes that involved a homicide. Knowing that reprisals were likely, they offered Jack's daughter, Julie, then about 10 years old, to a djabanangga man in return for a promise of aid in future fights. This man, however, did not honour his agreement in the next fight that occurred, so Jack declared the betrothal to be at an end. Other men criticized Jack's action. They stated that betrothal should be an unconditional arrangement and should lead automatically to marriage.

When Jack later moved to Hooker Creek, he took Julie with him. As he still feared repercussions from the earlier disputes, he offered the girl on the same terms to Peter djabanangga of Wave Hill. Peter was already married and refused to be entangled in Jack's quarrels. He suggested that Julie be offered to his "young brother", Robber djabanangga, who accepted the betrothal.

About a year afterwards, Robber arrived from Wave Hill and asked for Julie in marriage, guaranteeing to let her visit her mother at Hooker Creek frequently. Although Jack had had no cause to call on Robber for fighting aid, he handed the girl over without argument to avoid public disapproval of a second default. At the same time, he offered another young daughter to Peter in return for help in disputes, but Peter again declined.[42]

An example of how the stronger groups can exploit the weaker groups comes from the Yąnomamö. It also reflects how fighting over women can lead to the splitting up of a community and thus weaken its status in comparison to neighbouring communities:

The village of Patanowä-teri split during the last month of my first field trip. One of the young men took the wife of another because she was allegedly being mistreated by the husband. This resulted in a brutal club fight that involved almost every man in the village. The fight escalated to jabbing with the sharpened ends of the clubs when the husband of the woman in question was speared by his rival and badly wounded. The headman of the village, a 'brother' (parallel cousin) of Kąobawä, had been attempting to keep the fighting restricted to clubs. When the husband's rival speared his opponent, the headman went into a rage and speared him in turn, running his own sharpened club completely through the young man's body. He died when they tried to remove the weapon. The wife was then given back to her legitimate husband, who punished her by cutting both her ears off with his machete.

The kinsmen of the dead man were then ordered to leave the village before there was further bloodshed. The aggrieved faction joined the Monou-teri and the Bisaasi-teri because these two groups were at war with their natal village, and they knew that they would have an opportunity to raid their own village to get revenge. The Monou-teri and the two Bisaasi-teri groups accepted these new arrivals; they were kinsmen and would actively prosecute the war against the Patanowä-teri. The hosts, of course, took several women from the refugees, the price a vulnerable group must pay for protection.[43]

This extract exemplifies the fact that each community is made up of various, independent factions. A brother-alliance and other close kinsmen may leave its own community and join an enemy community to attack their 'own' community. Communities are organic rather than rigid. It also demonstrates how the weak can obtain assistance, but at a serious cost.

The most obvious manner in which the strong can manipulate the weak is by obtaining 'wrong' marriages or incestuous marriages. In the previous chapter, I outlined how societies were divided into four groups: by lineal relationship (brother/sister and husband/wife); and by generation (mother/father and father-in-law/mother-in-law). Which category people fall into is dependent on their relationship to other people.

Polygyny adds an element of chaos to this system. A man may have several wives, and his eldest and youngest children may be separated by twenty or thirty years. If a male has an eldest daughter who is forty, and a youngest daughter who is ten years old, then another male in the community may call these females 'mother' or 'mother-in-law' because this may be his genealogical relationship to the females.

Napoleon Chagnon comments on the implications of this for the Yąnomamö:

The net result is that the absolute ages of individuals gets out of synchrony with their generational identities. People will have brothers or sisters that are younger than their grandchildren. Since Yąnomamö kinship classifications utilize generational position, something must give. Girls are ready for marriage at puberty and boys when they are in their early 20s. No right-thinking Yąnomamö would sit patiently for 50 years until his sister's daughter is old enough to marry one of his sons. What 'gives' is the kinship classification which means that people *must* chronically re-classify some relatives in order to keep kinship classification more-or-less in harmony with ages and generational identities. They *must* break the kinship rules to make the actual marriage practices work.[44] (Chagnon's emphasis.)

Rerebawä once said to me, proudly predicting the future of Breakosi, his firstborn son. 'He is a real fierce little guy! So fierce that when he grows up he will probably commit incest!' What he was saying was that he would know the kinship rules and genealogies so well that if these got in his way he would break them and put someone into a marriageable category that didn't belong there... commit incest. You can only do this successfully and predictably in Yąnomamö culture if you are prepared to defend your rule violations, and your ability to do so depends largely on how much credibility of threat you can demonstrate, i.e., how 'fierce' you are. The most flagrant cases of incest I have in my records – men marrying parallel cousins or, in one case, a half-sister – are cases of men who are not only headmen but headmen with reputations for 'ferocity'. Breaking the rules to gain some personal advantage is always easier in any culture if you have power – acquired or inherited. In Yąnomamö culture it is partly acquired, via demonstrations of individual prowess, and partly inherited, by having lots of kin who will endorse your manipulations and rule-breaking.[45]

Interestingly, Warner also observed wrong marriages in the Murngin, but noted that the people attempt to 'right' the 'wrong':

> When children are born from a wrong marriage, the father calls them by the
> terms all other parents give their children (gatu). But everyone else calls the children by
> the term used if the mother had married normally. To use the native expression, "The
> father is thrown away."[46]

In other words, when a man marries a woman he was not meant to, the rest of the community
attempts to act as if the marriage never happened and to place the children in the category
they would be if she had married the correctly categorised man. Sometimes the fighting over
wrong marriages continues over generations:

> The Warumeri and Wangurri clans have until recently been bitter rivals for the
> possession of women of clans in the other moeity. Certain members of the Warumeri
> clan in the last generation married women who should have become the wives of
> Wangurri clansmen, since their fathers stood in the relationship of mother's brother
> to ego. The Wangurri organized a raid, killed many of the Warumeri men and took the
> women for themselves.[47]

In intermarried societies, people are often related in more than one way. Each of
these relationships is described as a 'link'. A man might be related to a woman through his
father's father's sister or his father's father's father's father's brother – two links. If the link
to the great-great-grandparent's brother involves the skipping of a generation, then there
could be two possible classifications for the woman relative to the man: for instance, 'wife' or
'mother-in-law'. When two patrilineages intermarry over many generations, what is created
is a spiderweb of relatedness, in which people are related to one another in many different
ways. This becomes useful in terms of reclassifying kin. If a man wants his son to marry a girl,
the girl might be related to his son in several ways, and the man might classify the girl in the
one that allows his son to marry her. So, even though the most obvious classification might be
'mother-in-law' to his son, there might be a more distant connection that allows a reclassification of the girl as 'wife', and thus the son can marry the girl – as long as the father can enforce
the reclassification. Other men might fight the reclassification, since they might want their
son or brother to marry the girl.

'Wrong' marriages are extremely common. In his work with the Yąnomamö, Napoleon Chagnon took data of 1326 people in 13 villages and found that of the 817 married
persons 274 were related with a 'wrong' or 'incestuous' link.[48]

Of the 274 marriages with an incestuous link, Chagnon found that 232 involved
generation-aberrant incest links, in other words, the 'wrong' part of the marriage was that
they had married someone who could be classified in a different generation. Of the 232,
88 marriages also possessed legitimate links; in other words, though the person could be
wrongly categorised, they could also be classified as being in the correct category by following
a different path of categorisation.

Thus 274 of 817 marriages, or 33 percent, involved 'incestuous' links between the spouses. There were 144 of 817 marriages, 17.6 percent, with generational incest links and no legitimate links. So a sixth of all marriages were 'wrong' and a third of all marriages could be interpreted as being 'wrong'. Incestuous or wrong marriages are thus a significant percentage of all marriages.[49]

The reason this causes so much conflict is that whenever a man gains a wife from a wrong category, it will always be the case that men in the correct category feel they have lost a potential wife. Thus every wrong marriage must be enforced against those men that hoped to marry the woman.

For the parties involved, a 'wrong' marriage may be very desirable; one man gets a wife, and perhaps another man gains a desirable ally. Or two intermarrying families may want to ensure that the bond is kept and agree to a marriage that is technically illegal. The desire for the alliance may overcome the illegality of the bond.

The situation of forming a cooperative society is complicated, far more complicated than obtaining the day's food or using a tool. Who marries whom is an endless source of potential conflict as each man attempts to maximise the number of wives he or his close genetic relatives obtain. The desire to obtain many wives and have many children must be balanced against the need to ensure the cooperation of the community. And all of this is directly linked to reproduction and cooperation.

Manipulating the system is not simple. It requires a man to be able to know all his ancestors, and the ancestors of others, going back several generations, and to be able to mentally consider who is related to whom without being able to write any of it down. While we might consider genealogy as simply a past-time, originally, genealogies were rulebooks of marriage and reproduction, and who was related to whom was something that could lead to lethal violence. It should also be noted that such manipulations occur years in advance. A man may begin to reclassify a girl so his son can marry her when she is only a baby – he is thus thinking fifteen years into the future, well beyond the life span of most large mammals!

Why is this important? This is the method by which humans cooperate and reproduce. This is how marriages are formed, how children are born, how societies are created. Any carnivore can hunt; any squirrel can gather; any chimpanzee can make tools; but only people can form cooperative long-term alliances and manipulate them. Humans must be able to see 10 to 15 years into the future, and be capable of creating or preventing complex manipulations that require knowing genealogies going back many generations. This is a plausible basis for human intelligence and large brains.

The importance of reproduction is not merely one of genes but also status. In these societies wealth is measured in terms of social allies, not monetary units. The wealthy of the 'stone-age' measured their wealth in people and not in possessions, since these were reliable allies (males) or allowed for the creation of allies (females). A multi-millionaire in a cooperatively-simple society might be the son of a man who had six wives, since the son will have plenty of social allies, and is in a strong position when it comes to bargaining with or intimidating others in the community or surrounding communities. This amplifies the importance of being polygynous.[50] As modern parents might hope to leave a substantial amount of money or property to their children, cooperatively-simple parents will hope to leave a substantial number of allies for their children. The monogamous male is thus leaving a smaller legacy to his children, and leaving them open to being manipulated and reducing their chance to bargain effectively. The importance of being polygynous is not a matter of sexual interest; it is a matter of a duty to one's children, to ensure that in a competitive environment, they are able to compete effectively. Many modern men simply see marriage in terms of pleasure and a few children because they live in societies with an army, a police force, and restrictions on marriage: for men in societies without extensive market economies, marriage was the manner in which a man created his own army or police force for his children, and if he didn't create such support, then his children were at a substantial disadvantage when in competition with men who did create such a level of support. In modern societies, status is measured in money rather than social allies, and parents attempt to maximise abstract status rather than reproductive status: in cooperatively-simple societies, the opposite is true, because money does not exist.

This gives a proper insight into why people in cooperatively-simple societies state that they fight over women.[51] Wives and children are directly linked to how much influence a man has in his community. Every extra wife he can gain increases his ability to create alliances within the community or between communities. If a man loses a potential wife, because she has been reclassified and married wrongly, he not only loses those genes he might have passed through to the next generation, he loses the influence in the community that he might have created through the marriages of his daughters. When men abduct women from other communities, the same is true. Wives and children, power and influence, are thus inextricably linked in these communities. Thus, if the men are fighting over women, they are fighting over power as much as they are fighting over their inclusive fitness (how many genes they leave in the following generation). Furthermore, his children inherit his biological legacy, and the larger his biological legacy, the greater the children's inherited biological wealth.

Male Behaviour – Aimed at Other Men, Not Women

Being a successful male in such an environment is thus much more difficult than simply finding a mate and feeding children. Men evolved in a complex society in which they need to succeed. Men in Western societies know that they need to find a position in a company or partnership and their ability to reproduce may depend upon their success in this environment. These companies or partnerships compete against other similar entities, and there is also competition within the company or partnership for promotions. Cooperatively-simple societies are no different.

In order for a man to be reproductively successful he does not need to be attractive to women: he needs to be attractive to men. Whether he gets married depends on whether other men will betroth their daughter or sister to him. Whether his sons get married depends on whether he can get other men to betroth their sisters or daughters to his sons.

From this perspective, we should expect that the majority of evolved male behaviour is aimed at other men, and not women.[52] Whether in the 'stone-age' or modern corporations, male behaviour is mostly aimed at succeeding in these competitive male environments, and men can expect to spend the vast majority of their time attempting to achieve this end. The amount of time spent wooing females, by comparison, will be small, and if men cannot succeed in the former, they'll be unlikely to succeed in the latter. Thus we observe that males prefer to socialise with other males, despite the fact that this appears to be a reproductive anomaly.

The success of a man will depend upon his status: the number of his allies, his own fighting strength and the fighting strength of his allies. The greater his status, the more attractive he will be to other men seeking his political support, and the greater his ability to form contracts for marriage and enforce them or to create contracts for marriage that are 'wrong'. Men with very high status may have numerous wives. Men with very low status may not be able to reproduce at all. For this reason, a man's first priority is not in finding a wife, but in finding male allies. In order to find an ally, he needs to be a desirable person to have as an ally, much like being a desirable employee.

Since men risk injury in giving their political support, we should also expect that men would prefer to make alliances with men that are less likely to cause problems. Some of the arguments in this text might suggest that all men want to be as violent as possible, but erratic, volatile men might be deemed too great a risk. In Papua, Karl Heider noted this aspect of violence as being both a positive and negative aspect of men's importance:

> In any case, aggressiveness and violence are not especially highly valued by the Dani. There is a category of men called *sinuknuk*. These are men who have acted too violently or rashly in warfare or within their own group. *Sinuknuk* implies something like "sometimes a little crazy," and people are wary of such men.
>
> Another attribute highly prized in anyone but especially characteristic of Big Men, is cleverness: *hat hotiak is* said to people in praise for something particularly clever which they have done. *Hotiak* refers to the greatest Dani virtue, which is a cleverness, or an extreme competence.[53]

Men who knew when to escalate a fight and when not to escalate a fight would be preferred. Men who were simply interested in fighting or seducing other men's wives and who were generally lazy might be viewed as undesirable. Men who refused to fight on another man's behalf would also be viewed as undesirable. There are likely to be many aspects of personality taken into consideration; Napoleon Chagnon mentions a few:

> Some Yanomamö men are in general more responsible, ambitious, economically industrious, aggressive, concerned about the welfare of their kin, and willing to take risks. Becoming an *unokai* is simply one of a number of male characteristics valued by the Yanomamö and an integral component in a more general complex of goals for which ambitious men strive. All the characteristics just mentioned makes some males more attractive as mates in arranged marriages and dispose some of them to take the risks involved in appropriating additional females by force. Both paths lead to higher reproductive success.[54]
>
> [*Unokai* = one who has killed]

Since the method of alliance also involves the creation of a marriage, it is not as simple as signing a piece of paper. An alliance will be consummated with a promise to marry a daughter, or a sister, so the implications of an alliance are not merely of friendship, but are intimately involved with one's immediate family. If a man in the community is known to be industrious or lazy, assertive or violently unstable, this man isn't simply a potential ally, but a potential husband for one's daughter or sister.

Surrounding these personal issues are larger issues. Two lineages may be intermarried over generations, so variations of personal behaviour may fade in comparison to the larger alliance. For instance, due to a short-life expectancy and the levirate, a man might promise his daughter to the hard-working elder brother, only to see him die and instead his daughter is passed to the lazy or violently unstable younger brother.

In this light, it is easy to understand Cicero's thoughts on friendship, that, for men, it is more important than anything else, because without allies a man cannot expect to reproduce, or possibly even survive because of competition with other groups of men. Alliances and cooperation are difficult; thus good ones are treasured. A man might love his sister's husband, as his brother-in-law, the father of his own children's future spouses (they are cross-cousins) and as a close ally who frequently comes to his aid and to whom he also gives aid. Friendship, reproduction, and an alliance can combine into an intense relationship. Of course, enmity can also emerge, such as if a man accepts a wife, for himself or a son, but refuses to give one in return. Trustworthiness is thus of tremendous importance, since a good ally or a bad ally can affect children and grandchildren.

The importance of the alliances within a community is most obvious when the alliance is placed at risk. In the following example, from Ronald and Catherine Berndt's study of Aboriginal sexuality in Western Arnhem land, a man unintentionally kills his own sister for adultery, and responds by killing his own brother's wife, so that the two patrilineages are equal, having both lost a sister:

> As we have seen elsewhere, it is unusual for a husband to exact the maximum penalty in the case of his wife's adultery, stressing the infidelity of the woman and ignoring her lover. This is the only case available in which a wife is killed because of her extra-marital liaison. There seems to have been a mistake made, as the brothers first decided to give the girl a hiding, which was quite normal, considering that the husband observed the act of coitus. The carrying out of sorcery by the younger brother did not meet with the other's whole-hearted approval and it seems that he failed to realize what was happening until after the deed was done. Then he reciprocated by killing his brother's wife, who was quite unconnected with the original cause of argument. The reason for his killing the brother's wife, instead of the brother, was probably motivated not only by kinship affiliations, but by the belief that it would be fair to do to one wife what had been done to the other, for a wife, apart from ties of affection, is not merely an economic asset, but a person who ministers pleasure of her husband.[55]

In the example a man meant to punish his own sister and instead killed her. This has seriously undermined the basis of the alliance between his kin and his sister's husband's kin. In response, the man kills his own brother's wife in order to ensure equality between the two groups, even though she had done nothing wrong. The alliance was more important than the woman's life, or even her personal child-bearing ability. The male alliance is thus more important than individuals or an individual's ability to reproduce; it is the total reproductive capacity of both groups that is of importance, and thus the individual is subordinate.

Village fissioning presents a second example. Chagnon found that if a Yąnomamö community breaks in two then a man may not be able to remain with his wife and his brothers, as his wife's brothers may demand she remain with them. The man thus faces the possibility of having to end the marriage to live with his brothers, or to live with his wife and leave his brothers. This seems a more likely scenario when enmity and the desire for revenge is the basis for the village having fissioned:

> Not all individuals are able to remain with the closest of kin at fission, usually because they are married to a person whose kin group elects to leave, and they have to go along or dissolve their marriages. A war between their new group and the old one puts such individuals in an ambiguous position. Such men often refuse to participate in raids against the group whence they fissioned, pointing out that they wish their close kin no harm. No stigma is associated with this, nor is such a man considered a legitimate target of vengeance by members of his current residential group. If one of his close kin in the original group is killed on a revenge raid by members of his current residential group a man may be moved by grief to the point of deserting his wife and rejoining his original group, filled with smouldering resentment and a concealed hatred of those co-residents who participated in the killing of his kinsman. This underscores the difficulty of interpreting Yanomamö warfare as a phenomenon that pits all the members of one political community against all the members of a different political community and makes clear why the village is not the most useful unit with which to analyze warfare in many tribes.[56]

The arranged marriage is a contract of cooperation; if the man ends the contract of cooperation, he loses his wife. The extract also reveals how communities are not distinct entities. Communities are over-lapping kin groups, allies and enemies, and a neighbouring community may include all three. Alliance and warfare must be viewed from the individual level: thus the individual may have allies and kin in an enemy community. An obvious example is where an enemy community has abducted a man's sister, thus his sister and possible nieces and nephews are now part of an enemy community – people he may wish to do no harm.

A man's world is dependent on other men, rather than his wife or any other factor. In order to gain a wife, he must please and impress other men, since all the marriages are arranged. If his close male allies become enemies of his wife's male allies, then he may have to end the marriage.

In terms of understanding a community as a group of distinct political entities, it is worth considering the final form of alliance, that which occurs between communities.

Inter-community Alliances

Ego	Brother	Cross-Cousins	Parallel-Cousins	Other Communities

I against my brother
I and my brother against our cousin
I, my brother and our cousin against the neighbours,
All of us against the foreigner.

A community of brothers, their cross-cousins and some parallel-cousins will not be enough individuals to form a competitive group.

Evidence of this comes from Helena Valero's document of her time amongst the Yąnomamö. At the time of this event she was in hiding from the Namoeteri. The headman of the Namoeteri had attempted to kill Valero after he had been informed that she had insulted him and so she fled. Her Yąnomamö name is 'Napagnuma'. Immediately after her escape a feast occurred, with the Kashoraweteri and the Mahekototeri visiting the powerful Namoeteri. It is important to remember that all the three villages mentioned are meant to be allies:

> What did those Namoeteri do? They told me later that when all the Mahekototeri and the Kashoraweteri who had come together had left, they began to say: 'Those young Mahekototeri left too early this morning. Perhaps they have caught Napagnuma on the road.' In the afternoon, they told me, the *tushaua* shouted loudly: 'I think those men, who left early, found Napagnuma in the forest yesterday and have left her near to where they stopped to paint themselves. That's why they left so early today, and why they ran.' The others, it seems, replied: 'Let's go back. That's what they must have done. We will not take Napagnuma, but their women!'
>
> They took bows and arrows and ran after those who had come to dance with them and had been their friends. It seems that towards evening they caught up with the Kashoraweteri and entered their *tapirí*. The Kashoraweteri grew afraid and their *tushaua* asked what they wanted. The Namoeteri, who were very numerous, did not answer and began to look among the *tapirí*. The Kashoraweteri *tushaua* shouted: 'Why do you come to search like this? You have entered without speaking to me. First the *tushaua* should have come in and spoken with me. I would have replied to whatever he asked.' The brother of the Namoeteri *tushaua* said: 'You must know. Those who left early this morning have found Napagnuma, who was a fugitive in the forest.' It seems that the chieftain replied: 'I saw no women with them. Four of our men, who had left at the same time as they did this morning, are here. They had left early, because, when we were coming to you, they had found an armadillo's hole and smoke it.' Then, it seems, he called the four men and asked whether they had seen women or women's footprints. 'No,' they replied; 'neither women nor tracks of women along the mud of the igarapé, only our own footprints.' Then the Kashoraweteri shouted angrily: 'You, proud Namoeteri, should not have behaved like this with us, threatening to kill us.'
>
> The Namoeteri went out without shooting and assembled along the *igarapé*,

nearby. There they decided to pursue the Mahekototeri. The *tushaua* Fusiwe ordered his men not to ask whether they had seen Napagnuma but to take the women, those young ones without children, and carry them out of the *tapiri*. They told me later that, when it was dark in the wood, they came to where the Mahekototeri had made their *tapiri*. They began to enter threateningly among the huts; the old men and women took fright and shouted: 'The enemy, the enemy, the Namoeteri, many of them!' So they took seven young women; one was even pregnant. They did not kill a single man, because, they said, the men had left their bows and arrows and fled into the woods.

At night I heard when the Namoeteri returned. The women were weeping; the men were saying: 'Don't cry; why do you weep for your men who are old and ugly? You, so young, live with old men whose hair is already white! Be happy! You will enjoy life with us, we are many, and we are strong; no one will be able to recapture you. You were never at peace with them, fleeing all the time for fear of the enemy.' I followed them and listened to them without letting myself be seen. Later I came close to the *shapuno*. The *tushaua* was saying in a loud voice: 'I have been on the attack; one of these women is for my brother. I have come from the Kashoraweteri; the old man said to me: "There's no woman here; Napagnuma is not here!" He was angry as he spoke to me. Does he think perhaps that I am afraid of him? If I had wanted my men to take all their women, I would have done it. I took pity on him, as he has other enemies, and I did not want to go on being his enemy.'[57]

Differences in fighting strength between communities make smaller communities vulnerable, even, as in the above example, to communities that are their allies. Indeed, such was the vulnerability of the weaker Mahekototeri that, despite the Namoeteri taking seven of their young women, the Mahekototeri did not take revenge and later even sought peace with the Namoeteri.

The example reveals three aspects concerning inter-community behaviour: firstly, trust between allied communities can be broken if the benefit is large enough (such as the abduction of seven women); secondly, that weaker communities can be dependent upon the larger communities, even if the latter exploit them; and, thirdly, and most importantly, the implications of failing to be able to defend oneself against enemies, or even friends, can be enormous. For the men of a smaller village such as the Mahekototeri, losing seven women would be an enormous reproductive cost. For the women who are captured, it can be equally dangerous, placing their children (born or unborn) at serious risk, and separating them from close male kin.

The previous example is one where no men lost their lives; yet, as the next example shows, the costs of warfare or deceit can be greater than simply losing women. Chagnon recorded the plight of the Bisaasi-teri who suspected that their allies, the Konabuma-teri, were using magic to kill their children – a common suspicion before the current knowledge of disease. The Bisaasi-teri killed one of the Konabuma-teri men, which led to revenge:

Ruwähiwä's group then set about to avenge this killing. They enlisted the support of a their village that was on friendly terms with the Bisaasi-teri and managed to get them to host a feast at which the Bisaasi-teri would be the guests of honor. They invited men from a fourth village to join them in hiding outside the village. The unsuspecting Bisaasi-teri had come *en masse* for the occasion: men, women, and children. Shortly after the feast began, and while the Bisaasi-teri men were lying motionless and helpless in their hosts' hammocks, someone gave a signal. The hosts suddenly set upon them with clubs, bowstaves, and arrows, attacking them in their hammocks. Many died immediately, but some managed to escape outside. There, they ran into showers of arrows from the hidden archers. More died and more were wounded, some badly enough that they later died. Approximately a dozen men were killed that afternoon. A number of women and pubescent girls were taken captive, never to be seen by their families again. The survivors retreated deep into the jungle, to the north, and hid for many days while the wounded recovered enough to move on. The survivors, depressed and anguished, sought refuge in a village to the north, Mahekodo-teri. They arrived early in the year 1951, a date recorded by James P. Barker, the first missionary to make a sustained contact with the Yanomamö a few months prior to this. He saw the Bisaasi-teri arrive at Mahekodo-teri, the village he had chosen for his mission station.[58]

Again, it is the allies as much as the enemies that can pose a threat. This is also true for 'allies' who offer aid to a newly weakened village:

> Within a week or so of the treacherous feat, the Mahekodo-teri, a visiting ally of Kąobawä's group, learned of the massacre and offered aid. The Mahekodo-teri headman himself visited Kąobawä's village and invited the entire group to his village to take refuge. They accepted the offer, and in January of 1951, after conducting one revenge raid on the enemies, moved to Mahekodo-teri.
>
> The Mahekodo-teri had been allied to Kąobawä's group a generation earlier, but after Kąobawä's group moved away from the Orinoco River, alliance activity had dwindled to just sporadic trading. True to Yanomamö political behavior, the Mahekodo-teri, being in an obviously stronger bargaining position, offered their protection and aid with gain in mind: They demanded and received a number of their guests' women. Again, the members of Kąobawä's group suspected further treachery from their new protectors and assiduously worked at establishing a new garden. ... When Mahekodo-teri later split into three factions, Kąobawä and his group learned from one of them the details of a plot in which the Mahekodo-teri were planning to kill the rest of the men and abduct the women. The only thing that prevented them from doing so was the development of a new war between Mahekodo-teri and another village, one that required the assistance of Kąobawä and his group.[59]

The picture that is gathered from these events are one of utter distrust of allies, and that the risks involved in forming alliances can be tremendous. I could present a similar picture of trust and mistrust in the wars of Western nations, such as in World War II, when Germany broke several agreements, including one with the Soviet Union, whom they would eventually attack despite promises of non-aggression.

Alliances between communities are difficult. Communities need other communities in order to increase their status and have an ally to turn to if they are weakened. On the other hand, as the previous examples reflect, allies can be the most dangerous enemies.

The method of alliance between communities is exactly the same as that between brother-alliances: the exchange of cooperation for reproductive benefits (women). The characteristics of a desirable ally will also be similar: trustworthiness, reputation, fighting ability, and similarity of status. A community should desire an ally that will be trustworthy, will increase their own status, and yet be of a similar status to themselves to diminish the opportunity for exploitation. If the levels of status are different, then one of the parties might be open to exploitation. Napoleon Chagnon found:

> One of the expectations and implications of alliance is that the partners are under obligation to provide shelter and sustenance to each other whenever one of them is driven from his village and garden by a powerful enemy. In some situations the beleaguered partner may be obliged to remain in the village of his host for a year or longer, approximately the length of time required to establish a new garden and productive base from which an independent existence is possible. Twice in the recent history of Kąobawa's group they were driven from their gardens by more powerful enemies and were forced to take refuge in the village of an ally. In each case the group remained with the ally for a year or so, moving away only after their new garden began producing. In both cases, the hosts demanded and received a number of women from Kąobawa's group without reciprocating in kind, a prerogative they exercised from their temporary position of strength. The longer the group takes advantage of a host's protection, the higher is the cost in terms of women, so visitors always make an attempt to establish their gardens as quickly as possible and move into them as soon as they begin producing. Without allies, therefore, the members of a village would either have to remain at their single garden and sustain the attacks of their enemies or disband into several smaller groups and join larger villages on a permanent basis, losing many of their women to their protectors. The jungle simply does not produce enough wild foods to permit large groups to remain sedentary, and the threat of warfare is such that smaller groups would soon be discovered by their enemies and victimized.
>
> Because of the ever-present risk of being driven from one's garden, no Yąnomamö village in Kąobawa's area can continue to exist as a sovereign entity without establishing alliances with other groups. Warfare is attended by a bellicose ideology which asserts that strong villages should take advantage of weaker ones and coerce them out of women; to prevent this, the members of all villages should therefore behave as if they were strong.[60]

Chagnon found a three-step method of alliance formation amongst the Yąnomamö: trading, feasting, and, finally, the exchange of women. Only when the latter occurs are the two villages bound by the exchange of cooperation for reproductive benefits.

Hence, in order for an occasionally nervous meeting of groups of men from different villages to evolve into a stable intervillage alliance based on the reciprocal exchange of women, the long and difficult road of feasting and trading usually must be traversed. Suspicion must give way to confidence, and this must develop into reciprocal feasting.[61]

The result is a difficult method of alliance formation:

> Members of allied villages are usually reluctant to take the final step in alliance formation and cede women to their partners, for they always worry that the latter might not reciprocate as promised. This attitude is especially conspicuous in smaller villages, for their larger partners in defense pressure them into demonstrating their 'friendship' by ceding women; the strong can and do coerce the weak in Yąnomamö politics. The weak, therefore, are compelled to advertize their alleged strength by bluff and intimidation and by attempting in general to appear to be stronger, militarily, than they really are, thereby hoping to convince their partners that they are equals, capable of an independent existence. By so doing, they also inform their partners that any attempt to coerce them out of women will be met with the appropriate reaction, such as a chest-pounding duel or a club fight. Nevertheless, each ally expects to gain women in the alliance and enters into it with this in mind; and each hopes to gain more women than it cedes in return.[62]

Darwin's theory of sexual selection thus occurs at a remarkable level involving, potentially, scores of genetically diverse males. In terms of status, the two communities will be most likely to enter into an alliance when they can prove that they are both capable of preventing exploitation.

The Yąnomamö also have a method of determining the strength of each group, using a chest-pounding duel or a club fight so that the actual strength of each group can be determined. This creates the rather odd situation of fighting for friendship – fighting for equality. The following comes from Helena Valero's time with the Yąnomamö, with two allies fighting one another after distrust was created:

> When they were half drunk with *epená*, those of the *shapuno* said: 'You are upset, we are upset, we must calm ourselves', and then they began to fight. Two men stood opposite each other; the first raised his bent arm and the man on home ground gave him the first blow with his closed fist, a very powerful blow on the chest. Sometimes they gave even two or three blows one after the other, then said: 'Now you,' and the other returned the blows. Some fell after three or four blows.

> They had prepared *nabrushi* which were thicker at the end which strikes and thinner at the end they hold in their hands. They wanted to fight in order to become friends again.

> They began in twos, but if one fell, his brother came to take his place, or his brother-in-law, or his father-in-law. If four, five, or six gathered around one man, the *tushaua* said, 'No, no, the contest is for two only; stand aside. He who has fallen must avenge himself.' They picked up the fallen man, threw water on his head, pulled his ears, wiped the blood, picked him up again and gave him the *nabrushi* once more. Then the

other leaned on his *nabrushi* and awaited the blow, lowering his head. ... While they struck they said to each other; 'I sent to call you to see whether you are a real man. If you are a man, let us now see if we become friends again and if our anger passes...' The other replied: 'Say so, say so, hit me, and we'll become friends again.' When one fell down and did not stand up again, the others carried him away. The women took fright, the old mothers shouted: 'Eat him, then, eat him! You want to kill him to eat him!' [Endocannibalism refers to consuming the ashes of the dead.]

After the *nabrushi* they took axes; they had seized them a long time before from a group of rubber workers. The *tushaua* gave two blows with the blunt end to the man who stood opposite him, on one side of his chest, and the other fell. He was replaced by his brother, who gave in return four blows on the *tushaua's* chest, but Fusiwe did not fall.

Then came Rashawe, the most *waiteri*, the most courageous, the man who had defended me. 'Now it is my turn,' he said, 'just try on me.' Already five young men had fallen under Fusiwe's blows. The *tushaua* struck him with his axe on one side and then the other, but Rashawe did not fall; Rashawe was indeed strong. Then he returned the blow. They strike each other and the blows increase in strength until one falls. I was standing to one side looking on with that other wife. In the end, Fusiwe sat down and vomited warm blood from his mouth.

When they had all finished beating each other, they remained friends and said: 'We have beaten you hard and you have beaten us hard. Our blood has flowed, we have caused your blood to flow. I am no longer troubled, for our anger has passed. If anyone dies, it is a different matter.'

As soon as the fight was over, the *tushaua* said: 'Now go your way, here is the meat, here is the rest,' and he had baskets of meat, of *papugnas* and of maize given to them. 'If you are tired and cannot walk, sleep near the *shapuno* or sleep tonight with us; you will then leave tomorrow. We have fought by hand, with the axe, with the *nabrushi*, not to be enemies, but to calm down our wrath. I sent to fetch you in order to calm you down with a fight. I had learned you were against me and I don't want to be your enemy, because we are the same race.'[63]

This example reflects how respect is gathered by showing equal amounts of fighting strength and courage. Only when the two groups have proven that they are a worthy opponent do they also prove themselves as a worthy friend, rather in the manner in which boxers hug each other after attempting to knock each other senseless. We see the same events in sport: if a team beats another team on penalties or over-time then there is little respect lost, but if a team defeats another by a large amount, this leads to derision and depression for the losers.

In order for men, as individuals, to give respect to one another they must prove they deserve respect. They must prove they are of equal status, and, therefore, not to be exploited or feared, but to be accepted on equal terms. If violence is an important method by which men prove their status, then the method by which men will earn each other's respect is by beating each other until they realise they are potentially equal. Once equality exists, men will become more trusting of each other because neither will fear exploitation. This will be the same for brother-alliances or entire communities.

Once two communities are on good terms, then the communities may visit regularly; the following from Chagnon:

> Friendly neighbors visit regularly, and the most commonly used trail from Kạobawä's village went south to the friendly Shamatari villages: Reyaböwei-teri and Mömariböwei-teri. Hardly a week went by during the dry season without someone, usually small groups of young men, going from one of the Shamatari villages to Bisaasi-teri or vice versa. Young men can make the trip easily in one day, for they travel swiftly and carry nothing but their bows and arrows. A family might also make it in one day if it kept on the move, but it would be a dawn-to-dusk trip if the women had to carry their babies or items they or their men planned to trade. Should the whole village decide to visit, it might be a two- or three-day trip, depending on how anxious they were to reach the village.[64]

Women will be exchanged between the strongly allied communities, and, over time, they will become increasingly genetically related. This will be especially true if the leaders of the two communities exchange daughters to marry their sons, and if their sons become leaders, thus making the sons direct first cross-cousins. The exchange of sisters or daughters between leaders is a common practice, and occurred in Europe between the royal families up until the twentieth century. Amongst the Tauade in Papua, Hallpike noted that 'It is always stressed that the chiefs marry one another's daughters and sisters, which helps to maintain a kind of solidarity among the leaders'.[65] Another researcher in Papua, Strathern, noted that 12 out of 18 leaders were the sons of a father who was also a leader ('big-man').[66] Since the leaders of communities are the most likely to be polygynous in their community, they may have the most daughters to exchange for the creation of political alliances. While we might assume that it is the daughters of lower status men that are the most likely to be exchanged into other communities, this would be an incorrect assumption. Since higher status men are looking to forge links with other communities in order to maintain their own status, it is they who may be most likely to exchange their own daughters, particularly with high status men in other communities. If two leaders of two communities are full cousins and genetically linked, then they may be more likely to continue the alliance between the communities, creating future marriages between their children, and having the political leverage to enforce cooperation and prevent deceit between the two communities. Anyone who has read of the politics amongst royal families in Europe over the last millennia will observe similar behaviour.

Yet, as Chagnon has shown, cooperation is difficult. Two communities that were once equal may become less equal with time. One may possess more enemies than the other, one may suffer more deaths from disease or war, one may fission into two communities, and in each case the weakened partner will be open to exploitation. The stronger community may renege on a promised marriage, and the weaker community may have little choice but to allow themselves to be exploited. Alternatively, if one community seeks refuge, there is a significant possibility for the refugee community to be exploited, as depicted above.

These issues of deceit and cooperation also reveal a benefit of the method of coop-eration, the arranged marriage. Since the marriages are arranged when the girl is very young, this allows for deceit to be punished by voiding the marriage before consummation. The possibility of deceit, whether between individuals or entire communities, can be taken into account by the long-term contract of the arranged marriage.

The dangerousness of the method of cooperation reveals a further point of inter-est: responsibility. While one might think that men would desire, always, to be the leader of their community or of their brother-alliance, this is in fact not the case. The following example comes from Chagnon concerning the aftermath of the killing of the Monou-teri's leader, Matowä:

> Instead of giving chase, as they ought to have done, according to Kąobawä and the others in Bisaasi-teri, the Monou-teri themselves fled into the jungle and hid until darkness, afraid that the raiders might return. Their awesome leader, Matowä, was dead and they were now demoralized.[67]

> Kąobawä, a classificatory brother to Matowä, then assumed the responsibility of organizing a revenge raid. Matowä's own brothers failed to step forward to assume this responsibility, and for a while there was no leadership whatsoever in Monou-teri. Finally, Orusiwä, the oldest and most competent member of the village, emerged as the *de facto* village leader; somebody had to take the leadership responsibility. He was related to the slain headman as brother-in-law, and their respective descent groups dominated village politics. Hence, leadership in Monou-teri shifted from one lineage to the other, equally large, lineage.[68]

Since the leader must also be the leader in seeking revenge, in enforcing an arranged mar-riage, in proving the courage of his community, one might expect that rather than individuals demand to be the leader, they would instead desire not to be the leader. Since cooperation is risky, those that organise the cooperation must also accept the responsibility if the coopera-tion fails. In order to gain many wives, a man must agree to cooperate in violent disputes. The more cooperative alliances he has, the more wives he gains, but the more risks he must take in order to be cooperative.

While most evolutionary thinkers today would instantly nominate food as the lim-iting factor for polygyny, Napoleon Chagnon found that alliances and risk-taking were the major limiting factors:

> In human societies, however, cooperation and sharing of resources are such that *individual efforts* at economic production are not the limiting factor insofar as acquiring resources is concerned. Among the Yanomamö, for example, sons-in-law, younger brothers, brothers-in-law, and other relatives contribute food, support, and aid to the mature males. The headman, for example, has a continuous flow of material and other resources – game, cultivated foods, wild foods, labor, services – to his family's hearth, and probably subsists better than most individuals in the village. In

brief, insofar as resources are concerned, he is in a good position to be able to invest in his mates' children and ipso facto in a good position to acquire multiple spouses. Reciprocally, he helps younger male kinsmen to secure mates, takes enormous and frequently mortal risks on behalf of his kin during periods of active raiding, and keeps order and relative peace within the village. His ability to acquire resources is predicated on the size and extent of his following, and this resource base likewise permits him to compete effectively for additional mates. As will be demonstrated... headmen *are* more successful in obtaining multiple spouses and siring more offspring than other men of comparable age.[69] (Chagnon's emphasis.)

Indeed, Chagnon compared twenty headmen to 108 males of 35 years or older and found that headmen had half as many wives again (3.6 to 2.4) and twice as many offspring (8.6 to 4.2). Interestingly, the headmen also had a less biased sex ratio with 110% sons to daughters compared to 121% sons to daughters for the 35+ males, suggesting that daughters are less likely to survive infancy amongst fathers of lower status. The headmen's wives also had higher average numbers of offspring (2.4 to 1.7) suggesting either earlier marriage or increased survival rates.[70]

In terms of assistance, Warner made the same observations amongst the Australian Murngin:

> Mutual help prevails between wawa and yukiyuko in giving frequent presents to gawel and mokul. When wawa is making presents for his first wives, yukiyuko gives him spears, game from the hunt and whatever wawa might like to present to his parents-in-law. When yukiyuko's turn comes, wawa is equally helpful.[71]
> [Wawa = older brother; yukiyuko = younger brother; gawel = father-in-law; mokul = father's sister.]

Food is not the main issue concerning a man's ability to reproduce, nor is any other resource. If a man promises a daughter to another man, then he can expect to be the beneficiary of the other man's labour, such as hunting. Thus a man with many daughters may never need to hunt since his future sons-in-law will be bearers of gifts. The main two factors which limit human behaviour are cooperation and reproduction (women), and risks are involved in obtaining the benefit of both.

In Western nations a similar system operates at a larger level, though there are differences. Money is used as a form of symbolic status for exchange, rather than reproduction. People's positions are more highly specialised: each man is not expected to perform the duties of a policeman, soldier, judge, farmer and hunter, but performs a single role as part of a larger society. Yet the hierarchies, the web of contracts, and the exchanges of status are much the same. Within each community (city), there are numerous brother-alliances (companies or partnerships) that combine the status of the individuals within each brother-alliance (companies or partnerships) to compete against other brother-alliances (companies or partnerships). The benefit of such competition in Western societies is largely status, since monogamy is enforced, limiting reproductive competition. Many evolutionary thinkers today speak about a 'stone-age' mind in a modern environment, but this view is incorrect: the similarities between the so-called 'stone-age' world and our own are far greater than the differences: the physical environments might be different, but the social environments are largely the same.

People cannot underestimate how vital the many above examples are in understanding human intelligence and behaviour. This level of cooperation and competition is unimaginable to the chimpanzee or gorilla. I will give one more example. It is important to keep in mind in the following example how the ceding and acquiring of marriages and women are all linked to both cooperation and reproduction.

The following is a further example of the chess game of the ceding and acquiring of women; it comes from Chagnon's study of the Yąnomamö. In this situation, Kąobawä's community, the Upper Bisaasi-teri, made an alliance with the Paruritawä-teri in order to engage in a revenge attack. The Paruritawä-teri are a Shamatari village, and the village they agreed to attack was another Shamatari village.

> This treacherous feast put the Paruritawä-teri in a very poor position. They were (and are) the westernmost Yąnomamö in the area, and their closest neighbors were the Monou-teri, Lower Bisaasi-teri, and Kąobawä's group. All their other neighbors, allied to the victims of the treacherous feast, relentlessly raided them. They were, in short, extremely dependent on Kąobawä's group and the two other villages related to Kąobawä's group.[72]

(Those two other villages are the Monou-teri and Lower Bisaasi-teri – the latter having split from a larger Bisaasi-teri to become the Lower and Upper Bisaasi-teri.)

> Kąobawä and his followers have astutely finessed this situation in two ways. First, they have managed to coerce the two Shamatari villages into ceding women without reciprocating an equal number of their own in return.[73]

Yet the two dependent Shamatari villages, the Paruritawä-teri and Reyaboböwei-teri, responded by organising their marriages in such a way as to create disunity amongst the Upper Bisaasi-teri and its allies. They did this by organising marriages with only one of the Bisaasi-teri villages, thus increasing their solidarity with one of the Bisaasi-teri villages, and weakening the solidarity between the Bisaasi-teri villages:

> The Shamatari are also shrewd in their alliance practices, and have succeeded in neutralizing their disadvantages by capitalizing on schisms between Kąobawä's group and Lower Bisaasi-teri. This has been done largely by confining their marriage exchanges to important men in the dominant lineages of Upper and Lower Bisaasi-teri. One of the Shamatari villages, Mömariböwei-teri, has restricted its exchanges to Lower Bisaasi-teri and Monou-teri, while completely ignoring Upper Bisaasi-teri. The other Shamatari village has given its women only to Upper Bisaasi-teri. This has resulted in the weakening of ties between the three villages of Upper Bisaasi-teri, Lower Bisaasi-teri, and Monou-teri, whose members are related agnatically to each other, and who share a common history. That is, they have lost solidarity as a network of allied villages

and, therefore, some of the advantages they enjoyed as wife receivers with respect to the Shamatari villages.[74]

This lack of solidarity eventually led to a club fight:

> Just before I arrived, there was a club fight involving all of the villages: The Upper Bisaasi-teri and Reyaboböwei-teri fought, as a group, against the Lower Bisaasi-teri and Mömariböwei-teri; that is, the groups that had entered into marriage exchanges with each other behaved as a unit and fought against their close agnates and erstwhile allies. A group of men from Mömariböwei-teri forcefully took a woman from the Reyaboböwei-teri, thereby precipitating a challenge to a club fight. The Monou-teri participated in the club fight as well, taking the side of the Mömariböwei-teri, from whom they had taken several women and to whom they had promised several.[75]

The remarkable aspect of this scenario is that the Monou-teri, Upper Bisaasi-teri and Lower Bisaasi-teri are now engaged against one another in a club fight. Historically, they were the same community:

> Kąobawä's group split in the process of abandoning the Kobou garden in 1954, a part of them taking the name Monou-teri. ... Kąobawä's group moved to the Orinoco again, to the site called Barauwä. ... They split into two factions after settling, half of them moving to the opposite bank of the Mavaca River. Thus Bisaasi-teri has an upstream and a downstream settlement, and the members of these two groups refer to themselves as the Upper and Lower Bisaasi-teri, respectively, Biaasi being the name of a grass commonly found around gardens.[76]

Thus group A has fissioned to become B and C (Monou-teri). Group B fissions to become D and E (Upper and Lower Bisaasi-teri, respectively). Group D (Upper Bisaasi-teri) forms a coalition with two Shamatari communities. The Shamatari communities are exploited and respond by organising their marriages with either group D (Upper) or group C (Monou-teri) and E (Lower). This then leads to a club fight between C, E and their Shamatari ally against group D and their Shamatari ally. Thus the original community has split into three sections, which are now active competitors using a club fight to resolve this competition.

The example reflects how using marriage alliances is rather like chess. Villages attempt to: form alliances with other communities, create enmity by preferring one ally over another, exploit other communities, and so on. Again, these marriages are all linked to reproduction, so success or failure has a direct bearing on genetic success. In these societies, how well people are able to play this highly political game has a direct bearing on their success in the gene pool. Since the game of chess is played with arranged marriages, people must make their move years in advance – promising marriages that will be consummated a decade into the future.

The reader should also note how remarkable this information is: the Shamatari communities that attacked the Paruritawä-teri, for instance, were uncontacted at the time – in other words, no Western person had met them.[77]

Those who wonder why we need intelligence need to ask themselves: could chimpanzees or gorillas engage in this behaviour? Those who attempt to explain human evolution through food production need to compare the difficulty of acquiring food with the above level of competition, cooperation and politics. Which is more likely to require a complex brain?

Location

A final aspect of competition is location. Obviously in order to possess a large group of cooperative people it is necessary to live in a nutritionally supportive environment. The final aspect of Napoleon Chagnon's work with the Yąnomamö required (at the time) modern technology, such as GPS equipment, in order to identify patterns of warfare in regard to location. He found that:

(1) Warfare in the low-lying areas is more intense than in adjacent foothills and mountainous zones, although the inhabitants of those regions recently lived in the lowlands. While I am still collecting relevant field data, the picture that is emerging is that populations in the lowlands are continuing to grow and fission, but the ability of newly fissioned villages to remain in the lowlands is limited by space and other reasons. Some villages are forced out, usually the smaller ones. These tend to relocate in the more defensible foothills and their members periodically attempt to relocate to the lowlands when their villages grow sufficiently large to compete effectively with the larger villages already there.

Some of the smaller, less viable villages are pushed further into the mountains and locate at higher elevations where resources are presumably less abundant and more energetically costly to produce or acquire. For example, many species of useful plants and animals are not even found there, such as caiman or tapir. Gardening is extremely difficult on the steep slopes and energetically very costly in comparison with lowland cultivation.

(2) Some villages that are expelled from one large lowland area find other, adjacent, large lowland areas that are unoccupied and colonize them. The process appears to be repeated in these areas, and the smaller villages are forced into more rugged terrain at higher elevations.[78]

It is obvious to anyone that some areas will be more difficult for survival than others. Rivers can be plentiful in fish, and soils can be more fertile and productive in some regions than others. This not merely affects access to fruits and vegetables, but also to the animals that feed off the more abundant flora. If people do not possess access to fresh water, then life will be extremely difficult. This is a concern for Australian Aborigines, particularly in the more barren centre of the country.

Napoleon Chagnon observed what occurred to Yąnomamö communities that were unable to remain competitive in fertile environments; the following concerns an on-going war between the Patanowä-teri and the Monou-teri. I have quoted this at length because it

includes much of what has been discussed in this chapter, while also revealing how movements by communities reflect their capacity to compete against other communities, rather
than any actual higher decision making in fighting for particular areas of land. If people
become uncompetitive, they are forced to move, and they will, slowly but surely, move into
less densely populated and less competitive areas with less fertile soil. It also reflects other
aspects, such as the importance that particular individuals play in making a community competitive. Not every man appears to be capable of initiating violent attacks on other communities or killing other men; only a small number of men in each community appear capable of
such violence, and these men are highly valued by the community and rewarded with more
wives, on average. Even if there are many men in the community, without men with a capacity for violence, a community becomes uncompetitive, and it can be forced out of the more
fertile, competitive environment.

> The Monou-teri had anticipated their raid by clearing a new garden site across the
> Mavaca River, where they hoped to take refuge after the inevitable revenge raids from
> Patanowä-teri began. They had hoped to complete their garden before the raids became
> intense, as the Mavaca River would have provided a natural obstacle to Patanowä-teri
> raiders. The Patanowä-teri, however, were infuriated by this killing and raided the
> Monou-teri immediately.[79]

After the revenge attack by the Patanowä-teri, the Monou-teri sought assistance from their
allies, the Upper Bisaasi-teri:

> The Monou-teri were afraid to return to their producing garden, so they
> divided their time between their newly cleared site, where they worked at cutting timber
> and burning it, and Kąobawä's village, where they took occasional rests to regain their
> energy. ... They returned to their old site only to collect plantains, which they carried to
> the new site. Kąobawä's group then built a new *shabono* and fortified it, anticipating the
> war they knew would be inevitable. Up to this point, Kąobawä's group, Upper Bisaasi-
> teri, maintained two small *shabonos* a few yards apart from each other: *he borarawä*. In
> anticipation of the war, they coalesced into a single larger group and moved into the
> new *shabono* when it was completed. The visiting Monou-teri also helped them work
> on the new structure.
>
> Meanwhile, the Patanowä-teri, knowing they would be raided by the Monou-
> teri and their allies, also began clearing a new garden... By this time the Patanowä-teri
> were in rather desperate straits. Their old enemies, the several groups on the Orinoco
> River, began raiding them with even greater frequency after learning that the Monou-
> teri and Bisaasi-teri were again at war with them. A few additional villages began raiding
> the Patanowä-teri to settle old grudges, realizing that the Patanowä-teri had so many
> enemies that they could not possibly retaliate against all of them. This 'ganging up' on
> a weak adversary is very characteristic of Yąnomamö politics and warfare is a solemn
> reminder to members of large villages to think twice about fissioning into smaller, more
> vulnerable ones.

The Patanowä-teri then began moving from one location to another, hoping to avoid and confuse their enemies. They spent the dry season in turns at their main producing garden, with the Ashidowä-teri, their only ally, and at their new garden. Each group that raided them passed the word to the other villages concerning the location of the Patanowä-teri. If they were not at one place, then they had to be at one of the other two. The raids were frequent and took a heavy toll. At least eight people were killed by raiders that year, and a number of others were wounded. Some of the dead were women and children, a consequence of the fact that the Patanowä-teri themselves sent a heavy volley of arrows into the village of one of their enemies and killed a woman. Females are normally not the target of raiders' arrows. Thus the Patanowä-teri were raided at least 25 times while I conducted my initial fieldwork. They themselves retaliated as frequently as possible but could not return tit-for-tat. They managed to drive one of their main enemies, the Hasuböwä-teri, away from their garden, forcing them to flee across the Orinoco. They concentrated on raiding this group until they had killed most of the *waiteris* (fierce ones). They were so successful at doing this that the Hasuböwä-teri ultimately withdrew from the war, about as demoralized as the Monou-teri were. Several of my informants claimed that they did so because their fierce ones were all dead, and nobody was capable of prosecuting the war any further against a village as ferocious as Patanowä-teri.[80]

When we consider competing for territories we tend to think in a Western sense, with people possessing maps and forming plans. The reality of competition for territories for cooperatively-simple societies is somewhat indirect, with the issue of whether communities can remain in a competitive environment depending on their capacity to compete with neighbouring rivals. The above example reflects this, with one of the communities, the Hasuböwä-teri, forced to cross the Orinoco River to escape their enemies. Thus, perhaps the Hasuböwä-teri moved into an environment that was less fertile, but the reason that they moved was not because others were coveting their territory, but rather because in order to maintain their position it was necessary to be able to compete with their neighbours, and they were no longer able to do this.

Competing for territories is more indirect than specific. No doubt the Yąnomamö know which areas of land are more fertile than others, but obtaining or maintaining a position in a fertile environment is dependent on the fact that fertile areas possess higher levels of competition, and whether the village is able to exist in such a highly competitive area.

For instance, Napoleon Chagnon found significant differences between populations residing in lowland and highland regions – with the highland areas being more difficult for horticulture. In the lowland areas compared to the highland areas, there was almost twice as many people (931 versus 499); a greater percentage of married men (90 versus 77 percent); and a greater number of polygynously married men (15 versus 8 percent).[81]

Chagnon also found that between the foothills and the mountain areas there was significant differences, with much higher percentages of unokai males (men who had killed a man) and abducted females in the foothills. Of the three foothill villages, the highest to lowest percentage of unokai men was 57.1 to 27.6, while in five mountain villages the highest to lowest percentage of unokai men was 21.4 to 0.0 percent.[82] So every foothill village possessed a greater percentage of unokai men than any of the mountainous villages.

Chagnon concluded:

> The political strategy appears to be to try to remain in the lowlands, an area that appears to be more desirable. But to remain there, villages have to grow large and remain large in order to compete militarily with neighbors. As villages grow in size, more fighting and conflict occurs within the village because villages become genealogically and socially more heterogeneous: formerly abducted women have offspring in the village, and unrelated or distantly related people from nearby villages migrate into the village. The Yanomamö will sometimes give as the explanation for a fission something like: 'We were too many in the village and were fighting all the time with each other. We got sick and tired of all the fighting and decided to split into two groups.' Eventually, the large villages fission to reduce conflict, but they attempt to remain in the same area and remain allied. Thus, they must claim and defend areas much larger than those required for subsistence reasons alone. They appear to do so in order to have unoccupied safe havens to retreat into when military pressures from their enemies and combined, allied villages of their enemies become particularly intense. When villages they have forced into the foothills or mountains attempt to move back into their unoccupied safe havens, the larger villages tend to react belligerently to them and harass them. This often takes the form of coercing them out of women should they tolerate their presence there at all, and the only way the smaller villages can remain there is to make peace with their larger neighbours and try to cultivate friendships and alliances with them. They may even be willing to tolerate the loss of a few women if that is the cost of occupying a more attractive lowland address. But each unchallenged act of coercion tends to lead to increased coercion: the strong exploit the weak if the costs are low and the benefits high. Eventually, members of the smaller group grow weary of the coercion and react, sometimes violently, and are forced out of the area again. A common example is when a group of men from the smaller village come to visit the larger village to demonstrate friendship and the men in the larger village forcibly appropriate all their women and send the men packing. A less common tactic is to invite the smaller village to a feast and then kill as many men as possible and take the women...[83]

The scenario depicted leads to several possible conclusions. Firstly, if men that live in more fertile regions have more wives (as Chagnon showed) and more children, then they will be more reproductively successful. This means that their progeny will compete with one another for space in the fertile region. In other words, not merely is competition between people who are similar, in the first place, but competition is likely to be between the progeny of those who were already successful. Brothers and full cousins will thus compete against men who are, most probably, fourth or fifth cousins or more related, rather than men who are very distantly related.

Secondly, movements appear to be based on weakness in the capacity to compete against other humans, rather than other needs. This suggests that human movements were more likely to be performed by less successful groups searching for new environments. Thus one might suggest that those that first moved out of Africa and into Europe, Asia, America, and Australia, may well have been uncompetitive humans that had been expelled from competitive environments. Needless to say, robust, healthy, competitive people do not move from the fertile environment that sustains their competitiveness. Movement may occur even if the new environment is itself inhospitable. A good example of this is amongst the San in the Kalahari Desert, where the environment is so poor that girls do not become parous until they are 19 and the height of the people is under five feet. It may well be that once people lose their capacity to compete, they may simply be better off searching for new lands, even if the new lands are extremely poor. Thus the movement of people around the world is probably one of refugees fleeing competitive environments and moving to less competitive environments, and those fleeing may be closely genetically related to those they are fleeing from.

There tends to be a belief that hunter-gatherers and -horticulturists are stagnant; the myths that the people create suggest that the people have always lived in the environment in which they now live. A good example of this is the use of plantains and bananas by the Yąnomamö, which they believe that they have been growing 'since the beginning of time'[84] when, in reality, they obtained them 'via diffusion from other Amazonian groups who ultimately obtained them from European colonists after 1492'.[85] Yet, the reality is that these peoples' histories, if recorded, would be as transitory as the recorded movements of peoples in Europe.

Concluding Remarks

Human cooperation is not easy. Being able to become a successful human being is not easy and cannot be achieved in isolation, nor with only a nuclear family, nor with only an extended family. Human beings need large numbers of supporters in the community and between communities in order to compete with other human beings. Males forge these large groups by making specific alliances with other males using arranged marriages. They do so primarily by forming alliances with cross-cousins. They may or may not benefit from the support of parallel-cousins. They attempt to form alliances with other communities through arranged marriages. People attempt to create a spider web of cooperation, with each thread an arranged marriage that directly links cooperation and reproduction.

The greater the number of supporters a man has: the more likely he will be able to prevent deceit or manipulation, the more likely he will be able to protect himself and his close kin, the more likely he will be to live in a fertile environment, and the more likely he will be to manipulate others and increase the number of wives for himself and his close kin. For those with few supporters, the converse is equally true.

The outcome of this is that we cooperate with close relations, people of lesser relatedness, and even with people living in different communities. We cooperate by forming long-term contracts involving betrothal, and thus can agree to cooperate for more than ten years into the future. We need a brain capable of forming cooperative links with up to several hundred people living in several communities and can form these cooperative links into the future. We need a brain capable of knowing these people, knowing our genealogical link between all these people, knowing how future and past people fit genealogically in relation to these existing people, and to organise arranged marriages to maximise our capacity to create predictable, trustworthy patterns of cooperation. In creating his own genealogies, Napoleon Chagnon commented on the headman who worked with him most closely:

> Kąobawä's familiarity with his group's history and his candidness were remarkable. His knowledge of details was almost encyclopedic, his memory almost photographic. More than that, he was enthusiastic about making sure I learned the truth, and he encouraged me, indeed, *demanded* that I learn all details I might otherwise have ignored. If there were subtle details he could not recite on the spot, he would advise me to wait until he could check things out with someone else in the village.[86]

> Once again I went over the genealogies with Kąobawä to recheck them, a considerable task by this time. They included about two thousand names, representing several generations of individuals from four different villages. ... Thus, after nearly a year of intensive effort on genealogies, Yąnomamö demographic patterns and social organization began to make a good deal of sense to me. Only at this point did the patterns through time begin to emerge in the data, and I could begin to understand how kinship groups took form, exchanged women in marriage over several generations, and only then did the fissioning of larger villages into smaller ones emerge as a chronic and important feature of Yąnomamö social, political, demographic, economic, and ecological adaptation.[87]

Without pen or paper, let alone a computer, a village headman was able to know up to 2000 names of living and dead relatives. How many people would consider such an issue of importance in the evolution of language and abstract thought? Furthermore, this knowledge concerned only four villages, and, as this chapter has shown, the Upper Bisaasi-teri, Kąobawä's village, was involved in alliance and conflict with far more than four villages – thus his knowledge would have gone beyond two thousand names. This is what our brain must be capable of doing: knowing perhaps two thousand individuals or far more, alive or dead, and know how they are related by marriage and birth, remembering that there are high levels of intermarriage over the generations. Only on acquiring this knowledge does a man know if a woman is a 'wife', 'sister', 'mother' or 'mother-in-law' and what other categories she might be placed into. This is the cooperative and reproductive system that binds the society. With that knowledge people know how the cooperative system is meant to operate, and who is attempting to cheat the system. Using and manipulating the cooperative system, people are able to forge large, competitive groups. This is why we need large brains.[88]

Today, many people talk about the focus on individuality in modern societies without fully comprehending what this means in a historical and cross-cultural perspective. Historically, individuals, as such, did not exist in a legal sense. Groups, whether a group of brothers, or an extended group of cousins, exist in a collective sense in terms of responsibility, simply because nobody could survive as an individual. Revenge is taken against the collective, rather than the individual; responsibility is held collectively, rather than individually; and it is arguable that reproduction is held collectively, rather than individually, due to the levirate and sororate. The human male can only be understood in his relation to other men, on whom he is completely dependent for his very existence, rather as an individual male within the nuclear family. Napoleon Chagnon summarises this:

> Adult brothers try to remain together for cooperation and defence, you can trust your kinsmen more than you can trust strangers. Brothers tend to be very cooperative and quick to defend each other. And without police or state or laws and courts, your only source of defence is your kinsmen. And the more closely you are related to your relatives, the greater is the probability that they will defend you, whether you are right or wrong. But they expect you to defend them, and kinsmen in general do defend each other, whether they are right or wrong.[89]

And if you don't have any kinsmen?

Then you are in bad luck.[90]

When there is no police force, kin and allies become everything.

Chapter 5

Marriage

Marriage is invariably viewed by Western society as a contract between merely two people: the husband and the wife. So far I have argued that human marriage concerns a contract between two groups of brothers: a contract of cooperation in exchange for genetic benefits.

This view of marriage did not end with cooperatively-simple societies, but can be found in western Europe until relatively recently. Margaret King found that:

> In England in the seventeenth century, former chief justice Sir Edward Coke forcibly abducted his daughter Francesca (and may have had her imprisoned and beaten) to compel her to consent to marry a mentally unbalanced brother of the Duke of Buckingham's: an opportune alliance was irresistible. Pressure to marry prudently could be exerted from beyond the grave. An English father in 1558 willed a hundred pounds in dowry to each of his daughters; but the sum was to be reduced to sixty-six "if any... will not be advised by my executors, but of their own fantastical brain bestow themselves lightly upon a light person." Another bequeathed his daughter some sheep and husband in a single sentence of his 1599 will: "To my daughter Marjorie, LX sheep, and I bestow her in marriage upon Edward, son of Reynold Shaftoe."
>
> Pressure to marry the parental candidate was surely most fierce in elite circles, where only the most determined and fortunate of heiresses could hope to choose her own marital destiny. Those completely free of property, and thus of any basis for negotiations in the marriage market, clearly had the most freedom to choose. But even among poorer families, family design and economic strictures dictated the marriage partners.[1]

Georges Duby found in the Middle Ages:

> The *Sponalia*, the ceremony in which the agreement was concluded between the two families and mutual consent was expressed, took place extremely early, and if the girl was too young to speak then a smile on her part was taken as a significant sign of her agreement. The wedding, too, was at a very young age. The moral code and custom of the day permitted that as early as the age of twelve a child could be taken from her cloistered world of women in her home, where she had been cocooned since birth, led with great pomp to the marital bed, placed in the arms of a greybeard whom she had never seen before, or of an adolescent scarcely older than herself, who, ever since he himself had left the care of women in about his seventh year, had lived only to prepare himself to combat through physical violence and the exaltation of male violence.[2]

Much of this is summarised in Stephanie Coontz's recent book *Marriage, A History: From Obedience to Intimacy or How Love Conquered Marriage*.

George Bernard Shaw described marriage as an institution that brings together two people 'under the influence of the most violent, most insane, most delusive, and most transient of passions. They are required to swear that they will remain in that excited, abnormal, and exhausting condition continuously until death do them part.'[3]

But only rarely in history has love been seen as the main reason for getting married. When someone did advocate such a strange belief, it was no laughing matter. Instead, it was considered a serious threat to social order.

When I studied history as an undergraduate in the 1960s, I was taught that people didn't begin marrying for love until the nineteenth century. In the 1970s many historians began dating the love match from the eighteenth or even the seventeenth century. Today many scholars trace the celebration of married love and companionship to the Protestant Reformation in the sixteenth century. A few even believe the basic contours of modern love took shape as far back as the thirteenth century.[4]

In her overview of the life of English women in the seventeenth century, Antonia Fraser concludes:

During this period, the emotion we should now term romantic love was treated with a mixture of suspicion, contempt, and outright disgust by virtually all pundits. From the Puritans in their benevolent handbooks of domestic conduct to the aristocrats concerned to see that society's pattern was reproduced in an orderly fashion, that tender passion which has animated much of the great literature of the world (including the plays of Shakespeare and the Jacobean dramatists familiar to audiences of the time) received a hearty condemnation. Nor was this a revolutionary state of affairs in seventeenth-century England, the arranged marriage as opposed to the romantic union having been preferred by most societies in the history of the world.[5]

Most people today wouldn't consider Shakespeare's *Romeo and Juliet* to be a revolutionary text that undermined the very foundations of society, but at a time when the arranged marriage was the main form of passing on wealth through the generations, there are few other ways of appreciating the text. Fraser continues:

Only at the bottom of society was some kind of proper independence enjoyed. Women of the serving or labouring classes were in theory subject to exactly the same pressures where love was concerned. 'This boiling affection is seldom worth anything' when making a choice of husband, wrote Hannah Woolley in her commonsensical handbook for 'the Female Sex' which included 'A Guide for Cook-Maids, Dairy-maids, Chamber-Maids, and all others that go to the service'.
Nevertheless, in practice these toiling females enjoyed a good deal more

freedom of choice where their marriage partner was concerned than their well-endowed sisters, simply because they lived below the level where such considerations as portions and settlements could be relevant. With freedom of choice came obviously the freedom to marry for love, if so desired, simply because no one else's interests were at stake. Richard Napier was a consultant clergyman-physician who kept notebooks of his cases between 1597 and 1634; they reveal, according to their editor, many instances of romantic love (and its problems) 'among youth of low and middling parentage'.[6]

One well-to-do young woman observed this difference in her letters:

> Dorothy Osborne described how she would walk out of a hot May night to a common near her house, 'where a great many wenches keep sheep and cows, and sit in the shade singing ballads'. When she walked over to them, she found their voices and appearance vastly different to 'some shepherdesses that I have read of'. But when she fell into discussion with them, she found that despite these deficiencies, they wanted 'nothing to make them the happiest people in the world but the knowledge that they are so'.
>
> If not the happiest people in the world, Dorothy's wenches were, in the single instance of their emotional independence, ahead of the majority of their female contemporaries.[7]

It should be noted that the women that did possess independence also had jobs. These freedoms may not have been available to women in earlier centuries since there would be fewer such positions. Indeed, while many people speak of the Enlightenment and the Industrial Revolution as being primarily concerning learning and finances, these two periods would have had a profound impact on the capacity for men and women to choose freely, merely because it enabled them to obtain work and extricate themselves from their families. Before these periods, few men and women would have been able to extricate themselves from family pressure. There must have been tremendous interest from young women to move to the growing towns and cities and away from their rural villages to be able to obtain freedom and independence, even if it meant living in a filthy town and performing mundane work.

Until recently, western Europe was not so different from the societies I have been focussing on, such as the Ya̧nomamö:

> Girls, and to a lesser extent boys, have almost no voice in the decisions reached by their elder kin in deciding whom they should marry. They are largely pawns to be disposed of by their kinsmen, and their wishes are given very little consideration. In many cases, the girl has been promised to a man long before she reaches puberty, and in some cases her husband-elect actually raises her for part of her childhood. 'Boys' seem to be more able to 'initiate' the process and have their older kin make the first marital inquiries, but since males marry later in life than females, these 'boys' are actually young men and the girls they are interested in are much, much younger, often just children. In a real sense, girls do not participate as equals in the political affairs of the corporate

kinship group and seem to inherit most of the duties without enjoying many of the privileges, largely because of age differences at first marriage and the increase in status that being slightly older entails.

Marriage does not automatically enhance the status of the girl or change her life much. There is no 'marriage ceremony,' and the public awareness of her marriage begins with hardly more than comments like 'her father has promised her to so-and-so.' She usually does not begin living with her husband until after she has had her first menstrual period, although she may be 'married' for several years before then. Her duties as wife require her to continue the difficult and laborious tasks she has already begun doing, such as collecting firewood and fetching water every day. Firewood collecting is particularly difficult, and women spend hours each day scouring the neighborhood for suitable wood.[8]

It is perhaps worth considering this as a thought experiment, to imagine oneself back at school, and, if a young woman, to imagine oneself married to a boy several grades higher than oneself, or any of the male teachers. Conversely, if a male, the opposite can be considered for younger girls. While some desirable outcomes can be imagined under these circumstances, it is easy to imagine many more undesirable outcomes. If a girl finds that her father has promised her to a violent man, a lazy man, an ugly or unintelligent man, a smelly or diseased man, then, no matter what her reservations, she will invariably marry that man. The same situation is true for the man who may find himself married to a young woman who might be either beautiful or ugly, industrious or lazy, strong-willed or meek, and so on.

For the man this presents some problems, which will be considered, but since he has the opportunity to marry more than one woman then a poor marriage may not be damaging. However, a woman can only marry one man and, unless death or divorce intervenes, this may impact significantly upon her success. Even if death or divorce do intervene, she will be unlikely to obtain any greater control over the choice of the second husband than she possessed in the first choice of husband. As part of the levirate, she will probably marry her dead husband's brother.

Helena Valero, the young woman kidnapped by the Yąnomamö, observed that the Yąnomamö possessed an instinctive understanding of the importance of the choice of father in infant survival:

Fusiwe's mother talked with me a great deal: mothers-in-law don't like to talk to their sons' wives, but she told me so many things. ... She told me of an evil thing which sometimes happens amongst them. When the first son is born and dies, and the second is born and dies, and the third is born and dies, if the woman changes her man, the son who is then born does not die. Sometimes the babies are born very yellow; then they say that the father has eaten the ashes of a dead man only recently prepared. For this reason, when the wife is pregnant, they do not eat the ashes of their dead.

One woman always had children who were born very yellow and then died. An old Hekurá inhaled *epená* and said to the husband that, if he wished to have children, he must send his woman with another man. The husband disapproved of this and refused, but I heard the old man say: 'I am not saying anything wicked; I am giving you advice for having children. You should not go with your wife any more, because it is your fault if the baby dies.' But that man refused to follow this advice and the children continued to be born yellow and afterwards to die. I remember on three other occasions when the husbands arranged with other men to have children. One woman did not want to go with the man chosen by her husband and came to weep with us, because her husband wanted her to go and beat her.[9]

[*Epená* is a hallucinogenic drug.]

Infant deaths in cooperatively-simple societies are common and a newly born child may not reach the age of ten. Napoleon Chagnon, in a study of 1466 Yąnomamö people, found that while children under the age of ten make up 37 percent of the community, children between the ages of 10 and 20 make up about 23 percent of the community, a significant decline.[10] (There is no record kept of births, and deaths are not spoken of privately or publicly, so counting living individuals is the only possible way of having some idea of death rates.) Women who wish to improve their chances of their child surviving infancy need to ensure that their child has an immune system capable of doing so. Even without a knowledge of genetics, the Yąnomamö people understood the importance of the father for the survival of the child. This is without taking into account other factors: intelligence, physical ability, or behavioural patterns. It may not be enough for a child to have the immune system necessary to survive the rigours of childhood; there is also the question as to whether the child will have the intellectual and physical capacity to become a success in the community. This is especially true of boys rather than girls.

Boys will face higher rates of violence in warfare and fighting; and they can be unmarried, monogamous or polygynous and thus have a greater variety of reproductive success. In Napoleon Chagnon's study death rates amongst males were much higher than females, with males dropping from 23% to 13% and then 8% in the 0-10, 10-20, and 20-30 categories. Females dropped from 14%, to 10% to 9% thus suggesting that female babies might be more likely to be the victims of infanticide or perhaps that the Yąnomamö possessed higher rates of male births[11] (it has been suggested by Robert Trivers that in times of war there is a higher male birth rate[12]). In any event, those girls that were allowed to live possessed a much greater chance of survival. Male death rates were such that any male child in the 0-10 age bracket had a 35% chance of making it to the 20-30 aged bracket; girls had a 64% chance. Thus the pressures on a woman to produce a healthy, vigorous male child are tremendous. She might expect that if she has three male children, only one might make it to the age of thirty. Even if she has three male children that make it to the 10-20 age group, she can only expect that 1.8 will make it to the 20-30 age group.

Furthermore for a woman, there is not merely the significant issue of whether her son will survive, but also the issue of whether her son may be polygynous. If a woman has a highly polygynous son she, in genetic terms, has hit the reproductive jackpot. A woman with one son who has five wives may have as many grandchildren as if she had given birth to five daughters. If her son has more than five wives, then the woman may, through that one son, have more grandchildren than she possibly could have had if she only gave birth to daughters (since the children are spaced by four year intervals of pregnancy and breast feeding, five children is about the maximum). However, sons are a greater risk since they are more likely to die in infancy due to disease and in adulthood due to violence. By having sons, a woman may end up having innumerable grandchildren if her sons are polygynous or no grandchildren if her sons die in infancy or early adulthood.

A woman's ability to control the health and competence of her children is restricted by the arranged marriage system that prevents her from being able to sexually select her husband. This is the sacrifice she makes in order to increase the number of cooperative males in the group. Women are thus placed in a difficult circumstance. If she seeks another father for her children, then she undermines the contract that binds her own brothers to her husband and his brothers. Alternatively, if her husband-by-arrangement is seriously lacking in terms of his immune system, his intelligence or his physical ability, then she may be seriously undermining her long-term genetic success if she does not attempt to find a different father.

This is the battle of the sexes. Rather than two loving, cooperative people, who have selected one another, instead we find two people part of a much larger system, pushed together to serve a greater need.

The Battle of the Sexes

Marriage is partly cooperative and partly combative: the husband and wife-by-arrangement may or may not be in accord as to whose children she wishes to produce. While the husband and wife-by-arrangement may produce food, shelter, and other essentials together and may copulate regularly, there is the possibility for the woman to undermine her spouse's reproductive success. All it takes is for a woman to copulate with a man on several occasions and for the next four years her husband-by-arrangement may find himself caring for, protecting, and fighting on-behalf of, another man's child.

The implication of this is jealousy; the following comes from Burbank's *Fighting Women*, which concerns a study of an Australian Aboriginal society on a mission:

> Spending time alone with a member of the opposite sex or accompanying a man or a woman into the bush or an unoccupied house is suspect behavior. For example, after interviewing a white settlement official for several afternoons in the rare privacy and stillness of his air-conditioned office, he told me that one of his Aboriginal colleagues had asked him if he was thinking of taking me as a second wife. Walking alone may even arouse suspicion, for lovers who wish to tryst signal their intentions to one another or send a messenger naming the spot, then leave camp from different directions, meeting later at the prearranged location. Watching a person of the opposite sex also signals sexual interest.[13]

Jealousy may not merely be aroused in the husband, but in his extended family; the following is from Hallpike's study in Papua New Guinea:

> An instance of this excitability was very aptly illustrated to the writer while at the

Kataipa rest house. A girl was talking to another man who was not her husband. Her husband's father, on seeing this, was under the impression that she was flirting with the man, and as such he did not like it. He rushed into a house and brought out an axe and was intercepted by the police only just in time. (R. Hill, P.O., Patrol 3/58-9, Tapini.)[14]

The marriage alliance system would reflect that the husband's kin would support his ability to ensure paternity; since the alliance is between the two kin groups, the woman's own kin group may ensure her husband's paternity as well. If a woman was copulating with a man other than her husband-by-arrangement, and her brothers felt that this placed their contract for cooperation with their sister's husband's family at risk, then they might attack their sister for engaging in the extra-marital sexual selection.

What is known in modern parlance as 'domestic violence' is ultimately the attempt by a man to control a woman's reproduction. Violence between spouses in cooperatively-simple societies can be found in every study, and usually the anthropologist only bothers to mention the most extreme examples involving burning, mutilation, or killing.

Warner, in his study of the Murngin in Arnhem Land found:

> If a wife continues to be unfaithful she might be killed by her dué and members of her own family. Usually the dué would depend upon a magician to accomplish this...
>
> A beating is the usual punishment for a wife's adultery. Garawerpa, an old man of the Daiuror clan, put fire on his wives' vaginas, as did Binindaio when their galles copulated with Willidjungo, the medicine man. With the beating goes a severe tongue lashing.[15]

Napoleon Chagnon found amongst the Yąnomamö:

> Most physical reprimands meted out take the form of blows with the hand or with a piece of firewood, but a good many husbands are more severe. Some of them chop their wives with the sharp edge of a machete or ax or shoot them with a barbed arrow in some nonvital area, such as the buttocks or leg. Some men are given to punishing their wives by holding the glowing end of a piece of firewood against them, producing painful and serious burns. The punishment is usually, however, more consistent with the perceived seriousness of the wife's shortcomings, more drastic measures being reserved for infidelity or suspicion of infidelity. It is not uncommon for a man to injure his sexually errant wife seriously and some men have even killed wives for infidelity by shooting them with an arrow.[16]

Though no statistics are ever taken of violence between spouses in cooperatively-simple environments, which is unfortunate, statistics from cooperatively-modern societies give a hint of the level of violence. It is important to remember that these statistics deal with men and women who *selected* one another, and that the levels of violence would increase if men and women were in arranged marriages. In *Naked Motherhood*, one of the few realistic books concerning what actually occurs to a woman when she becomes a mother, Wendy Leblanc explains some of the statistics and their reproductive basis:

As statistics on violence against women are gathered, the magnitude of the problem is becoming increasingly apparent. In the United Kingdom in 1989, 32% of all violence against women was classified as domestic and in 1990, 43% of all female murder victims were killed by their husbands or lovers. A study in 1985 by Finklehor and Yllo in the United States found that 10% of wives are sexually assaulted by their husbands at least once in their marriage. In Australia in 1995, 50% of homicides were domestic.

Worse still it has been found in the United States that 60% of battered women were pregnant at the time of their partner's attack. The sickening prospect that the onset of physical abuse may occur when a woman becomes pregnant with her first child is borne out in a study by Webster, Sweett and Stolz at The Royal Women's Hospital in Brisbane. They state that:

> Pregnancy may be a stimulus for the first episode of domestic violence or may prompt an escalation in an abusive relationship.

This study of 1,014 women revealed that 18% experienced their first abuse by their husbands after they had become pregnant for the first time, and 24% said previously established abuse escalated during their pregnancy. Punching, kicking, choking and assault with knives and other weapons was not uncommon and the abdomen and reproductive organs regularly became the targets of abuse.

> Characteristically, the abusive man controls his partner in multiple ways... Pregnancy may threaten his dominance over the woman; he may also see the foetus as a rival. Blows to the abdomen may be a direct attempt to terminate the pregnancy.[17]

In Canada in 1991-92, of 438 female victims of homicide, 236 were killed by a spouse, ex-spouse or boyfriend (compared to 54 of 745 male victims).[18] Men are also far more likely to kill a wife who was no longer co-habiting and if the woman was younger.[19] A meta-study of 13 African studies found that 15.23% of pregnant women were victims of violence from their partner (2% to 57%).[20] A comparison of three studies in Western nations which examined the timing of violence found that violence ranged from 10.4 to 23.4% overall, and 6.6 to 8.3% during the pregnancy. In the two smaller studies (n = 548 and 290), violence during the pregnancy began after it had never previously occurred 12.5 to 13.9%, while in the larger study (n=1243), 88.0% of violence began during the pregnancy after it had not occurred in the previous 3 months.[21] A study of 1897 pregnant women in Guatemala City found that 10% (181) suffered physical violence from their partner, and that women had a greater risk of a miscarriage if they experienced such violence rather than no violence (15.5% to 9.2%). Women who experienced sexual violence had the highest rate of miscarriage (19.7%), while even verbal violence was correlated with a higher rate of miscarriage (14.2%).[22]

The object of violence by a husband against a wife is control. Rape is a similar method of control, an overt statement of sexual authority. The following example from Meggitt's study of the Walbiri reflects this:

> Yarry djabangari, aged about 40, had two wives and five children. He was devoted to the children and whenever they were involved in a misadventure he angrily berated and thrashed the wife who, he thought, had failed to look after them properly. His beatings became so frequent that the junior wife, Polly nambidjimba, decided to leave him. She moved to the widows' camp and for several days ignored his orders to return home. Yarry thereupon stormed naked through all the camps, waving his boomerangs and swearing obscenely at Polly. Onlookers were deeply shocked and withdrew to avoid witnessing Polly's humiliation. Yarry trailed her to her hiding place, raped her, and then thrashed her. Thoroughly cowed, she went back to his camp. Other men criticized his behaviour among themselves, but none of Polly's close kin present in the camp came forward to avenge the insult. Yarry's size, rage and demonstrated fighting ability had intimidated the men, while at the same time Polly's declared intention of leaving him had permanently cost her their sympathy.[23]

Rape by a husband-by-arrangement of a wife is an extreme method, since it would obviously seriously affect any small affection that a woman might have left for a husband. Until the 1980s, in virtually all Western nations and probably all other nations, a husband was legally entitled to rape his wife.[24] Even in societies with selected marriages, rape in marriages may be common. Diane Russell found that 87 out of 930 in a random sample of women in San Francisco stated that they had experienced a rape or attempted rape by their husband or ex-husband (44% reported a rape or attempted during their lives).[25] In a small sample of women in Maine, David Finkelhor and Kersti Yllo found that 9 women out of 133 reported that they had forced sex with a husband or live-in partner.[26] Diane Russell quotes an unpublished German study in 1976 that involved 332 married women interviewed by seventy-two trained female interviewers. They found that 18 percent of married women were forced to participate in marital intercourse against their will; of these 52 percent selected categories that involved physical force: 33 percent chose he 'grabbed me and attacked me' and 19 percent chose 'he held me against my will, so that I couldn't move'. In response to the question of whether a husband had forced himself on the interviewee when she was not in the mood, 7 percent responded 'frequently', 57 percent responded 'sometimes' and 35 responded 'never' (the husband was not necessarily the woman's current husband).[27] It should be noted that the wording (and translation) obviously plays an enormous role in the response since questions regarding being pressured to have sex, being forced to have sex, being physically forced to have sex, and being raped, may elicit different responses. In a survey conducted in 1980 by an Australian women's magazine, when the word 'rape' was used, one percent of the 30,000 respondents to a questionnaire indicated that they had been raped by a husband.[28] One might expect that forced sex in societies with arranged marriages would be much greater, based on the fact that there was no choice to begin with. In societies with arranged marriages, the basis of the legal entitlement to rape a wife-by-arrangement may be reflected in the contractual nature of the relationship. Her husband-by-arrangement is given a legal entitlement to reproduce with her, and rape, as a method of presenting sexual authority, reflects that legal entitlement (beatings were also a legal entitlement,[29] and continue to be largely ignored). In cooperatively-modern societies, where women are free to select their long-term sexual partners, such an entitlement has little logic to it, yet, remarkably, this entitlement has only recently been rejected.

Invariably, a woman's kin group, especially her close brothers, will attempt to protect their sister from extreme behaviour. Amongst the Yąnomamö:

> A woman can usually depend on her brothers for protection. They will defend her against a cruel husband. If a man is too severe to a wife, her brothers may take the woman away from him and give her to another man. It is largely for this reason that women usually abhor the possibility of being married off to men in distant villages; they know that their brothers cannot protect them under these circumstances.[30]

Yet, the expectation is that the woman will always be cooperative, and such intervention may only occur if the husband is likely to kill a man's sister. This raises a serious difficulty, as what was previously an alliance may become a blood feud, as the brothers seek to avenge their sister's death.

Whilst this might create the picture of complete antagonism, Phyllis Kaberry refuted such an idea in her study in the Kimberleys in northern Australia – though it must be remembered that the community she was studying was affected by a Western government:

> From time to time, the camp would be disturbed by outbursts of quarrelling, but it would be wrong to assume that it existed in a permanent state of tension and distrust, in an atmosphere of jealousy resembling the conditions, for instance, amongst the Dobuans. The women did not pass from one extra-marital affair to another; the majority lived in peace with their husbands, and much depended on the individual temperament, the degree of assertion and submission. If a woman had an affair and was discovered, then it temporarily disrupted the union: but where it remained secret, or only happened once or twice, it probably did much to reconcile her to a marriage with a man who was no longer or perhaps never had been sexually attractive to her: it presented a possible source of compensation for the lack of freedom in choice in the first place.[31]

The size of the community may also play a role in the level of hostility between the husband and wife-by-arrangement. Both men and women are surrounded by potential partners yet are effectively prevented from selecting at all and the larger the community, the greater the number of potentially attractive sexual partners. This situation obviously has volatile elements. Men too may be placed in an awkward situation, being expected to control a teenage girl, probably much younger than himself, and to enforce the marriage alliance between his kin and her kin. Considering how assertive many women can be, the expectation that her husband-by-arrangement will fight any actual or potential lover of his wife may be a dangerous obligation for the husband. Furthermore, he may not remotely like her. Catherine and Ronald Berndt summarise this:

> But if arranged marriages make for personal discontent or dissatisfaction in individual cases, this affects men as well as women. On the formal level, a man may be just as

passive a figure while the negotiations go on around him. Nevertheless, for him the issues are simpler. He does not have to choose between one woman and another when it comes to marriage. If he likes, he can take another wife, although it makes things easier if there is no opposition from the wife or wives he already has.[32]

Men and women, I argue, possess instincts to enforce marriage rights, using whatever means they think are necessary, and that this exists in order for the arranged marriage system to work effectively. This carries over to our own day where many women accept serious violence from their spouses and have no instinct to leave, violence which may be perpetuated for years. If women evolved in societies where long-term partners were selected, one might expect the women would leave at the first sign of violence, but they do not. Indeed, women often have numerous instinctive responses to their partner's violence, usually based on self-blame. It is a first instinct, it seems, that when a partner becomes hostile women become very self-aware of their own behaviour to mitigate his violence, rather than simply to leave the relationship – whether or not there are children. In a society with arranged marriages such an instinct would make perfect sense: since women had no opportunity to end the marriage, all they could do is to try their best to minimise the harm to themselves and this would be their first instinct. If women evolved with Western-style marriages, this response would be nonsensical.[33]

In cooperatively-simple societies, women are observed to have their own methods of imparting some control over the relationship, usually involving psychological abuse. Though 'nagging' is not regularly associated with psychological abuse, any perusal of the anthropological literature would suggest that this is the main aim of nagging. Mervyn Meggitt found such abuse more common amongst 'virtuous' wives in the Walbiri:

> Many women resent the operation of the double standard. Some cannot understand why they should be more heavily penalized for adultery than are their husbands; others cannot see why, when they themselves remain chaste, their husbands can fornicate with impunity. As public expression of their grievances merely antagonizes their husbands, the women adopt other courses of action. Virtuous wives become almost insufferably so and privately nag at their husbands almost without pause. Anyone who has heard a Walbiri woman nag will readily appreciate the effectiveness of her technique.[34]

Berndt and Berndt comment:

> In a number of areas men say that it is only by violence that a nagging wife can be silenced or disciplined: her weapon is her tongue, his a spear or a club – and if an exasperated husband hits his wife too hard, he must bear the consequences.[35]

Psychological tactics used by men, and probably, though to a lesser degree, women, include verbal insults in front of other people. It may actually be to a man's benefit to lower his wife-by-arrangement's self-esteem by verbally abusing her in public, and attacking her appearance, intelligence, and general behaviour. The benefit might come in lowering her self-belief, which would be coupled to her ability to select other men, and to attempt to reduce other men's estimation of her. A woman that attempted the same tactic would obviously be less likely to succeed, since her husband-by-arrangement might quickly escalate the argument to violence. Verbal abuse does appear to be common; in one study over 90 percent of U.S. college or New Zealand university students reported verbal insults by sexual partners.[36]

Ronald Berndt and Catherine Berndt summarise marriage amongst Australian Aborigines (a 'sweetheart marriage' is a selected marriage):

> Relations between husband and wife, as persons, do not rest on such formal matters as how the marriage was arranged, and whether or not it is monogamous. Even when there are several wives, this need not exclude affection between the two of them – just as a 'sweetheart marriage' is not inevitably a happy one. This is merely a different way of looking at the institution of marriage, as not simply an exclusive relationship between two people. But it *is* a relationship between two people, even though it brings in others as well. Whether they merely tolerate each other for want of something better, or from sheer inertia, or have a more positive affection for each other, depends to some extent on their personal experience together. If a woman neglects her husband or children, or is always looking for a new erotic experience: if he is so rough and hasty-tempered that she is afraid of him, or so weak that she cannot rely on him in a crisis – that marriage may perhaps last for a long time, but largely as a matter of habit: and it is likely to come to an abrupt end if the opportunity arises. On the other hand, a marriage need not be on the verge of a rupture because husband and wife are constantly quarrelling: in fact, they may prefer it that way.[37]

Alternatively:

> There are marriages which persist into the old age of both partners, with every appearance of devotion between them.[38]

If we accept the evolution of human reproduction as based on selected marriages, then it is difficult to understand why people find married life so difficult – or, for that matter, why marriage even exists as an institution or why fathers 'give away' their daughters at the ceremony. However, if we consider the basis of sexual reproduction in humans to be the arranged marriage, then the basis for marriage becomes obvious, and the difficulty between women and men in marriage also becomes obvious. All our instincts are based on the notion that one partner may have little or no interest whatsoever in the other partner or, from the woman's point of view, may face producing infants that die in infancy or are ineffective as adults.

From this perspective: physical or psychological violence, mate-guarding, jealousy and rape may all be seen as methods of attempting to control an uncooperative partner, whether from the woman's point of view, or the man's. Equally, cooperative, harmonious unions may also be possible. One might expect suspicion to be normal, something to be overcome by trust. Only once a husband-by-arrangement trusted his wife might he refrain from using controlling tactics and such trust might be hard won. For instance, women have been found to assist their husband in a fight with another man in Aboriginal society; Meggitt observed in the Walbiri:

> A person whose own spouse is attacked may often retaliate by falling upon the aggressor's spouse as well. Thus, when one man is struck by another, the victim's wife first tries to belabour the assailant from the rear with a club; if she is prevented, she is likely to turn on his wife. Although women are not positively enjoined to assist their husbands in this way, they are usually quick to do so. Such subsidiary engagements are often far more sanguinary than are the quarrels in which they originated, and in a matter of minutes they may transform a half-hearted altercation between two men into a wild brawl that involves half the camp.[39]

Sexuality

Many female primates, such as gorillas, present a genital swelling when ovulating, but women's ovulation is so well hidden that women resort to thermometers and calendars to attempt to predict ovulation. Evolutionary theorists have attempted for decades to come up with a logical reason why a woman that evolved in a Western-style marriage system would hide her ovulation from her long-term, sexually selected partner. This, of course, makes no sense: why would a woman hide her ovulation from a man who will be required to invest decades of work into her children and whom she selected? She wouldn't, and, amongst primates such as gorillas, where the male spends years protecting a female's infant, the female doesn't.

However, if women evolved in societies with arranged marriages then women would have no alternative but to hide their ovulation from their husband. If a woman did not hide her ovulation from her husband-by-arrangement, then he would merely need to mate guard her during her ovulation to ensure that any child was his own. Since the marriage is arranged, this would end all female sexual selection amongst humans. We should expect that women would evolve to ensure their capacity to sexually select, and that they would hide their ovulation from their husband-by-arrangement so that he could never predict if she was ovulating.

A second confusing aspect of human sexuality is regularity of copulation. Humans tend to copulate very frequently, something rarely found amongst other primates with long-term sexual bonds to a particular partner. If humans evolved with Western-style marriages, then we might expect that she would give a signal as to when she was ovulating, and the pair would need to mate only a few times – this is the case for gorillas. Evolutionary theorists have suggested all kinds of arguments, such as bonding, to explain the human prevalence to waste large amounts of time and resources in needless copulation.

Alternatively, a natural consequence of arranged marriages and hidden ovulation will be that the husband-by-arrangement will want to maximise his chances of being the father of his wife's children. Thus, we should expect that he will want to copulate regularly with her, as often as possible, especially during times when she might be ovulating. We should expect that in response to a woman ensuring her capacity to sexually select by hiding her ovulation, she should passively accept to copulate with her husband-by-arrangement whenever he so desires. (Alternatively, if she wanted to copulate with her husband-by-arrangement, because she wanted him to be the father of her children, then she should actively seek to copulate with him.)

The female should present confusing signals of interest and no sign of ovulation; he should copulate with her regularly to increase his capacity to ensure paternity. She thus accepts her part of the deal in cementing the cooperation of two groups of men by passively copulating with her husband-by-arrangement; however, by never presenting a definite signal of ovulation, she allows herself the capacity to limit his chances of fertilisation. Men in some countries have responded to this insecurity of fertilisation with extreme methods, such as, infibulation, clitoridectomy or ostracising their partner from society. Some women are prevented from leaving their home's walled enclosure, even to go shopping, and male guests are entertained in an area separated from the women. Unni Wikan noted of women in the Oman: 'Epitomizing women's seclusion and the narrow horizon of their world is their own reference to these wards and villages as "my country" (baladna: literally, our country), a usage that the men find amusing.'[40] 'Every compound has a roofed platform at the entrance, just inside the gate, where the menfolk may receive their unrelated friends, without bringing them into the inner areas occupied by women.'[41] 'Every item in these homes has been bought by the men; the women have only a vague idea of the range of selection from which items are picked, as they have never set foot in the market.'[42]

Either way, a woman ovulates for only 12 to 24 hours. Her vaginal mucus will regulate how receptive she is to sperm during and before this period. Even though sperm can exist for three to five days in fertile mucus, this depends upon the presence of fertile mucus. The mucus in a woman's vagina alters according to whether she is fertile or not: during times of infertility, the mucus provides a barrier to sperm; during times of fertility, the mucus provides an environment in which sperm can survive and move. Sperm die without the presence of fertile mucus. Factors such as stress can reduce the presence of fertile mucus.[43]

Sexuality is fundamentally important. In terms of the gene pool, nothing is more crucial than the issue of fertilisation. There is a small window of opportunity for both the husband-by-arrangement and his wife. Children born of extra-marital sexual selections will be a minority. What defines human beings is the arranged marriage system of cooperation and for the genes underpinning this system to exist, most children need to be born of arranged marriages. This argument is reflected in Warner's study of the Murngin of north-eastern Australia:

> The reader may assume from this account of illegal unions that marriage is a very unstable institution. Such is not the case. Although love affairs are common, a lifelong marriage either of a galle mielk to one dué, or to him and his younger brothers through the levirate, is the normal thing. Murngin society could not exist were this not true.[44]
> [Galle mielk = mother's brother's daughter or wife; dué = father's sister's son or husband, they are cross-cousins.]

If most children were born outside of marriage then the genes underpinning the arranged marriage system would slowly disappear and so would human cooperation, thus most children must be born of the system. A small percentage might be the product of women selecting outside the arranged marriage system. It should also be noted that Darwin's theory of sexual selection involves both female selection of males and male competition with other males; thus, either way, sexual selection is the basis for reproduction.

A Wife's Jealousy

I have considered the question of male jealousy and that males will use whatever means necessary to enforce their paternity. There is also the question of a wife-by-arrangement's response to her husband's behaviour.

It is interesting to consider under what circumstances a woman might be jealous of her husband. It is often suggested, as would be expected if humans evolved with Western-style selected marriages, that women are protective of their partners to prevent their partners leaving them. This might be true for Western societies with a state that enforces monogamy, but humans normally allow polygyny, and no man would leave his wife under any circumstances if he could be polygynous. In societies with arranged marriages all the women are married immediately after puberty and none are free to select another man. There are no free women to select, and, even if there were such women, men would simply marry the women, monogamously or polygynously.

The woman cannot lose her husband-by-arrangement, since they are contracted to remain together to bind their two brother-alliances in a cooperative agreement. A man will not leave his wife-by-arrangement for another woman, because the other woman will not be able to leave *her* husband-by-arrangement. The arguments normally applied to human behaviour assumes that people are free parties, but in cooperatively-simple societies there are no free parties.

There are good reasons for jealousy on the part of the wife-by-arrangement. The consequence of a husband copulating with another man's wife-by-arrangement is that the husband-by-arrangement will be expected to attack the lover. If a woman finds that her husband-by-arrangement is copulating with another man's wife-by-arrangement, and she likes her husband, then this creates a serious problem for her. If there is a fight between her husband-by-arrangement and the woman's husband-by-arrangement, her husband could be injured or killed. For those women who like or do not want to lose their husband-by-arrangement, we should expect that they should be protective of them. Conversely, if her husband-by-arrangement is extremely violent, then she may be unconcerned by his activities, and, indeed, might enjoy him being violently assaulted.

A woman has two responses to such a situation: attacking her husband or attacking her husband's lover. Both are used by women; violence appears to be a regular method of preventing a husband-by-arrangement's sexual selection of other, potentially dangerous, women. Warner found amongst the Murngin that:

The fights usually take place between two women, almost always because a young woman has seduced another's husband. Occasionally more than two women are involved, sisters taking side with their kin; or several wives of one man may attack his sweetheart, in which case the latter usually calls upon her own relatives for help. The women's ironwood digging sticks are the usual weapons, although if a man's club is within reach the women often use it. They strike each other on the head. The fights are of such intensity that bloodshed always results.[45]

Verbal attacks can also occur; Warner, again:

An outraged wife who has caught her husband in a sexual relation with another woman resorts to public abuse of her mate for his infidelities. Her obscenity and abuse are usually more proficiently and much more adequately expressed than a husband's.

At noon one day Bruk Bruk, the young and attractive wife of Lika, who had inherited her from an older brother, accused Djolli, a man from a more distant clan, of trying to seduce her in the bush. A tremendous noise was made in the camp. All the relatives of the parties concerned talked at once, and the two men armed themselves with spears, spear-throwers, and clubs and charged at each other, exchanging curses.

Djolli was angry because he claimed to be falsely accused; and so was Lika because he knew something was wrong, for either his wife had defended her virtue against Djolli, or she had succumbed and then accused him. In either case his self-esteem had been injured. Djolli's wife, an older woman, no longer attractive, stood by, and instead of helping her husband, the usual thing for a wife, screamed at him: "You belong to me. I am your sexual partner. You are like a dog. You are incestuous and sleep with your own mothers and sisters. Why don't you keep your penis where it belongs – in me, not other women?"[46]

[The 'mothers and sisters' are classificatory rather than actual.]

A wife will never expect her husband-by-arrangement to leave her since he may have waited ten years to copulate with her after their betrothal. But if she discovers that her husband-by-arrangement has been selected by a woman with a dangerous husband then she should attempt, with whatever means necessary, to prevent the sexual selection occurring. If a fight did precipitate, and her husband-by-arrangement was injured or killed, then this could mean that she could be passed on to one of his brothers, undermine the security of her children, and reduce the strength of her husband's brother-alliance.

Jealousy may not merely be created by a husband's potential lover, but also by a second, third or fourth wife. Violence between co-wives of a husband is a common theme in cooperatively-simple societies. Other wives can obviously create difficulties: a young wife might engage in a sexual selection that leads to the husband-by-arrangement fighting her lover or the wives may feel that they have lost their husband's support if he shows a preference for his new wife, and there may be arguments as to the sharing of work. Alternatively, some women appear to be thankful for the assistance of another wife who can share the work load, or the wives may be sisters, in accordance with the sorority. Jealously at a younger, preferred, sexually attractive wife also appears to be a common if not universal theme.[47] Warner observed one such situation:

> The wives live in a fair degree of amiability, but there is sometimes jealousy over a husband.
>
>   Dorng, a man past middle-age, but still very active, had three wives, of whom Opossum was the youngest. He complained one-day:
>
>   "My wives growl at each other all the time. The two old ones are good friends, but they are jealous of Opossum. They say to her, 'Why don't you get another man? This man is our husband. He made our children come. He belongs to us. He doesn't belong to you.' Opossum says, 'Where am I to go? My other dué are too old; I'd be the same as dead if I lived with one of them.' The two old ones say, 'Your first man is dead already.'" [Dorng's brother died and he got Opossum through the levirate.] His old wives were very pleasant and kind to him, but Opossum was a scold. "When I want to sleep with one of the old sisters at night," he said, "Opossum won't go to sleep as she should. I wait and wait; sometimes morning comes, and that woman watches and won't go to sleep. All right, I wait no longer, I play with her. When I do, Opossum growls at me. The old sisters go to sleep, but sometimes they laugh at me when I play with Opossum."[48]

Meggitt observed of the Murngin:

> Consequently, sororal polygyny is approved, for actual or close sisters are thought to be less likely to quarrel. My observations revealed that most co-wives do in fact live together amicably and that a strong bond unites them. Simple affection is one element of this tie, which seems also to be based on a common solidarity of the women arising from the shared, intimate knowledge of the husband's foibles, faults and virtues, and reinforced by the shared responsibility of rearing his children.
>
>   The attitude is a more intense form of the tenuous solidarity that most women display in the face of assertions of male superiority. Between co-wives the bond is most clearly expressed when one defends the other from the verbal or physical attacks of the husband, or lies to conceal the other's omissions or misconduct. Such behaviour is neither a duty nor an explicit convention; it follows from friendship. Nevertheless, relations between close sisters are not necessarily smoother than those between unrelated co-wives. The sisters may exhibit a greater affection, but (perhaps because of this) the quarrels that do occur are often very bitter.[49]

A long-term issue may be the issue of which wife's sons benefit from any future arranged marriages. The more co-wives, the more sons that will compete for marriages amongst those organised by the husband. Though this is obviously a serious area for conflict, Mervyn Meggitt noted: 'Finally, a man should not discriminate among the children of his various wives, and I saw no evidence of such prejudice.'[50] Alternatively, folk stories such as Cinderella, of step-mothers attempting to harm or derogate the former wife's children, suggests that competition between women on behalf of their children may have occurred.

Manipulating the Husband-by-arrangement

Since conflict is expected between a husband-by-arrangement and his wife's par-amour, if a woman does not like her husband-by-arrangement then she can use the expected violence between her husband-by-arrangement and a potential paramour to her own advantage. (The quote above, from Warner, concerning Bruk Bruk, gives an example.) A woman can place her husband-by-arrangement in serious danger if she sexually selects a very dangerous man. While this might mean she could be violently attacked by her husband-by-arrangement, she can place her him in danger if he feels compelled to attack the other man.

Flirting and actual or fabricated sexual selections of other men can thus be extremely dangerous to the man who is expected to fight any man his partner sexually selects. Mary Batten summarises the situation amongst the Yąnomamö:

> Napoleon Chagnon reports that a Yanomamö woman chooses her lovers and may precipitate fights between husband and lover in which the husband may be severely beaten. In Yanomamö society, the husband of an unfaithful wife is expected to challenge her lover in a club fight during which the husband may sustain serious blows to his head. If he doesn't fight to preserve his honor, he loses status. Eventually the husband becomes weary of this situation, and will "throw away," or divorce, his wife, freeing her to marry again. Of course, she risks being violently beaten by her husband for her infidelity, but she may regard this as a small price to pay to get rid of him and be with the man of her choice.[51]

It may not be likely, however, that the woman may get any choice, but she may get a different husband:

> Kąobawä gave another wife to one of his other brothers because she was *beshi* ("horny"). In fact, this earlier wife had been married to two other men, both of whom discarded her because of her infidelity.[52]

Following the levirate, she may be given to a younger brother or close parallel-cousin or 'brother'.

Modern theorists seem to interpret flirting in front of a partner as an attempt by a woman to present their attractiveness to their partner.[53] Yet, any man who has observed flirting in the presence of a male partner will readily appreciate that a major factor is violence. Most men, if their partner flirts with another man, will attempt to mitigate any violence she may create – avoiding eye contact with, or physically pulling her away from, the other man. He may, of course, attempt to intimidate or attack his potential opponent. By contrast, if a woman was actually attempting to find an alternative partner then she would do so without the presence of her partner. If her partner is present, she is surely attempting to intimidate her partner through another man. She may not be able use violence to defeat or intimidate her partner, but another man might be able to do so, if she can gain that man's interest. She can thus utilise others to engage in 'domestic violence' on her behalf.

Third parties may well be her best allies in winning any wars against her husband-by-arrangement. Considering that some men in cooperatively-simple societies may be known to have killed one or more men, it is easy to imagine that most men would be seriously concerned if their wife-by-arrangement began to flirt with a man who has killed several men with little more than spears or bows and arrows.

Parental Investment

Along with the conflict concerning paternity and control within the marriage, there are other issues such as nutrients, shelter, and children. While theorists using the Western-styled marriage emphasise the married couple as the major participators in the raising of children, the cliché that 'it takes a village to raise a child' is closer to the truth. In the previous chapter concerning male cooperation I emphasised Napoleon Chagnon's work in identifying that the nuclear family is not the basic reproductive group, but that the brother-alliance is the basic, reproductive group.

Robert Trivers pioneered the idea of parental investment in evolution. He noted that where two sexes existed in a species, the lesser ornamented of the two sexes placed the greater amount of resources in the infants they produced. The peacock, which aids the pea-hen with little more than his sperm, is far more ornamented than she is. Where males placed greater resources into the young, it was the female that was more ornamented.[54]

In humans, it is the female that is more ornamented, with developed breasts and less body hair. The reasons for this will be considered in a later chapter, yet, considering that in cooperatively-simple societies women will be pregnant and breast-feed their child for at least three years, as well as obtain a reasonable share of the food (depending on the environment, more or less than fifty percent), obtain firewood, water, and do a large amount of the cooking and cleaning, then the fact that she is considered to be more attractive than the male seems absurd. Considering her investment, it would seem odd that she is more ornamented rather than the male, and more concerned with her physical appearance.

This would be true if we were to consider that food, shelter, firewood, and other essentials were the sole basis of parental investment. While the men will provide food and shelter, their contribution would seem less than the woman's. This is summarised in the following extract; Phyllis Kaberry has summarised portions of Malinowski's 1913 survey[55] of Australian Aboriginal society:

> "It seems beyond doubt that in the aboriginal society the husband exercised almost complete authority over his wife; she was entirely in his hands and he might ill-treat her, provided he did not kill her." "By closer analysis we find that ill-treatment is – in the primitive state of the aboriginal society – in most cases a form of regulated intra-family justice; and that although the methods of treatment in general are very harsh, still they are applied to more resistant natures and should not be measured by the standard of our ideas and nerves." "He had well-nigh complete authority over his wife; that he treated her in harmony with the low standard of culture, harshly, but not excessively harshly;

that apparently the more tender feelings of love and affection and attachment were not entirely absent. But it must be added that, on these last two points, the information is contradictory and insufficient." "The husband had a definite sexual 'over-right' over his wife, which secured to him the privilege of disposing of his wife or at least exercising a certain control over her conduct in sexual matters." "Some caution must be used in accepting the above-mentioned cases of absolute faithfulness and chastity required from the woman."

Judged in terms of the facts at Professor Malinowski's disposal the woman would appear to have had the worst of the bargain. She provided the major part of the food, and received apparently a grudging portion in return. She reared his children, but the privilege of disposing of them in marriage seemed to be vested in him; and only in cases of excessive ill-treatment might she appeal successfully for the intervention of her own kin. If she were killed, then they would avenge her death. In other words the emphasis is placed on the rights of the husband, whilse those of the woman only seem to emerge in the limitations to the bounds of his authority. Instances of her assertion tend to read as acts of defiance rather than as those based on law or in conformity with social norms.[56]

In her work, Kaberry found a superior existence for women than Malinowski detected in his summary of the material available at the time, yet the tenor of Malinowski's work is reflected in much of the anthropological material: the women appear to perform more of the work. Even what work the men do perform can become subsumed into the larger political structure; Catherine and Ronald Berndt observed:

Food-collecting by women is more of an individual affair. Even when a group of women goes out together, each collects food for her own family: herself, her husband and children, and perhaps her parents or some near relative. When several men go hunting together, more than one of them may have rights in the kangaroo or other animal that they kill; but in any case a large animal is usually divided up, even to the extent that the main hunter may keep almost none of it for himself. In other words, a man is more intimately caught up in a network of kin and ritual obligations, which involve giving as well as receiving both food and other goods, and specific services.[57]

Indeed it is this basis of the political structure that allows polygyny to occur. While a Western view of marriage might suggest that a polygynously married man might need to work harder to feed his family, in fact, the exact opposite may be the case. In the previous chapter I quoted Napoleon Chagnon observing a similar scenario for the Yąnomamö, that polygynously married men are owed a bridal service from their future sons-in-law, and therefore receive a constant supply of meat and other resources.[58]

Since males are expected to give resources to those upon whom they are politically or reproductively dependent, we cannot simply consider a human economic and reproductive group as being the husband, wife, and children. A man with many wives-by-arrangement may do little hunting, but have a good supply of meat, supplied by those men who expect to marry his daughters.

We can go one step further. Love, care, food, shelter, firewood, water, and breast milk are all worthless without safety. Perhaps the most significant and difficult to quantify of all parental investment is security and safety. How would we quantify the men's ability to provide the members of their community with a secure environment, considering that all it may take is a single, successful attack by another group of men and several years of caring for a child may be extinguished in a single day? An infant may lose an abducted mother or even be killed. We might compare this to gorillas. A male gorilla gives no food and modest attention to his children, but without him, they would die since other males attempt to remove females and can attack either the male or his infants. So how would we quantify his ability to protect his children? How could this be compared with the female's investment? If the male gorilla is physically twice the size of the female, does that mean his investment is twice as large?

If the above were placed in a *hypothetical* pie-graph, the results could be either of the following, according to the level of male investment due to protection.

Graph representing current assumptions of social investment in children

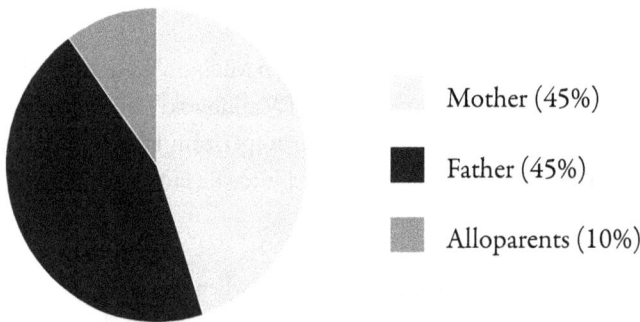

Mother (45%)

Father (45%)

Alloparents (10%)

Figure 1: it is the current assumption that the only participants in the investment in children are the mother, father, and childcare assistants.

Hypothetical graph of parental investment in cooperatively-simple societies

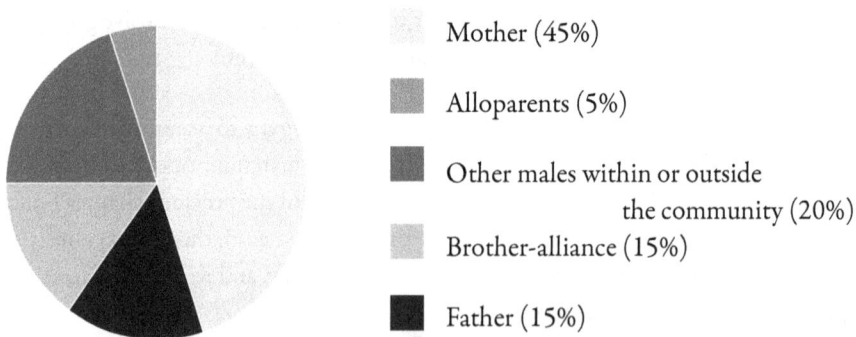

Mother (45%)

Alloparents (5%)

Other males within or outside the community (20%)

Brother-alliance (15%)

Father (15%)

Figure 2: the anthropological literature would suggest that other males, besides the father, invest heavily in each other's children through protection and nutrition; the other males being brothers and half-brothers of the father (the brother-alliance), plus other males in the community and males in other communities. It should be noted that the father invests in his own children, and also invests in the children of other males, so his total investment may be comparable to the mother in total.

Alternative hypothetical graph of parental investment in cooperatively-simple societies

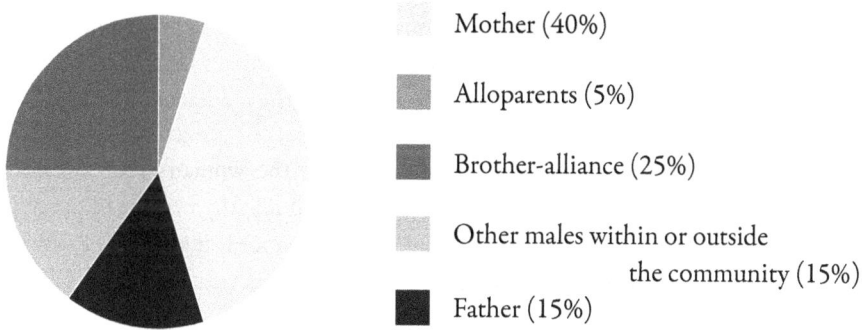

Mother (40%)

Alloparents (5%)

Brother-alliance (25%)

Other males within or outside
the community (15%)

Father (15%)

Figure 3: it may be the case that the males in the community and other communities place almost as much investment in the children as the mother. (In modern societies, women have entered the realm of the 'other males' and 'brother-alliance' who create extra status (money) that can be used for nutrition, conflict, protection and regulation. The modern corporation, farming bodies or corporations, government, and so on, would thus equate to the 'brother-alliance' and 'other males within or outside the community' which women are now heavily investing in.)

The father's investment in his children may be important, but his contribution may be simply one individual amongst a larger group of males: the brother-alliance, allied with the other males in the community. This may be true in terms of protection and nutrition.

Parental investment in humans is not simply the nuclear family with grandparents, parents and children. In cooperatively-simple societies where violence between groups and within the group is common, much of the parental investment may be in the form of male protection. This becomes particularly obvious when we consider such things as the levirate, where if the father dies his brother will become the new 'father' of the children. Such protection is equally important as breast milk, and a child cannot grow to become a successful adult without both. A father's direct input into his child's growth may be relatively small, but the combined effort of various men in the community into their own and other men's children by protecting the entire community may increase the total male input to almost the equal, the equal, or more than the mother's input. Considering that a quarter of the adult male population may die in violence this is not an unreasonable estimate. The input of the mother is obviously significant. Since the female is generally deemed to be more attractive than the male, it is worth considering these issues of parental investment are more complex than a simple observation of Western society and home life. Indeed, since the female is the more attractive, we might assume that the collective male investment may be the greater of the two.

While suspicion of cuckoldry may lead to violence, for the most part, a child appears to bring peace to the volatile relationship between the husband-by-arrangement and wife.

Kaberry found this amongst the Australian Aborigines in the Kimberleys:

> There is little doubt that the presence of children makes a marriage more stable, for the occasional divorces in the Kimberleys were in the main limited to childless couples. Moreover, until the child is able to crawl, sexual intercourse is forbidden between the mother and her husband; and this, together with the fact that a child is rarely out of its mother's arm, limits her opportunities for a casual affair and ensure a certain amount of domestic peace. The common interests they share in their children would seem, on the face of things, to be one of the elements which consolidate the marital relationship.[59]

Rather than the married couple battling each month over the woman's reproduction, it would tend to be an infrequent occurrence, interspersed with lengthy periods of pregnancy and breast-feeding. This, however, creates a further difficulty, since it means that many men, even if married to one or more women, may not actually have a sexual partner. Chagnon found amongst the Yąnomamö:

> Finally, there are taboos against coitus during pregnancy and for more than two years after the birth of a woman's child while the woman is nursing her infant. A polygynous headman, for example, might have as many as four or five wives, but it can happen that all four are either pregnant or nursing a new born – and the husband may not have sexual relationships with any of them. He must therefore satisfy his sexual desires by seducing the wives of other men. The situation is even more restricted, sexually, for men that have only one wife – or no wife at all. It would be fair to say that most adult male Yanomamö have healthy sexual appetites and are usually alert for opportunities to engage in sexual affairs with any woman who is not otherwise prohibited as a sexual partner because of incest proscriptions.
>
> Indeed, even in the latter situation, the particularly "fierce" men do not seem to be especially concerned about engaging in incestuous sexual liaisons strongly identified in Yanomamö beliefs with male aggressiveness and "waiteri" qualities.[60]

Even in a large community, men may have no sexual partner even if a man has several wives-by-arrangement. The few women in the community or visiting communities that are not breast-feeding or pregnant may thus attract a great deal of male attention.

Men spend very little time actually handling their children; the reason for this should be obvious – a woman who is constantly handling a child is very restricted. We should expect that men would take almost no interest in handling their children, because every hour that his wife is caring for the child is an hour when she cannot sexually select another man. This may have little importance in the first months or year after the birth of the child, but significant importance in the second or third year. Men should, if anything, have an instinct to *avoid* caring for their children, in order to force his wife to care for the child and prevent her from any sexual selections. Men should have an *aversion* to caring for children for anything more than a few minutes, even their own children.

This is not to suggest that they do not love their children, it is simply the kind of tactic that increases a man's chance of siring the next child. Indeed, we might even expect that men would generally possess the tactic to let his wife-by-arrangement do as much of the work as she can, and that he should do as little as possible. The more work his wife might do; the more tired, the more stressed his wife-by-arrangement might be; the less opportunity she has to seek a sexual selection. If a wife-by-arrangement spends all her time looking after her husband and children's needs then this is to his advantage; the more time his wife-by-arrangement has to consider and seek pleasure, then this is to his disadvantage.

One might expand upon this, and suggest that a husband-by-arrangement might even dislike others caring for his children other than the mother, including her close relatives. Some writers have commented upon the importance of grandparents, yet, to the husband-by-arrangement, if his wife's mother cared for his wife's children for lengthy periods of time, this would increase her ability to sexually select alternative genes. We should expect that a husband-by-arrangement might ostracise his wife's family to increase the time that his wife cared for her child. Opposing such an argument, in societies that are cooperatively-simple, the father would expect his son to marry his wife's brother's children. Thus the bond between a man and his wife-by-arrangement's family are expected to be good, since his children can be expected to marry her brother's children. It would be unlikely that he would attempt to ostracise members of a family when his son or daughter would be expected to marry a member of that family. However, considering that many women will either marry a husband-by-arrangement who lives in a different community, or perhaps become separated from her own family due to the community fissioning into smaller communities or due to abduction, a woman may well end up in a community with few relatives and her husband may well look poorly upon others caring for her child. A very jealous man might also attempt to limit assistance to his wife from her relatives if they lived in the same community.

Despite such potential altercations, a husband-by-arrangement will be expected to defend his wife and children against others, and assist brothers and half-brothers in their defence of their children against others. Meggitt observed what is commonly found in cooperatively-simple societies:

> A man immediately retaliates when another attacks his wife, and this is his duty rather than his privilege. Indeed, some of the angriest Walbiri I have seen were those whose wives and children had been injured in this way.[61]

Napoleon Chagnon and Timothy Asch filmed the outcome of an attack on a man's wife in the aptly named film *The Ax Fight*.

Concluding Remarks

While many theorists today argue that women and men evolved marriage in order to form a cooperative union to raise children, why would women then hide their ovulation from their chosen partner? Why is the cooperative union so frequently uncooperative involving jealousy, violence, and, until recently in probably all societies, legalised rape.

By contrast, the difficulties that we observe in marriage are explicable if humans evolved with arranged marriages. The regular copulation – desired or passively accepted, – the violence, the hidden ovulation, the potential to cause a disturbance through flirtation, the potential for a harmonious cooperative relationship, and so on, all exist so that either the male or the female can ensure their own particular desired paternity of the child. Considering the high infant death rates, the very high mortality rates for males, and the potential for polygynous males, the importance for females to produce healthy, strong, intelligent males cannot be underestimated. Equally, the importance for the male in ensuring that the child carries his own genes, a child he may risk or lose his life to protect, is paramount. Those males that could not ensure paternity would obviously leave a lesser presence in the gene pool. For females it is an issue of survival and quality – dead infants, monogamous males and females or polygynous males. For males, the main issue is cuckoldry.

Some will, no doubt, point out that humans have not stopped being cooperative despite the fact that arranged marriages no longer exist in Western societies. However, we must observe that humans only began to select their long-term sexual partners once a profoundly extensive market economy was in operation, and, indeed, that this only occurred in the last five hundred years in western Europe.

How did this change from arranged marriages to selected marriages occur? It appears that it is a bottom-up occurrence. It appears that girls in the poorer families were able to escape arranged marriages and become servants or other workers. This was only plausible on a large scale as individual protection and wealth increased. Even then, as any description of the lives of servant girls would suggest, girls faced serious dangers, such as rape.[62] Evidently, the risks were worth it to escape an arranged marriage in which they might never experience affection for their long-term partner. One has only to observe the death threats aimed at some women of Middle-Eastern or Asian heritage that now live in Western nations by their own families, to see that families have enormous trouble maintaining the arranged marriage system once women are able to obtain self-sustaining employment. The last women to be able to escape the arranged marriage systems were those who were wealthy. The princes and princesses of Europe continued to utilise arranged marriages in the late nineteenth century. There is thus a gradual process towards selected marriages, rather than a sudden change. It is a change that has occurred hand-in-hand with the evolution of Western Civilisation itself. One might view this alteration as perhaps the most important change in Western nations, where a market economy is now so extensive that it alone is able to bind members of the community together, and arranged marriages have become unnecessary. The market economy has become so extensive and the human memory of history so poor, that most people, nowadays, have little or no understanding of the part that marriage plays in cooperation, and see it as little more than an act of love, sentiment and reproduction. History would suggest that people in Western nations, particularly wealthier people, once viewed marriage as vital, since it played such a major role in alliance formation and the increase of individual and familial status.

Chapter 6

Female Sexual Selection of Males

So far a number of issues concerning sexual selection have been considered, with the focus on competition between groups within communities and competition between communities. To this extent, sexual selection has occurred in the sense that groups of males use cooperation to compete with other groups of males. To cooperate, small groups of closely related males, the brother-alliance, use arranged marriages in an exchange of cooperation for reproduction, to form larger, more competitive groups. At the individual level, very able men are desirable political partners, obtain more wives-by-arrangement and have more children. At the group level, very able small groups of closely-related males are able to obtain more wives-by-arrangement for themselves and near kin and leave more descendents (inclusive fitness). Humans evolved as cooperative animals, with those most competitively cooperative benefiting to the greatest degree, reproductively speaking.

The implications of this are that humans cooperate; that women gain more men to protect themselves and their children; and that reproduction occurs largely without female sexual selection. Since there is no selection in a marriage the man attempts to ensure his paternity with regular sexual intercourse and coercion. He is not alone in ensuring his paternity. His wives-by-arrangement will be observed by himself and his close kin – particularly parents (if alive), brothers, and half-brothers. Since the marriage is a contract for reproduction and cooperation, a wife-by-arrangement may also be observed by her own kin, who will want to ensure the contract is fulfilled. However, a woman may prefer to select elsewhere if her husband-by-arrangement is unacceptable or with him she produces unfit babies.

A wife-by-arrangement is in an awkward situation. She is watched by many pairs of eyes, and her husband can use extensive violence against her if he wishes; she could even be killed. As stated in the previous chapter, even being alone can be enough to instigate violence against her by her husband-by-arrangement, and talking to another man can instigate a violent attack.

A woman may be able to speak openly with other women and speak in public situations, where it is obvious that there is no cause for the husband-by-arrangement to be concerned. Yet, if she was found speaking to a man, with no one present to vouch for the subject matter of the conversation, this could lead to an extreme response from her husband-by-arrangement. This may be the case even if she is speaking to her husband's own brother. An example of this comes from Helena Valero's time with the Yąnomamö when one of her husband's wives is absent while her child is crying. In a fury, Valero's husband, Fusiwe, immediately orders Valero to give the absent wife her hammock and to remain with 'whoever she is with, my brother or anybody else'.[1] Eventually, Valero takes the hammock to the wife. The wife, Shabotami, is found speaking to Fusiwe's younger brother, and the brother is given the hammock.

> The woman remained silent. The man became serious and shouted: 'I don't want her.' Then he threw the hammock on the ground and said: 'Take it back to him and tell him that I don't want the woman!' I picked up the hammock and replied: 'No, you like the woman; if you didn't like her, you would not have kept whispering, but you would have said: 'Go back at once, for your son is crying'; I then threw the hammock on the woman.[2]

Later the mother comments to her son that he is jealous of his younger brother:

> The mother answered: 'But I think you are jealous of her and that's why you have sent to give her the hammock.' 'If I had been jealous, I would not have sent the hammock, but I would have struck your son and the woman with my *nabrushi*. I am certainly not jealous of my brother; I think, poor brother, he has no wife and I have so many, I'd better give him one of mine. For days I have been thinking about it and now today I have my opportunity.'[3]

Even talking to a husband-by-arrangement's own brother can thus cause conflict.

The problems in sexual selection for women are self-evident. Any verbal communication with another man could lead to violence. If she cannot speak intimately to other men, then she certainly will not be able to touch or caress them. Physical contact will be impossible. A woman is thus isolated from other men, and her husband-by-arrangement and his close genetic relatives will do their best to ensure that this isolation is sustained.

So how does she then sexually select? And if she is to sexually select, what is she looking for in a potential mate?

Violence and Protection

The most crucial element of any potential sexual selection is violence. Let's suggest that a woman is one of four wives to a very successful, violent and jealous man. Let's suggest that he once had five wives, but on discovering an infidelity, killed her. One might suspect that the other four wives would be very unlikely to consider sexually selecting another man, even if the husband was infertile. In Papua New Guinea, Hallpike noted of one chief: 'Other chiefs, such as Kire Keruvu, have had several wives, but none of them has borne children, presumably because the man is sterile.'[4] The chief in question was known to have killed: 'Much later Kire Keruvu of Goilala killed B- Ki- of Laitate and fed his body to the pigs...'[5] Evidently the women were so fearful of their husband-by-arrangement, that they never considered selecting another man, even if it meant never having children.

The first and foremost issue in any sexual selection will be safety, and her greatest enemy will be her husband-by-arrangement. If she is to sexually select, she must be able to ensure her own life is not placed in danger. The best method of achieving this is to sexually select a man who can defeat her husband-by-arrangement in combat or politics, and to ensure that the sexually selected man will prevent attacks upon herself.

The ultimate example of this comes from Chagnon's observations of a headman, Kaobawä, who:

> ...has had six wives thus far – and temporary affairs with as many more women, at least one affair resulting in a child that he has publicly acknowledged as his own.[6]

For a woman considering selecting a man other than her husband-by-arrangement, the best outcome would be that the selected man could protect her and the child, even if the selection became publicly known.

The logic is simple, if counterintuitive to those who assume that the husband and wife will have selected one another. If violence from her husband-by-arrangement is the largest limiting factor in terms of sexual selection, then a woman's best method of sexually selecting is to prevent or minimise such violence. Since the marriage is arranged, it is important to remember that the wife and husband-by-arrangement may have no feelings for each other at all, of love, desire, or intimacy. They may dislike each other and fight regularly. A woman who is regularly beaten by her husband-by-arrangement may enjoy observing her husband viciously beaten by another man. She may not be upset if he was killed. This was observed by Helena Valero; her husband had shot one of his wives in the leg with an arrow out of anger, and when her husband was himself killed Valero remembered the following confrontation with the wife who had been shot:

> We, four wives, tied our hammocks up around the fire. The wife who had been shot asked me: 'Shall we now all be separated?' I answered: 'I don't know, you only thought of getting him killed, now you can be happy! You have a father and a mother, and so you wanted his death; now you are happy and you may look for a man who will not maltreat you.' She began to cry and said to me: 'I do not want to be separated from you.'[7]

It is imperative to appreciate this fundamental difference between a selected marriage and an arranged marriage. One only has to read some of the examples of women being beaten, mutilated, burned, or shot with an arrow, to appreciate that many women might be very pleased to see similar things occurring to their husband-by-arrangement. The same is true in modern marriages, where, despite a sexual selection occurring, violence may be so great that enmity may be the overwhelming feeling between the couple.

Observations, let alone statistical data, of 'domestic' violence in cooperatively-simple societies is limited. In most studies, the only time an anthropologist bothers to mention domestic violence is when the violence is extreme. From my readings, there does not appear to be any study that documents variations in violence against women.

Consider the following two outcomes of women's sexual selections in cooperatively-simple societies. In the first example, it should be noted that the two co-wives sexually selected two brothers, men that they considered would unite to protect them against their husband. The example comes from Meggitt's study of the Walbiri:

Ginger djabangari had two wives, Marcie and Liddy nambidjimba, who were actual sisters. For some months the women had been involved in a liaison with two bachelors, Willy and Johnny djabangari, who were not Ginger's countrymen. Ginger knew this but because of Marcie's poor reputation, was in no hurry to make an issue of it. Eventually Liddy became pregnant, so retired to the bush one day, where Marcie helped her procure an abortion. Somehow the news reached the camp and some well-meaning friends told Ginger, not only of the abortion, but also of the names of the women's lovers. This forced him to take action. He enlisted the support of several countrymen who were close brothers of his wives and then attacked the two women and their lovers. He speared Johnny through the knee and Liddy through the arm; Willy he stabbed in the back with a long knife, and only European intervention prevented him from cutting Marcie's throat. At the same time his djambidjimba countrymen thrashed all four of the offenders with clubs and boomerangs.[8]

Marcie and Liddy made a poor selection; their selected partners were unable to combat their husband-by-arrangement and one of the women might have been killed. It should be noticed that the women's own brothers attacked the women and assisted their husband, as the marriage alliance would dictate.

By contrast, consider the following from the Yąnomamö:

Although Rerebawä has displayed his manhood and fearlessness in many ways, one incident in particular illustrates his character. Before leaving his own village to take his wife in Bisaasi-teri, he had had an affair with the wife of an older "brother" (probably a parallel cousin). When the affair was discovered, the brother attacked Rerebawä with a club. He responded furiously, grabbing a single-bit ax, soundly beating the brother with the blunt side, and driving him out of the village. The brother was so intimidated by the thrashing and the promise of more to come that he did not return to the village for several days. I visited this village with Kaobawä shortly after the fight, and Rerebawä was along to serve as my guide. He made it a point to introduce me to the man. He approached the brother's hammock, grabbed him by the wrist, and dragged him out on the ground, exclaiming, "This is the brother whose wife I screwed when he wasn't around." A deadly insult like that was usually enough to provoke a bloody club fight among the Yanomamö. This man only slunk sheepishly back into his hammock, shamed but relieved to have Rerebawä release his grip.[9]

Unfortunately, Chagnon does not tell us of the level of domestic violence that the man's wife received when the defeated man returned from the forest. Yet women's sexual selections, according to Chagnon, reflect the desire for men like Rerebawä:

...it should be clear that competition among males for mates can become a severely disruptive force in the internal ordering of large, heterogeneously composed villages. One expression that this takes is a tendency for men to represent themselves as aggressively as possible, indicating to potential competitors that affronts, insults, and cuckoldry will be immediately challenged and met with physical force. In addition, displays of masculinity, such as fighting prowess and "*waiter*" (ferocity) are admired by Yanomamö women, and particularly aggressive men have an advantage both in soliciting sexual favors of larger numbers of women as well as depressing the temptation of other men to seduce their wives. For example, it would be hazardous for a young braggart to try to seduce the wife of one of the village "fierce ones," for it is certain that the young man would be severely beaten in a club-fight by the older, fiercer men – and perhaps even killed. On the other hand, many of the older, fiercer men will openly flirt with and try to seduce the wives of younger men, and in some cases, they even do so to the wives of their younger "brothers."[10]

The preference for excellent fighters also occurs in Papua New Guinea; Hallpike notes:

> While it is true that a man attracts women sexually by his reputation as a killer, and that chiefs play a significant part in society as war leaders, it still seems to be the case that general social status is more important in acquiring wives than a reputation for mere bloodshed.[11]

> Superficially the women appear to be regarded as being merely useful for making gardens, looking after pigs, etc. In reporting damage or raids the native almost invariably places pigs first, belongings second, and women last. Actually the women wield a great deal of influence which although invisible is very real. Their partiality for men who have the right to wear the 'paka' [Moto = 'homicide emblem'] is probably the indirect cause of many murders in the past and present. Women are by no means slaves and have no hesitation in leaving their husbands if ill-treated or dissatisfied. (Annual Report of the Goilala Sub-District 1946/7; reply to Questionnaire TA/62/1.)[12]

One of the few definite examples where the domestic violence that wives have received is lowered is in an extended example in Meggitt's study of the Walbiri. Louis djuburula attempted to keep the wives of his younger brother, Romeo, who has gone insane. Louis, who is older, and already has three wives, is not meant to receive the wives, and he becomes embroiled with Yarry djabangari who attempts to have Louis relinquish the women. The dispute leads to numerous fights and Louis's wives take advantage of the situation:

> Meanwhile, Louis was having domestic troubles. Emboldened by his general unpopularity, Charlie, Paddy and Jim Tulum djuburula took every opportunity to copulate with his wives – Charlie with Milly, Paddy with Minnie, and Jim with Yma. Louis could not watch all the women simultaneously and, as he knew what was afoot, he daily grew more morose. The fact that everyone was aware of the situation and did not trouble to hide their smiles added to his discomfort.[13]

Louis, with no supporters, and confronted by superior numbers can do nothing to intervene to stop his wives copulating with other men.

The duel, whether with clubs or guns, is a common outcome of a woman's sexual selection, whether in cooperatively-simple societies or pre-twentieth-century Europe. Amongst the Yąnomamö, Chagnon observes:

> Most duels start between two men, usually after one of them has been accused of or observed *en flagrante* trysting with the other's wife. The enraged husband challenges his opponent to strike him on the head with a club.[14]

> The 'ideal' rules are that one man exposes his head and allows his opponent to strike him as hard as he can on the top of his head with a long, wiry pole, usually some six to eight feet long. He then gets to deliver as many blows back to his opponent as he received. The blows invariably cause bloodshed, and once the blood starts flowing, free-for-alls often follow and everyone starts clubbing opponents on any part of the body they can hit – bashing skulls and sometimes breaking arms and shoulders.[15]

These fights can be fatal; this is from Helena Valero's statement:

> We were already on the journey when a young man took another's wife by one arm; the husband saw them and they had a fight with *nabrushi*. I was at a distance with the *tushaua's* other women; we heard the shouts and went closer. The young man had a great wound on his head and blood was coming out of his mouth and nose. He had fallen on the ground and could not get up again: they then made him comfortable in the hammock, which they tied to a tree-trunk, and carried him. When the *tushaua* saw him, he said: 'What has happened? A snake-bite?' 'No,' they replied; 'it was a vendetta; he has been struck on account of jealousy.'

> The next morning the young man was dead: when we reached the *tapirí* where the relatives were with the dead man, we found the women weeping and collecting wood to burn his body. ... The next morning they burnt the body; when the pyre began to burn, the heat made the white part of the brain come out of the hole in the head. Then they said: 'Put more wood on, cover him up well.'[16]

There are also examples where people take advantage of the general confusion caused by disputes to gain a sudden advantage on someone not directly involved in the original dispute. The actual events that occur in brawls are rarely recorded. The following are two examples where a male takes advantage of other disputes; however, on these occasions it is to attack a former lover. The events are taken from Meggitt's work with the Walbiri and involve Joan, who, at the time, was recently widowed:

> Next day, when Windy again told her that he intended to marry Joan, Kitty upbraided him and belaboured Joan with a club. Alec II, angered by the loss of his ex-mistress, also joined in and split open Joan's head.[17]

Later, Alec II makes a second attack:

Indeed, Milly herself did not wish to join Charlie. When she said so, Netta, the wife of Jack djagamara and the close mother of Charlie, at once attacked her for not favouring Charlie. She broke Milly's hand and received a split head in return. Alec II promptly took advantage of the general confusion to thrash his ex-mistress, Joan, once more to punish her for deserting him. Windy retaliated but did no serious harm.[18]

In disputes involving many people, there would be ample opportunities for people to attack others, so that a paramour can attack an ex-lover or the wife's husband-by-arrangement, thus potentially giving the wife protection.

One of the few modern studies on the subject of what a woman wants, and what men think women want, in an extra-marital partner emphasise the importance of protection. In a study by Heidi Greiling and David Buss, women were asked to rate the desirability or undesirability of 139 characteristics in a potential affair partner with 1 being highly undesirable and 9 being desirable. They found that the women gave 'protective of her' a 9.0; 'physically fit' received 8.2; both 'strong' and 'muscular' received 7.8; and athletic received 7.7. David Buss writes on what women wanted:

> In our in-depth interviews with women about what they wanted in an affair partner, most vehemently denied that they were looking for someone with the physique of Arnold Schwarzeneger or a Sylvester Stallone. They found those bodies overblown and worried that muscle-bound men were overconcerned with their physical appearance. But they did want a man who could handle himself physically with other men – a man in good shape, fit, strong, athletic, well toned, and willing to step in when the situation called for it.[19]

David Buss has interpreted this as 'mate insurance':

> The backup mate can serve many functions when the regular mate is not around. He can defend against predators, fend off aggressive men who might otherwise exploit the absence of a regular mate, or provide food during a food shortage.[20]

I am dissatisfied with these answers. Do women in modern societies face predators or starvation? This is highly doubtful. As to the dangerous male present, one would expect a woman would call the police and not the affair partner; if a woman did call her affair partner, then this would dramatically increase the likelihood that the sexual selection would be discovered. In my opinion, evolutionary theorists must be particularly careful in attributing instincts that existed for cooperatively-simple societies to cooperatively-modern societies. Human beings, like many animals, are remarkably adaptable. If women are stating that they are looking for certain traits in an extra-marital partner, then these traits should, firstly, be expected to reflect their current existence, not an entirely different existence that even their ancestors have not experienced for millennia.

So, why would a woman want protection and from whom? If one were to look at the statistics, they would suggest that the most likely perpetrator of violence against a woman will be her partner. The most likely person to kill a woman will be her male partner. In fact, her greatest physical threat is her partner. This is reflected in the Buss and Greiling study. Out of 35 choices, as 'The most beneficial aspect of short-term extra-pair mating for women', 53 women and 45 men chose 'Received protection from abusive family through sexual partner.' Women gave a slightly higher rating than men (4.26 as against 3.76 out of 5). For men, the second highest choice was 'Sexual partner protected her/his children from steady partner,' while this was the fifth choice of women.[21] These results fit the prediction.

Considering that many men will become violent if confronted by the sight of his partner with another man, what kind of man would a woman want to be with if caught by her partner? Probably a man that was 'protective of her', 'could handle himself physically with other men', and was 'willing to step in when the situation called for it'. It seems remarkable to me that David Buss did not come to this conclusion. The above extract comes from his book *The Dangerous Passion*, a book concerned with jealousy, and particularly violence and murder due to jealousy. It astonishes me that, in a book concerned with domestic violence, he would ignore this issue when considering the main reason for women's choice of an extra-marital partner. Furthermore, whether in cooperatively-simple societies or modern societies, the situation remains the same – a husband is dangerous, and if the studies of the Yąnomamö are taken into account, far more dangerous to women than the jaguars of the Amazon.

Other modern studies have confirmed women's preferences for aggressive or asser-tive men. In Bruce J. Ellis's overview of women's preferences, when dominance traits were expressed in film, pictures or writing, women reacted positively:

High-dominance males in the Sadalla, Kenrick, and Vershure study were rated considerably more sexually attractive, but significantly less warm, likable, and tender – qualities that presumably offer cues to a male's willingness to invest in a woman and her children.[22]

Of height, Ellis writes:

The most salient criterion concerning height that women apply to men appears to be the "male-taller norm," which is so prevalent that it has been called the "cardinal principle of date selection".[23]

Of dominance, Ellis writes:

Perceptions of social dominance in humans, as well as in nonhuman animals, are affected by physiognomic traits. In particular, traits associated with physical maturity (proportionately thin lips and eyes, receding hairline) and physical strength (wide face, square jaw [Guthrie, 1970]) are linked to perceptions of social dominance in people (Keating, 1985, 1987; Keating, Mazur, and Segall, 1981). This correlation emerged among a majority of observers in at least 10 of the 11 diverse cultural settings that Keating et al. studied. Consequently, if females find social dominance attractive in men, then women should rate mature male facial features more favorably than immature ones on measures of sexual attractiveness.

Keating (1985) constructed mature and immature facial composites from identiKit materials... On a scale of 1-to-7 (1 = very unattractive, 7 = very attractive), women gave immature and mature male composites mean ratings of 2.48 and 4.40 respectively. In short, female perceptions of male attractiveness increased in response to morphological enhancements of dominance signals.[24]

These types of selection methods have left psychologists like Bruce J. Ellis to simply describe the method of selection as the 'mate choice paradox'. It is expected that women will prefer good providers, but instead they prefer dominant men, even when they know they will be less tender, warm, and likeable. The results simply do not reflect the idea that women should want men to rear children; instead they generally want dominant, cold men.

Perhaps the most obvious example of the preference for dominant men comes from women's fantasies. In the U.S., romance fiction accounts for 14 percent of all fiction sales and in 2009 was the single largest category for book sales.[25] If women did want a good provider, we might expect that the plot would follow such a design, with a woman slowly getting to know a caring, loving man, the kind of man who would want to care for children. We might expect women to be fantasising about caring, loving providers, and that the plots suggest that women are sexually aroused by men who show an interest in children.

In a book written by romance writers, *Dangerous Men and Adventurous Women*, it is revealed that there are two basic plots in romance fiction: a woman is forced into a 'marriage of convenience' or a 'shotgun wedding' to a violent and dangerous man. In one plot she teaches the violent man to treat her well and love her, and in the second plot, she finds an even more dangerous man to get her out of the marriage.

Romance writer Jayne Ann Krentz outlined four aspects of romance novels that feminists in the publishing industry have attempted to change:
  (1) that all the male heroes are powerful, aggressive, and violent 'alpha' males;
  (2) that the heroine is aggressively seduced by the hero (often after a marriage of 'convenience');
  (3) that the heroines are usually virgins; and
  (4) that all the plots are identical.[26]
This is, of course, the exact situation in which a Yąnomamö or Australian Aboriginal woman would find herself:
  (1) her husband will most likely be a powerful, aggressive, and violent 'alpha' male type;
  (2) her marriage will be arranged; and
  (3) she will be a virgin.
In the books it is imperative that the man be violent and dangerous, and not merely to other people, but to the heroine:

The hero in a romance is the most important challenge the heroine must face and conquer. The hero is her real problem in the book, not whatever trendy issue or daring adventure is also going on in the subplot. In some way, shape, or form, in some manner either real or perceived on the heroine's part, the hero must be a source of emotional, and, yes, sometimes physical risk. He must present a genuine threat.

The hero must be part villain or else he won't be much of a challenge for a strong woman. The heroine must put herself at risk with him if the story is to achieve the level of excitement and the particular sense of danger that only a classic romance can provide.

Any woman who, as a little girl, indulged herself in books featuring other little girls taming wild stallions knows instinctively what makes a romance novel work. Those much-loved tales of brave young women taming and gentling magnificent, potentially dangerous beasts are the childhood version of the adult romance novel. The thrill and satisfaction of teaching that powerful male creature to respond only to your touch, of linking with him in a bond that transcends the physical, of communicating with him in a manner that goes beyond mere speech – that thrill is deeply satisfying.[27]

If women were selecting men to care for babies, is this what we would expect would excite their instincts? Wouldn't we expect women to be fantasising about men who brought food, built houses, and held children? On the other hand, if women evolved in violent societies with arranged marriages, this is the exact, challenging situation women would find themselves in where they would need to gain the confidence of an extremely dangerous man.

In the 1970s, Kathleen Woodiwiss led something of a revolution in romance novels; hers depicted violent altercations, rape or near rape of the heroine (even by the 'hero'), and explicit sex scenes that often ran for twenty pages. In *The Flame and the Flower*, her first romance, the heroine is forced to marry a plantation owner in the south of the U.S. and in their first sexual encounter he rapes her. In *The Wolf and the Dove*, Woodiwiss depicts the scenario that many women would have found themselves in for millennia: the heroine is captured by a victorious enemy and forced to become the man's wife; in fact, she is initially chained to the bed. In both books the heroine eventually finds her new husband to be desirable, falls in love with him, and earns his love in return. Obtaining the love and respect of a violent man is the basic plot of each novel.

While no modern woman would elect or desire to be placed into an arranged marriage with an extremely violent man, the fact that such a situation elicits a strong erotic response, in such a manner that the plot in each book is virtually identical, would tend to suggest something of women's instinctive responses. We could say a similar thing of men who enjoy watching violent movies: that they instinctively react to the thrill of combat. Many people enjoy horror movies, which elicit the strong feelings of fear that, historically, would have occurred in our lives. This isn't to say that men want to be in a war or that people want to be hunted by psychopaths, but these plots elicit instinctive responses within us that cause our pulses to race in a safe environment. In romance fiction, the instincts concern arranged marriages, violent husbands, violent societies, and selecting within those societies. Children rarely get a mention. These instincts are exactly the same instincts we would expect women to have if they evolved in societies with arranged marriages.

In *Dangerous Men and Adventurous Women*, another romance writer, Linda Barlow, specifically states the preference for male coldness and aggression over tenderness in fantasy:

During the period when I was writing my historical romance novel *Fires of Destiny* I had a recurring dream in which Roger Trevor, the hero, would appear on my doorstep, saber rattling, to complain that I was allowing my feminist sensibilities to subvert his original rough, tough, macho character.

"The more you perfect your bloody little fairy tale, the more of a lily-livered wimp I turn out to be," says he.

"You're becoming a more mature, more civilized, and sympathetic hero," I assure him.

"Yeah, well I liked it better when I dragged the heroine off to my ship at knifepoint and ravished her."

Truth is, despite my feminist qualms, so did I.

I'm not ashamed to admit that I've always been one of those die-hard fans of the old-fashioned, hard-edged romances which feature a feisty heroine who falls into love and conflict with a dangerous hero with sardonic eyebrows and a cruel but sensual mouth. In the romances I most enjoy, as well as the ones I write, the intensity of excitement I feel while reading is directly proportional to the level of emotional hazard the heroine experiences as her relationship with the hero develops. When he stalks her, carries her off, besieges her honor, and finally makes love to her with a passion and determination that would unnerve me if I ever encountered it in real life, I greedily turn the pages, finding within such scenes a catharsis of the essential impulse or desire that led me to pick up the book in the first place.[28]

Although I have read several fascinating essays on romance novels as political or cultural documents, I see them as psychological maps which provide intriguing insights into the emotional landscape of women. The various elements contained in them function as internal archetypes within the female psyche. This includes the hero, whom I see not as the masculine object of feminine consciousness but as a significant aspect of feminine consciousness itself.[29]

Most people will appreciate that, where sex occurs in a dangerous setting, the sexual responses of a person may be heightened. What is most interesting are the lengths to which women's romance novels will go in order to create such a setting, whether the sex is explicit or merely alluded to.

The alternative plot to the woman being forced into marriage and creating a loving relationship with her husband is the woman finding a man to take her from that marriage. The other man is invariably even more dangerous and violent than the man she is taken from. Perhaps the most famous romance in literature is *Romeo and Juliet*, a story written by many writers, but most famously by William Shakespeare. The story uses a standard romance plot of a woman escaping from an arranged marriage with another high status man. Though rarely portrayed in this manner, it is obvious that Romeo is an extremely violent young man who is prepared to kill other men. While Romeo is usually depicted as a sweet, tender young man, he might equally be depicted as a cold-hearted, violent thug with occasional self-doubt.

Elopements rarely occur in cooperatively-simple societies and usually only where the circumstances allow. In Australia, due to the vast expanse of the landscape, and the lack of reliance upon horticulture, elopements appear to be more common. Baldwin Spencer noted the outcome of elopements in his 1899 study *The Native Tribes of Central Australia*. In this case, elopement involves the attempted escape of a man with a woman from a different community.

> Not infrequently the elopement of a woman with some man is the cause of serious trouble between the members of different local groups. When an elopement takes place and the man succeeds in getting safely away, some time may elapse before the aggrieved husband takes any action, though at times the eloping couple are at once followed up and then, if caught, the woman is, if not killed on the spot, at all events treated in such way that any further attempt at elopement on her part is not likely to take place. If the man and woman succeed in getting away to a distant part, then the chances are that sooner or later the original husband of the woman will, accompanied by his friends, go in search of her and the man who has run off with his property. As a general rule the upshot of the matter is a fight between the two interested parties; but at times the result may be that the friends get restive and interfere, in which case the fight becomes more serious and leads to a general quarrel between the two local groups, the men of the resident group, to which the man who has taken away the woman belongs, making common side against the men of the other group.[30]

In the above example, if a woman were to select a man and decide to elope with him, it is understandable that she would desire him to be able to defend her. Above, in chapter four, I gave one example from Chagnon's study of the Yąnomamö where a man attempts to remove a wife-by-arrangement from her husband, which only leads to the paramour dying, the woman's ears being cut off, and the village fissioning in two.[31] The woman chose poorly, and the man could not protect her from her husband-by-arrangement.

My expectation would be that if a study was taken of women in cooperatively-simple societies, women who chose men that were more dominant than their husbands-by-arrangement would suffer less violence because of successful intimidation of their husbands or be more likely to successfully elope from their husbands. Alternatively, in modern societies this might involve financial assistance and protection to leave their partner.

If women evolved with arranged marriages, then they might possess almost no instincts to respond to nurturing signals in a man, since they evolved without selecting their husband-by-arrangement, and would be unlikely to be able to select another. All women's instincts should be concerned with the issue of dealing with her husband-by-arrangement's violence. She should want a high status male who can protect her. This doesn't necessarily mean the best fighter in the community. The status of a man is reflected not merely in his own fighting ability, but the fighting ability of those whose assistance he can call on.

That women rarely select men below their own status, even in modern societies, is something that is commonly known. Experiments by Townsend and Levy performed experiments where the same group of men and women, either high or low in attractiveness, were photographed wearing different clothing.[32] The clothing reflected high status (including a blazer and Rolex watch); medium status (plain cream shirt); and a fast-food chain uniform. Models were pre-rated as to their level of attractiveness by a separate group of participants yielding two groups of either high attractiveness or low attractiveness. Male and female undergraduates and law students were then shown the photographs of the models, and asked whether they would be willing to engage in coffee and conversation, a date, sex only, serious involvement with marriage potential, sexual and serious with marriage potential, and marriage.

The results showed that women preferred the high and medium status/high physical attractiveness males, but preferred the high status/low physical attractiveness male compared to the low status/high physical attractiveness male. The preference for medium status/high physical attractiveness over high status/low physical attractiveness was small.

(Interestingly, the results for males should be baffling for most mammals, but were unsurprising for humans. The men preferred 'high physical attractiveness' women in every category. For 'sex only,' 'low attractiveness' women scored only 2.9 out of 5 as against 1.55 out of 5 for 'high attractiveness' (1 being highest). This seems paradoxical since one would expect the males to be very willing to copulate with any female. The status of the women made little difference. In fact, interest in low physical attractiveness women was poor regardless of her status. Interest in high physical attractiveness women varied only with the seriousness of the relationship, with interest in marriage increasing with status.)

The results for women are not surprising and merely reflect what has been known for generations. Many theorists will automatically invoke ideas of survival for such results, but we can assume that starvation or protection from predators are not issues for undergraduates and law students. Any woman should be perfectly capable of producing successful children with a high, medium or low status male. A good genes hypothesis – at least in terms of appearance – is equally unsuccessful since low status attractive males are ignored for high status, unattractive males – while males always preferred attractive females.[33]

If we were to invoke the idea that women evolved with arranged marriages these conclusions would make perfect sense: women would always instinctively select a man of their own status (their husband's-by-arrangement) or above (the women in the Townsend study were undergraduates and law students). Women should possess an instinctive aversion to men of lower status, even if they consciously admit that their aversion makes no sense in a modern environment. In modern societies, like so many other behaviours, female sexual selection seems paradoxical; however, in a cooperatively-simple society, it is perfectly logical.

This argument will be counterintuitive, particularly to those that may well wonder why a woman would sexually select in this manner in a Western society when she possessed no partner. The lack of a partner is immaterial; women should possess an aversion to men of lower status to themselves. Instincts will always be better than trial-or-error, particularly when the matter may well be life or death. Many women are killed by their husbands, even in societies where women get to choose their husband. Whenever an anticipated event is a matter of life and death, it is more likely that a strong instinct will exist for that event. We all have an instinctive aversion to potentially dangerous animals such as snakes and spiders. We should expect that women would possess an instinctive aversion to those males that traditionally could not protect them from possible death: men of lower status than their husband-by-arrangement. Thus we should expect that even if women know that a lower status man can provide for themselves and their children in terms of food, they should possess an irrational, inexplicable aversion to these men. And they do. So strong is this aversion, that many women of middle and upper-income backgrounds would rather not marry at all than marry a man with a low income.

Men also respond to these impulses. Many men in modern societies spend enormous amounts of time in the gym doing weights, despite the fact that their day job requires little more to lift than a pen. These same men may spend enormous amounts of time producing high levels of status, despite the fact that the resources needed to create such levels of status are well beyond what is needed for mere survival.

In understanding human behaviour we must decide how much of our behaviour is applied to the new environment of Western Civilisation and what environment the behaviours were originally intended for. We cannot observe someone driving a luxury car and compare this to meat. If women select heavily on the basis of status, even if there are lower status men that are better looking and more entertaining, then there must be a basis.

Women's selection methods and fantasies reflect a society with arranged marriages: a preference for status and an aversion to men with lower status than their own.

Reticence

If a woman is to select another man then the most preferable outcome is that her husband-by-arrangement should never know. While a woman may prefer a man who will be dangerous to her husband-by-arrangement, it is better if her sexual selection is never revealed. We should thus expect that women will prefer men that are silent and emotionally controlled. Women should reject men who are more likely to reveal the sexual selection; a reticent or stoic demeanour should be desired.

In her study of Aborigines in the Kimberleys, Phyllis Kaberry observed:

> One woman who had given up her lover said to me: "Me finish longa him; too much all about make 'em row along me – all day him talk-talk." A more stable union, to judge by what happened in the camp, did not mean the cessation of extra-marital adventures nor a rigid curtailing of her liberties. Still it was true that a woman would have to deal with the jealousy of her husband; she generally availed herself of the opportunity created by his absence from the camp, and whether she had an affair or not the rest of the community was not inclined to give her the benefit of the doubt.[34]

Stoicism is invariably an attribute of most male heroes in film and literature. Most of the major male actors are stoical: rarely are heroes portrayed as home-makers. Effusive and voluble men are usually portrayed as either homosexual or the community clown.

Carefully chosen behaviour should extend beyond speech to physical demeanour. A man should be careful with his words, his eyes, and his interaction with the woman who sexually selected him. Men who were known to stare at women, or be unable to control their emotions would be avoided. Women will be attracted to men that are severe, unemotional, and perhaps even cold in public.

Reticence and emotional control should be the most desired characteristic in men, after equal or higher status, since the most desirable outcome for a woman is that nobody learns of the sexual selection.

Emotions

I have already commented that violence from the husband-by-arrangement is the greatest difficulty for a woman or man, which presents the problem that no person in their right mind would engage in a sexual selection. Depending on a woman's husband-by-arrangement, both parties may be killed, seriously injured, or severely beaten due to the sexual selection.

There is a very high negative motivation involved in a sexual selection. In order for a person to actually engage in a sexual selection, there must be an internal motivation that is greater than the possible violence. The positive motivation must be great enough to overcome the fear of a husband-by-arrangement attacking his wife or the sexually selected man with a club or axe. This motivation is obviously 'love'. In order to engage in a sexual selection one would, at times, almost need to be insane, and this is what love supplies, a positive, overwhelming drive. Love is an emotion that overcomes all reason and logic.

The longer the duration of the sexual selection the greater the chance that the husband-by-arrangement will discover the sexual selection, so we should expect that the overwhelming, positive motivation should last only long enough for the woman to become pregnant. While the overwhelming positive motivation might disappear, we might expect that it might not disappear entirely, but be reduced to a lower secondary level. This would be beneficial for the man to ensure that the husband-by-arrangement's wife is protected during her pregnancy. A man's feelings for a woman should be reduced to a lower level, but one where he would still be concerned for her. A woman's feelings for a man, by contrast, might not disappear so easily. We should expect that after one child she might again attempt to sexually select the same man a second time.

The level of motivation should be higher in women. Men have the potential to copulate and reproduce with many women, while women can only produce their own children: for women, the issue of finding an alternative to her husband-by-arrangement is of greater importance. While a man might be able to decide to ignore a woman's interest for fear of her husband-by-arrangement, with the expectation that other opportunities might arise later, a woman has no capacity to wait. A woman will possess a small avenue to sexually select before she eventually becomes pregnant with her husband-by-arrangement, and we should expect that during this time a woman's positive motivational drive to sexually select should be very high. Since she may only have a maximum of four or five adult children in her life, the difference in her success between sexually selecting the genes of a brilliant man and reproducing with her less-than-average husband-by-arrangement may be significant. With each period of ovulation, with each window of opportunity, we should expect that the positive motivation would be extremely high. Thus we might expect the experience of love would be greater in women than men.

It is worth considering the contrary idea that people evolved with Western-style marriages. If women and men evolved to select another person for a long-term relationship, one that would last for decades, why would the following emotions, here summed up by Shakespeare in *Romeo and Juliet*, be helpful?

> Love is a smoke rais'd with the fume of sighs;
> Being purg'd, a fire sparkling in lovers' eyes;
> Being vex'd, a sea nourish'd with lovers' tears:
> What is it else? a madness most discreet,
> A choking gall, and a preserving sweet.[35]

Emotionally, if humans evolved with Western-style marriages, then any affection between the couple should be slow in developing. True affection between a male and female in such a relationship should only be reached after a child has been born, and nursed past the most dangerous period of infancy. We should expect the emotions to be small at the beginning and at their peak after a child is successfully past infancy. Instead, sexual therapist Theresa Crenshaw concluded after years of private practice dealing with couples' sexual problems: 'Contrary to popular belief, in the very beginning your relationship is usually as good as it is going to be.'[36]

In many societies, marriage and love are not related. Yąnomamö men made this obvious to Napoleon Chagnon:

> I am sometimes asked what the Yąnomamö idea of love is. They have a concept, *buhi yabraö*, that at first I thought could be translated into our notion of love. I remember being very excited when I discovered it, and I ran around asking lots of questions like these: "Do you love so-and-so?" – naming a brother or sister. "Yes!" "Do you love so-and-so?" – naming a child. "Yes!" "Do you love so-and-so?" – naming the man's wife. A stunned silence, peals of laughter, and then this: "you *don't* love your wife, you idiot![37]

This is not an odd remark by the Yąnomamö men, it is normal. Husbands and wives in arranged marriages are not lovers, and nor should they be expected to love each other.

For instance, in David Buss's study on international mate preferences in 37 countries, involving 9,474 people, no differentiation was made between countries where arranged marriages were normal and those where selected marriages were normal. In Western nations 'Mutual-attraction – love' was the preferred or second most desirable trait in a mate out of eighteen choices for both men and women. In Nigeria, however, love was rated as the fourth most desired characteristic for both men and women, while amongst the Zulu in South Africa, love fell to tenth and fifth for men and women respectively. In China, another country where at the time arranged marriages would be expected, love was fourth and eighth for men and women respectively as the desired characteristic in a mate. In China the women rated 'ambition' and the men rated 'good health' as more important than 'mutual-attraction – love'. Amongst the Zulu in South Africa, 'good cook and housekeeper' was the second most preferred trait amongst the men after 'emotional stability'. Such results seem perplexing to a person from a Western culture, but the combination of love and marriage are a recent occurrence, and love evolved for a very different purpose than to initiate long-term relationships. For people expecting an arranged marriage, expectations would be lower and people would choose those traits that they might have a reasonable chance of experiencing. In that context, the desire for 'emotional stability' seems more understandable.

The very word 'love' suggests extra-marital sexual relations. The word 'lover' has always suggested a person outside of a marriage, while the object of the love affair, the paramour, is 'a partner in an illicit love intrigue; a mistress,'[38] and comes from the French *par amour* – through love. The Roman poet Ovid wrote *The Art of Love*, a detailed account of how men could have love affairs with other men's wives, and how women could attract them. For women, Ovid advocated make-up to attract a short-term sexual liaison, not a husband. Marriage for the Romans was a matter of arrangement.

Consider how humans initiate sexual contact. Is it with a long conversation, or through lengthy grooming? All it takes is a glance; humans make sexual contact with their eyes. Sexual therapist Theresa Crenshaw comments:

> As important as scent is in the early dynamics of love, lust and romance, when the potential object of your affections is out of the range of scent, sight has a special impact all its own. The term *eye contact*, touching with your eyes, takes on new meaning. Humans can determine whether or not they want someone with just a glance – even if there is a glass partition making scent impossible. Lower animals can't.[39]

Humans do not need to touch, smell or speak to one another to experience intense desire. This would be expected for women in arranged marriages who were unable to communicate intimately with men, or perhaps communicate at all. A woman in a community that was being visited by another community could, merely through eye contact, initiate a sexual selection with a man from the visiting community. Since the visiting community might only stay for a few days, this would give her the opportunity to select genes from any visiting community. Phyllis Kaberry in her 1930s study of the Australian Aborigines of the Kimberleys reported the following concerning women's sexual preferences (outside of marriage):

> There is also love magic, with which I shall deal later, that of the men being different from the type practised by the women. The latter have very definite ideas as to what are desirable qualities in a husband or a lover. They include a good physique, an elegant coiffure, bodily adornment, and prowess in dancing and hunting. The appearance of some of the young men visiting the tribe used to arouse even the middle-aged and soberly married to overt admiration and enthusiasm.[40]

The idea of a couple falling in love and living happily ever after is a recent idea. The very word 'love' always meant the feelings involved in an extra-marital affair. People in marriages never expected to love each other because they could not choose each other.

If humans evolved to sexually select each other for long-term relationships in which children took fifteen years to mature we would expect the emotions of sexual desire to emerge slowly. That sexual desire emerges suddenly, sometimes taking a mere glance at the opposite sex, cannot be adequately explained through Western-style marriages. That this sexual desire, love, is often equated to near insanity makes even less sense. However, if humans evolved with arranged marriages, then the requirement for sexual desire to reach levels of near insanity makes sense since all the sexual selections possess the risk of extreme violence. Thus we should expect the evolution of emotions that are so positively overwhelming that they nullify the possible risk – even if the risk is death. Now, in Western societies, these emotions seem paradoxical.

Lack of Insulating Hair

If a woman wishes to sexually select, since she did not sexually select her husband-by-arrangement, and she cannot touch any of the men around her, then she has the difficulty of being able to determine their genetic quality and health. If men were covered in hair, as are all other primates, then a woman would not be able to tell if the men around her were diseased or possessed a history of disease. If a female chimpanzee grooms a male she will be able to smell him and observe the level of parasites. She may also be able to determine a level of historical disease that, from a distance, may not be obvious, but is evident upon inspection.

The response of female hominids was to select males that were able to demonstrate that they were not diseased and possessed little physical evidence of disease. The only way males are able to demonstrate this is to lose their body hair. Females were able, at a glance, to observe whether a male was diseased and the effectiveness of his immune system. Considering that many children die in infancy, and some women have one baby die after another, the immune system is extremely important. Considering that a woman may risk her life in sexually selecting another male's genes, she will want excellent evidence that his genes are worthwhile. Once our ancestors began using fire or animal skins and could keep warm despite a lack of insulating body hair, this allowed males to give females a record of their immune system, unimpaired by hair.

There is an argument that humans lost most of their body hair in order to reduce parasites, put forward by Mark Pagel and Walter Bodmer.[41] This argument is unlikely as there is barely another animal that walks the earth that is not covered by a protective layer of hair, feathers or scales. This is testimony to the fact that parasites are a small burden compared to the benefit of a layer of hair that can protect the individual from the cold and insulate from the heat. Since humans are surrounded by others who can remove parasites then it would seem odd that humans would have such an urgency to lose a protective layer of thick hair, indeed, social primates with opposable thumbs would be the least likely to shed their hair. That humans can face a serious threat of death through exposure without either clothing or fire would suggest against such a hypothesis.

Another argument is that the loss of hair allowed for increased uptake in vitamin D. Cholecalciferol, naturally occurring vitamin D, forms in the skin when affected by ultraviolet light, but then, so it also forms in fur and feathers. Animals with fur and feathers obtain cholecalciferol by licking their fur or preening their feathers. Vitamin D is essential for bone formation, so if hairlessness was essential for vitamin D uptake one would expect that all other primates must be suffering from rickets – but this is not the case. To present a benefit is of little use if we then observe that almost all other mammals have taken a different path and do not appear to be deficient and simply use a different method to obtain the same end. Simply because humans can walk and run is not the explanation of why they use two legs, since what we are attempting to explain is why humans use two legs rather than the four used by every other primate. Unfortunately, too many scientists are easily pleased and look the other way when it is pointed out that there happens to be an exception rate of over 99 percent in the rest of the ground-dwelling mammalian species.

Another argument is that a lack of covering in hair allows for the use of sweat as a cooling system.[42] Again, if this were the case then all mammals would be hairless and they would sweat. The fact that almost all other mammals, and certainly all other primates, are covered in hair and usually prefer to use a different cooling mechanism suggests that there are minimal benefits to be found in minimal hair coverage and the capacity to sweat. There are those that argue that such a cooling mechanism is essential for hunting large animals. Yet, this is false, as hyenas, wolves and hunting dogs hunt large prey over great distances and have no need for copious amounts of sweat. Furthermore, sweating requires copious amounts of water to maintain: a human exercising in a hot environment uses two litres of water an hour. One might suggest that sweating is possibly the most ridiculously luxurious cooling mechanism in the animal kingdom, and that our ancestors were able to evolve such a cooling mechanism only because they emphatically dominated their environment so that their access to water was unquestioned by any other animal. No doubt, they would have had access to consistent sources of fish, and the abundance of food produced by plants also attracted to the water rich environment. If humans were able to sweat, then they must have been able to dominate rivers and water holes, and possessed consistent access to the richest nutrients available. Most probably, big game hunting would be an optional extra for these omnivores – while wolves, hyenas and African hunting dogs depend on meat as their main source of food.

It is all very well to suggest a benefit, but it is unfortunate if we then find other mammals with similar behavioural patterns to be covered with thick hair. Since hairlessness is ignored in all other ground-dwelling mammalian species, we must follow Darwin's path and see that functionality cannot be the answer, we are looking for something unusual. We are looking for something very unusual, so particular that no other animal engages in this behaviour. Arranged marriages and the prevention of female selection of the male gives this unique behavioural pattern. The fact that females must select males under dangerous conditions and that they are incapable of direct communication, thus requiring selection at a distance, gives the basis for hairlessness. Naked skin allows females to diagnose the health of males by being able to observe pimples, blemishes, rashes, spots, colour changes (such as yellowing) and so on. Disorders often leave scars so that the skin gives a remarkable historical record of disease. One can gaze at another person's skin and find a reasonable record of a person's recent physical and mental health (since stress can cause physical disorders), particularly if they happen to be virtually naked, so that even genital diseases might be able to be observed in males. When females may be risking their life or health in order to engage in an extra-marital sexual selection, that they would consistently select males that were naked in order to be able to obtain a clearer account of their health would seem logical. Between the hair-covered male that might be diseased or not and the naked male with clear and unblemished skin, selecting the former is always a risk, particularly in an environment where less than half the population survive to become adults. No other argument presents a behavioural pattern in humans that is different to other hair-covered mammals. Since humans are unique in their appearance we must have an explanation that presents a unique function: nutrition, vitamin D, or any other argument pertaining to natural selection can never be an answer, since other mammals require all these things and yet possess a common, different appearance. This argument is similar to Darwin's views: 'The females apparently first had their bodies denuded of hair, also as a sexual orna-ment; but they transmitted this character almost equally to both sexes.'[43] The basis for female hairlessness will be dealt with in the following chapter.

There is also the question of colouration. In terms of human evolution, the issue of colouration is less important, since it simply presents a variation of the same model – we are attempting to explain the model, not the variations.

The basic human appears to be various shades of brown, and people with brown skin represent the vast majority of the world's population – the Americas, Asia, Africa and Australia (two billion in India and China alone). Europe is the only exception to this other-wise universal skin colouration. This skin colouration is found in every environment, from the Amazon to Alaska, to the deserts of Australia and the northern reaches of eastern Russia and China. Outside of Europe, whether people lived in igloos or in deserts, they tended to be brown – until mass migrations occurred.

Currently, there is an assumption that any variation is due to interaction with ultra-violet light. It is argued that darker skin protects against UV light and that lighter skin is superior for vitamin D production in limited UV environments. Ultraviolet light is thus seen as the cause of both darkening and lightening of skin colour.

It is worth pausing for a moment to remember exactly what natural selection means. Natural selection means death or inability to reproduce. The argument is that without very light skin, people could not exist in Scandinavia since they would experience a vitamin D deficiency and thus rickets, as vitamin D is essential for the calcification of the bones. Equally, the argument is that without very dark skin, people could not exist in parts of Africa due to the ill-effects of UV radiation. The argument is not that these people suffered more problems on average, but rather that they either died or were incapable of reproducing and disappeared completely from these environments, replaced by those that possessed genes for lighter or darker skin. This is not a minor flaw, but a major, catastrophic flaw that led to their own death or the death of their descendants.

The issue of the evolution of skin colour and the vitamin D argument has been cov-ered in Ashley Robins' *Biological Perspectives on Human Pigmentation*. The arguments sup-porting the need for vitamin D above are countered by the arguments that (1) rickets is a disease of civilisation, and that before dense housing rickets did not exist; and (2) that people with darker skin are capable of obtaining enough vitamin D since vitamin D can be stored and these summer stores are enough to last someone throughout winter. Robins concludes:

The upshot of all the arguments raised here is that the hypothesis of Murray (1934) and

Loomis (1967) does not survive critical scrutiny. Even assuming for the transmission characteristics that black epidermis would always cause a shortfall in vitamin D formation compared to white epidermis, Beadle (1977) calculated that Negroids would still produce a total annual amount of vitamin D that fell within the normal recommended range right up to the latitudes of the Arctic Circle – and this estimate was based on an exposure of only 10.5 per cent of the body area (head, neck and hands). Furthermore, there was no synthesis of vitamin D in Caucasoid skin specimens exposed to sunlight in Boston (42.2° N) from November to February (inclusive) and in Edmonton (52° N) from October to March (inclusive) (Webb *et al.*, 1988). However, during the same period of exposure in Boston (November to February), there was a significant photodegradation within the specimens of the existing vitamin D stores (Webb, De Costa & Holick, 1989). The major thrust of the hypothesis is that skin depigmentation evolved in the north to utilize the winter sunlight for vitamin D synthesis. As 'white' skin during the northern winter appears not only incapable of such synthesis but also likely to promote *depletion* of vitamin D, the hypothesis is severely undermined.

In summary, rickets is a disease of industrialization, urbanization and overpopulation; it is a disease which stems from a deprivation of sunshine. ... Under such circumstances there is not a scintilla of palaeontological or experimental evidence that rickets (or osteomalacia) ever did or would have manifested, regardless of skin colour.[44]

Robins notes the archaeological record, which finds that only 1 percent of skeletons in Swedish and Danish cemeteries (AD 1100-1550) show a sign of rickets, and that there is no trace of rickets in medieval cemeteries in East Anglia.[45]

The disease gradually increased during the post-medieval era of urbanization and rapidly escalated with the advent of industrialization, so that by the eighteenth and nineteenth centuries it was common in skeletal surveys. This again highlights the fact that rickets is not a 'natural' disorder but arises from abnormal environmental conditions.[46]

It should be noted that it is early studies from the turn of the twentieth century that appear to impress those who accept the vitamin D argument. In Britain, for instance, rickets was observed in at least a third of children tested between 1868 and 1935;[47] however, in a 1944 study of 5,000 children that used a radiological diagnosis 'the incidence of rickets diagnosed radiologically in children between 3 and 18 months of age is 2½% before 6 months, 4% in the first year, and negligible after that period',[48] though this could have been reduced due to recent vitamin D supplements. Furthermore: 'The rate of incidence of active rickets varied from 61% in Sheffield, a smoky, working class, industrial area, to nil in St. Albans, a well-to-do, suburban district.'[49] Once radiological testing is used, the number of actual cases of rickets tends to be significantly reduced, often to a tenth of the level compared to clinical observation (see the Bangladesh example, below). There is the question of what the actual impact of the epidemic in rickets had on the population. An archaeological investigation of

135 adults in a Birmingham cemetery found deformities associated with rickets were common, though they suggest that it may not have had adverse social implications.[50] By contrast, a study of 234 post-medieval subadults found no impact from rickets on growth.[51] The actual impact of childhood diagnosis of rickets during this period does not appear to be well-documented.

While the rates of rickets, above, seem remarkable, it needs to be noted that this was suffered amongst a pale-skinned population in an unusual environment. In the United States, according to Census reports, the African-American population had increased: in New York from 31,320 in 1800 to 70,092 in 1890; in Massachusetts from 6,452 to 22,144; and in Connecticut from 6,281 to 12,302.[52] These increases were lower than for the Caucasian population (which had significant increases due to immigration), but they occurred at latitude levels comparable to Geneva. If people in northern latitudes suffer from an incapacity to maintain acceptable vitamin D levels, how did these people survive and increase in number?

People may cite modern studies to suggest that darker skin people have higher rates of rickets. One recent meta-analysis, a review of reported cases of children with rickets in the U.S.A. between 1986 and 2003, found that 166 children (4 months to 54 months, with seventeen studies less than 30 months) were reported with rickets, and of these children eighty-three percent were African-American.[53] Yet thirty cases in ten years were reported in North Carolina (all were African-American) and another nine reports were from Dallas, Texas (eight were African-American), both of which have average winter temperatures of at least 10° C. Whether in colder climates, such as Connecticut, or warmer climates, such as Texas, people with darker skin were more likely to suffer from rickets. Rickets are also a problem in Bangladesh, where rachitic deformities were found in 0.26% of 21,571 surveyed children in 2000 and in 0.12% of 10,005 surveyed children in 2004. In the Chittagong region, 8.7% of children surveyed had at least one clinical finding indicative of rickets, and 0.9% had confirmed rickets (positive physical features, raised alkaline phosphatase and positive radiology).[54] The cause of rickets in these children has been argued to be calcium deficiency[55] and antenatal and exclusive breast-feeding by vitamin D deficient mothers may play a role.[56] In any event, even with supplements, the numbers of African-Americans that suffer from rickets is very low.

The time required to obtain a healthy amount of vitamin D is not extensive:

> It has been suggested by some vitamin D researchers, for example, that approximately 5-30 minutes of sun exposure between 10 AM and 3 PM at least twice a week to the face, arms, legs, or back without sunscreen usually lead to sufficient vitamin D synthesis...[57]

Even if the amount of time is increased six-fold for dark skinned people, as has been suggested,[58] this is not an overwhelming amount of time – approximately thirty minutes to three hours, twice a week, at a time when people needed to procure food in traditional outdoor methods. Added to this is the opportunity to eat fish, another source for vitamin D, which would be a natural part of most human's lives, since they would need to live near a fresh water source. Fifty grams of sardines is equivalent to seventy percent of daily requirements.[59]

The evidence supporting the vitamin D theory is far from overwhelming. We are

informed that an epidemic of rickets occurred in industrial England, while at slightly southern latitudes an African-American population more than doubled. The indigenous population in North America has not altered as the prediction would predict: indigenous people in
North America remain relatively brown compared to their white counterparts at similar latitudes. If we found that people with darker skin living in New York in the nineteenth century
significantly reduced in number over time or very pale skin amongst the indigenous North
American population at higher latitudes, then this would be positive evidence, but there is
no such evidence. Supporters of the vitamin D argument use evidence such as large numbers
of people with dark skin being deficient in vitamin D levels in the northern hemisphere, but
there is no evidence that this significantly impacts their ability to survive and reproduce.
When natural selection is invoked, we must expect evidence of early death or inability to
reproduce successfully; this is quite different to sexual selection, which is selection of a trait
within a healthy population. There is no credible evidence that darker skin in the northern
hemisphere is so detrimental that it could invoke natural selection.

The argument concerning protection from UV is equally questionable. While
people with light skin are more likely to get skin cancer, historically, few people lived long
enough to experience this problem, since few people lived until the age of fifty. In Australia,
which has the highest rate of melanoma in the world, the average age of diagnosis is 60 for
males and 40 for females.[60] The higher rate for females is most likely due to sunbathing and
greater levels of self-inspection. In 1997 skin cancers were the 19th leading cause of death,
below suicide (10th), and have increased slightly to be 16th in 2006.[61] Historically, it does
not appear to have had a catastrophic effect based on life expectancy data. In 1884, average
life expectancy for a male was 47.2 at birth and 63.6 at age 30, and even higher for females.[62]
Skin cancer was not listed as a leading cause of death, so actual rates are unknown; however,
considering the visual evidence of advanced melanoma (large, black growths on the skin) it
could not have been missed by physicians. An average life expectancy of 63.6 for a man once
he reached thirty does not suggest that cancers from melanoma had a significant impact on
the Caucasian population. It should be noted that the tendency of Europeans to deforest the
landscape further increases outside workers' exposure to the sun.

The skin cancer argument would require that significant numbers of fair skin people
would die due to skin cancer before the age of thirty, or even earlier. A continent such as
Australia, should have been impossible to colonise if the skin cancer argument was correct,
since cancers to the face, forearms, hands, and other regularly exposed regions should have
eliminated Europeans in high numbers at a young age, particularly in males working outdoors. Skin cancers are more common in people of European origin, but the argument of
natural selection requires death at young ages, and, even with young women sunbathing, this
does not appear to be satisfied. By the age of 40, a woman would expect to be a grandmother
under normal circumstances of human life. If the argument for darker skin in areas with high
UV levels was correct, then we should expect that Australia would satisfy a test based on
natural selection, yet lighter skin people in Australia have thrived. Nineteenth and twentieth
century immigrants from Europe form the vast majority of Argentineans, and they have also
been unaffected. If lighter skin is so detrimental, requiring darker skin pigmentation for survival, where is the evidence from experiments in Australia, Argentina and Southern Africa?

Most critically, the largest experiment, taking thousands of years, the immigration of people into North and South America, has not yielded the diversity of skin pigmentation found in Europe and Africa – people remain a medium brown with minor variations. For those in the northern region of North America, this is often explained away by suggesting that the population, such as the Inuit, obtained large amounts of vitamin D through whale and seal fat;[63] however, indigenous people were spread throughout the area, including the plains, such as the Cree and Blackfoot, at latitudes comparable to London and Antwerp. These people have lighter brown skin, comparable to people in Latin America, and hunted buffalo, not seals.

This leads us back to the question of why people would have variations in skin colour if it cannot be linked to questions of natural selection or survival. The variations of skin colour are simply an extension of the original reason for hairlessness itself. Pale skin allows for greater ability to observe the skin – particularly important in environments where people wear clothes. By preferentially selecting pale males, females were able to increase their likelihood of ensuring a lack of disease, which, again, is vital in an environment with extremely high mortality rates for infants and children. Even the smallest blemish or rash is easily observable on light skin, while it is far more difficult to observe on darker skin. In societies where people wear few clothes, such disorders are obvious, even on darker skin, but in societies where people wear clothes, the face, hands, ankles and neck may be the only observable skin on which to detect disease.

The benefit of brown and darker skin is observable in people with light skin: women wear mascara around their eyes in order to frame the eye, as did actors in black and white, silent movies. Since all selection is at a distance, the capacity to make eye contact at a distance is vital. Light skin around the eye makes eye contact more ambiguous. Light skin is viewed as unhealthy, perhaps even anaemic, and many people with light skin now go to extraordinary lengths to darken their skin. The impact of the sun on skin cannot be ignored, since, though few die of skin cancer in Australia at an early age, it does rapidly increase lining on the face, giving a person a crinkled, weathered appearance at a young age. Darker skin is more resilient to high levels of UV, and maintains skin condition.

Human skin reflects health and ageing, and selection should operate between these two features: the preference for skin that remains youthful, yet presents visible evidence of disease. Lighter skin should be preferentially selected, except where the skin begins to be damaged. In Australia, Caucasians who spend large amounts of time outside will begin to show significant levels of lining in the face, even in their early twenties. Males that could maintain youthful skin would be selected in preference to males that presented damaged and ageing skin for a longer part of their life. In environments where skin damage was reduced, females would select for lighter skin for greater presentation of disease.

Any attempt for uniformity with regard to region and appearance is rendered impossible due to migrations which have occurred over millennia. These migrations undermine arguments based on natural selection, but not sexual selection.

If females only selected males at a distance and preferentially selected males whom they could see were not diseased because they had lost their hair, then eventually all males' skin would become visible. Light skin aids the capacity to observe a history of disease. The visual importance of skin is reflected by the female presentation of unblemished skin, removing hair, wearing minimal clothing and attempting to cover blemishes with make-up. If skin were merely beneficial in terms of the capacity to sweat or for UV light uptake, then we would expect there to be no interest in the state of a person's skin, and we would be unlikely to observe a multi-billion dollar industry for its preservation and presentation. If one were to create an experiment where a person with pale skin had their skin slightly yellowed and then sent the person out into the public, the result would surely be that many people would suggest that the person should go to hospital. Furthermore, we need also question why humans, unlike other mammals, use visual rather than scent cues, to create sexual interest. If all sexual selection in the human species occurred between individuals that were incapable of touching one another and barely able to smell one another, then the requirement for such visual cues makes sense. By contrast, mechanical arguments make no sense whatsoever since, if they were essential, we should expect them to be reflected in the other mammalian species – which they are not. Those that wish to continue to argue the vitamin D argument, or lice argument, or any other natural selection argument, need to explain why this does not apply to the rest of the mammalian ground-dwelling species.

Pubic and Underarm Hair and Smell

It has been suggested that pheromones and olfactory cues are an important component in sexual selection. It has been suggested that this is an important method for females to assess genetic variation in males, with greater genetic diversity possibly improving the defence mechanisms of the immune system.[64] Considering the levels of infant mortality and that perhaps half of the number of infants do not make it to adulthood, the importance of the immune system cannot be underestimated. In many animals there is a specific organ, the vomeronasal organ, to detect pheromones,[65] though it has been inactivated in apes.[66]

Recently, however, the evidence for pheromones has been doubted.[67] There seems to be good reason for doubting the importance of pheromones, due to the lack of a specific organ for olfactory communication and, secondly, the fact that human scent is not so much human odour as bacterial odour. Underarm odour is caused by the bacterial breakdown of secretions from the apocrine glands – a gland that does not play a role in creating sweat, which is odourless; the greater the number of bacteria, the greater the smell.[68] Before modern methods of medical diagnosis, scent played a role, with particular diseases causing a definite scent (scrofula, stale beer; typhoid, freshly baked brown bread; rubella, plucked feathers; diphtheria, sweetish).[69] This link was originally identified in ancient medicine, and has since been added to by conventional medicine: diabetes, some hepatic diseases,[70] and gonorrhoea ('putrid').[71] It appears to be the case that human secretions and bacteria may vary according to an individual's state of health.

Other primates have the capacity to groom one another, but if humans evolved with arranged marriages, then men and women in cooperatively-simple societies would be unable to touch one another. Since disease and smell are linked, a woman would want to smell a desired man. Women report that scent has a stronger impact on sexual interest than do men.[72]

Humans, by any standards, are particularly odorous. A man who has exercised can quickly fill a room with his scent. Physically active people that do not clean themselves regularly, after several days, will produce a strong odour. In cooperatively-simple societies, men and women cannot touch each other or easily smell each other. If smell is an important part of sexual selection, then one might expect that women might be more likely to select men that they could smell – from a distance of several metres.

Underarm hair and pubic hair increases the surface area that can spread the scent. The level of odour is particularly important in an environment where there are many competing smells, such as other men and women (who do not use soap), dogs or other animals, decomposing matter, dried or cooked food, and possibly regurgitated food, excrement, or other human waste (infants are present). Napoleon Chagnon commented on his first experience of a Yąnomamö village: 'Then the stench of the decaying vegetation and filth hit me and I almost got sick,'[73] comments which may have also been uttered by people entering Western towns or cities before the existence of sewers. Significant levels of odour may have been particularly valuable to women who were considering sexually selecting men from other communities who may only be present for a short amount of time. A woman would thus be given every opportunity to smell the man without actually coming close to him.

The prevalence of men to be odorous and to have hair to increase the surface area for dissemination of their scent makes sense in an environment where selection is at a distance. If selection occurred in a Western manner, with the possibility of close contact, then the need for strong odour and increased surface area would make less sense. If human odour becomes altered with disease, then a strong 'normal' odour may be attractive, while a strong foul or unusual odour may be unattractive. There remains the question, though, of what it is that people are actually capable of smelling.

Head and Facial Hair

Head hair allows for greater display, allowing a man or woman to present more information as to their health. One only needs to consider those undergoing chemotherapy to appreciate that hair loss is seen as a sign of ill health. Receding hair or balding can also suggest that the man is maturing.

While in Western nations a vigorous display of head hair is normally associated with women, one has only to peruse the anthropological photographs of people living in Papua New Guinea and Australia to see that it is men that use their hair for display in cooperatively-simple societies. It is often long, thick and shaped. Much like Western men in the music industry, it appears that once conventions and restrictions are removed, then men are as adaptive as women in utilising hair as a display.

Hair can also be a signal of weakness, with greying hair a symbol to younger men of an older man's weakness. As men age, they present signals of a lack of display. This may increase their longevity if they are not viewed as a serious threat by other males. Once males reach a certain age, reproduction may be solely via the arranged marriage system, as against the selected extra-marital selection method, since the latter is so dangerous.

Facial hair appears to be more variable. Amongst some peoples in humid climates, facial hair is largely non-existent. In others it allows a man to further increase his display, and, in Papua New Guinea and Australia, men often use their facial hair as imaginatively as they use their head hair. Like receding hair, facial hair also reflects maturity, as few younger men are able to grow long facial hair.[74]

Similarity

Finally, one feature that women should find highly attractive is that the men are similar to themselves in appearance. If a woman is to seek a sexual selection, then she obviously takes a tremendous risk if the man has features different to both the husband-by-arrangement and herself. Women should find men with similar faces and bodies to themselves to be attractive and those with differing features to be less attractive.[75] If the husband and wife-by-arrangement both have red hair and pale skin, then it is particularly dangerous for the wife to select an alternative male that has black hair and dark skin. This would suggest that humans should not necessarily follow the assumed path of out-breeding to increase genetic diversity, through either marriage – where cousin-marriage is normal – or selection outside marriage. One recent study of 92 women selecting from 55 males did find that women preferred similar males with a similar genetic make-up, which would go against the expectation of a preference for different genes.[76] Another study found that observers were able to predict whether two people were actual couples or randomly selected people used to create couples solely by the similarity of the people.[77] Alternatively, one study found that people prefer facial images that have been manipulated to appear average rather than similar.[78] Explanations for this preference of similarity have followed the ideas of sexual imprinting in birds; however, it makes more sense in the context of selection within arranged marriage systems, that selection must involve an element of similarity between the female and the male of her choice to reduce concern by the husband-by-arrangement.

An implication of this is that female selection should thus operate to increase the physical similarity of all those in the community. If one man was particularly successful, and everyone in a community and the surrounding communities shared his genes, then such a man might also become the blueprint for further female sexual selection, as women preferred men who appeared like themselves. If the particularly successful man possessed blonde head hair rather than brown head hair, and this was a feature that a female grandchild shared, then she might actively select against other kinds of men that did not share this feature. The same would be true for skin colour, the shape of the face, or any trait that would be easily distinguishable in a child. Women should actively select against any man who differed significantly from herself, since the sexual selection would be more obvious, and the child would be more likely to suffer violence from the woman's suspicious husband-by-arrangement.

Female sexual selection, in order to hide any actual sexual selections, should increase the likeness of a community over time, and non-intermarried communities might become increasingly different with time. The closer the resemblance of a man to herself, the greater the attraction. This might well account for the lack of bi-racial marriages in cooperatively-modern societies.

Concluding Remarks

How important is female sexual selection of men? Occasionally, a woman may be extremely successful through her sons. Napoleon Chagnon, in his study of the Yąnomamö, has closely followed the genetic legacy of a man named Matakuwä, or 'Shinbone'. This man produced 43 children with 11 wives. His 43 children produced 231 grandchildren; the 231 grandchildren produced 480 great grandchildren; and there were, at the time of report, some 295 great, great grandchildren. In seven villages, 756 individuals were descended from him.[79] Even allowing for some cuckoldry by non-relatives of Matakuwä this is a significant number. Chagnon found similar success for Matakuwä's father: 14 children, 143 grandchildren, 335 great-grandchildren, and 401 great-great grandchildren.

Since there are believed to be about 20,000 Yąnomamö, if one were to only consider the great grandchildren of Matakuwä, the chance that any Yąnomamö could claim to be related to this one man is 480 in 20,000, or about 1 in 42. If there were more than 480 living relatives (children, grandchildren, great grandchildren and so on) at any one time, which is possible, then the probability would be higher. Considering that there was no high-level social organisation available to this man, as it was for some highly polygynous Chinese emperors, this is quite a remarkable feat of human reproduction.

By contrast, a man who married monogamously and produced six children would be fortunate if half of these children reached adulthood. Napoleon Chagnon, in his study of 1466 Yąnomamö, found that 37% of the society was aged between 0 and 10. The group aged 10-20 consisted of 23% and the group aged 20-30 consisted of 17%.[80] So, if a man married monogamously and had six children in the 0-10 region over a period of time, the chance that they would reach the 20-30 range would be roughly 46% or the equivalent of 2.76 of his offspring. If this monogamous success were repeated by these children then he might have about 7.6 grandchildren and 21.0 great grandchildren, somewhat fewer than Matakuwä – and this is assuming probably a best outcome scenario for a monogamously married lineage, since 2.76 is above replacement levels. Thus this man's genes would, in the Yąnomamö gene pool, represent 0.105% at the great grandchildren stage (pop. = 20000). Matakuwä's 2.4% to a hypothetically monogamous man of 0.105% gives some suggestion that though most men might be married monogamously, polygynous men can have a tremendous individual impact on a gene pool. Napoleon Chagnon records this influence in regard to Matakuwä's father, which he records as individual 1222:

> The total contribution of individual 1222 to the entire population (all ten villages) is startling: 45 percent of the entire population are descended from him in some way. If multiple relationships are considered, the picture is even more startling. Of the some 1,500 living people in the ten villages, 45 percent are related to him at least by *one* link, but some people are related in several different ways. There are nearly 1,000 total relationships between the present living population and individual 1222![81]

Thus, in some manner 675 living people were related to one individual, and some of those people were related to that individual in more than one manner – because of intermarriage, they could 'link' their descent through more than one line of descendants.

It is easy to imagine that if Matakuwä (or his father) possessed a certain trait – such as hair colour – it may well become very common; by contrast, the monogamously married man would be unlikely to leave an obvious trait. If one were to consider that humans have existed in excess of 100,000 years, then one must assume that we must all be related to many Matakuwäs.

While the mother of Matakuwä's father could be jubilant at the genetic success of her child and grandchild, as Helena Valero witnessed, other Yąnomamö women might watch child after child die in infancy. Not only do women thus have a tremendous incentive to ensure that their children are born healthy, but they also have a tremendous incentive to ensure that their children are competitive in a very competitive environment.

For women, quality is essential for both survival and success. Women should be particularly careful as to who is the father of their child. We should also expect that her husband-by-arrangement should do his utmost to prevent any sexual selection of other men.

I should add that though my suggestions are generalised, actual data-based studies of extra-marital selections in cooperatively-simple societies are obviously almost impossible to gather. Since the selections are meant to be unobserved and entirely secret, this makes any systematic study almost impossible. This appears to be particularly the case since most anthropologists are men, who had little ability to gain evidence of sexual selections unless they became publicly known. This is most obvious in the cases of studies in Australia where female anthropologists or the wife of a male anthropologist were able to gain more access to information of sexual selections, which women were more likely to know about and discuss. In these studies, sexual selections receive a far more detailed treatment than probably any reports solely by men. From my readings, it must be the case that many sexual selections occur without public knowledge.

The arranged marriage system is something to be manipulated. Most of the children will be born of arranged marriages, and it is up to a woman to attempt to manipulate the system to her best advantage. If she sexually selects a man outside of her own marriage, then she is attempting to produce a son who is capable of surviving and then manipulating the arranged marriage system better than if she reproduced with her own husband-by-arrangement.

If manipulating the system to her best advantage means considering an alternative father for her child, then she will want a man of similar appearance who will be sexually selected in such a manner that the sexual selection will not be discovered; if it is discovered, she will want a man who can protect or her from her husband-by-arrangement or remove her from him; and, she will want a man with traits that reveal an excellent immune system, such as skin without discolouration. Dominance and status are extremely important: an intelligent, physically gifted, facially attractive low status male, may be attractive, but be ignored by females either in fear of their husband-by-arrangement or looking for evidence of success in a competitive social environment.

Chapter 7

Male Sexual Selection of Females

Many of the aspects that are relevant to female sexual selection of males will obviously also be relevant to male sexual selection of females, so some of the previous chapter will be equally true, but from the point of view of the opposite sex.

The Scenario

Perhaps one of the most perplexing aspects of human beings is that females are deemed more attractive than males.

In the animal kingdom, Robert Trivers showed that if a female places the larger amount of the resources into her children then she will be choosy and the male attempts to attract her interest.[1] In humans, females are the ones with specific features to attract males, such as prominent breasts. Considering that women will breast-feed for more than two years and be pregnant for nine months, then the fact that women would need to attract a male would appear against evolutionary principles.

Current evolutionary theorists argue that this is because women need to attract a lifelong mate. However, I have already argued that in simpler societies, women engage in arranged marriages, and have almost no capacity to select their husband. The capacity of women to choose their male partner only began several hundred years ago in Western Europe.

If women cannot select their long-term partner, then their attractive physical features – such as enlarged breasts – can only exist to attract a short-term partner.

Attracting a short-term partner, as I have explained previously, is dangerous. In sexual selections, violence is the limiting factor, and the violence will emanate from the husband-by-arrangement and his supporters. If a woman's husband-by-arrangement is an extremely dangerous man, with a capacity for extensive violence, then any man that shows an interest in the woman is taking an enormous personal risk. Men are placed in a dilemma as to sexual selections. If a woman shows interest in him, then whether he responds may depend upon her husband-by-arrangement's capacity for violence.

For males the risks can be high, and I have already given examples where paramours were beaten or killed. Furthermore, the woman's husband-by-arrangement may be regularly copulating with her, so there is no guarantee to the paramour that his risk-taking will produce a child. The husband-by-arrangement may also attempt to kill or abort any child if he suspects it is not his own. William Buckley noted this in his three decades of living with Australian Aborigines in south-eastern Australia:

They are in general, very kind to their children, excepting the child is from any cause, believed to be illegitimate; and, again, when a woman has been promised to one man, and is afterwards given to another; in such case, her first-born is almost invariably killed at its birth.[2]

The risks are high and there may be no benefit. Most men may be better off waiting for a wife or a second wife than even considering being sexually selected by a woman. However, women cannot wait. A woman will only produce four or five children, assuming they reach adulthood, and she is limited purely by quality rather than quantity. If her children are dying in infancy, her reproductive situation may be dire.

If a woman is to attract a man for a sexual selection she needs to offer excellent evidence that the short-term risk is worthwhile for the man.

Invariably, in the natural world, it is the male that provides excellent evidence of his own genetic impressiveness; however, women are the only female primates who find themselves with a partner without selecting him, and are placed in a unique situation.

She cannot leave her husband-by-arrangement since the marriage exists in order to knit together two groups of cooperating people; therefore, almost all sexual selections are short-term. As I argued in the previous chapter, this is why love is a short-lived emotion, this is why sexual contact can be made solely with eye contact, and so on.

So what might induce a male to take the enormous risk of copulating with another man's wife-by-arrangement?

Primogeniture and Age

In the chapter concerning male alliances I noted that it is the eldest brother who is most likely to benefit from the exchange of women with other brother-alliances. If a man betroths several daughters to the sons of another man, then it is likely that it will be the eldest son who receives most of the women.

This means that the eldest son will be the son most likely to become polygynous. He will be the eldest male, with several extra years of experience and growth compared to any younger male siblings. Occasionally, it may be the case that a younger brother is more assertive; however, we should expect that the eldest brother will be the one to benefit most from any betrothals organised by his father or close male relatives.

This means that a woman's chance of producing a polygynous son diminishes as she gets older.

Furthermore, as a woman ages her responsibility increases. Once she has a child or two, she may deem that the needs of her children are more important than the risks involved in sexually selecting other men. If she selected another man and her husband-by-arrangement killed or maimed her in anger, then this would affect or end her capacity to care for her children. Thus the benefits of sexual selection will decrease as her responsibilities to living children increase. By contrast, if her children repeatedly die in infancy then her desire to take risks may increase.

From the woman's point of view, this means that the most important time to sexually select a man will be early in her sexual life, a time when she will possess the least responsibility, can take the most risk, and when she is most likely to give birth to a son who may one-day become polygynous.

Attractiveness should thus occur at the beginning of a woman's life. It should slowly disappear, particularly after the birth of a second child and certainly a third child. Becoming less attractive with time may actually be to her advantage, since it may not merely reduce her capacity to select, but reduce other men's advances and interest.

From the male point of view, any man considering an extra-marital selection can be killed in a duel. He should be more willing to engage in such a selection when the woman is most likely to be at her reproductive peak: when she is likely to produce an eldest son. As she ages, even though she can still have children, her future sons will be less likely to benefit from any marriage contracts and an extra-marital selection might only produce a monogamously married son or a girl. From a man's perspective, the benefits of an extra-marital selection decrease as women age. Men should see older women as less attractive, even though they are still capable of having children. The risk outweighs the benefit.

We tend to think that women are attractive to men when they are young because the woman is attempting to attract a lifelong mate. This may seem logical in societies with enforced monogamy and extensive market economies, but cooperatively-simple societies are polygynous and possess arranged marriages. Since the marriages are arranged, appearance should be unimportant. Thus, women's attractiveness is not aimed at husbands, but extra-marital selections. This fact is well represented in middle-eastern nations where women are clothed so that other men cannot tell their age or their physique.

Visual signals

In the previous chapter on female sexual selection of males I considered that since men and women in societies with arranged marriages were unable to come into intimate contact, then they would need to sexually select at a distance.

In her book *Naked Motherhood*, a rare attempt to give a reflection of the true implications of becoming a mother, Wendy LeBlanc gives details of the enormous changes to women's bodies after the birth of a first child:

> By the time we have weaned our babies, returned to normal bras or none at all and are ready to show off our prized breasts – alas, they are no more. The brief display of big-breasted womanhood is now sagging and shrinking but we're left with the protruding nipples.

> With exercise we can lose the extra weight, trim down the waist and tone up the muscles again but it is unlikely we will push it all back into the exact same shape of our girlhood days... For some women it takes only one child to irrevocably alter her shape. For months after childbirth many women are aghast when they see themselves in a full length mirror and, for the majority, a bikini never quite looks the same again.

Hips which once looked like those of a boy may fill to matronly proportions and never return to their former slender shape. Thighs may thicken and show stretch marks, too. Bottoms become rounder and maybe they drop just a titch. Some ankles and calves grow larger. Varicose veins developed during pregnancy do not necessarily vanish after the birth and in some cases grow worse as the mother is on her feet most of the day carrying a growing baby's weight as well.[3]

Even the face and hair is not immune to the effect of one child:

The changes to a woman's body throughout pregnancy are obvious. Essentially we expect to grow a fatter tummy and bigger breasts. What many women do not realise is that this additional roundness extends to her face, smoothing out many of the little lines. This is the 'glow' of pregnancy which affects the older mother more profoundly than the younger one. Because this bloom grows gradually it is quietly incorporated into her self-image and she does not realise it is a temporary phenomenon.

After pregnancy the lines which appear on some women's faces are truly astonishing. Fluid which has built up in the body during pregnancy rapidly returns to normal levels, withdrawing its plumping support from every cell – notably in the face.

As the exhaustion and the drudgery begin to take their toll most of us age noticeably. This is especially distressing in contrast to the fullness of face we experienced during pregnancy.[4]

During pregnancy a woman's hair grows more vigorously all over her body....

Within weeks of baby's birth our flowing, curling locks may begin to wilt. The shine fades away. The once-prized hair begins to fall out with every washing or brushing, producing drains and brushes filled with hunks of lank hair....

In time the vast majority of us cut our hair shorter, usually under the pretext that it is easier to look after in our busy schedules. Few of us admit to ourselves that this is an attempt to lift our sagging faces and rid ourselves of hair which no longer looks particularly attractive.[5]

Since most women in Western societies delay childbearing, often into their thirties, the impact of pregnancy and feeding are delayed. However, women in cooperatively-simple societies will experience their first pregnancy in their mid-to-late teens, and, by their early twenties, will probably be breast-feeding their second child. In *The Native Tribes of Central Australia*, Baldwin Spencer gives an unsentimental view of the impact that children have on women's bodies in cooperatively-simple societies:

Naturally, in the case of the women, everything depends upon their age, the younger ones, that is those between fourteen and perhaps twenty, have decidedly well-formed figures, and, from their habit of carrying on the head *pitchis* containing food and water, they carry themselves often with remarkable grace. As is usual, however, in the case of savage tribes the drudgery of food-collecting and child-bearing tells upon them at an early age, and between twenty and twenty-five they begin to lose their graceful carriage; the face wrinkles, the breasts hang pendulous, and, as a general rule, the whole body begins to shrivel up, until, at about the age of thirty, all traces of an earlier well-formed figure and graceful carriage are lost, and the woman develops into what can only be called an old and wrinkled hag.[6]

Napoleon Chagnon reported a similar situation, though more tactfully: 'The women tend to lose their shapes by the time they are thirty, because of the children they have borne and nursed for up to three years each and because of their years of hard work.'[7]

The above statement from Spencer is exactly what should occur, women's bodies should become less attractive to men as they age, because the benefits of a sexual selection reduce with age. By the age of 25 she is far less attractive, and by 30 she is completely unattractive to men – this is despite the fact that a 30 year old woman may still be able to produce two more children. Despite being fertile, women in their late twenties and certainly women in their thirties will be viewed as unattractive. However, by the age of thirty, a woman will have three children under her care, and probably have had at least one male child. Thus, the benefits of a sexual selection are not worth the risks, to either herself or a potential paramour.

What we should thus expect is that men will be attracted to traits that suggest a lack of children. Indeed, women's bodies should emphasise any trait that suggests a lack of children since it is to their own advantage to have the opportunity to gain a better father for her eldest son if she should want to do so.

Breasts are the obvious example. As soon as a woman has a child, her breasts hang from the milk of several years of breast-feeding. A man will be able to tell whether a woman has breast-fed a child, and, potentially, how many children she may have fed, from a distance of twenty metres, whether or not he has any knowledge of her history. Thus a man who visits a community for the first time will be able to make an accurate assessment of which women are likely to bear their first son, and which women are likely to produce the most children in the future, without hearing a single word of their history. Before pregnancy and breast-feeding a woman's breasts will be pert with small nipples; these will change after one child. It is interesting that in some states of the United States, and historically in men's magazines, women's nipples are deemed to be so capable of inciting men that, even though a woman could show her entire breast, the nipple needed to be covered with a pasty.

Large breasts allow a woman a greater sized signal. The fact that women can afford to layer their chests with excess fat reveals what a luxurious environment in which women evolved, in terms of nutrition. Women would never have evolved with breasts if nutrition was a significant issue.

Catherine and Ronald Berndt note about Aboriginal ideas of beauty:

On the Arnhem Land coast, and in the Western Desert too, to call a woman 'bony' or 'skinny' is to use a term of abuse; but this is a relative matter.[8]

To the assets of a healthy body, and clear skin, they add such extras as a good crop of hair and straight nose (parts of Arnhem Land coast), or a nose broad and strong across the nostrils (parts of the Western Desert), or moderately long legs and arms (Western Desert). And so on.

But whatever the area, and however frankly or otherwise this is expressed, as far as physical attractiveness is concerned the greatest stress is on youth. A girl who has just reached puberty is most desirable as a spouse – one with small, rounded breasts, not yet drooping after years of childbearing; so is a young man newly initiated. This quality of youth, like physical features which conform with local standards of what is good in a face or a body, is an advantage in love affairs: and it is undoubtedly an asset in marriage, insofar as it puts a person in a better bargaining position in any dispute – although this may be counterbalanced by the jealousy it provokes. In a situation where women are regarded as old at menopause, or even before, the period of youth for them is obviously all too short; and even though they may take various measures to enhance their appearance – rubbing their hair and bodies with fats and red ochre, decorating their hair with gum-nuts, wearing an elegant nose-stick on special occasions – these, of course, like the songs in their love-magic series, have only a limited effectiveness.[9]

This preference for young women with physical signals is reflected in some of the songs of the Australian Aborigines in Western Arnhem Land:

Girls take up their strings, to make 'cat's-cradles', girls from the Burara and Gunwinggu tribes...
They take their breast girdles, twisting the string. We saw their breasts, young girls of Goulburn Islands...
They are always there in that camp, at the wide expanse of water...
Making string figures, leaning back on the forked rails of the hut, the sea-eagle nest...
We saw their breasts, and their hands moving – Goulburn Island girls, clans from the Woolen River...
Their breasts in the cold west wind, as they flutter their eyes at the men:
Swaying their buttocks, speaking in Goulburn Island language.[10]

They saw the young girls twisting their strings, Goulburn Island men and men from the Woolen River:
Young girls of the western clans, twisting their breast girdles among the cabbage palm foliage...
Stealthily creeping, the men grasp the cabbage tree leaves to search for their sweethearts.
Stealthily moving, they bend down to hide with their lovers among the foliage...
With penis erect, those Goulburn Island men, from the young girls' swaying buttocks...
They are always there, at the wide expanse of water...[11]

They seize the young girls of the western tribes, with their swaying buttocks – those Goulburn Island men...
Young girls squealing in pain, from the long penis...

Girls of the western clans, desiring pleasure, pushed on to their backs among the
cabbage palm foliage...
Lying down, copulating – always there, moving their buttocks...
Men of Goulburn Islands, with long penes...
Seizing the beautiful young girls, of the western tribes...[12]

Ejaculating into their vaginas – young girls of the western tribes.
Ejaculating semen, into the young Burara girls...
Those Goulburn Island men, with their long penes;
Semen flowing from them into the young girls...[13]

What men find attractive in women are simply signs of a lack of childbearing. We
tend not to think of them in this manner, rather seeing them as signals of attractiveness.
Yet, the only actual signal that women present to men are their breasts. All other signals
simply reflect a lack of the impact of childbearing. For instance, a woman with a small waist
compared to her hips is not sending a signal, this is simply a natural reflection of her bone
structure. Though the fact that fat is not stored on the waist suggests the body does naturally
emphasise this contrast, a woman with a small waist, wide hips, stretchmarks and significantly
drooping breasts will be less attractive to men. Since childbirth appears to increase the width
of a woman's hips, and women in cooperatively-simple societies perform large amounts
of physical work (thus reducing fat storage) this lessens the importance of the waist and
hip contrast. No matter what a woman's actual physique, her breasts will give an accurate
reflection of whether she has had children.

Women do tend to show off a lack of childbearing in other ways, such as limited fat
storage on the waist and less body hair than men. The fact that women possess less body hair
than men tends to suggest how much more important sexual selection is for women than for
men.

The statement being made to men is: 'I am worth the risk of my husband's violence
against you.'

Some may argue that youth is an obvious factor in terms of attractiveness for all
animals, but this is not true. In chimpanzees, for instance, females become more attractive
as they become older. This is probably because brothers tend to team up and monopolise
matings with females, thus as females become older and more likely to produce a second
male infant, they become more attractive.[14] From a male chimpanzee's perspective, signs of
age and previous childbirth actually make a female more attractive: the opposite to human
beings. This is most obvious in Jane Goodall's studies in Gombe, where the old female Flo
was extremely attractive to males: 'Sex appeal, that strange mystery, is a phenomenon as
inexplicable and as obvious among chimpanzees as human beings. Old Flo, bulbous-nosed
and ragged-eared, incredibly ugly by human standards, undoubtedly had her fair share of it.'[15]
At one point, Flo has mated so many times that (note that a swelling indicates the female is
potentially ovulating and therefore receptive to all males in the community):

On the eighth day of her swelling Flo arrived in camp with a torn and bleeding bottom.

The injury must have just occurred. Within another couple of hours her swelling had gone. She looked somewhat tattered and exhausted by then and we were relieved for her sake that everything was over. At least we thought it was over, for normally a swelling only lasts about ten days. But five days later, to our utter astonishment Flo was fully pink again. She arrived, as before, with a large following of attendant males. This time her swelling lasted for three consecutive weeks, during which time the ardor of her suitors did not appear to abate in any way.[16]

Frans de Waal noted in *Chimpanzee Politics* 'the males are most interested in the mature females'.[17]

One could imaginatively alter aspects of human behaviour to create different outcomes. If primogeniture did not exist, and, in fact, it was the youngest son that was most likely to become polygynous, then we should expect women would become more attractive as they aged. Alternatively, if each son had a very high chance of being monogamously married and no son was polygynous, then women should remain attractive throughout their life. Yet this is not the case. Not merely are women less attractive when they are in their thirties in cooperatively-simple societies, but they are *unattractive*, despite the fact that they can have more children. The only reason a woman might be unattractive to a male, despite the fact that she is still ovulating, would be if there were a negative factor that overcame the positive factor of a potential child; the husband-by-arrangement's violence supplies that factor.

Permanent Selections

While a temporary selection may be highly desirable, if she can permanently select a man who can take her from her husband-by-arrangement, then this may be a superior outcome. The benefits are twofold: she may be able to select a man that is superior in intelligence, behaviour, physique or immune system to her husband-by-arrangement to be the father of all or most of her children; or, she may be able to move from one brother-alliance to a more powerful brother-alliance, in which her children will be more competitive due to the greater number of kin supporting them. (The power or status of the brother-alliance is comparable to the monetary wealth of a family, in modern terms.)

The latter of the two ideas is perhaps the most important. A woman married to a lone man may produce two or three sons, but if they are not connected with a larger group of males, their status may be small. In competition, two or three brothers may be unable to compete with a man, his brothers, his half-brothers, and first parallel- or cross-cousins.

While a man is locked within his brother-alliance, needing to remain with closely related males, a woman can move between brother-alliances. Women can move from a lesser group of men to a stronger group, and increase the ability of their sons to compete against other men. This is also true in more complex societies where families create a dowry, a monetary gift, which they use to attempt to marry a daughter to a family higher up in the hierarchy. By contrast, their sons are stuck in their family's position in the hierarchy. Enforced monogamy can be a boon for daughters and a tremendous limitation for males.

Permanent selections are obviously dangerous since the husband-by-arrangement is losing his wife. While a temporary selection might lead to a non-lethal duel, an attempt to permanently select may lead to a duel that approaches lethal proportions. Such selections are rare, but they do occur. The following is from Catherine and Ronald Berndt's summary of the Australian anthropological literature:

> Elopement seems to be more common that the ideal picture allows for, and in many instances, in actual practice, accepted as a more or less legitimate way of affecting a marriage, or at least as a possibility not to be condemned out of hand. This potential tolerance may have been reinforced by outside pressures, modifying the indigenous systems; but certainly it is latent in love-magic songs and rites...
>
> The runaway couple almost invariably make for another area, preferably one not easily accessible; but in most cases a revenge party will follow them, and perhaps kill or injure either or both. Among the Aranda a woman caught in the act of eloping is severely punished, if not actually killed.[18]

Meggitt found amongst the Walbiri of central Australia:

> Should the couple elope, the wife's kinsmen and the husband and his close agnates are obliged to follow them. Other close kin of both the spouses may join the pursuit, to ensure not only that the offending man receives his deserts but also that the wife is not seriously injured. The aggrieved husband and his agnates should thrash the enticer and spear him, although not fatally. The wife should also be beaten by her brothers and then speared lightly in the thigh, so that she will "sit down quietly" in her husband's camp. Some men said that the husband could spear his wife, but most disagreed with this. They asserted that his action would certainly lead to a fight with his wife's father.[19]

Note, it is the woman's own brothers who are expected to enforce the arranged marriage. Of the Murngin Warner observed:

> Five varieties of extra-legal sexual relations between men and women are recognized as possible, i.e., the tribe would not condemn them as unpardonable offenses against customary law: (1) relations when the legal spouse of either or when both spouses are carrying on an affair without the mate's full knowledge; (2) runaway matches when the man and woman go to a distant clan to live – usually a true love match, but not always; (3) a union when a man steals a woman from her husband and takes her to his own clan; (4) a union when warring clansmen kill off the husbands of the women and keep them for themselves; and (5) the union of a daughter whose father has given her to a relative without a legal claim to her in another tribe.
>
> Each of the above is considered illegal and condemned by all people, yet each is practiced to a considerable degree among all the tribes in this region, and each frequently leads to a permanent union that has full tribal recognition.[20]

Lovers' unions often develop into permanent ones. If there is a dispute about a girl between two young men, one of whom has been living with her, it is likely that, all else being equal, the lover would get the girl, for, say the old men, "They were sweethearts." If the woman should be given to the other man, the lovers frequently run away, and the tribe later recognizes the union.[21]

However, Warner also observes that the eloping couple may be followed by the husband and wife-by-arrangement's families to return the wife to her former husband. Warner notes the difference between the abduction of a woman and an elopement, in which only in the former may a man expect support from his close kin, since it is invariably part of warfare and rivalry, while the latter is an individual decision:

A runaway marriage differs from a union caused by a man's stealing a woman in that the woman gives her consent; such an arrangement has the consent of neither her people nor his; whereas in the case of theft, a man's own people frequently support him, particularly if there has been hard feeling between the two clans and a conscious rivalry for the women of certain groups. The runaway union usually ends in his and her people attempting to send her back to her true dué [husband], because it would lead to armed trouble between the man's people and the members of the right dué's clan, and also for the woman's people if they did not condemn the act by helping the true dué regain his wife. However, no man's clan would desert him in such a situation to the point of allowing him to be killed or attacked by the injured man's group. Ordinarily, such a situation ends in a "growling" match, in which the two sides stand armed behind the two opponents, the lover allowing the injured husband to have the better of the cursing. Sometimes such a runaway marriage leads to a permanent union.

After a couple have been lovers for some time they may be so fond of each other as to risk everything and leave their own kin and go to a distant clan. Her people, the husband's gawel and galle, immediately help the injured spouse to regain his wife. The people of their groups generally frown upon the lover's attempt. Further, her people consider it an affront to them by the people of her lover's clan, and it is necessary for them to protest that they had nothing to do with the matter and show good faith by helping to get her back. A woman or a man can take the initiative in a runaway marriage, but most of the cases recorded were due to the influence of the woman.[22]
[Gawel = mother's brother; and galle = mother's brother's son.]

In the fourth chapter I also mentioned a situation where a Yạnomamö man attempted to take a woman from her husband, only to be killed by the headman after inflicting a serious wound on the husband. The risks are thus high. The Walbiri of Australia told Meggitt of elopements: 'The people could not recall any instance of a married man's eloping with another woman; and it was their opinion that only an insane man would thus jeopardize his own marriage.'[23]

It appears that elopements are more common in Australian Aboriginal society than others, and there are few similar statements to be found amongst the Yạnomamö. Either way, in the Australian Aboriginal literature, though the implications of an elopement were known, reported elopements were rare. This is not surprising since a failed elopement may lead to serious injury or death.

Yet, if a woman's only method of permanently changing her husband is to elope or to select a man to defend her against her husband-by-arrangement and move to another community, then the risk may be worthwhile under certain circumstances.

It is often difficult to discover whether the attractiveness of the woman plays a significant role or not. Though there are numerous examples from literature, it is, however, difficult to find exact examples in the anthropological literature where a woman's physical appearance, as against her youth, played a role in the lengths to which a man will go to be either temporarily or permanently selected by a woman. As to one of the many fights that William Buckley observed over women during the three decades he lived with a community of Australian Aborigines, he commented upon one:

> They then set up a dance, the men remaining as spectators, encouraging them with cheers, and all sorts of noises. This diversion finished, as usual, with a regular fight, beating each other about with their clubs most unmercifully. I afterwards understood this quarrel to be occasioned by a woman having been forcibly carried away by another tribe: one of those with us. She was living with the man who had taken her, and, as the man and woman were then both present, they wanted to chastise her for not returning to the tribe to which she belonged. In the skirmish this woman was felled by a heavy blow; seeing this, the men began to prepare for a fight also; one man threw a boomerang amongst the women, when they all ran away. The native who had stolen the girl, then came forward by himself and told them to take revenge on him, and began to sing and jump and dance, upon which her father went up to him. They both remained quiet for some time, when the men called out to the father, telling him to let him have her, as the man she had been promised to was not worthy of her. Eventually the girl returned to her father. She appeared to be about fifteen years of age, and certainly was no beauty to fight about.[24]

Young women tend to be fairly scarce, so the attractiveness of a woman may not necessarily be of great importance. Considering that many men will wait until their mid-twenties before they can marry, and, thereby, engage in coitus for the first time without taking extreme risks, it is understandable that the years of frustration might lead a man to take extreme risks in either copulating with another man's wife-by-arrangement or by attempting to forcibly remove her or elope with her – whatever her appearance. This is particularly true if warfare is endemic and men's lives are short.

In any event, force is thus a legitimate basis on which to allow a woman to change husbands, but only if the man she has sexually selected is more powerful than the first; Ronald Berndt:

> Apart from so-called 'wrong' marriages (not necessarily with tabooed relatives or within the same moiety), which did occasionally take place, there were several

acknowledged forms of sexual association which, while not officially desirable, were recognized as ways in which a wife could be obtained. Nevertheless, such cases caused a breach of the peace and could bring about a protracted feud. This often occurred with an elopement ... or an abduction. In the latter case, a man might steal a woman by force from her legitimate husband, spearing him in the course of fighting and taking over his wife or wives. Such occurrences are said to have been relatively frequent in the past in north-eastern Arnhem Land. In 1946-47 they took place only occasionally.[25]

There is not enough evidence to suggest that women with small waists, pert and large breasts with small nipples, attractive hair, a boyish face, and excellent symmetry are more likely to be chosen for permanent selections than women who are merely young. It is almost impossible to find a single instance of men fighting over a woman because she was deemed to be attractive. This may reflect the diversity in women's appearances.

Being Permanently Selected

A second form of sexual selection is when a man forces a woman to be his partner, against the woman's wishes. Though sexual selection is usually deemed to be something where both parties desire the outcome, this is not the limit of Darwin's argument. There is the issue as to whether women have been selected by forced sexual selections.

The Old Testament, in Book 5, Deuteronomy, supports forced selection:

DEU:021:010 When thou goest forth to war against thine enemies, and the LORD thy God hath delivered them into thine hands, and thou hast taken them captive,

DEU:021:011 And seest among the captives a beautiful woman, and hast a desire unto her, that thou wouldest have her to thy wife;

DEU:021:013 And she shall put the raiment of her captivity from off her, and shall remain in thine house, and bewail her father and her mother a full month: and after that thou shalt go in unto her, and be her husband, and she shall be thy wife.

The emphasis being on a 'beautiful woman' leads to the question as to whether a woman's physical appearance affected sexually selected traits in women. For instance, if a group of women were taken captive and if one woman was deemed to be more attractive than the others, would she be more likely to become one of the leaders' wives? This might affect the success of her children if she was married into a successful lineage. By contrast, less attractive women may have been forced to marry into less successful lineages.

Helena Valero noted during one of the large abductions of women she experienced, that the desirability of the abducted women varied. At the time, Valero was still only a young girl and pre-pubescent, and this may have played a role in her desirability, since she would need several years of care before going through puberty:

Sometimes those who had taken the women did not want theirs and said: 'I took this woman, but I don't want her any longer; who wants her?' Some others replied: 'Give her to me.' Then the one who didn't want her any more gave her to the other and so he passed in front with those who were leading the way and had no women. The young man who had taken me, brother of that girl who was always with me, said to her: 'I want to give this woman to another man.' 'No,' answered his sister, 'she'll stay with us, I'll take care of her.'[26]

Xoxotami took me to her mother in her tapiri. 'Who's this?' asked the mother. 'Where does she come from?' Xoxotami replied: 'She is a Napagnuma. She's that white woman whom our men captured together with the Kohoroshiwetari, the one whom the Kohoroshiwetari did not want to give up. This is why our men have become their enemies and have gone to fight them. Xoxotami continued: 'She's pretty thin; no one wanted her. I have helped her and brought her here; now she'll stay with us.' The mother looked at me and answered: 'There are lots of young men here who have no women. This white woman will stay with me; then nobody shall come to take her away.' Then she said in a loud voice: 'Let everybody listen. There are so many men without women. Soon this one will become a woman and then nobody must take her away from me.'[27]

The evidence of forced sexual selections being based on specific physical features or behaviours is scarce since only a few warring peoples in cooperatively-simple societies have ever been studied. Yet, there can be found definite examples where age made a significant difference in the way women were treated. Helena Valero observed this on two occasions:

The *tushaua* Fusiwe was in his hammock. Then the sister of the dead *tushaua* Rohariwe came up and said amid tears: 'I am old, I am no good for living with a man; I will stay here three days, I will wash the blood, I will burn my brother's body, and then I will go away. If I had been young you could have kept me, but I am old and I am useless; let me go.' She spoke and wept. Fusiwe replied: 'Stay here, go away, do as you please.' The wife of the *tushaua* Rohariwe, on the other hand, tried to escape with her little girl. A man ran after her, took her by one arm and brought her back to the *shapuno*: she was still a young woman.[28]

Two men continued to hold me by the arms. Others also took Xoxotami. The Shamatari warriors were looking for the women, who were hiding in the corners of the *shapuno*. There were two enclosures of palm leaves: inside were two young girls 'of consequence'; that is, in puberty. An old Hekurawetari woman said 'Don't touch my grand-daughter! It's only three days that she's been of consequence. If you take her, *Gnaru* (Thunder) will kill you.' The girl began to cry, but the old woman continued: 'Don't cry; it's bad to cry when you are like this. If you cry, you die.' They opened the shelters and dragged the girls out.[29]

Only on one occasion was physical attractiveness a factor:

They also took one of Hekuraw's daughters. She was beautiful, pleasant and nice; the nephew of the Shamatari *tushaua* took her. Then the *tushaua* saw her and asked: 'Who has taken this woman?'

    'It was your nephew.'

    'This woman is for me.'[30]

Thus the woman is moved from the nephew to the *tushaua*, the headman – an important move up the hierarchy in the community. Valero's text is an extremely rare first-hand account of what exactly happens to women that are abducted or captured in warfare. While such evidence is very limited, it does suggest that physical attractiveness may play a role, and could increase the likelihood of a woman marrying a dominant man in the community hierarchy. This is an important difference for the woman's children, since the dominant men are likely to be polygynous, and they may gain many half-brothers and half-sisters to act as supporters or to exchange in marriage, a tremendous advantage in a community where cooperation and success are heavily dependent on kinship. In being selected by men, even against the woman's will, the attractiveness of the woman can aid her in moving up the hierarchy.

    Previously I have documented that Napoleon Chagnon showed that men who had killed other men (unokai) had more wives than men who had not killed and that in villages with higher percentages of unokai men there were higher percentages of abducted women. He also showed that lowland villages possessed higher percentages of polygynously married men than highland villages, higher percentages of married men, and larger village sizes than the highland villages. These details suggest that it is likely that there are larger communities in territories where there are higher nutrients, and it is these communities that will contain more married men, more polygynous men, more men that have killed other men, and more abducted women. Therefore, abducted women would be more likely to be married to high status men in high status communities.

    A similar situation of violence between men over a woman arises where a woman has been betrothed to more than one man. Catherine and Ronald Berndt observed:

Betrothal arrangements did not always run smoothly. A girl's father might make betrothal commitments with several different families, so that several men might claim her at or just before puberty. Or an irregular union on the part of a girl's parents might mean that she had more than one set of eligible betrothal partners, and this might lead to conflict and feuding which could delay her marriage. Under these circumstances, even if she did go to one man, others might continue to feel they had claims to her and go so far as to press these. Trouble was bound to ensue if a girl's parents behaved avariciously, trying to betroth one daughter to several different men. In such cases, the men to whom the girl had been promised had a right to claim her – especially since they had probably ratified the contract by making presents to her relatives. Or parents anxious to improve their economic position might give their marriageable daughter to a certain man but, if he failed to live up to their expectations, take her away and give her to someone else who would, they hoped, be more generous. But this juggling with a potential husband and wife's affections could be dangerous, particularly if a girl was pregnant or had borne a child from the first man. It was usually attempted only when she was young and had not conceived: but even then it could lead to fighting. Conventional betrothal arrangements might be upset, too, should the girl have a sweetheart to whom she was deeply attached, or if the betrothed husband already had one wife who strongly objected to his taking another. However, consistent jealousy on the part of a first wife was rare. [31]

The seriousness of a multiple betrothal can be found in the following example from William Buckley's time with the Australian Aborigines in Victoria:

> Having erected their bark huts near ours, they remained peaceable enough for several days, hunting and enjoying themselves at length, the Putnaroos suddenly surrounded our people, and without any previous altercation speared a young man about twenty years of age. The cause stated to be was that the murdered man had been promised a girl who his assailant wanted for himself.
>
> Poor fellow, when he was speared, he ran only a very few paces, and then dropped down dead. Our tribe expostulated with the others against this assault, but were answered by the threat, that if they said much about it they would serve us in a similar manner; so we, being by far the weaker party, were obliged to appear to be satisfied.[32]

The response by those afflicted reflects the typical method of cooperation – competitive cooperation:

> When they had consoled themselves a little, the father summoned all the tribe and other friends he could muster; they came in considerable force, and having pipe-clayed and ochred themselves all over, they set off, prepared for battle. This however was evaded, as the Putnaroo invaders had taken to their heels, on seeing the great numbers to which they were opposed.[33]

Catherine and Ronald Berndt also noted the possibility of extreme consequences:

> Then there is marriage by capture. Spencer and Gillen (1938:103-4, 554-5) speak of this as one possible procedure, although among the Aranda a very uncommon one. Occasionally a man seizes a woman from a member of his own group, in which case he may or may not be able to persuade some of his kinsfolk to help him. In north-eastern Arnhem Land this may happen when a man feels that a particular woman he wants has, unfairly, gone to someone else. Or he may simply be attracted to her, and abduct her, killing her husband in the process – without paying any compensation for her unless her father or brothers are strong enough to enforce it; and the same holds good for the relatives of the man he has killed.[34]

There is thus the question of whether men will be more inclined to take greater risks to enforce a betrothal if the woman involved is deemed to be attractive. Since women are scarce, it may well be the case that no matter what her appearance or behaviours she will be fought over.

In summary on the issue of the sexual selection of women through abductions or enforced marriages, age appears to be the major issue. A young woman might be fought over, while an older woman might be ignored. Young women, no matter what their shape or appearance, will be presenting numerous signals that men find attractive. If there are only a few women in the community presenting such signals, then such women will be highly desirable. If she were betrothed to more than one man then they might fight over her. If she were captured by another community, she might be more likely to find herself married to a high status man. This would be the case even if she were not attractive.

This is reflected in the conclusions of Napoleon Chagnon:

> These data raise a broader and more intriguing question about sexual selection theory and the extent to which marriage/reproduction in the environment of evolutionary adaptedness (EEA) and in contemporary tribal societies with prescriptive rules of marriage are the result of female choice and female preferences for older men with higher amounts of "material" resources... or whether differential male reproductive success is measurably affected by male preferences for high symmetrical face, clean complexion, or a sexually appealing waist-to-hip ratio. My data show that the vast majority of Yąnomamö babies result from arranged unions in which men produce offspring with women they classify as *suaböya*, and much of the male-male competition has to do with getting as many females into this category as possible, regardless of facial symmetry, complexion, or waist-to-hip ratios of the little girls they are legitimately able to marry or those they sometimes classify to achieve this end.[35]
>
> [Many theorists use the term 'EEA' to describe the environments in which humans evolved.]

All young women appear to be highly desirable. However, Valero's testimony suggests that attractiveness may play a role in the status of the man the captured woman is forced to marry, with beautiful women forced to marry higher status men – which is to the woman's children's advantage. The importance of attractiveness in modern societies appears to be accentuated by the enforcement of monogamy; in societies with polygyny, then physical attractiveness appears to be less important. The lack of uniformity of female appearance reflects the inherent desirability of all women due to the children they can produce, which are directly linked to competitiveness (sons) and cooperation through exchange (daughters).

One might conclude that in terms of male sexual selection of females, that the same aspects that make females attractive to males for short-term selections will attract males for long-term selections, whether the female has any input into the selection or not. Fat levels, such as large breasts that have not been affected by breast-feeding, may well be the only physical signal women actually present. Such signals may play a part in women not merely being able to attract a male for an elopement, but her marriage if abducted. Other aspects of the female form seem more based on a lack of evidence of child-bearing than an actual signal; thus the hour-glass figure is more the reflection that the woman has yet to bear a child, rather than a specific signal in which her body has evolved to produce a particular effect, though this may be countered in that fat deposits are not placed on the abdomen.

Emotions

Women's emotions will be affected by two factors: her husband-by-arrangement and her need to sexually select.

For women, sexual selection is more important than for men since a man can be polygynous while a woman can only produce her own children and, at best, a polygynous son. Quality is the important issue. Since women will be violently attacked by their husband-by-arrangement for sexually selecting, this is a serious negative in sexual selection.

Unlike her potential paramour a wife-by-arrangement has an intrinsic value to her husband. He would be highly unlikely to kill her, particularly if she is his only wife. If he killed her, he might have to wait years to marry again. By contrast, her paramour may have no intrinsic value to husband-by-arrangement, and serious violence may be exercised.

This means that men will need to be more careful with their emotions than women. This is particularly true since a woman is limited in only being able to have her own children. She is ovulating immediately, while a man can wait for the first of what might be several marriages. We should expect that the positive emotions associated with sexual selections are more powerful in women since women have a more immediate need to create a sexual selection. In other words, love should be a more powerful emotion in women; while, for men, love might be something they can ignore.

One only needs to look at entertainment to see that this is true. Women's romance fiction is a large seller, while many men's novels are largely devoid of romantic attachment. In fact, love and romance seem to be the most constant and important aspect of women's entertainment.

For women, love should be an overpowering emotion, something well beyond the fear of violence. Only if a woman's husband is so violent that she is sure he will seriously harm or even kill her should the fear of violence overcome the potential for a sexual selection. For men, the issue of violence should be more overpowering than the emotion. If her husband-by-arrangement is a known killer, then he would do well to simply ignore a potential sexual selection with the man's wife.

By contrast, if women evolved with Western-style marriages, then the opposite should be true. A woman's emotional range should be lower than a man's. She should be slow to show interest, picky, and relatively unemotional. By contrast, he should be excitable, quick to show interest and fall in love.

Women are also more endowed with the capacity to present physical manifestations of their emotions. Blushing or tears allow women to present to a man, at a distance, her true interest. Thus, even without speaking to him, a woman can blush in order to show her embarrassment at her own feelings, or cry, to present her negative emotions at his presence (sadness). Men are less capable of presenting such emotions, and facial hair can even undermine the capacity to send a strong signal.

Again, if women were choosing long-term partners in a Western manner, then the need to send long-distance signals of their emotions makes no sense – after all, she should be able to simply walk up and talk to all men. If her sexual selection is short-term, extra-marital and clandestine, then the need for physical manifestations of her feelings make sense.

Concluding Remarks

There is the interesting question of which sex is actually the more attractive. While one might conclude that women's breasts suggest that women are indeed the more attractive, this is complicated by the fact that women sexually select men, and then it is up to the man as to whether he will accept the sexual selection. In other words, women choose men, but women must attract men. For any other animal such a situation would be absurd, but the existence of the husband-by-arrangement creates a need for women to choose and attract. She must prove to the man that she is worth the risk of her husband-by-arrangement's violence.

To that extent, since women are selecting, one might conclude that it is actually males that are the more attractive of the two sexes. Thus, if there is a very attractive male in the community, all the females that are not closely related can sexually select the same male, as with any other species. However, she is restricted by the husband-by-arrangement and thus she must prove to the selected male that the risk is worth taking. Thus women's breasts and relative hairlessness, compared to males, is misleading. Men are thus still the more attractive of the two sexes; it is simply that there is a third factor.

Alternatively, one could equally argue that the husband-by-arrangement is such a significant force that females truly are the more attractive sex and that males are the choosier sex. In other words, females might attempt to choose many males, but only several males might – due to the husband-by-arrangement's potential violence – be able to accept her sexual selection. Of course, if the male chooses and the female doesn't then this is rape.

There would appear to be no simple answer to this question. In modern society, one could observe female groupies of rock bands to see that it is the male that is more attractive. Rock bands could thus be viewed as a type of lek, where females gather to select individual males, and that the same males are often repeatedly selected. Or one could equally look to women who carefully select between several males in something closer to Jane Austen's *Pride and Prejudice*.

Ultimately, it is probably best to simply observe these aspects as minor issues within the more complex web of human cooperation. Female attractiveness is the lesser issue compared to the potential for violence of her husband-by-arrangement; if her husband is an extremely dangerous man then she will, no matter what her appearance, be unattractive to men (the femme fatale). If her husband-by-arrangement is less dangerous, then she may be quite attractive whatever her appearance. By contrast, male attractiveness is dependent not merely on his appearance, but his cooperative status. If a male has many supporters who will come to his aid in a dispute, then that male may be more capable of seeking out women who sexually select him than males who have few or no supporters. The physical appearance of the male is not vital for the female either: people are restricted by the larger circumstances.

In any event, humans thus do not fall into the usual simplistic idea of the choosy female and the impetuous male. Both sexes can be choosy and attractive depending on the scenario.

Males are faced with the fact that all sexually active females will be married, or betrothed, to a male. This male will attack them if they attempt to copulate with the female. Thus males are attracted to features that reflect the possibility that the female will produce a first son. They are attracted to signals of a lack of childbearing. They have also selected in females many of the same traits that exist in themselves to aid sexual selection at a distance, while adding a few others: lack of body hair, odour and methods of transmitting this scent through underarm and pubic hair, eye contact that also present emotions with tears and redness, blushing, and fat deposits on the chest that are only firm when unaffected by years of breast-feeding. Without touching a woman or speaking to her, he can observe how many children she has had, her immune system, her smell, and communicate with her emotionally through tears and blushing. It is difficult to imagine how these features could evolve without arranged marriages.

Chapter 8

Sexuality

Female sexuality

From the beginning, it has simply been assumed that people's sexuality should reflect a simple model where man and woman choose each other, and experience entirely fulfilling sexual and reproductive lives. Men and women should both experience positive emotions during intercourse, a woman's sexual desire should be timed with her ovulation, her menstrual cycles should be regular, and that sexuality should perfectly reflect positive emotions and increased likelihood of impregnation. The couple should meet, fall in love, both experience pleasure during sex, she should quickly fall pregnant, and this pattern should continue for their entire lives.

However, none of this has been found to be the case. For instance, women's sexual desire has not been found to equate with ovulation:

> Schreiner-Engel et al reviewed 32 studies of the distribution of sexual activity through the menstrual cycle. Sexual activity was found to be increased during the premenstrual phase in 17 studies, in the postmenstrual phase in 18, during menstruation in 4, and around the time of ovulation in 8.[1]

The inconsistency of these findings could also reflect the inconsistency of methods used to discover the stages of the participating women's menstrual cycles.[2] Women's timing of ovulation is so well hidden that obtaining a correlation between desire and ovulation is incredibly difficult.

Then there is the issue of women's pleasure. Did women orgasm almost every time they had sexual intercourse as men do?

There has been found no correlation between sexual intercourse and women experiencing orgasm. Alfred Kinsey and Shere Hite's studies reported that between 40 and 60 percent of women self-report orgasm with coitus. Sexual therapist Theresa Crenshaw commented on the numbers of women experiencing orgasm:

> Some surveys today indicate that fewer and fewer women are actually orgasmic with intercourse. Is this true? As time goes on, are women becoming less sexually responsive? Or, are they perhaps becoming more open, accurate, and honest in talking about themselves?
>
> There is no way to be certain, but given that current statistics available in research and impressions from my clinical practice, it appears that estimates of the percentage of women who have orgasms with intercourse are lower than formerly

believed, and estimates of the number of women being orgasmic through masturbation and other methods involving clitoral stimulation are probably not high enough.[3]

This problem of women failing to experience orgasm during sexual intercourse was viewed as an opportunity for pharmaceutical intervention. After the success of the male impotency drug viagra, the pharmaceutical company Pfizer spent eight years attempting to create a drug to overcome women's incapacity to experience orgasm during sex:

> Dr. Joe Feczko, president of worldwide development with the company, said in the statement: "FSAD [female sexual arousal disorder] is an emerging area of research and is far more complex than male erectile dysfunction. Diagnosing FSAD involves assessing physical, emotional, and relationship factors, and these complex and interdependent factors make measuring a medicine's effect very difficult."

> Dr. John Bancroft, director of the Kinsey Institute at Indiana University, Bloomington, said: "I am not surprised by the negative results with Viagra in women." ... "The recent history of the study of female sexual dysfunction is a classic example of starting with some preconceived, and non-evidence based diagnostic categorisation for women's sexual dysfunctions, based on the male model, and then requiring further research to be based on that structure. Increasingly it is becoming evident that women's sexual problems are not usefully conceptualised in that way."

> In a recent survey of women, his group found that indicators of physiological response during sexual activity, including genital response, lubrication, and orgasm, were not predictive of whether a woman was distressed about her sexual life, whereas more emotional and subjective aspects of the sexual relationship, plus her mental and physical health, were predictive (Archives of Sexual Behavior 2003;32:193-208).[4]

Another of the Pfizer researchers commented:

> Dr. Mitra Boolel, the leader of the company's sex research team, says: "The brain is the crucial sexual organ in a woman." While a man's arousal almost always led to a desire for sex, there was no such obvious corresponding factor with women, he said.
> "There's a disconnect in many women between genital changes and mental changes," he explained. "This disconnect does not exist in men. Men consistently get erections in the presence of naked women and want to have sex. With women, things depend on a myriad of factors."[5]

One might thus suppose that women's sexuality is fundamentally different to men's, but this doesn't appear to be the case. For instance, Alfred Kinsey, after his study of 8,000 women in 1953, concluded:

> Some 45 per cent of all those females in the sample who had ever masturbated reported that they usually reached orgasm in three minutes or less, and another 25 per cent in

something between four or five minutes. The median for the whole group was a few seconds under four minutes. Many of those who took longer to reach orgasm did so deliberately in order to prolong the pleasure of the activity and not because they were incapable of responding more quickly.

These data on the female's speed in reaching orgasm provide important information on her basic sexual capacities. There is widespread opinion that the female is slower than the male in her sexual responses, but the masturbatory data do not support that opinion. The average male may take something between two and three minutes to reach orgasm unless he deliberately prolongs his activity, and a calculation of the median time required would probably show that he responds not more than some seconds faster than the female. It is true that the average female responds more slowly than the average male in coitus, but this seems to be due to the ineffectiveness of the usual coital techniques.[6]

Kinsey's research would appear to suggest that women and men, at a basic level, appear to possess similar potentials for sexual pleasure – at least, when they are giving pleasure to themselves.

The simple model of human sexuality is not reflected in the findings. Instead we discover that women's sexuality is extremely complicated. Desire doesn't coincide with ovulation. Menstrual cycles are irregular and ovulation hidden. Though women are as capable of bringing themselves to orgasm as men are, perhaps as many as fifty percent of women, or more, don't experience orgasm with intercourse.

Another issue is frequency of copulation. Sexual reproduction should be simple, and in some animals it is. Female horses, for instance, give an exact sign of when they are ovulating. Stud farms can thus exist, where mares that are ovulating can be taken to a stud farm, have a single act of intercourse with a male, and become pregnant. Female gorillas also give a signal of when they are ovulating in the form of a genital swelling. All that is required for a female to become pregnant is a few acts of sexual intercourse.

By contrast, humans have many acts of sexual intercourse. In Western societies, this is amplified by contraception. Again, if we initiate the simple model of men and women evolving in a Western-style of marriage, this doesn't appear to make sense. Surely, we should be like horses and gorillas. The couple should meet, fall in love, there should be a few acts of sexual intercourse, and she should become pregnant.

Some have suggested that the point of repeated intercourse is for the couple to pair-bond, but other animals don't need repeated sexual intercourse to be able to bond. A male gorilla will risk his life to protect his female partners and their infants, despite the fact that he will only have intercourse with the females on a few occasions.

There also appears to be no correlation between how well a man and a woman are bonded and how much they copulate. This appears most evidently in societies with arranged marriages and polygyny. A good example of this concerns Napoleon Chagnon's observation of a headman's two wives.

When I first met him he had just two wives, Bahimi and Koamashima. Bahimi had two living children then – others had died. She was the older and enduring wife, as much a friend to him as a mate. Their relationship was as close to what people in our culture think of as love as I have ever seen among the Yąnomamö. The second wife, Koamashima, was a girl of about twenty years with a new baby boy, her first child. There was speculation that Kaobawä was planning to give Koamashima to Shararaiwä, one of his younger brothers, who had no wife; he occasionally allowed Shararaiwä to have sex with Koamashima, but only if he asked in advance.[7]

Kaobawä thus was deeply attached to his elder wife, and thinking of giving the second to a younger brother; yet who did he have coitus with the most? Koamashima.

She is one of Kaobawä's younger wives, and because of her youth she enjoys his favors more regularly.[8]

Meggitt noted the same preference amongst the Walbiri:

A man is usually most attentive to his latest wife and sleeps with her to the exclusion of the other women for the first few months. After a reasonable time, however, this partiality should cease, and the sexually active co-wives should also share his attentions and his bed.[9]

Sexual frequency appears to be based on the youth of the woman not on the bonding. (The reasons for the importance of youth have been dealt with previously, being that the first son of each woman is more likely to be polygynous than later sons.)

We are thus left with a conundrum in that human sexual behaviour and Western-style marriage patterns do not appear to fit. Alternatively, if we instead assume that men and women evolved with arranged marriages, then since the woman didn't choose her partner we should expect that she would hide her ovulation from him, otherwise she has no opportunity to choose. Furthermore, her ovulation would become even more obscure if her menstrual cycles were irregular. We should expect that she would not necessarily experience pleasure with intercourse, since she did not select her partner and she may not want him to be the father of her children. We should expect that men would desire to have frequent sexual intercourse with their partners to increase the likelihood of pregnancy. We should expect that a man's desire to copulate with his partner should be dependent on her attractiveness to other men rather than on how well bonded he is with his partner. Indeed, if he is completely trustful of her, they may copulate infrequently. We should also expect that women would abide by their partner's desires, otherwise she would be undermining the contract for cooperation that unites their families.

A further point of ignorance is the assumption that men must be the more sexual of the two sexes. It is assumed that since men can impregnate many women, men must want to have sexual intercourse with many women. Men are consistently presented as always thinking about sex and that it is men that are the more sexually aroused.

Yet there is one tremendous difference in the sexuality of men and women: women can experience multiple-orgasms and men cannot. Alfred Kinsey noted that while most men are similar in their sexuality, women are wildly divergent. In his 1953 study he found:

> Because there is such wide variation in the sexual responsiveness and frequencies of overt activity among females, many females are incapable of understanding other females. There are fewer males who are incapable of understanding other males. Even the sexually least responsive males can comprehend something of the meaning of the frequent and continuous arousal which some other males experience. But the female who goes through life or for any long period of years with little or no experience in orgasm, finds it very difficult to comprehend the female who is capable of several orgasms every time she has sexual contact, and who may, on occasion, have a score or more orgasms in an hour. To the third or more of the females who have rarely been aroused by psychologic stimuli, it may seem fantastic to believe that there are females who come to orgasm as the result of sexual fantasy, without any physical stimulation of their genitalia or of any other part of their body.[10]

Perhaps the most remarkable aspects of female sexuality is the capacity for orgasm without physical contact – merely through thought alone ('self-induced imagery').[11]

Multiple-orgasm is perhaps the most obvious difference between the extent of male and female sexual pleasure. I doubt that there is any man who has experienced a score of orgasms in an hour or an orgasm without physical stimulation. Theresa Crenshaw defines multiple orgasms in women in the following manner:

> Women have the potential to have more than one orgasm. They can have several orgasms in a row, without stopping. Multiple orgasms can be measured as a cluster of uterine contractions (three to eight) interrupted briefly, and repeated at intervals. Some women define multiple orgasms as more than one orgasm occurring during the same sexual encounter, even if the orgasms occur at one-hour intervals. Most, however, consider these individual orgasms and define multiple orgasms as those occurring in rapid succession (within seconds or two to three minutes).
>
> After the first few orgasms, the intensity of each response generally diminishes. In between orgasms, a woman's clitoris might become momentarily hypersensitive, before being able to continue stimulation. The capacity to be multiorgasmic is delightful for most women, with peaks of orgasmic tension occurring in rapid succession until stimulation is stopped, or until satisfaction or exhaustion terminates the experience.[12]

Using the Western model, if human sexuality reflected man and woman choosing each other, falling in love, and having children, why would women possess a sexuality where they can possess a multiple orgasmic potential well beyond any man's? Women are meant to be carefully selecting a lifetime partner, one that can invest in her children. One might expect that women's sexuality should be fairly low. That women should be very slow to possess any interest in a male. That women's sexuality should be dependent on a male's intent on investment.

Instead we find that women's sexual potential is off the scale, compared to men's. Linked to love, an intense emotion often compared to insanity, one must wonder why women, who are meant to be selecting men for lifetime relationships, would be capable of experiencing insane levels of affection and for this to be supported by orgasmic levels many times greater than a man's. Furthermore, women hide their ovulation from their partner and possess irregular menstrual cycles.

By contrast, if we assume that women evolved with arranged marriages, love and multiple orgasm make perfect sense. Women in arranged marriages need to take large risks in order to copulate with a man other than their husband-by-arrangement. Helena Valero, in her time with the Yąnomamö, observed:

> There was another married woman who loved a man younger than her husband. She waited for her husband to go hunting a long way off and sent to tell the youth to meet her in the *roça*. When the husband came back he got to know of this and gave her so many beatings, and burnt her with live firebrands on the chest; her skin remained attached to the brands. But the woman who had this weakness said, 'He has beaten me; now I'll behave even worse.'[13]

What might motivate a woman who has been so badly beaten and burned to continue to engage in an illicit relationship? In cooperatively-simple societies, such treatment of wives-by-arrangement that have affairs is common, and sometimes the woman can be killed, as has been documented in previous chapters. With such consequences at stake, what would it take for a woman, who never selected her husband-by-arrangement, to desire to have sexual intercourse with a man other than her husband? She would need a positive experience that was much stronger than the negative experience. Since a woman in a society with arranged marriages did not choose her husband-by-arrangement, if she is going to choose a man to be the father of her children, then she needs to take large risks. She needs an extreme positive inducement to take those risks. Multiple-orgasm is such an inducement. Rather like love, a short-term intense emotion, multiple-orgasm is a positive inducement to act in an extreme manner. As Shakespeare put it in *Romeo and Juliet*:

> Being held a foe, he may not have access
> To breathe such vows as lovers us'd to swear;
> And she as much in love, her means much less
> To meet her new-beloved any where:
> But passion lends them power, time means, to meet,
> Tempering extremities with extreme sweet.[14]

A further important reason for multiple-orgasm is that women, unlike men, cannot wait. If a woman is in an arranged marriage then she only has a short time until her husband-by-arrangement will impregnate her. Thus when a woman is ovulating we would expect that women will possess a tremendous drive to seek a potential alternative father to her husband-by-arrangement whom she did not select, and that upon selection there will be a tremendous positive experience upon having sexually selected. Since she may only have a few months, at the most, until impregnation occurs with her husband-by-arrangement, we would expect that during this time she might take tremendous risks to copulate with another man, and that her body would reward her for taking those risks with a greater pleasure. Men, on the other hand, can wait. If a man views an affair with another man's wife as too risky, then he may decide that it simply isn't worth the risk. A woman cannot wait. She may only have five children in her lifetime, since she will be pregnant and breast-feed for up to three years for each child. She will only have a small number of opportunities when she will be able to decide the quality of her children. Since many children die in infancy and her sons can be polygynous, then the difference in quality of her children, particularly male children, can be enormous. The difference between selecting a vigorous healthy man as the father of her child, but doing so illicitly, or reproducing with her sickly husband-by-arrangement, could be the difference between children that die in infancy and a strapping son who obtains four or five wives. It can be the difference between leaving a dynasty that dominates her community and surrounding communities for generations or no legacy except children that do not reach adulthood. Multiple-orgasm and love are a combination that ensures women have all the motivation they need to select a different father for their child, no matter what level of violence their husband-by-arrangement is willing to subject them to.

The desire for pleasure, particularly orgasms, was found by David Buss and Heidi Greiling to be a major factor in why women in cooperatively-modern societies seek extra-marital sexual selections. David Buss summarises his findings:

> When Heidi Greiling and I began to interview women about their affairs, one prominent theme emerged repeatedly: sexual gratification. In our very first study of women's perceptions of the benefits of affairs, sexual gratification was at the top of the list. Our initial sample of 90 women judged it to be "highly likely" that a woman would receive sexual gratification from an affair partner, more likely than any of the 28 potential benefits in the initial study. When we asked women to evaluate which circumstances would be most likely to open them up to an affair, sexual unhappiness with regular partners loomed large: "my current partner is unwilling to engage in sexual relations with me"; "sexual relations with my current partner have been unsatisfying for a long time"; and "sexual relations with my current partner are too infrequent for me."
>
> Moreover, women rated experiencing sexual gratification with an affair partner to be one of the most important benefits of an extramarital affair. Orgasms in particular seem especially important, as women rate these as more beneficial than merely receiving sexual gratification.[15]

A further aspect of sexual selections is that danger should heighten pleasure. Numerous books from D.H. Lawrence's *Sons and Lovers* to a recent book by Catherine Millet *The Sexual Life of Catherine M.* show people in modern societies regularly seeking out risky situations in order to heighten the pleasure of sex, such as a public environment. The element of danger also appears to be a common theme in women's romance literature. In cooperatively-simple societies almost all sexual selections would have a strong element of danger, and this danger should heighten the excitement and pleasure involved in the selection.[16]

Some might argue that women in modern societies also need high levels of motivation in order to sexually select an alternative father. But this is not true. Women in modern societies select their husband, and will have lower levels of motivation to select an alternative. Enforced monogamy means that women need to keep and maintain a partner for their entire lifetime in order to gain his investment that increases the status of her children. If a woman in a society with arranged marriages is caught having an extra-marital affair, then she may be beaten by her husband. In a modern society, she may be divorced, and become a single mother living on welfare, with the lowest possible status in society. For women in modern societies to possess intense levels of sexual pleasure and emotion makes no sense: they chose their long-term partner, so why put it at risk? Multiple-orgasm and love are a hindrance to the formation of long-term pair bonds, not a benefit.

Many have pondered the point of orgasm beyond a positive motivation. Is there a secondary purpose?

When women orgasm there are powerful uterine contractions, similar to those experienced before giving birth to a child, and there has been tremendous speculation as to the importance of these contractions.

The most comprehensive experiments concerning these contractions were performed by William Masters and Virginia Johnson. Masters and Johnson observed hundreds of men and women performing sexual acts to determine the physiology of sexual activity in humans. The importance of their work in sexual physiology is comparable to Kinsey's work in sexual behaviour. Before their study it was generally accepted that the purpose of women's orgasms were to increase sperm levels.

> There have been numerous references to a sucking effect developed by the uterus and directed toward the seminal-pool content. Usually the concept is expressed that during orgasm the uterus develops some form of pressure and sucks the seminal fluid through the external cervical os into the cervical canal, and ultimately even into the endometrial cavity. Thus, in theory, uterine activity mechanically would shorten the transportation interval and the migratory distance for the spermatozoa elevated from the vaginal seminal pool by this reaction.
>
> Evidence assembled during the past decade raises grave questions as to the authenticity of this concept, so well established in biologic thinking. As described earlier in the chapter, corpus contractions start in the fundus, work down through the midzone, and terminate in the lower uterine segment. In other words, orgasmic uterine contractions are expulsive, not sucking or ingesting in character. Even if a negative pressure could be established in the uterus with its normal direct tubal connection into the abdominal cavity, corpus contractions would be expected to work from the midzone up toward the fundus in order to establish a sucking effect.[17]

In other words, the contractions of the uterus is toward the vagina. Masters and Johnson concluded that the role of uterine contractions were expulsive. Theresa Crenshaw, who studied under Masters and Johnson, details the aftermath of their investigations:

> Some fascinating procedures were done to determine what happened to the uterus during orgasm. If sex was indeed for reproduction, the uterus must have sucking

contractions to pull the sperm into the tubes for fertilization. If the contractions were expulsive instead, they seemed to be meaningless and counterproductive.

When the original conclusions were published showing that contractions were expulsive, they provoked such an uproar that Dr. Masters performed additional studies to refine his data. First, he studied women during orgasm under a special kind of X-ray equipment. A cap filled with dye was placed over the cervix (the opening to the uterus). If, during orgasm, the dye was sucked into the uterus, it would show on the X-ray and demonstrate that the uterine contractions were sucking in nature. In fact, none of the dye was found in the uterus during or after orgasm.

In a second study, he had women stimulate themselves with a doctor's metal speculum in place so that he could watch what happened to the cervix during orgasm. All of the women studied were having their period. If the contractions were expulsive, it was felt, blood would be forced from their uterus and would be observed coming out of the cervix. As it turned out, the uterine contractions were so forceful that blood was expelled from the vagina, splattering the white lab coat Dr. Masters was wearing.[18]

Masters and Johnson describe the second experiment:

As stated previously, 50 women cooperated with investigation of uterine response to sexual stimulation during menstrual flow. Thirty-one of these women were parous and 19 were nulliparous individuals. At research request, the women subjectively selected the period of their heaviest flow for evaluation of uterine response. With a speculum placed in the vagina to provide full view of the cervix phase tension release by automanipulative techniques. Seventeen of these women desired more than one orgasmic experience during the experimental sessions. During the terminal stages of orgasmic experience or within the first few seconds of the resolution phase, menstrual fluid could be observed spurting from the external cervical os under pressure. In many instances the pressure was so great that initial portions of the menstrual fluid actually were expelled from the vaginal barrel without contacting either blade of the speculum. It should be recalled that an indwelling speculum holds the cervix high and far from the vaginal outlet.

These observations provide further clinical evidence of uterine contractile response to effective sexual stimulation. A strong expelling force must be created in the corpus to extrude menstrual fluid under the degree of pressure evidenced in these 50 clinical observations. Since the menstrual flow was extruded in spurts rather than in continuous flow, the expulsive force can be presumed contractile rather than spastic in character. It should be recalled that corpus contractions initiated by orgasm have been recorded physiologically to start in the fundus and work toward the lower uterine segment. There now is objective clinical observation to support many women's subjective contention that sexual activity during menstrual flow markedly increases the flow on a temporary basis during an immediate postcoital or postmanipulative time sequence. Additionally, these observations support the concept of expulsive rather than ingestive reactions of the corpus to effective sexual stimulation.[19]

Why would women's orgasm be expulsive? If a woman is copulating regularly with her husband-by-arrangement and copulating irregularly with another man, then this gives her ample opportunity to rid herself of the unwanted sperm by literally ejecting it from her uterus. In no way is a woman, or any other female, benefited from having desirable and undesirable sperm in her reproductive tract. We should expect that all females would possess mechanisms of disposing with undesirable sperm and keeping desirable sperm. Expulsive orgasm yields such a mechanism. If a woman copulates with an undesirable male and then a desirable male, orgasm will expel the unwanted sperm and allow the desirable man's sperm to take its place. If the woman copulates with a desirable man and then copulates with an undesirable man, then the desirable man's sperm also takes precedence. This system also allows women to become pregnant whether they experience orgasm or not. The contrary position is nonsensical, not merely physiologically, as Masters and Johnson have proven, but logically. If orgasm assists sperm to enter the cervix and uterus, then why do women become pregnant without orgasm, even with rape? Women will have fertile mucus that transports sperm with or without sexual desire.[20] If orgasm increases sperm levels, how does this assist the woman to choose, since the sperm that is first in line will be first to fertilize the egg? Increasing sperm levels does not assist a woman to choose the father of her child, since it is a question of first to the egg not quantity to the egg.

In biology, there is a great deal written about 'sperm competition' but much less is written about 'sperm manipulation'. Robert Trivers argued that if the female is the one that places the largest investment into her infants, as against the male partner, then she should be the more particular concerning which male fertilises her large investment. We should expect that this would include selecting sperm. If a female places far greater resources into an infant than the male, then she should also place greater importance on being able to manipulate the sperm of males, and reducing the capacity for the males to manipulate her reproductive tract. William Eberhard's *Cryptic Female Choice* considers the information for female post-copulation manipulation of sperm across various species and is a rare attempt to emphasise the importance of female control.[21]

One of the reasons for scepticism concerning female control of sperm is that it is generally believed that the female must be able to actually tell the difference between different sperm from males.[22] Thus the female's body would be able to tell the difference between sperm from two different males and preferentially select one as against the other. This is highly unlikely, since there is no inherent difference between sperm, and the quality of the DNA is not reflected in the quality of the sperm. Thus it is wondered how the female can control fertilisation if she cannot tell the difference between the sperm of different males. Yet, the fact that females do not possess any such control is reflected in the fact that rapes do lead to fertilisation. On the other hand, the female does possess eyes, ears, a nose and a brain; thus the female can see, hear, smell or understand the differences between males and use mechanisms to affect fertilisation. Upon copulating with a desirable male, a female can eject any unwanted sperm or can increase levels of fertile mucus. Thus the mechanisms are not particular to each sperm, but are generalised responses aimed at affecting sperm. Specific responses to each individual sperm would be highly unlikely.

Many have speculated concerning the implications of the visible region of the clitoris, yet there are obvious further benefits from an expulsive implication of orgasm. If a woman experiences unsatisfactory sex she can masturbate and rid herself of unwanted sperm. Even if a woman is not actually sexually selecting another male, by masturbating she can rid herself of unwanted sperm to reduce the likelihood of pregnancy.[23] If women experience very satisfying intercourse, it is illogical that she would then wish to masturbate afterwards; it makes a great deal of sense that unsatisfying sex would lead to masturbation.

Masters and Johnson documented the implications of masturbation after coitus by examining a prostitute who had experienced 6.5 hours of sexual intercourse without orgasm:

> A decade ago during the prostitute phase of the sex research program, investigators first were alerted to the extent that pelvic vasocongestion may be developed by long-continued sexual stimulation. One individual underwent repeated pelvic examinations during a six-and-a-half-hour working period and for six hours of observation thereafter. During the working period multiple coital exposures maintained the woman at excitement-phase levels of response. There were five subjective plateau-phase experiences superimposed on maintained excitement tension levels, but orgasmic relief was not experienced.
>
> Toward the end of the working period the uterus was increased two to three times the unstimulated size; the broad ligaments thickened with venous congestion; the walls of the vaginal barrel were edematous and grossly engorged; and the major and minor labia were swollen two to three times normal size. Pelvic examinations and coital activity became increasingly painful toward the end of the six-and-a-half-hour working period.
>
> During the six hour observation period gross venous engorgement of the external and internal genitalia persisted – so much so, in fact, that the woman was irritable, emotionally disturbed, and could not sleep. She complained of pelvic fullness, pressure, cramping, moments of true pain, and a persistent, severe low backache. After the termination for the observation period automanipulation brought immediate relief from the subjective pelvic distress and the low backache. The objective findings also disappeared rapidly. Pelvic vasocongestion was reduced by an estimated 50 percent in five minutes and had disappeared completely ten minutes after orgasmic experience.[24]

Though obviously this is a dramatic example, women feel congested if they have intercourse for long periods of time and do not experience orgasm (as do men). If they masturbate, the congested feeling disappears as does the visible evidence. If orgasm is expulsive, then some or much of the semen would also be removed from the unsatisfying men.

Masters and Johnson studied 43 women to observe the effect of vasocongestion and changes in the size of the uterus during coitus and masturbation:

> The study subjects were stimulated to plateau-phase levels of sexual tension. Automanipulation or coition was continued until orgasm was judged imminent by the study subjects. Pelvic examinations were conducted before onset of each stimulative session and immediately after arbitrary cessation of sexual stimulation late in the

plateau phase.

In every instance the 31 parous study subjects demonstrated a significant increase in uterine size. Usually the uterus increased in size from 50 to 100 percent over that described immediately prior to the onset of sexual stimulation.[25]

When the uterus obviously was enlarged, the deep vasocongestive response to sex tension increment developed both with automanipulative and coital stimulation. The longer sexually stimulative activity was continued before late plateau-phase levels of tension were achieved, the more severe was the deep vasocongestive response of the pelvic viscera.[26]

The nulliparous women lost all demonstrable vasocongestion from the pelvic viscera within 10 minutes after orgasm, while the multiparous study subjects needed from 10 to 20 minutes after orgasm before all evidence of uterine vasocongestive size increase was dissipated. When orgasm was not accomplished, clinically obvious uterine size increase frequently remained for 30 to 60 minutes.[27]

In other words, with sexual excitement the uterus increases in size from 50 to 100 percent, and, without orgasm, remains in an increased state for up to an hour after sexual excitement. With orgasm, the uterus returned to its normal condition invariably within 10 to 20 minutes after orgasm. Considering the evidence of the prostitute above, the increased time of congestion and increased uterine size would suggest that women that experience sexual excitement without orgasm will remain in a congested state for up to an hour after sexual activity is ended. How many women follow the prostitute's method to reduce the feeling of congestion? I'm sure most men understand the congested feeling of sexual excitement without orgasm and the need for orgasm.

Furthermore, considering that coitus without orgasm leads to longer periods of congestion, then one would consider that unsatisfying sex would dramatically increase the likelihood of masturbation after intercourse. This would not be a matter of sexual desire or fantasy either, but merely of a desire to reduce discomfort. If orgasm were expulsive, it would make perfect sense that a woman would wish to orgasm after unsatisfactory sexual intercourse: it makes no sense if orgasm existed to increase sperm levels, since then women would feel congestion after unsatisfying coitus, and increase sperm levels from an undesired partner. If a woman masturbated every time after unwanted coitus with a husband-by-arrangement, then it seems plausible that with expulsive orgasm she might reduce the chance of pregnancy for some amount of time.

The benefit of an external clitoris becomes obvious in such a scenario. With the presence of a clitoris, women can sexually stimulate themselves to rid themselves of physical discomfort experienced from unsatisfying coitus, and coincidentally rid themselves of unwanted sperm. That men, in some societies, remove the clitoris through a clitoridectomy suggests a deep, instinctive mistrust of women's sexuality by men.

While Masters and Johnson's numerous studies have been accepted as vitally important to the understanding of human sexuality, those involving female orgasm have been ignored. Nowadays, it is rare to find a paper or book concerning female orgasm that even refers to Masters and Johnson's results, despite the fact that they would probably be considered to have completed the most authoritative work on human sexual physiology. For instance, in *The Science of Orgasm*, Barry R. Komisaruk, Carlos Beyer-Flores and Beverly Whipple do not even mention the above experiments by Masters and Johnson, yet they do record the subsequent attempts to prove that female orgasm is sucking in nature. Thornhill and Gangestad's *The Evolutionary Biology of Female Sexuality* also fails to mention any of the experiments. In her overview *The Case of the Female Orgasm: Bias in the Science of Evolution*, Elisabeth Lloyd refers to Masters and Johnson's findings of an expulsive nature to female orgasm, but only to negate the 'upsuck' concept. A rare example of Masters and Johnson's experiments being properly documented is Mary Roach's humorous popular science book *Bonk: The Curious Coupling of Science and Sex*. Yet nobody appears to consider that an expulsive mechanism to female orgasm could be an evolutionary adaptation. Masters and Johnson's studies have been criticised,[28] but they have never been disproven experimentally. The study in which women were experiencing orgasm during their menstrual flow should be easy to replicate (allowing for volunteers), and could be replicated over the duration of the menstrual cycle by an alternative liquid being placed into the uterus (this is done in procedures such as saline enhanced sonography). It does seem remarkable that Masters and Johnson have had 50 women experiencing an orgasm with a menstrual flow, that menstrual fluids were visibly expelled, and the results ignored.

An expulsive nature to female orgasm is the only possible explanation that makes sense. Sperm can be replaced by more desirable sperm, allowing for female selection, while also allowing for impregnation without orgasm or desire. That female masturbation as a response to vasocongestion after coitus without orgasm could also reduce the chance of pregnancy is also logical. The contrary argument for increased sperm levels possesses no logic, since we should expect all women to desire to masturbate after orgasmic coitus and not after non-orgasmic coitus, when the contrary is true. We should expect limited capacity for impregnation without orgasm, when the contrary is true.

Male sexuality

Men's sexuality has received much less attention from researchers than women's sexuality. Men's sexuality appears to be relatively simple, or at least, it has been presented this way. A man sees a naked woman, he wants to copulate, and he can do so and ejaculate.

Yet, there appear to be somewhat aggressive aspects of men's sexuality. Much of this appears to stem from a deep mistrust of women, that approaches a generalised level of misogyny. Some examples of this are institutional clitoridectomy or infibulation where a woman's clitoris is mutilated and her labia are mutilated and sewn together. This is an extreme way of preventing a woman from being sexually active and occurs in some societies with arranged marriages.

There are other examples of men's fear of women. Much of this appears in men's pornographic material. One might expect, according to the Western ideal of human sexual relationships, that pornographic websites would present men and women meeting, falling in love, and having sexual intercourse. Yet, anyone that peruses these websites would discover that love and romance are almost non-existent. Instead, we are presented with films of women being humiliated and sometimes violated by men. Many men appear to be extremely aroused by seeing footage of men engaging in sexual acts with women where not merely are women's feelings not accounted for, but where women are specifically subjected to pain and degradation. One might thus suggest that only degraded men view pornographic websites, but this is not true either, since investigations show that the majority of men view these websites and these men cross the political spectrum.[29]

As women are fantasising about arranged marriages in romance literature, are men watching the natural conclusion of their fantasy of an arranged marriage – domination of a partner? Romance fiction appears to be the attempt to rise above such treatment, to invoke in the male proof of fidelity and the invoking of tenderness. Men's pornography appears not to bother with a positive outcome, and present repetitive examples of male domination of women.

One might conclude that Western nations are somehow debasing women. Consider the following extract from the Yąnomamö myth 'The Origin of Copulation':

> Yoawä went fishing one day. He tied bait to his fishing line and cast it into the river. As he fished, he came upon a beautiful maiden. She was Raharaiyomä, daughter of the giant river monster, Raharariwä. There were no women at that time and Yoawä wanted to capture her so he could copulate with her. She was in the river, spinning cotton into yarn – and very sexy. 'Wow! A real female! She has a beautiful vagina! I'd like to try it!' he said to himself. He knew she would not cooperate, so he decided to change into a small bird and lure her up to the surface where he could catch her."[30]

Yoawä captures her with the aid of his brother Omawä.

> Their nephew lived with them – Howasi, the white (Cappuchin) monkey. He saw her vagina and the provocative pubic hair and was immediately overcome with lust. 'Let me have her first! Let me have her! I'm horny as hell, and just look at that fantastic vagina!' They let him have the first turn. He passionately mounted her and stuck his penis into her vagina with a mighty pelvic thrust – and immediately screamed in agony, withdrawing his bloody stump. She had piranha fish inside her vagina and they bit the end of his penis off. He screamed and howled and fled into the forest, holding what was left of his bloody penis.
>
> Yoawä removed the fish from her vagina with a barbed arrow and then mounted her. He was consumed with lust, and made long, passionate pelvic thrusts. His penis went in deep, and came back out rhythmically, and made a foul, disgusting noise: 'Soka! Soka! Soka! Soka!' Omawä threw up his arms in despair, for such clumsy copulating, such foul noise, would anger people who might overhear it. He demanded that Yoawä cease and dismount so he could show him how to copulate in the proper manner. He then mounted as Yoawä observed. He copulated with slow, discreet pelvic

thrusts and made no noise as he proceeded. 'See? That is how to copulate properly, so nobody can hear your penis going in and out!' From that time on, people have been able to copulate discreetly.[31]

This story is in many ways sexual education, and, it might be added, presents a different method of copulation from the hard thrusting manner often depicted in modern pornography. Yet, even so, in this story, the only consistent description of the woman is of her genitals. There is no consideration of the discomfort due to long thrusts, only the noise such thrusts might make to be heard by others. If the 'Origin of Copulation' were made into a pornographic film, would it be any different to western pornography? The general theme appears to be the same, male pleasure and no consideration for the female. I have also quoted some of the love songs of Arnhem land in the previous chapter, where the men 'seize the young girls of the western tribes', 'Young girls squealing in pain, from the long penis', 'Girls of the western clans, desiring pleasure, pushed on to their backs among the cabbage palm foliage'.

As I have previously shown, in societies with arranged marriages men do not expect to love their wives. A dominated, humiliated, subordinate wife-by-arrangement that accepted her husband's every whim, even if it were painful to her, may thus be the desired outcome. Such a dominated partner would be more likely to have his child. A free, strong-willed, dominating wife-by-arrangement would, by contrast, feel more likely to choose a father for her child of her own preference and to act on these intentions.

Other aspects of male sexuality appear not to have been researched. One example is the lack of sexual interest that many men show their partners after several years in a relationship. While men are consistently presented as being continually and easily aroused and that it is the woman that is generally refusing him, one has to only read sexual therapy columns in newspapers (or to hear women's personal complaints) to conclude that the contrary position – little or no sexual interest from the man in his partner, even where she is interested in him, – is extremely common. For every study focused at assessing the level of interest in women's sexual desire with ovulation, there appear to be very few assessing the aspects of male desire in long-term relationships of more than one year. Since dwindling sexual interest can lead to affairs and divorce, one might wonder why this issue has not received significant interest from researchers.

Do men feel greater levels of pleasure in intercourse with new partners by comparison to long-term partners? Do men feel greater pleasure with more attractive partners? Is there much variation in terms of pleasure for men? Considering that women can experience multiple orgasms, one might conclude that male pleasure has a smaller level of variation. This would perhaps reflect the long-term method of male reproductive success, with males depending on polygynous marriage for their genetic success, in comparison to women, for whom quality is more important, and thus each child remains vital. Males, therefore, gain less from extreme sexual desire and inflated levels of pleasure, which might undermine their long-term capacity to accumulate wives. Since most children can be expected to be born of arranged marriages, men may not be benefited from experiencing large differences in pleasure with different partners; however, there is no doubt that men are willing to engage in risky behaviour to copulate with women.

Concerning the male, there are perhaps only three major morphological aspects that specifically concern sexuality; the size of the testes, the size of the penis, and the lack of a penis bone.

Much comment has been made on comparisons of men's testes with those of the gorilla and chimpanzee.[32] Female gorillas only mate with the resident male silverback, and there is no competition with other males. Female chimpanzees mate with numerous males, and there is a tremendous amount of competition between males to fertilise a female. In comparison to body size, unsurprisingly, male gorillas have small testes containing small amounts of sperm, while male chimpanzees have very large testes with large amounts of sperm.

Men have testes somewhere in between. The basis of this would be, firstly and primarily, repeated attempts to ensure the paternity of the children of a wife or wives. Since a husband can never be sure of when a wife-by-arrangement ovulates, he will copulate with her regularly when she is not either pregnant or breast-feeding. Therefore, while a male chimpanzee has large testes in order to copulate with many females repeatedly, in men, the testes exist largely for repeated copulations with the same female. Even if a man has several wives, it is highly likely that most or even all could be pregnant or breast-feeding and so the likelihood that he would need to copulate with more than one wife-by-arrangement at any particular time would be small. In cooperatively-simple societies, there are few opportunities to copulate with women, since there are few women to begin with and since women are often pregnant or breast-feeding children. Thus the requirement for large testes to be able to copulate with many women is unlikely. It has been suggested that men's testes are large due to multiple matings with various females; however, it is far more likely that men need a large amount of sperm to copulate repeatedly with the same wife or wives.

A further trait is the large size of the male penis. There are certainly enough jokes concerning the size of men's penises in both modern society and in the anthropological literature to assume that size is important. Certainly I have heard, in conversations between women, comments made where women were somewhat outraged at the lack of a man's penis size.

Dixson suggests that penis length is linked to vaginal length in primates:

> Selection has presumably favoured efficient placement of the ejaculate close to the cervical os, in order that spermatozoa may migrate (or be actively transported) into the uterus. The human vagina is approximately 5-6in. (12.5-15cm) long (Dickinson 1949) and is capable of dilation to accommodate the penis during intercourse (Masters and Johnson 1966). Among the primates as a whole, there are only nine species for which both vaginal lengths and (erect) penis lengths are known, including *H. sapiens*. For this small sample there is a highly significant correlation between these two genital measurements... There is nothing exceptional about vaginal or penile length in human beings.[33]

Interestingly, vaginal length varies, with the largest primate, the gorilla, having a shorter vaginal length than humans, which is shorter again than that of chimpanzees. Yet, penis length is longer in humans than either chimpanzees or gorillas.[34] The extra length of the chimpanzee vaginal length could be enhanced by a significant genital swelling in the chimpanzee, which does not occur in gorillas to anywhere near the same extent. In any event, similarities in vaginal and penile length appear to be fairly conclusive, with a very obvious correlation existing between the two.

As to the lack of a penis bone, throughout this book I have presented one continuous idea that a man's interest in other men's wives-by-arrangement is extremely dangerous. All it takes is for a woman to speak to a man other than her husband-by-arrangement to provoke serious violence.

Possibly in response, most men restrain their penis to ensure that an erection is not possible to view in public. For instance, the men in Papua New Guinea place a phallocarp – a large gourd, usually brightly coloured and decorated, - over their penis. The following is from Jared Diamond:

> In response to my question as to why they wore phallocarps, the Ketengbans replied that they felt naked and immodest without them. That answer surprised me, with my Western perspective, because the Ketengbans were otherwise completely naked and left even their testes exposed.[35]

The Yąnomamö men tie their penis to their body using the foreskin and string. The Yąnomamö give the same response to the idea of allowing their penis to be loose, despite otherwise being utterly naked:

> It takes time to stretch the foreskin to the length required to tie it adequately and securely, and until that happens, it keeps slipping out of the string, much to the embarrassment of its owner and much to the mirth of the older boys and men around. Sometimes older boys and men accidentally get 'untied', causing great embarrassment – for it is like being completely naked. A penis string is not comfortable. Take my word for it. Wherever possible, as at mission posts, the men become rapidly accustomed to wearing short pants or loincloths and these become popular trade items very quickly. Men who wear pants stop tying their penises. It should not be assumed that all customs are enjoyed or liked by the practitioners, a topic of anthropological research that deserves far more attention than it has thus far received.[36]

A Papua New Guinean man is being modest by wearing a large and coloured phallocarp, because nothing will be as immodest and dangerous as an erection in the presence of other men's wives-by-arrangement. Equally with the Yąnomamö, a piece of string could be the difference between being sexually attracted to another man's wife-by-arrangement and presenting no signal, or presenting a very strong signal of attraction to another man's wife-by-arrangement and initiating a violent club fight. That the penis string is painful may also prevent erections and be beneficial.

The greatest instant killer of libido is fear, in both men and women. Women who are married to dangerous husbands-by-arrangement will fear risking a sexual selection and may be depressed by the constant violence that is aimed at them by their husbands. Such women are unlikely to have any libido, and, indeed, to have any libido could be dangerous. Theresa Crenshaw observed, in her years as a sexual therapist: 'Sex is the most vulnerable of all natural functions. If an individual is depressed, the first appetite to go is sex.'[37] The same is true for men. Theresa Crenshaw explains what fear does to men:

> Loss of erection actually results from a chemical within our body triggered by fear. An emotion (fear) causes a chemical (adrenalin) to instantly appear in your bloodstream. *Our mind causes chemical and physical reactions in our body.* This is not just theory, it is fact! It is well known that stress contributes to ulcers, heart disease, and impotence – each of which in turn increases the stress on the body.
>
> Until recently, the mechanism involved in impotence, not caused by trauma or disease, has not been well understood. The adrenalin reflex, however, explains it in a clear and sensible fashion. Most people know that when they are suddenly frightened when they have almost been run over by a car, adrenalin – the survival chemical – instantly surges through their system, preparing the person for "fight or flight." Adrenalin results from an emotion and appears instantly in the bloodstream.
>
> Adrenalin causes specific physical reactions within the body. It causes impotence by directly constricting the blood vessels in your penis – just as it does those in your skin – and the erection is lost. The more upset you get about it, the more adrenalin pumps into your system, and the more impossible it is to become erect again.
>
> Another effect of adrenalin on sex is to stimulate ejaculation. That is why an anxious man will ejaculate faster than one who is at ease. Extreme fear can cause ejaculation. Men who are about to be executed have ejaculated – hardly an erotic experience. Some men report ejaculating during physical exertion, such as a man who said: "When I was sixteen I was climbing a rope trying to beat everybody else, and I ended up ejaculating instead."[38] (Crenshaw's emphasis.)

The implications of the combination of fear and impotence cannot be underestimated:

> Impotence isn't contagious, rarely fatal, but can be life-threatening. It affects men at any age and devastates them. Because of impotence, a man can become depressed, and sometimes suicidal.[39]

In an environment where sexual selections can be potentially dangerous, even lethal, if a woman sexually selects a man and the man is frightened he will not be able to maintain an erection. Men have no bone in their penis to maintain an erection under duress. Considering that many men are rendered impotent in modern societies simply by virtue of fearing the loss of an erection, it is perhaps worth considering the difficulty in maintaining an erection in a social environment where sex with another man's wife-by-arrangement could lead to extreme violence.

The combination of fear and the ability to lose an erection is not a handicap but a benefit, since the inability to gain or maintain an erection might end the sexual selection and save the man's life, if he had good reason to be in fear. Male hominids with a penis bone may have been more likely to engage in extremely dangerous copulations, and been more likely to be killed in the subsequent violence. This is a feature of human behaviour, that those males with the longer-term approach, that cooperate and wait for the benefit of arranged marriages, are favoured over those that take shorter-term, riskier approaches.

Premature ejaculation may also be a half-way measure between impotence and confidence, since it minimises the time of copulation. This is especially true for younger men, who can have great difficulty in preventing premature ejaculation. For younger men, every moment spent in a sexual selection could be very dangerous, and premature ejaculation may greatly assist them in minimising the danger, though probably it would lead to a shortening of the relationship. Also, very young men, in their teens, often get erections for no apparent reason; this may also assist them, if the younger man has an erection in the presence of an older man's wife-by-arrangement, since the young man can get erections without sexual interest the older man may give the younger man the benefit of the doubt.

A large penis in coitus with a woman is of little benefit during coitus since a woman's vagina opens out to a size larger than a penis. Theresa Crenshaw comments:

> The wall of the inner two-thirds seems to lose feeling while the outer third gains feeling. In actuality, reduced sensation is an illusion. It is just the result of ballooning of the vagina. As the vagina expands, it becomes larger than any penis would ever want to be. The surface of the inner two-thirds of the vagina is usually not in contact with the penis, except perhaps at the deepest part.[40]

Indeed, girth might be expected to be commented upon by women as more important than length, since only the first few centimetres of the vagina touch the penis.

Male Homosexuality

The final aspect of male sexuality that I will consider is homosexuality (female homosexuality will be dealt with in a later chapter). The question of how genes that allow for homosexuality could be beneficial is easily explained: homosexual men simply marry women and have children like other men. In societies with arranged marriages, whether one is homosexual or heterosexual has no impact on whom one marries or whether one marries; marriage is a matter of duty. As people have no choice as to their partner they also have no choice as to the sex of their partner. (This also explains how genes allowing for female homosexuality are passed on.) If a man has two wives and he dies, then his wives will be passed on to his younger brother, following the levirate. If his younger brother happens to be homosexual then this has no bearing on the outcome, though it might be plausible he could pass the women onto another brother.

Though male homosexuality is often presented as being a matter of men being incapable of copulating with women, this is nonsense. Oscar Wilde, for instance, one of the most famous homosexual men of all time was married and had children. Indeed, it would be interesting to discover how many homosexual men do marry and have children.

Sue Joseph's *She's My Wife, He's Just Sex*[41] is one of the few books to deal with the issue of actively homosexual married men. She interviews Bruce Gibbs, the administrative officer with GAMMA (Gay and married men's association):

> GAMMA is what we describe as a self help group for men who are married or have had an ongoing relationship with a female partner but who have had feelings of attraction to men. They may have actually practised having sex with men or they may have actually conducted a sexual relationship while they have been with a woman.[42]

As to these men's sexuality, the following question is asked:

> 'You are walking down the street on Saturday morning doing the shopping and a couple the same age as you are walking towards you. Who do you look at?' 'Oh I look at the man.' 'OK, what do you do? Where do you look?' 'I look at his face then I look straight at his dick.'[43]

Some men do not identify at all with homosexual behaviour and argue that they are 'straight'. Bruce Gibbs and other 'beat' researchers observed that these men are normally from more conservative 'ethnic' backgrounds, and these construe most of 'the other class' of non-gay self-identifiers:

> It happens in the ethnic communities. They will go out and have sex and not identify [with being gay] in any way. They just have these sudden impulses. Do things and they are unsafe.[44]

This impulsive behaviour has important health impacts: in India it has been found that married men who have sex with men have twice the HIV rate of non-married men (see below). In terms of male sexuality, Gibbs placed men on an arc with one side being straight and the other gay:

> Some people will fit right down in the 10 degrees angle of being totally straight. Never thought of another man. Never touched another man. Never had sex with a man. Others move up there a bit because they had some sort of sexual contact as a teenager. As they move through life and start thinking they tend to move along this arc and approach the 90 degree line up the top. At this stage they would have had sex with men and they are deciding which way they want to go. And some of them will move back. They might move back to the 45 degree and they might live a married life and go and have sex with a man once a year. Just somewhere they happen to be and the occasion arises and they enjoy it. Others just move into the other sector and somewhere in that sector they will settle and you sort of get the various classes of homosexual people and you'll get the ones who wear high heels and flirt and carry on and scream right out 90 degrees that way and others who lead a more 'normal' life will be somewhere in the upper arc. Sometimes you'll get gay men who go and have sex with women because they like it.[45]

None of this is new. Alfred Kinsey observed in 1948:

> Males do not represent two discrete populations, heterosexual and homosexual... Not all things are black nor all things white, It is a fundamental of taxonomy that nature rarely deals with discrete categories. Only the human mind invents categories and tries to force facts into separated pigeon-holes. The living world is a continuum in each and every one of its aspects. The sooner we learn this concerning human sexual behavior the sooner we shall reach a sound understanding of the realities of sex.[46]

Perhaps because of the politics, or perhaps because of regimented thinking, people have tended to view men as either homosexual or heterosexual. It would be understandable for political reasons if homosexuality in males was viewed as invariable, since, historically, there has been the belief that men can be altered 'back' to heterosexuality. Yet one must wonder what percentage of homosexual men are actually married and have children.

As to these men's feelings towards their children, Bruce Gibbs of GAMMA observes:

> We haven't got anyone who has come along and said: 'Well, I just got rid of the wife and the kids, thank goodness.' Whether it is because of the type of guy who is gay is more caring, I don't know. Their greatest concern is to have access to their kids. They can't live without their kids.[47]

Sue Joseph quotes another worker in the area, Phillip Keen, Sydney Outreach Co-ordinator from the Beats Project, Western Sydney, who claims:

> My personal estimation is that there are 10s of 1000s of men in Sydney beats on any one busy day. There are 60 active beats in and around Sydney, so that figure is a minimum. And 40-60 per cent of the men who use them are non-gay identifying. And many of them are married or de facto, or simply sleep with women, too.

> A beat is a public place which is known to men who go to it as a place where there are other men going there for the purpose of meeting each other for sex. They can be anywhere, but usually are toilets, public parks or a section of a public park, isolated roads, beaches or car parks.[48]

> These guys often don't even speak to the men they have sex with – there is a code of etiquette. There are signals and understandings that only they know, and each beat can be different.[49]

> In the hetero world sex means penetration and penetration means sex – other stuff is fooling around and often non-gay identifying men describe their sex with men in the same way, which is a barrier to HIV education.[50]

Educating these men about AIDS was an important part of the reason of Phillip Keen's work, though even talking to these men is difficult:

> What seemed to be the most effective – yet ethically problematic for us – was to pretend we were cruising in order to get close enough to say who we were. That never felt good – either for us or the client. It wasn't actually a definite come-on, but near enough and at the earliest opportunity we would say who we were.[51]

> I've had rocks thrown at me and I've been thrown down a set of stone steps. For the guys, particularly the non-gay identifiers, we represented being caught and believe me, these men do not want to get caught. They want to remain hidden away, forever. They have a lot to lose.[52]

The female partners appear to take the revelation harshly. Sue Joseph interviewed Phillipa C. who runs the Women Partners of Bisexual Men Peer Support Group:

> But when she started the support group she was in fact quite surprised by the response – most of the women had found out after the relationship had started that their partner was bisexual.[53]

> Many of the women walk away from their relationships doubting whether their partner ever loved them at all; whether they were just being cynically used from day one. He needed a woman in his life. Maybe he wanted kids too and she was just an instrument for him to achieve those things. And when he didn't need her any more, he walked away. Spat her out.[54]

> This is very sad because they feel their whole married life has just been a lie – even a joke – and they don't know whether to believe their husband when he says he really did love her, and maybe still does.[55]

> Two scenarios: Wife Number One, married for 20 years, is told by her husband that he has fallen in love with another man and is planning to leave her. He says that he has always known he was gay, even before they were married. She is in complete shock, the marriage has had its ups and downs, but she thought they were basically happy.

> He takes a wedding photo off the mantelpiece and points out how the best man had his hand on the groom's behind on their wedding day.[56]

> Wife Number Two is told by her husband that he is experiencing some disturbing feelings, that is, a sexual attraction to other men. He might even have acted on these feelings once or twice, perhaps when he was drunk. He is very confused because he still loves his wife dearly, but he wants to be honest with her and, hopefully, work things out with her support as well as allow her the right to have some input and control in what is, in effect, a family crisis.[57]

Very few of the women kick their husbands out immediately. Most of them will stick around to try and work it out for a while and then leave or ask their husband to leave. Where there is a knee jerk reaction is more often when the man says: 'OK, I'm out of here.'[58]

Research on this issue, mostly in regard to HIV infection, has recently increased, particularly in non-Western nations. This issue in these countries is important because of cultural expectations of marriage, and a different perspective on homosexual behaviour. The authors of one Indian study note:

> Cultural norms in India ensure that there are predetermined roles for women and men that impact on sexuality. Women are raised from an early age to repress sexual desires and adopt the role of the obedient wife, whose primary responsibility is to reproduce. No such restrictions are placed on male children; masculinity is not defined by sexuality, but rather by fatherhood. Further, in Indian culture, close physical contact between individuals of the same gender is not considered inappropriate. Close contact between men of the same sex often begins in adolescence and, in some cases, evolves to sexual contact between men. Most men would not consider this behaviour to be inappropriate, nor would they identify themselves as "homosexual", especially when this behaviour occurs within the expectation or reality of marriage and fatherhood. Indian societal norms allow large numbers of men, who may or may not self-identify as homosexual, to have sex with men, while at the same time being married to women.[59]

Further, the majority concurred that the primary reasons for getting married were due to parental pressures and the fear that if they did not get married, their younger siblings would also not be able to get married, a situation that is customary in India.[60]

Duty to the family overcomes individual desires (MSM = men who have sex with men):

> In India, MSM are a diverse group of individuals who may not necessarily be associated with an overriding homosexual identity. Social norms proscribing premarital and extramarital heterosexual sex are strictly enforced in a manner that tacitly encourages men to have sex with men; because women are not easily accessible to men outside of marriage, men may release their sexual tensions with other men. At the same time, men who seek sex with men are frequently married and engage in marital sex as a social duty in response to strong social norms and intense familial and community pressure for procreation.[61]

In one study that screened 10,785 men who were at a sexually transmitted infection clinic in Pune, India, 708 (6.6%) were men that had sex with men, and of these, 253 (35.7%) were married (of these 13.4% were HIV positive).[62] Reflecting the need for secrecy: 'Most (95%) married MSM self-identified as bisexual. While nearly all (97%) had disclosed their same-sex behaviour to other MSM, virtually none had disclosed their behaviour to their wives (2%), other family members (6%), and health care professionals (15%).'[63] In another study that sampled 1,950 people in Mumbai slums, 46 were men that reported having sex with other men, and of these 26 (57%) were married.[64] Both these studies benefit by being samples from a wider population, rather than individuals recruited from a specific sub-population. A study of 2,910 rural Indian men aged 18-40 found that 72 (3%) married men had unprotected sex with other men in the previous 12 months and had more female partners than male partners.[65] A study in four Indian states of 4,597 men who had sex with men, 18.4% to 37.1% were married, which is fairly high when the median age was 23 to 26 across states. Of those who described themselves as bisexual, 61.2% were married. India has various descriptions for men who have sex with men, including Hijras, who dress as women. Of Hijras, 23.1% reported they had been married though only 29.3% were living with their spouse.[66] Married men who have sex with men have also been found to have double the rate of HIV infection as compared to men that are not married.[67]

Studies of men engaging in sex with other men in other countries have reported much lower rates of marriage: 5% in Latin America, 7% in Eastern Europe and the Central Asia ranges, 14% in Nairobi and Malawi reporting 34%.[68] However, many of these studies obtained their sample directly through networks of men engaging in homosexual behaviour rather than being samples from wider populations, which may reduce the likelihood of more clandestine participants, such as married men. Other findings have been: 93.8% in Timor-Leste; 37.8% in Kathmandu, Nepal; 36.3% and 39.3% in Bangladesh; 29% and 26% in the Maldives; and 7.6% in Sri Lanka.[69]

This data explains how men with homosexual or bisexual behaviour can be reproductively successful: they get married to women and have children. This can be the case even if the men dress as women, as reflected in marriage rates amongst Hijra in India. How is this possible? The fact that men's capacity for sex is based more on functionality than emotion is reflected in the findings by Pfizer in their research with the drug viagra, quoted above. If men were more like women, emotional than functional in their sexual behaviour, then a homosexual or bisexual man getting married and having children would be less plausible since a man needs an erection to initiate sex. For men, attraction is not the key issue; the capacity to function is fundamental, regardless of attraction.

Even if this is the case, how could men's homosexuality be beneficial? Previously, I have stated that in cooperatively-simple societies children are produced in two ways, through arranged marriages and extra-marital selections. The first of these methods requires patience, since most men marry in their early-to-mid twenties. The second method requires risk-taking, since copulating with another man's wife-by-arrangement may lead to an attack from the husband and possibly his supporters. All sexual females are married and all pre-sexual females are betrothed. Most children are the product of arranged marriages.

Homosexuality thus provides a benefit to men who are unlikely to marry. Since arranged marriages are based on primogeniture, the youngest of several brothers is the least likely to marry early. Warner observed of the Murngin:

> The junior levirate is a prominent mechanism. When an older brother dies the brother next in age and consanguinity receives his wives and becomes father to the children. This is not only a privilege, but a duty. Frequently wives thus acquired, being past the age of bearing children or gathering food, are really an economic liability to the

heir; yet he must take them and look after them.[70]

Finally, an older brother always helps a younger brother to acquire a wife because it settles the younger man and tends to prevent his getting into trouble with more remote wawas, yukiyukos, and other kin, by copulating with their wives, thereby forcing the wawa to defend his brother from their vengeance.[71]

[Yukiyuko or Wawa = younger or older real or classificatory brother.]

There are three situations where men might be unlikely to marry: when they have many older brothers, they have no sisters or half-sisters to exchange, or have a weak group of supporting male relatives and lack the political weight in the community to attract alliances with other men – his kin are more likely to concede girls in marriage exchange in the need for protection. Men with strong sexual desires towards women, that live in an environment where all women are either married or betrothed, and cannot expect to marry until well past their mid-twenties, face a serious problem since any attempt to initiate sexual contact with a woman could seriously endanger his life. An obvious response to this is to have a preference for males, yet, when the possibility of marriage arises, to be able to marry and reproduce with the woman.

Studies have found that birth-order affects the likelihood of homosexuality, and accounts for 1 in 7 cases of homosexuality, and that each older brother increases the likelihood for homosexuality by around 33 percent.[72] In societies with arranged marriages where the younger brothers are less likely to be able to take advantage of any such arrangements and may even need to wait for an elder brother to die, it makes sense that his sexual attention may be better diverted elsewhere. By contrast, an assumption of human evolution along Western-style marriages, where elder and younger brothers are equally capable of attracting female partners, then such a trait for homosexuality makes no sense.

Though there is no data to support the suggestion, that I have seen, isolation and weakness may be other triggers for homosexuality. These could include high levels of stress, as either a child or youth, which might suggest that the boy is growing up in an environment where he has little protection, and thus has few male relatives and is unlikely to marry early. If he is politically weak, with few protectors, then as a man, he would be ill-advised to take an interest in other men's wives-by-arrangement (the only available women) since he would have few other men to support him. Indeed, an isolated man might be benefited by creating alliances with other isolated males or younger brothers of large, politically strong kin groups through a common sexual bond, which might increase his political influence amongst other men and his likelihood to marry. It is interesting that many gay men in modern societies present either an overtly muscular image or a lithe and sometimes overtly feminised image, suggesting protection and weakness.

Another consequence of male homosexuality is that men should be either homosexual or heterosexual. It would not benefit men to be interested in both men and women, since this would mean he would be interested in other men's wives-by-arrangement. That men are either predominantly homosexual or heterosexual has been shown in studies, as Dean Hamer explains:

The difference is that most men are at one end or the other of the Kinsey scale – either they're straight or they're gay – while women are much more spread out along the scale – many of them show some degree of bisexuality, either in their behavior or, more frequently, in their thoughts. An example is a study of 4,903 twins in Australia in which the majority of both the men and the women identified themselves as predominantly or exclusively heterosexual (Kinsey 0 or 1). However, for the men there was a "valley" in the distribution in the bisexual score range of 2 to 4, then a "peak" in the predominantly or exclusively homosexual range of 5 to 6. But for the women there was a gradual tapering off of the distribution in the higher Kinsey score range. There wasn't any peak at all in the predominantly or exclusively lesbian range; most of the nonheterosexual women were in the intermediate, bisexual range.[73]

The reason why women are in the bisexual range will be considered later.

I argue that the benefit of homosexuality is that it prevents weak males from engaging in competition with strong males where competition is dangerous. In human terms, weakness is measured not merely in individual terms, but group terms (how many brothers), and in terms of birth-order (likelihood of marriage).

That homosexuality might be common, particularly amongst males, in the animal kingdom, would not be surprising. Reproductive sex is often competitive and sometimes lethal.

Concluding Remarks

Human sexual behaviour only makes logical sense within the context of societies with arranged marriages. That women are specific in their sexual response to their partner makes logical sense if one assumes that women evolved with arranged marriages. Women thus possess a dual sexuality with either dutiful acceptance or extreme desire.

The ignorance of the factor of violence in humans not merely undermines people's capacity to understand human cooperation, but also human sexuality. Multiple orgasm and love cannot make sense unless viewed as an extreme positive feedback, required to overcome a strong negative feedback. Unless it is assumed that women could be subjected to violence for an extra-marital selection, multiple orgasm and love will be viewed as implausible behaviours for the creation of a long-term pair-bond. These aspects of female sexuality only make sense within the context of arranged marriages, as does the expulsive aspect of orgasm.

Male sexuality is also easier to understand within the context of arranged marriages. Male pornography makes no sense within the context of the Western ideal of selection and love of a partner. Neither does premature ejaculation, regular copulation, or the lack of a penis bone. Confidence and fear appear to be a large part of male sexuality.

Human sexuality appears paradoxical and unlike the sexual behaviour of any other animal. This sexuality can only make sense within the context of the arranged marriage, where marital hostility, extra-marital selections, and violence create a volatile mixture reflected in human sexual behaviour.

Chapter 9
Female Alliances

I have considered the idea that humans evolved with the arranged marriage system, and that sexual selection, in terms of females selecting males, was made more difficult by the husband-by-arrangement and his supporters attempting to prevent his wife selecting other men. I have suggested that humans evolved with numerous visual signals that enable for sexual selection to occur at a distance: naked skin; strong scent; eye-contact initiating sexual selections; blushing and tears to present emotions in a visual manner; and so on. We thus have males and females with the ability to select the opposite sex on the basis of visual signals, they possess the ability to communicate emotions visually, and possess the internal reward mechanisms – love and erotic pleasure – to motivate the individual to sexually select or be selected.

Yet, there remains the difficulty that the female and the male remain at a distance. How is the sexual selection consummated? The female cannot simply go to the male and organise a meeting place, and nor can the male: if they are observed by the husband-by-arrangement, or one of his allies, then extreme violence could result, and the sexual selection would be prevented.

The Intermediary

In order for a woman to sexually select a man she thus requires an intermediary. She requires another person to organise the sexual selection between herself and the selected man. She is limited in choosing such a person. The only males to whom she can speak are either close male relatives or her husband. Her husband-by-arrangement obviously will not help her, and nor might her brothers, since she is the cement that cooperatively links her own brothers to her husband and his brothers.

Since she cannot select a male to act as an intermediary, the only other people she can communicate with freely are women. The chosen woman will have similar difficulties, though, since she will only be able to communicate freely with her brothers and husband-by-arrangement. Women will thus form an alliance with another woman in order to select that woman's brothers or half-brothers. The man's sister acts as the intermediary.

Evidence of a trusted messenger in cooperatively-simple societies is as rare as long-term studies by women anthropologists. Karl Heider noted the difficulty of studying women amongst the Dani of West Papua:

> But I was never able to have real discussion with women or to use them as informants. Men, yes; boys, usually; girls, often; but women almost never. I am still not

sure why. Whenever I tried to ask a woman a question, she would usually giggle, shake her head, and say she did not know.[1]

In 1967 Paul Ekman worked with Heider and the Dani in examining the universality of emotions, at a time when the social sciences were arguing that the emotions were based purely on culture and were not universal – thus the Dani were examined to see if they expressed happiness by narrowing their eyes or dropping the corners of their mouths (such was the state of the study of humans at the time). Ekman observed: 'I inadvertently made a woman angry by looking directly at her in public. No one should have difficulty determining which woman in Fig. 4 is angry.'[2] In societies where men do not trust their wives, even a direct glance between the male and female is unacceptable, and perfectly understandable from the female perspective. Contact between male anthropologist and potential female informant is thus extremely limited.

Fortunately there is some definite evidence from female anthropologists. Phyllis Kaberry noted amongst the Aborigines in the Kimberleys, in northern Australia:

> There is a recognized procedure in courtship, in which either the man or the woman may take the initiative. The girl may send a gift of tobacco to a boy through her mate, djalindja:ru. Bulagil, who finally eloped with her tribal son-in-law, was said to have done this. If the man is agreeable he will keep the tobacco and later make a similar gift, and this continues until a meeting is arranged and the two have an affair.[3]

The djalindja:ru is explained as:

> The grouping resolves itself first of all into that of hordes; when these are examined individually, they are found to comprise a man and his family separated from other horde members by his own blood relatives on his father's side. Generally we find that if two women have been foraging together, they either share the one fireplace as co-wives or are mother and unmarried daughter; or else they are neighbours and affinal relatives. Sometimes they may be married to men of the same horde, and youth, temperament, and common interests may forge a bond of companionship between them. They are djalindja:ru or "mates".[4]

The djalindja:ru appears to be the equivalent of a best friend, as she would be described in English. Kaberry's definition of the djalindja:ru is as varied as one might expect in reference to each woman's situation, particularly in reference to age. The previous reference to the participation of the djalindja:ru in sexual selections suggests that while older women may have a close friend as a daughter, many of the younger women have a close friend as another woman, and, explicitly, not a sister.

Warner noted a similar method in sexual selections amongst the Murngin:

> Sometimes a man has a woman's friend – always a woman – drop a certain type of string in the loved one's basket. If the woman belongs to the Yiritja moiety and the man to

the Dua he puts an opossum fur string into it; if he is Yiritja and she Dua, he uses a red parrot feather string. This, of course, occurs only after he feels fairly confident that his advances will have a favorable reception. She replies through her woman friend, and arrangements are made for a time and place.[5]

Sometimes a woman attempts to seduce a young man into joining her for a lovers' assignation. She usually starts by sending him food. After these presents have been given for a while and he is sure of their meaning, he usually tells the woman who brings them for her friend to tell the other woman where he will meet her. If they have an assignation, he usually makes presents to her. Sometimes young men are so bashful that it takes much persuasion to force them into such a union.[6]

Jane C. Goodale is the only anthropologist who specifically notes that the intermediary is a relative of the male; in *Tiwi Wives* she notes:

Go-betweens, usually close siblings of the lovers, aid them in arranging meetings in the bush, for the two should not be seen talking together. When lovers walk together, the boy is careful to walk in the girl's footprints, obliterating them so that her husband cannot follow her.[7]

It is worth reflecting on the implications of this. Firstly, a woman that is sexually selecting another man will be potentially risking her life. Secondly, if she needs an intermediary to organise the sexual selection, then she needs an intermediary whom she can trust will not reveal the information to others. Thirdly, if certain men are the most desirable in the community, then their sisters and half-sisters will be the most popular choices as potential intermediaries. Thus we should expect that women will possess a hierarchy in terms of the desirability of the women's close male relatives. Fourthly, as part of this hierarchy, we should expect extensive competition to exist in the formation of potential intermediaries. Fifthly, women without close male relatives present in the community (due to, for instance, abduction from an enemy community in which all her close male relatives reside) should have tremendous difficulty in becoming an intermediary, simply because they have no male relatives. Sixthly, due to the importance of the information given to the intermediary, women should engage in a long relationship before they trust a woman to act as an intermediary.

Considering the importance that I have placed on sexual selection in this book, then if women's sexual selections of men are entirely dependent on the intermediary to establish communication between the man and woman, then the intermediary must be fundamental in human evolution. Indeed, one must expect that women would place more emphasis on the intermediary than the sexually selected man, because without the former they cannot select the latter. Therefore, before women begin to consider sexually selecting a man, they must already have formed a close bond with his sister or half-sister. Once a woman has formed a close bond with a woman with several brothers or half-brothers, then she possesses the ability to be able to communicate with these men through her closely bonded ally.

Unfortunately, the study of women's friendships is a relatively recent area of research (though superior to the study of male friendships, which is virtually non-existent), and so most of the material concerning their relationships must be taken from modern studies. Most studies examine girls in high school. Terri Apter and Ruthellen Josselson give extensive examples in *Best Friends*. In this book, the most common theme is that of the importance of trust in a friend in their ability to keep information secret:

> We reveal our secrets and then we set up structures to protect ourselves after we make ourselves vulnerable: "Promise you won't tell," we plead, and confidentiality becomes part and parcel of what we mean by friendship. Over and over, as we interviewed women and girls, we heard that the worst betrayal of friendship is to "tell."[8]

> The safety net promised by friendship talk is constantly broken. By the time girls are fourteen, most say they have already been betrayed by a friend – betrayed by having a "secret" or a "problem" or "something private" revealed behind their backs.[9]

> Betrayals are as common as dust, yet each one comes as a shock, since each girl believes her system of control will be effective. "But she *promised*," "But she *swore*," "But I was sure she'd never tell," girls protest, even as they see the promise broken. Some girls are traumatized by these betrayals, which can lead to depression and the conviction that they will "never trust anyone again," and "life isn't worth living." ... Those girls and women who do have friends note the conflict and danger but insist, "I can't give up my friends."[10]

> Even as mature women, the dilemma we experienced with our girlfriends during adolescence recurs: To be absolutely safe from betrayal, we must be alone, but being alone is lonely, and so we seek friends. Our paradigm of friendship involves exchanges of confidence and self-exposure – and hence vulnerability.[11]

One study of 60 males and 60 females aged 10 to 15 years found that girls same-sex friendships were of significantly shorter duration than boys', that if their current same-sex friendship ended they would feel significantly worse than boys and that their life would change more significantly than boys. Despite the shorter duration of the friendships, girls also reported that their current closest same-sex friend had already done something to hurt their friendship. Girls also reported that they had more same-sex close friendships with whom they were no longer friends than boys. Age differences had no bearing. If friends in general were considered, then there were no sex differences, it was only the closest friendship that revealed sex differences.[12]

The importance that these girls are placing upon secrecy makes little sense in modern societies – what would they have to fear? Yet, in cooperatively-simple societies, where a woman trusts another woman to act as the intermediary with her sexual selection, then such information is extremely dangerous. A revelation could lead to extreme violence. The importance of trust and secrecy must be paramount in such circumstances, and these girls, I would argue, are acting in a purely instinctive fashion, learning how to trust other girls, to prepare for selecting an intermediary. This reflects the intense levels of emotion in girls' friendships.

The consequences of failure were documented by Mervyn Meggitt and his wife; they found in the Walbiri:

> Adulterous wives generally combine in a conspiracy of silence, so that their husbands are uncertain of what is occurring in their absences. But female solidarity is a rather brittle bond, and a woman never knows when some minor dispute with another woman will lead to tale-bearing.[13]

Gossip is a consistent theme and linked to violence. Catherine and Ronald Berndt, in their study *Sexual Behaviour in Western Arnhem Land*, concluded:

> Natives of this region (like those of others) are confirmed gossips, and take much pleasure in interfering in the affairs of others – with which, moreover, they seem to be fairly well acquainted. It is often this inquisitiveness, and this tendency to meddle in the affairs of others, which directly accentuate the seriousness of arguments and quarrels.[14]

Though some may consider aspects such as assistance in the gathering of food or child care that may have been important in the evolution of women's alliances, I would argue that if the gathering of food or child care was so important, then women would be more interested in behaviours other than the exchange of secretive information – such as mutual assistance and physical ability. Yet, in women, the most important part of their alliances appears to be the ability to keep information confidential, and, without this idea of sexual selections being paramount, it is difficult to imagine for what reason such confidentiality could be important.

Status

If a woman is to form an alliance with another woman, then we must expect that the timing for the creation of such an alliance will be early in her reproductive life. Girls passing through puberty, who will soon expect to produce their first son, the son most likely to be polygynous due to primogeniture, will place the greatest importance on the formation of alliances. As women age, then the importance of female alliances can be expected to fade, as sexual selections become less likely or important.

A girl's status will depend upon how many brothers or half-brothers she possesses who may either live in the community or are frequent visitors to the community. Thus, even if a woman marries out of the community, if her brothers' and half-brothers' are in an allied community then they may pass into her community often enough to make her an attractive candidate for an intermediary. Napoleon Chagnon observed:

> Villages that are allied, moreover, send young men to each other to look for mates. These men are always suspect, since they try to seduce the young girls in the village and, when caught, they flee for home. If the alliance is a particularly critical one for either or

both villages, the culprit might be able to return after a cooling-off period and resume his bride-service for a promised wife. If the alliance is not critical, relationships between the two villages deteriorate.[15]

Chagnon also observes the importance of a woman's brothers in terms of her own status, particularly in her brothers' ability to prevent extreme violence from her husband-by-arrangement:

> The women who are exchanged are the ones who like it least of all, for in their new village they will have no brothers or other kinsmen to protect them from a cruel husband. And often their parents share the feeling.[16]

A woman's capacity to be protected from extreme violence is thus dependent on her brothers:

> In Village 14, for example, there is one individual who has no relatives whatsoever; she is a captive from a distant village. Attractive to a larger number of males than most women in the village (she has no kin, and therefore all men may copulate with her), she is more often exposed to attempts to seduce her. Her husband is a jealous man. The mere suspicion that she has been unfaithful enrages him: on one occasion, he shot her with an arrow in the stomach, and she nearly died. His violent actions are largely unrestrained, since she has no brothers to protect her.[17]

Interestingly, Eugene Kanin found a correlation between the presence of an older male sibling and a young woman's ability to avoid sexual aggression in modern environments.[18] Following the usual sociological view of the importance of education, Kanin argued that this was due to male instruction and education of sisters, though I wonder if it isn't due to the potential for male sibling violence that protects their sisters.[19] One wonders how many young men would rape a woman if they knew she had two or three older brothers, and if the presence of vigorous brothers is a greater deterrent than the police. While the police, judge or jury might doubt a woman's statement of the fact she has been raped, it's doubtful her brothers would. Equally, under modern law, a violent response by her brothers might fall under the category of provocation, suggesting limited criminal liability.

In cooperatively-simple societies, those women with brothers in the community will be the ones most confident of being able to sexually select, since their brothers will protect them from a husband-by-arrangement's violence. Women with brothers in neighbouring, friendly communities may be less confident of brotherly protection. Women with no kin in the community or neighbouring communities may be the least confident in their capacity to sexually select, since no-one will protect her from her husband. This hierarchy will be the same in terms of a woman's ability to act as an intermediary for other women. Women with the most brothers and half-brothers in the community will be the one's most able to act as an intermediary; those with brothers or half-brothers in neighbouring communities will be less able to act as an intermediary; those with no male kin either present, or as visitors, will be incapable of communicating with men other than her husband.

Along with brotherly protection and attraction, this hierarchy will possess a third element. Those women that marry within their community will remain with girls with whom they grew up and formed close bonds. Women that marry into another community will need to form new alliances in a community that they may have only ever visited. Women that are abducted need to form alliances with women that are complete strangers.

Women face other difficulties. Their community may fission into two or three smaller groups, thus separating them from a close, life-long ally. Women have no control over these events; and even whether a woman remains in her natal community, or moves to another community with her husband-by-arrangement, depends on the strength of her brothers. In his work on village fissioning, Napoleon Chagnon noted that men's reproductive success (inclusive fitness) depended on their ability to keep sisters in their community and men in smaller brother-alliances were less able to control this outcome:

> Preliminary results of analyses of cross-cousin marriages (which would be the result of systematic sister-exchange over several generations between members of two descent groups) in large descent groups suggests that the inclusive fitness of males depends to a high degree on their ability to get their own offspring married off to the offspring of their sisters – which means that they must try to keep sisters with them. The extent to which they are able to do this depends on the prominence of their descent group – the number of social allies they have. Men whose sisters have married into larger descent groups are less likely to be able to keep their sisters with them if their descent group is small and their own wives come from descent groups that fission away from the larger groups.[20]

Brothers and sisters face similar difficulties. Brothers need sisters to exchange for wives; and, more importantly, their sister's children will be cross-cousins to their own children and thus the most natural and stable choice for marriage in political terms. Sisters need brothers to attract other women's interest for sexual selections and for protection from their husband-by-arrangement. An attractive girl with lots of brothers in her community should thus desire an ally who is in exactly the same situation: an attractive girl with lots of brothers in her community. Girls in high status positions should thus want one type of ally: a girl who is exactly the same. Girls in lower status positions should desire a girl in the same situation or higher. Because of genetic relatedness, very attractive women will possess, most probably, very attractive brothers. Very intelligent women will possess, most probably, very intelligent brothers; physically robust women will possess physically robust brothers.

> Girls learn to find and take a place among the multiple regimes of power that exist among them. Standards of beauty have been well explored as conferring higher prestige, but there are other aspects of the political world that girls create to differentiate among themselves which have not been well understood.[21]

Again it seems odd why beauty would be important in alliances if the alliances existed for reasons of mutual assistance, such as childcare. Yet, if women needed to form alliances with other women in order to sexually select the only men that allied women could freely communicate with – her brothers – then it makes sense that she should want to ally herself with beautiful women who will probably have beautiful brothers.

One study that compared 46 pairs of female friends (aged 19 years) found a high correlation in their attractiveness (with the attractiveness judged by 9 naive students), though the women themselves did not judge themselves as being similarly attractive. There was also found to be a high level of correlation between self-judged differences in attraction and rivalry within the friendship: with those that felt less attractive reporting higher levels than those that felt more attractive, who in turn reported more rivalry than women who reported similar levels of attractiveness.[22]

I argue that a girl or woman's status will be reflected in her brothers' status. Those in the most powerful brother-alliance in the community will be at the top of the hierarchy; those in a less powerful brother-alliance in the community will be lower in the hierarchy; those with a brother-alliance in a neighbouring community will be lower still; and those women with no resident or visiting brothers will be at the bottom. This will certainly be true from other women's point of view. As stated, this hierarchy has four aspects: brothers to attract other women's interest; brothers to protect her; likelihood of remaining in her natal community where her lifelong female allies exist; and, her brothers' ability to ensure their sisters stay with them if the community fissions. (In modern societies, such status will be reflected in monetary terms.)

I would argue that genetic quality (attractiveness, intelligence) is an important aspect of alliance. However, a girl's status will be reflected in the status of her brothers and this should be more important than genetic quality. There is no point forming an alliance with an attractive woman if she cannot communicate with any men other than her husband-by-arrangement. One might thus conclude that women are extremely aware of status issues in their alliances, and that status would count before all other matters.

The Formation of Alliances

The process of relationship building appears to be similar to the intense feelings involved in sexual selections; in other words, girls fall in love with each other, then they fall in love with men. In their study, Apter and Josselson found:

> Admiration – and its close cousin, envy – is often what draws us to a potential friend. Often a girl "falls in love" with a friend – first admiring her from a distance, perhaps imitating her in a manner or dress, and then managing to draw close to her as a friend.[23]

> Many girls begin to idealize a friend and their love for her, just as they will later idealize a romantic partner and their romantic feelings.[24]

This is reflected in the bracelets that are often sold to girls, with the love heart broken into two pieces. Though on first instance, it might seem obvious that they would be for sexual couples, they are aimed at girls:

In any preteen boutique in any mall, there will be a large rack of broken necklaces sold together, two halves of a heart etched with letters that spell out "best friends." For girls just entering adolescence, having one half of one of these hearts means that she has a best friend, that she has been chosen by somebody as most important.[25]

The depth of the emotions involved appear to be the same as that felt between sexually selected men and women, as one girl in the study observed:

I fell in love with her and we tell each other everything.[26]

Girls and women appear to consider their ability to form alliances in much the same way as they might consider their ability to attract a man:

Much as girls may later learn to regard themselves as potential marriage material ("I could never get a guy like that!"), they first learn to evaluate their assets in being able to attract friends.[27]

Girls evaluate themselves in terms of who they can attract as friends. Their decisions about how to behave within a friendship are shadowed by their assessment of their own options within the friendship market.[28]

We learn as girls about our drawing power in the friendship market: Who can we attract to us and keep with us? And how do we do that?[29]

The method of creating the alliance is based on the exchange of secrets. Deborah Tannen, who wrote the linguistic classic *You Just Don't Understand* concerning the differences in the manner in which men and women speak, notes:

Not only is telling secrets evidence of friendship; it *creates* a friendship, when the listener responds in the expected way.[30]

Intimacy has generally found to be a consistent part of girls and women's conversations, but less so for boys and men. For instance, one study that compared 62 men and 74 women's conversations with a close same-sex friend found the following topics discussed frequently by women but not men: personal problems (45% to 14%), doubts and fears (46% to 16%), family problems (47% to 26%), and intimate relationships (26% to 8%). Men discussed sports far more frequently than women.[31] A review of 50 studies found that in the ages of 6-8 years, there were no differences in self-disclosure, studies involving ages 9-11 years showed 59% greater disclosure by girls, and of the studies of 12-20 years, 89% showed greater disclosure by girls.[32] These differences with age suggest that self-disclosure is linked to important requirements linked to puberty and reproduction. These differences appear to be cross-cultural. For instance, a study involving 66 girls and 45 boys in Barbados with an average age of 15, girls were found to describe a close friendship with a same-sex friend as having exclusivity and warmth/closeness significantly higher than boys, while conflict was similar. Girls also chose prosocial behaviour, admiration, similarity and affection significantly more than boys.[33]

Alliance formation appears to be a matter of exchanging information. Probably unimportant secrets are exchanged, followed by more vital secrets, and increasingly personal information. Girls and women will test each other, waiting to see if any information is divulged to a third party. If it is, the relationship may deteriorate; if not, they may divulge increasingly important information. Women will slowly develop a level of trust with another woman. I argue that with this knowledge, a woman will decide whether a woman can be trusted to act as an intermediary for her. If a woman trusts another woman implicitly with information, she may not hesitate to ask her to pass a message to the desired man.

Perhaps it is also likely that men will attempt to pass on messages to other women through his sisters. It is, after all, very much a part of Western human behaviour that a man can be expected to ask. A woman can sexually select a man through flirting – eye-contact, blushing, and so on – and he will then attempt to request to meet her through his sister. I suspect, however, that sexual selection would be female oriented in direction rather than male; if females depended on males to pass information to the sister, then there would be no preference for a best friend; simply knowing or being an ally of the male's sister would be enough. The existence of the best friend suggests that sexual selections are female directed, and this appears to be the case in the above examples quoted concerning Australian Aborigines. Phyllis Kaberry also notes:

> The woman as we have seen is capable of taking the initiative in courtship. She has her ornaments – headbands, fringes, cicatrices, necklaces, and nowadays perhaps one or two dresses, skirts, and red handkerchiefs. Occasionally she uses red ochre and grease; and when she desires a lover, she very practically sends a gift of tobacco by some intermediary. As such gifts are usually made to blood or affinal relatives, the implication is obvious.[34]

Furthermore, a woman needs an intermediary she can trust in order to organise several or numerous meetings. If a man sends a message through his sister that he wants to meet a woman, then that woman might decline the invitation because she views the intermediary as untrustworthy, even if the man is attractive. The trusted intermediary defines which extra-marital selections are possible, rather than any other factor.

As with men, proof of an alliance may also be reflected in violent cooperation; Meggitt observes one incident in the Walbiri:

> There had long been bad blood between Doris nagamara, the strapping young wife of Jack djabaldjari, and Beryl nagamara, the usually even tempered wife of Alec djabaldjari, and the women often quarrelled violently over trifles. The two men, who were good friends, although not countrymen, kept out of these disputes. One day, an argument between the women flared into a bitter fight with clubs. As the two were evenly matched in strength and skill, the onlookers did not intervene; but, just as Chloe nagamara, Doris' close friend, arrived, Doris suffered a severe scalp wound. Swinging

an enormous club, Chloe ran to her aid, and the pair were rapidly reducing Beryl to impotent screams when Clarry djabaldjari, Chloe's husband and Alec's half-brother, appeared. Energetically using the flat of a boomerang, he flogged Chloe the 400 yards back to their camp and furiously berated her for "double-banking" his brother's wife. Elsewhere in the camp, Jack and Alec elaborately ignored the affair.[35]

Apter and Josselson also comment on the importance of continuous support:

> What girls and women say they want is for someone who is "always there" for them. A true best friend is someone who is ready to console you when you are upset, to listen to you when you need to talk, to take your side in any dispute, to try to meet your needs. A best friend looks out for you.[36]

I would argue that girls and women form friendships with similar levels of intensity or 'love' as they will in sexual relationships; that these relationships are based upon secretive information and a lack of revelation of that information; and that girls analyse female friends in much the same way that they will analyse men. Girls want a physically attractive, intelligent female ally, much as they would want an attractive, intelligent man to sire their children. In societies with arranged marriages, I would argue that a woman's only link to other men is through a desired man's sister, and this creates the market for female alliances.

Women's alliance formation would encompass a gradual exchange of increasingly personal information or secrets. This would allow women to possess varying degrees of trust with other women according to the level of importance of the information exchanged.

Information

Anne Campbell in her book *Her Own Mind* summarises women's friendships:

> There is a high degree of exclusivity in these friendships and friends are acutely concerned with maintaining them. To spend too much time with a third girl is to threaten the bond between friends and much of the gossip that takes place concerns who is best friend with whom and whether that relationship is in jeopardy. Because girls invest so much of their private lives and personal feelings in the revelations that they make to one another, the loss of a best friend threatens not only loneliness but also the betrayal of the confidences that have been made. The importance of trust runs through the literature on girls' and women's friendships and the greatest betrayal is the abuse of trust. It is trust that distinguishes a best friend from other friends...[37]

Deborah Tannen, in *You Just Don't Understand*, observes:

> For most women, getting together and telling about their feelings and what is happening in their lives is at the heart of friendship. Having someone to tell your secrets to means you are not alone in the world. But telling secrets is not an endeavor without risks.

Someone who knows your secrets has power over you. She can tell your secrets to others and create trouble for you. This is the source of the negative image of gossip.[38]

Again, in a cooperatively-modern society, one must wonder what girls and women would lose by such revelations. Yet, in societies with arranged-marriages, where sexual selections are extremely dangerous, and where women depend upon other women as the messengers to men, secrets are important. A woman, in such a situation, does have power over another woman. A messenger could reveal a sexual selection, and the woman in the sexual selection could be severely beaten, burned, or even killed. In such circumstances it is not surprising that women would take the betrayal of a confidence so seriously. This instinctive behaviour continues in women in cooperatively-modern societies.

An example of how dangerous information can be comes from Helena Valero's experiences with the Yąnomamö. In an environment where men fiercely protect their reputation, Valero states that she called a dog 'dirty, bad', but it was relayed by another woman to the *tushaua*, the headman, that Valero had called the headman 'dirty, bad'. The *tushaua* then proceeded to attempt to kill her and shot two poisoned arrows at her, forcing her to run and hide in the jungle for an extended period of time. The entire event appears to be based on enmity between Valero and the other woman:

> Along the path at that moment came the *tushaua* Fusiwe, returning with the other men; behind them came a huge dog. The children said: 'Here comes that bad dog!' Then I exclaimed: '*Shami watiwe*' (dirty, bad!). I threw myself into the water, I was so scared, and stayed in the middle of the *igarapé*. A woman, a cousin of those with whom I had lived, heard me and told the *tushaua's* mother that I had called him 'dirty, bad', and that I had said: 'I will not go and live with that scoundrel who has so many women!' It was not true. I had said nothing, I had said those words only to the dog. That woman probably wanted to see me killed. The mother called the *tushaua* and – they told me afterwards – spoke to him as follows: 'You are a grown man, the strongest among all these men here; do you allow yourself to be called dirty and wicked by a woman? You are not *shami, watiwe*, you are the handsomest man here. Kill her at once!'
>
> I went back; I knew nothing of it. I started to prepare *pupugnas*. I saw the *tushaua* approaching me with bow and arrows in his hands. But I was innocent and suspected nothing. He looked at me intently and I turned my head. 'Here's the woman who says I am ugly and dirty!' I turned round and saw that he was pointing his arrow in my direction: I made a leap over the woman who was squatting down near me. The other women threw themselves on this side and that, and the small son ran with his bare feet into the fire. I parted the palm branches of the roof and jumped over the palisade. I don't even know how I did it, I was so frightened. They told me when I came back: 'But how high you jumped!' While I was running, there came an arrow. It did not enter my flesh, it just hit me, tac, on the arm, like the blow of a stick. The arrow fixed itself into a *pashuba* trunk: it was a poisoned arrow. They told me he shot two arrows.[39]

She eventually married the *tushaua* and they had three sons. Valero does not explain why she thinks the cousin wanted her dead.

Information or gossip can be extremely dangerous. If a woman tells another woman about her ally's sexual selection, then she is giving that third party the ability to harm her ally. This, I would argue, is the very basis of the 'best friend'; true safety only occurs when the sexually selected woman knows of the sexual selection, and if the messenger knows. The more people that know about the sexual selection, the more likely it might be revealed.

Yet, since information or gossip can lead to extreme violence, this makes it valuable. As there is a marketplace for allies, we can also expect a marketplace for information. Apter and Josselson comment:

> "Juicy" gossip, with its overtones of secrecy, is tinged with excitement, marking it as information out of the ordinary. Women and girls know the delicious anticipation that attends a friend coming to them with a gleam in their eye and the words "Have I got something to tell you!" The very secrecy of information makes it sparkle.

> Gossip is something we depend on a friend to have. It's part of what makes her fun. We thrill to our own status when we have a rich piece of gossip, something that is a source of pleasure or excitement or entertainment to others ("Wait till you hear this!"). Our insider knowledge raises our social standing, especially when the information comes from high-status sources: If you are close to a popular girl, then you, too, are popular. You can prove that you are close to her by knowing her secrets, but you can only prove you know her secrets by betraying them, so friend talk and betrayal – those seemingly utter opposites – often fit together like hand in glove.[40]

There are numerous women who might want to end another woman's sexual selection. A wife-by-arrangement of an attractive man might be extremely jealous of him, and attempt to prevent any woman sexually selecting him, possibly by revealing the sexual selection. A woman that desires a man and discovers that another woman has sexually selected him can end the sexual selection by revealing it publicly. Women can control other women's sexual selections by being informed of the sexual selection, and revealing it to those who will inform the husband-by-arrangement. A sister might be able to end a sexual selection of a brother, because she finds the woman involved unworthy of him. On each occasion, the threat of disclosure may be enough, and the information may not even need to reach a husband-by-arrangement. Alternatively, violence may also be used; the following is from Warner's study of the Murngin:

> The galle has another recourse. She can attack and abuse her husband's mistress. A wife usually feels that it is the other woman's fault anyway.

> Balliman, a young man noted for his fighting and dancing abilities, was Mumulaiki's lover, so camp gossip had it. Balliman's wife (he had inherited her), who was much older than he, took her digging stick and, with the aid of much cursing, gave Mumulaiki a thorough thrashing. Both women were in fainting condition when the fight was stopped by others. Their heads were cut open and they were covered with

bruises and cuts.[41]

The following from Burbank's study of Australian Aboriginal women, *Fighting Women*. This concerns a community of 400 people in a community in Arnhem Land, which Burbank calls 'Mangrove'. The following is one example of a complicated dispute concerning a young woman's interest in a man. In this case a young woman is interested in another woman's male partner. The latter woman has a child by that partner. This mother's extended female allies attempt to intimidate the young woman's relatives (note that almost no men are mentioned):

> Yesterday, shop time, [a woman] was in the office and [another woman] saw her and went into the office and they were growling—because [the first woman's sister's daughter] was teasing [the second woman's daughter]. Then just after supper, just before sun go down time, the whole lot [the second woman, two of her sisters-in-law, her mother-in-law, and another woman] came marching down to [the first woman's mother's] house. [The second woman and her mother-in-law] had sticks and [the second woman] hit a trash barrel. But they only growled at [the first woman's sister and her husband—the parents of the girl who had been "teasing"—and mother]. [The first woman] wasn't there. [The second woman] said, "You don't let your children be cruel to my children when I go away." [The first woman's sister's husband] was growling too, and so was [her sister]. Her "family," her mother [and two of her mother's half-sisters] were trying to stop her, but she just kept growling. I don't know how she didn't get that temper and [start fighting physically]. She used to be like that. [One of the sisters-in-law] said, "Don't growl at me, I have a pretty leg." And then she just walked away. That means she knows someone is sleeping with her brother [real or classificatory]. I don't know if anyone is. [The first woman's mother] picked up a rock, but she didn't throw it. [Another of the second woman's daughters] had a baby for that man; they don't like [the first woman's sister's daughter] to go after him. [A young woman] was angry at [the first woman's sister's daughter] for that and that's why she was teasing [the second woman's daughter]. [The first woman's sister's daughter] was trying to get that boy.[42]

Rather than a fight between schoolgirls, concerns over sexual selections can involve intimidation between large numbers of their extended families, though mostly between women. Men appear to only get involved when they must, perhaps a reflection of the higher levels of violence that occur in disputes between men. In cases of aggression where women were the instigators, Burbank found, in fights involving women, that about half of all fights involved women against other women (52%: 147 of 285); a quarter involved women against men (22%: 64 of 285); and another quarter involved aggression against a single woman (26%: 74 of 285). This reflects the social aspect of disputes, and that women tend to team up, rather than act individually.[43]

Indeed, Burbank found that women were involved in aggression more often than men: 'Women are among the initial actors in 495 of the 793 cases of aggressive behavior that I collected. In 285 of these cases, women are clearly the instigators of aggressive behavior; in

147, women are the victims of other women'.[44] In another 74 a single woman was the victim. While the women being studied were in a community heavily affected by Western influence or products (such as alcohol consumption and petrol sniffing), the arranged marriage system continued to exist. Burbank was able to identify a reason for the aggression in 217 of the 285 incidents of female initiated aggression. While elsewhere I have documented that men fight over marriage arrangements, Burbank noted only three cases where marriage arrangements led to aggression instigated by women. Burbank observed that in the 113 cases of aggression between groups of women, 23 involved pre-marital sex, and 17 involved jealousy; by contrast, only 5 involved essential resources, and 14 involved the misbehaviour of others (not working or helping). When aggression involved women and men, the most common reason was essential resources (8 out of 55 occurrences of aggression). If the aggression was aimed at a solitary woman, the most likely cause was gossip or sorcery (aggression) (10 out of 49) and disputes concerning children was the second most likely (9 out of 49).[45] When the aggression escalated to fights, jealousy caused the highest number (20 out of 49 fights), and a further 7 involved pre-marital sex; four involved children and only two involved essential resources.[46] Fights between women led to seventy-five injuries: fifty were caused by fights due to jealousy or pre-marital sex; by contrast, four were caused by a fight over essential goods, three injuries involved disputes regarding children, and one involved the arrangement of marriages. Swearing led to four injuries, gossip or suspected sorcery (aggression) led to five injuries, misbehaviour (not working or helping) led to two, and delinquency (theft) led to two, suggesting that issues of status and discipline are as significant as essential goods or children.[47] It needs to be kept in mind that the people studied were heavily affected by Western culture (such as receiving welfare and living in houses); however, this is a rare example where data concerning aggression and violence involving women living in a society with arranged marriages has been collected. The fact that two-thirds of injuries in fights between women were due to sex outside marriage gives at least some indication of what motivates women to escalate disputes. Once women learn of the extra-marital sex that is occurring within the society, it appears that such information is the most likely cause of violence between women; gossip or sorcery (which might be defined as 'lies') was marginally more likely to lead to injury than essential goods.

The more information a woman has about what is going on in the community, the more she can control what occurs. This perhaps leads to a further example of status. Previously I have commented on a woman's brother-alliance's strength, or her physical appearance; a further aspect of status will be knowledge. The more a woman knows, the more other women will want her information, and the more popular she will be. Women may well fear a woman who knows she has sexually selected a man as much as she might her own husband-by-arrangement if he finds out.

In recent years it has become more accepted that girls and women are equally as aggressive as men, but that most of the aggression involves indirect aggression.[48] However, Anne Campbell has found that young women are more likely to engage in violence in lower socio-economic environments,[49] and in the anthropological literature there certainly does seem to be a greater willingness by women to engage in physical violence.

As to indirect aggression, in a Finnish study involving 389 8- to 15-year-olds it was found that the use of indirect aggression develops at around age 11 in girls, and that girls are more likely to use indirect aggression and withdrawal, while boys are more likely to use direct verbal aggression or physical aggression.[50] Amongst 15-year-olds, methods more likely to be used by girls were: trying to irritate the other so as to cause a loss of temper, trying to make the other look stupid, backbiting (gossiping), spreading vicious rumours as revenge, attempting to defame the other over a long period, and gathering other friends to one's side. Group sizes of friends also differed, with 49% of girls most likely to be named by at least 30% of peers as being in a friendship group of two, while, for boys, the most common were four or more (27%) or alone (40%). A later study involving Finland, Poland, Israel and Italy confirmed these direct and indirect methods amongst boys and girls, respectively.[51] Other studies have shown that indirect aggression begins in pre-school ages of 3-6 year olds and has a positive correlation with the number of friends.[52] Relational aggression has also been positively correlated with popularity and physical attractiveness in both boys and girls (aged 13 years).[53] A three year longitudinal study of two schools involving 480 students in early adolescence found that students nominated as friends students who they deemed to be popular rather than unpopular, and also nominated these popular students as engaging in indirect aggression, but not physical aggression.[54]

This leads to a further aspect of women's alliances: inclusion and exclusion. Apter and Josselson note:

> While boys' bullying systems center on the question "Who is stronger?" girls' teasing defines other girls as acceptable or unacceptable, "in" or "out." The question is not "Who's dominant?" but "Who's included?" For girls, power is the authority to exclude.[55]

A woman who is not a part of the exchange of information is not merely unable to control other women's lives, but unable to know if they can control hers. Girls' and women's analysis of exchanges of information are, thus, unsurprisingly minute in detail; Apter and Josselson report one girls' definition of her friendship:

> We spend hours talking. Just talking. Sometimes it's about just one thing, like what happened last night, or yesterday at school. I tell her what happened, and then she'll ask a hundred questions. "What did she do then?" or "How did she say it?" and "Who else was there?" And it goes on and on...[56]

Deborah Tannen speaks of men's inability to converse in the same manner:

> Yet another woman recalled a time that her best friend's husband tried unsuccessfully to take part in one of their conversations. Breaking with tradition, he tried to tell about an experience that he thought was similar to the ones they were discussing. The two women plied him with questions he couldn't answer about exactly what had been said and how and why. He backed off from the story, and didn't try to tell any more. Perhaps he was wondering to himself why the women were interested in all these unimportant details.[57]

The formation of cliques is the best example of inclusion and exclusion. A clique is a centre of information sharing, and girls are either in or out of the clique. A clique allows girls and women to pool information and exclude others from the information. Furthermore, in cooperatively-simple societies, it enlarges their group of close allies who can assist in sexual selections. Apter and Josselson note:

> Many adult women still name someone as their "best" friend, while others name several close friends but don't feel that any qualify as a "best friend," seeing each friend within a wider network. And even as they think in "best friend" terms, girls often form friendships in threesomes and foursomes, and establish a circle, a culture with its own values and rituals.[58]

Even if she trusts one woman more than another, by possessing a larger circle, a woman increases her ability to sexually select, since she increases the number of women's brothers with whom she can communicate. Since such relationships are based on the exchange of secrets, she also increases the pool of available information. She thus increases her ability to manipulate others' lives and to sexually select. By contrast, excluded women are excluded from both the information and the potential to sexually select the brothers of the girls in the clique. Without the basis of assistance in sexual selection, it is difficult to appreciate exactly why women would need to form such relationships, nor why information was so valuable.

In cooperatively-modern societies, this method of trust and exchanges of confidences can be counter-productive. One woman in Apter and Josselson's study noted the difficulty of the exchange of secrets between women in the modern business environment:

> It is easier to be friends with the men, even though they're often more difficult. The thing is – you see the competitiveness. It's up front, and you know that if they have something to use against you, when the time comes, they'll use it. With women, there are mixed messages. I can't afford to be what I call friends with a woman I work with, even though many of the women here are great, and I miss like hell that closeness that I could have with them.[59]

One example given was of a woman who learned later that her workplace friend had cancer but was not informed because the latter feared it would affect her chances for promotion.[60] Another comments:

> I have to be more like a man when it comes to relationships at work. I have to keep a watch on myself.[61]

By contrast, most men are fully emotionally prepared for such an environment. Most men will be used to the idea that even a good friend may not inform them of negative events in their life, because this is to inform other men of his weaknesses. In a society with arranged marriages, a man that told others of his weaknesses would be opening up other men's interest in his wives-by-arrangement, and could lead to their taking advantage of his sudden weakness by initiating fights they knew they would now be able to win. The idea that a man would keep knowledge of his cancer from others in a competitive workplace would be, to many men, common sense.

Yet women did not evolve for such an environment. Women's exchanges of confidences may be counterproductive in the workplace, all she does is expose herself with no apparent benefit. This is not because of men, or because men have created the workplace environment; the criticism levelled in this situation is from women about other women and women's behaviour in the workplace.

Though men lack the intimacy, men immediately recognise that their success is based on their ability to build networks with other men, over the long-term, even if they do not necessarily like the actual man. Men's competitive survival is based on their ability to form a large, cooperative group that can compete against other cooperative groups of men. Grievances, anger, and hostility are smoothed over for the main task of cooperating. Women, by contrast, do not need such levels of cooperation. Women may be more cooperative with a single woman, but they are less cooperative with women generally. Apter and Josselson remark on the difference between men's rooms and women's rooms in business:

> Preparing to enter a difficult world in which they would have to compete to succeed, many women were disconcerted to see how men often did not have to compete with one another. Instead, men befriended one another, mentored one another, supported one another. What many women experienced was not the cutthroat environment they had expected, but a cozy club.[62]

> Once women did try to form a corresponding "women's room," they found that, far from its being the place of comfort they had envisioned, the rivalry and backbiting in some groups often seemed to threaten the fabric of women's progress. Women found themselves unwittingly re-creating the "clubs" of junior high, with punishments for those who didn't conform, competition for closeness to the women who seemed most popular (and therefore powerful), and a grapevine of rumors about who said what about whom behind her back.[63]

Workplace bullying has only become an issue for research very recently. Studies of female dominated workplaces, such as nursing, report that high levels of bullying do occur. In one British study of 1100 NHS staff, in a workplace where 84% of staff were female, nurses reported bullying in the previous twelve months at a higher level than other staff (44% to 35%). Males were over-represented as bullies (26%), 66% of bullies were female, and 8% involved both sexes.[64] The most extensive and longest studies come from Norway, where in one survey of 7,986 workers, 85% in public organisations, 8.6% stated they had been bullied in the last 6 months, with 4.5% experiencing serious bullying. Those that were bullied labelled males as the perpetrators in 49% of cases, and 30% were female perpetrators.[65] Men reported another man being the bully in 90% of cases. In another study, of a nearly all-male large industrial organisation, revealed that 89% had been harassed in the last 6 months.[66]

With such evidence, perhaps the anecdotes presented above reflect an idealised expectation of women in the workforce, and an idealised view of men's relationships. However, it does appear to be the case that women do engage in interpersonal behaviour that is demeaning, undermining, or antagonistic. Evidence of women's inability to engage in normal exchanges of personal information with colleagues with whom they have a positive relationship and fearing any information may undermine their position in the workplace appears not to have been thoroughly researched and so it is difficult to assess its true nature. Since it appears to be a natural part of women's behaviour to share personal information, it would not be surprising if this was found to potentially lead to conflicting situations in a competitive environment. Women appear to be less likely to enter competitive environments than men even when they were higher achievers,[67] and more likely to state that a positive social environment will affect their decision to remain in a workplace than men.[68] There is a question as to whether women avoid workplaces where their normal behaviours will be less suited to the environment.

The basis, I would argue, of women's confidences has nothing to do with work – whether gathering food in a cooperatively-simple society or in a modern office, – it has solely to do with sexual selection in an entirely different environment.

Finally, it is instructive to consider the reasons of why women stop engaging in secretive exchanges. Tannen notes that a lack of information, or secrets, can lead to the disintegration of a friendship between two women. This is often likely to occur when one of the women is happy with her partner:

> Because telling secrets is an essential part of friendship for most women, they may find themselves in trouble when they have no secrets to tell.[69]

If women's need to engage in such sharing of secrets was based on sexual selections, then we should expect that a desirable marriage should end a need to share secrets. Even if a woman wanted to form a close relationship with another woman, she simply lacks the tools to do so. In order to be able to receive, she needs to be able to give.

A secondary reason for no longer engaging in secretive exchanges appears to be depression: 'Closing the door on girl talk is a symptom of depression – a sign that a girl believes she is beyond help.'[70] One study of 13,465 adolescents in a longitudinal study found that the biggest predictors of suicidal thoughts for girls were 'friend attempted suicide', 'intransivity index' (friends do not name one another as friends), and 'isolation'. By contrast, for boys, isolation and a high intransivity index were low predictors of suicidal thoughts. The biggest predictors were whether a friend or family member had committed suicide. For both sexes, these were bigger predictors than 'depression', 'homosexual attraction', 'parental distance' or 'forced sexual relations' – commonly cited reasons for suicide.[71] It would seem remarkable that whether a girl was in a tight-knit friendship group or a loose friendship group would affect suicide ideation, yet this reflects the overwhelming instinctive requirement for girls to have close friendship ties.

Alliances and Men

If women's alliances concern sexual selections, then two further conclusions can be drawn. Firstly, while unallied women will be competitors for the sexual selection of a man;

allied women will be cooperative. Allied women should assist each other to select men and not compete. Secondly, since women's alliances concern sexual selections, men should possess a strong suspicion of women's alliances.

Apter and Josselson note:

> Girl's and women's interest in men is more likely to make them allies than enemies. Men, and the special kind of interest we have in them, become important points of contact, significant conversation themes, as girls and women share their observations and descriptions of men, analyze their character and their value. They watch a friend's love life intently, learning from her trials, experimenting with life and expanding through a friend's experience. In some respects, they learn to love and to idealize men in one another's company. Girls see their personalities as "interesting" or "cute." They find "depth" in even the most brutally abrupt young man. They guess at – or create – a wonderful inner life on his behalf: "He's sweet. And he's really sensitive. *I* can tell."[72]

> Throughout their lives, females offer one another love advice. Together they plot strategies to interest a romantic partner and scheme revenge on unfaithful lovers. Attraction to a boy can act as cement in girls' friendships: It is something to talk about, something to plot and plan, to analyze, to exult in or despair over.[73]

Phyllis Kaberry observed amongst the women she studied in the Kimberley region their tendency to speak together openly about their relationships with men, even if the relationships were forbidden:

> Women in these matters are ready to laugh at their own delinquencies, gossip about those of others and as a rule condone laxity unless they happen to be near relatives, when they showed some shame and embarrassment.[74]

Kaberry documented how women's ceremonies in the Kimberleys concerned not merely the passage from girlhood to womanhood,[75] menstruation,[76] and childbirth,[77] but also black magic in which she may attack a lover, a husband, or another woman.[78] It is in the secret women's corroborees,[79] however, where women seek to control their husband while they were away,[80] or – reflecting the usual double standards of both sexes – engaged in singing and dancing that was intended to increase the chances of one or more women in the group to gain a lover. The latter of these corroborees was not a simple matter, with two hours spent on body painting alone:[81]

> The younger women would express a desire to have the corroboree and the maŋanbara:l would then decide, perhaps in consultation with the other old women, when it would take place. They might wait till the next inter-tribal meeting, but often there were sufficient women living on the larger stations to make a more frequent performance possible. The women when alone discussed the event eagerly, describe ones they had seen, teased one another about their lovers. They enthused over the designs painted on the dancers, and continually stressed the fact to me that this was "proper *lubra* business;

blackfellow can't see him. This one play belonga *djibönir* (sweetheart). *Lubra* no more want him old ŋuːmbana (husband) him get 'em young boy." A statement that can be readily understood if it is remembered that some of the younger women were married to men considerably older than themselves. ... Now in the songs the tribal son-in-law is cited as a particularly desirable lover – a man, in short, whom the women should have been most assiduously avoiding. They giggled when they explained this and seemed to delight in the illicit character of the dance and the fact that they were breaking one of the strictest tribal taboos.[82]

Summing up one of the songs, Kaberry writes:

> Summed up briefly, the songs refer to the painting, the other forms of bodily decoration, and to the male and female genitals. The women constantly used in explanation the word "sing", which in pidgin-English has the significance of magic directed towards or into a particular object or person. So they would say: "Me sing that one *djibönir*; me sing that one *mari* (penis)." The lover is pictured as sitting in the camp. He thinks of the girl and trembles with sexual excitement. The woman then "sings" his penis, so that it will grow long, so that she can say "that my penis now". The boy cries for the girl, and his penis grows so long that it falls to the ground like an umbilical cord. The girl trembles with desire and goes to him. He opens her legs and sees her clitoris. The boy is symbolized by his headband, he becomes dizzy with desire, and the woman says "that my headband now". The woman "sings" her own genitals, and the man ejaculates. She makes a bed of leaves. The song now reverts to the use of the *miliri* which is rubbed between her legs in the dance "to make her genitals long; to make them 'poison', to make them good fellow alonga boy". Finally the boy opens her legs and they have intercourse.[83]

In that passage it is interesting that the women themselves appear to acknowledge their own illicitness in sexual behaviour, similar to the Adam and Eve or Pandora's box myths. It should be noted that the actual dance, however, is simple and not erotic:

> Their steps alternate between a kind of shuffle and a half-skip. Some of the other movements are more graceful – the swaying of the body from side to side from the hips, the gestures with the arms, or with sticks carried in both hands and the response to the rhythm of the singing.[84]

The *Yirbindji* does not confer sexual rights on a woman of which she did not illicitly avail herself before. But it does give her the moral or a-moral support of the other women; it does generate additional confidence in her powers to arouse desire in a lover or a husband; it does sheathe her in a protective magical armour; and it is an opportunity for display and recreation corresponding to that enjoyed by the men.[85]

The corroborees have two desires:

...they represent a sanctioned method of partially circumventing some of the disadvantage of a law that prescribes the relationship of those who shall marry, irrespective of mutual desire. The dance is thus a safety-valve; a means of mitigating the dissatisfaction of the younger women (and indirectly that of the men).[86]

For a woman who is satisfied with her husband, there is probably the fear that he may be unfaithful or that if she is approaching middle-age, that she may cease to attract him. The dance may thus be a means by which the marital tie is strengthened or it may be a means by which temporary sexual satisfaction is found elsewhere, so that the marriage can be endured for the other advantages it offers – a settled stable existence in conformity with tribal law.[87]

Amongst the communities in the Kimberley region, the women not merely join with another woman or clique in order to talk and plot, but rather create entire corroborees that set up a formal theatre to present their interest, either general or for a specific man or to maintain a relationship with a desired husband. Women are thus communal in their support for each other's sexual desires, and, in these communities, formally so.

A woman's ally is thus not a competitor, but an assistant. One can immediately contrast this with men, amongst whom, even good friends may be rivals.

As might be expected, men are instinctively aware of this and suspicious of women's alliances. The most obvious example of this is the manner in which women's conversations are often referred to as 'gossip' which is seen as something that can damage society.

Men understand all too well the radical nature of female friendship. Their fear that women friends are laughing at them can be heard in the way woman talk is demeaned as "natter" or "gossip."[88]

In a society where women can secretly pass information without men knowing and create disorder through revelations of extra-marital affairs, or create suspicions of extra-marital affairs, then it is obvious that men would view women's conversations negatively.

A further aspect of concern for men is that a wife's conversations with other women undermines the care with which they prevent their own weaknesses becoming common knowledge. Tannen writes:

Many men resent their wives' or girlfriends' talking about their relationships to friends. To these men, talking about a personal relationship to others is an act of disloyalty.[89]

Such strong reactions support the claim by anthropologist Jill Dubisch (writing about Greek culture) that talking about family matters to nonfamily members is taboo because it destroys a sacred boundary between inside and outside, taking outside the home what properly belongs inside it.

Dubisch also points out the symbolic connection between verbal and sexual pollution: Allowing strangers into the house by telling the family's secrets is like "illicit sexual penetration." This seems to capture the dilemma of widows in Greece as seen in a line from a lament that Caraveli recorded: "Widow in the house, gossips at the door." The widow is confined to the home, because if she steps outside, anything she does will expose her to the sexual accusations of gossip.[90]

If a man is suffering from occasional impotence, for instance, it is immediately understandable that he would be concerned at his wife-by-arrangement's conversations with other women. The same would be true if he was suffering a physical or mental injury, or if he was depressed; these are things that he would not reveal and his own brothers might not know, but, considering the physical, sexual relationship with his wife-by-arrangement, she might. Weaknesses that he might hide, afraid other men would take advantage of them, could be revealed and become common knowledge.

We might expect that men would not merely be suspicious of a woman's allies, he might attempt to separate her from them, and end her alliances with other women. In any environment, the better a man can isolate a woman, the more confident he will be of his paternity. This would be especially true in environments where men and women have arranged marriages.

Romantic Alliances

There is no doubt that female homosexuality exists in cooperatively-simple societies. Helena Valero observed, during her time with the Yąnomamö:

> There were some women who were friendly with each other, but the others found this very repulsive. One day some girls were saying to one: 'That is your woman.' The girl's mother listened and said nothing; it was a girl who did not want men and fled from them. Some time later the same girl was asking her girl friend to go with her to collect wild *caju* in the forest; the other girls wanted to go, but this one said: 'No, we are going on our own, because you never collect anything.' The others answered: 'Go on your own, then, for you are husband and wife!' Then the mother took a stick, ran after her daughter, knocked her down and gave her such a rain of blows.[91]

From modern studies homosexuality in women appears to be exactly the same as any alliance, but with sexual activity between the women. In *Lesbian Friendships: for ourselves and each other*, Weinstock and Rothblum and their contributors argue that the most important aspect is the friendship between the women, that a homosexual relationship is something of a romantic friendship. Women in homosexual relationships often consider their lover to be their best friend and their sexual partner;[92] they are often friends first and then become lovers;[93] and women that were homosexual lovers also tend to remain friends once the sexual relationship has ended.[94] I think it is also interesting that homosexual women may refer to their homosexual friends as their 'family'.[95]

The difference between homosexual women and best friends appears to be that they are very similar, but the former includes sexual activity. This appears to be a matter of distinct confusion for homosexual women:

> Jill touches me, looks at me closely, compliments me just like a lover – but no way
> would she ever be my lover. There's a code I'm trying to crack. Most women have it pat
> because they don't need it. They feel this love for a woman friend, and everyone knows
> it's not sexual. So how do you go about reading desire in another woman? That's what
> I'm trying to figure out each time I feel drawn to someone.[96]

> I've asked Jill, "Do you love me?" And she says yes, and says that she loves several women
> – maybe ten or eleven – and she even described something that I would call "falling in
> love." She meets a woman, and there's something about her – maybe something strong,
> like the ways she stands up to a colleague, or it can be some show of weakness, like being
> self-conscious about her hips, or it can be physical: "I like her voice. I like her gestures. I
> like the way she looks." And then Jill says, "But I'd never even think of going to bed with
> her or developing that kind of relationship with her." And that's where I get confused.
> Because it's so very different for me.[97]

The major question thus concerning homosexuality in women, is why is there any
necessity for this sexual interest in other women, when the difference between being best
friends and sexual partners might be very small.

Women in cooperatively-simple societies who are homosexual to any degree will
simply get married as part of the arranged marriage system. As other women have no choice
as to their husband-by-arrangement, whether a woman is homosexual or not, she will have
as little choice over her sexual partner's identity as she will over the sexual partner's sex. She
will get married and have children, no matter what. The same is true for homosexual men.
Men whose father has chosen a bride for him will marry her and have children with her;
a man whose brother dies will inherit his wives and copulate with them. This will be true
even if he is sexually attracted to men rather than women. To comprehend this, one need
only consider the 'honour' killings that occur in many 'traditional' societies or the various
examples I have given of women being abducted by force. Indeed, one need only consider the
history of Western nations to appreciate that one hundred years ago, the concept of 'coming
out' in a homosexual relationship was implausible in Western society.

Evidence also suggests the fluidity of women's sexual orientation. In a study of 3,963
adolescent females, 21 identified as lesbian: 4 had only male partners while 12 had male and
female partners. Of the 113 who were 'not sure' of their sexual orientation, 91 had only male
partners and 19 had male and female partners. Of the 163 who were bisexual, 66 had only
male partners and 91 had male and female partners. Of the 3,666 who were heterosexual, 65
had only female partners. In fact, of the 3,963 adolescent females, only 79 had female partners
only, and 82% of these adolescents stated they were heterosexual.[98] While this may be in part
due to experimentation and learning, almost all female adolescents, no matter what their
stated sexual preference, had male partners. One study of 1,304 women with an average age
of 40 had a much higher ratio of heterosexual, lesbian and bisexual women (49%, 40%, 11%
respectively). Despite their sexual preference, lesbian women had more male sexual partners

in the previous year (2.4) than bisexual women (2.2), while heterosexual women had the fewest male sexual partners (1.4).[99] These studies show that women self-identifying as lesbian have male sexual partners in adolescence and in later life. Two studies investigated the fluidity of women's sexual preference. One study of 81 non-heterosexual identifying women aged 18-25 years observed the changes in sexual identity before and after 5 years: lesbians (39 initially, 22 after five years) bisexual women (27, 12), unlabelled (25, 7). Of the changes, 9 now identified as bisexual, 10 as lesbian, 12 as unlabelled, and 10 as heterosexual.[100] A similar study over two years of 80 non-heterosexual identifying women aged 16-23 years, found that 44% of lesbians had previously identified as bisexual, and 25% of bisexuals had previously identified as lesbian.[101]

Female homosexuality appears only to be a slightly different intensity to being best friends, and homosexual women will marry a husband-by-arrangement and reproduce like other women. The only factor that I believe could benefit women with homosexual and heterosexual preferences is that women with homosexual preferences could initiate female alliances at a faster rate than women with solely heterosexual preferences. This might be beneficial for women that marry exogamically, to a man outside of her community. If this were true, then isolation should be linked to homosexual preferences. The benefit would be that a woman with homosexual preferences who married exogamically would be able to create a new network of friends faster than a woman without homosexual preferences. With the alliances intact, it would then be a matter of using those alliances to select heterosexually. That women with homosexual preferences continue to engage in sexual relations with men is reflected in the above data.

However, in order for this to be true then one would expect that there would be a link between isolation and separation during puberty and homosexuality. I have never read of such a link. Since female homosexuality is so similar to female alliance formation, I can only suspect that the two are very closely linked, and that homosexuality is simply a manner of creating allies. That women who engage in homosexual activity tend to be both interested in men and women, amplifies the possibility that a woman could create a best friend in a sexual relationship, and later consider an extra-marital selection. Since, above, I have argued that women depend upon allies in sexual selections in societies with arranged marriages, that homosexual behaviour might form the groundwork for later sexual selections is somewhat plausible.

Menopause

The final alliance that a woman makes is during and after menopause. With menopause a woman enters a period of time, sometimes lasting decades, where she will not reproduce. There have been various suggestions as to the reasons for this, and the reason most regularly used is that a woman gives up her fertility to assist her daughters. While a mother would, no doubt, assist her daughter, I doubt that this is the reason for menopause.

If a woman is to give up her fertility in her forties then she is giving up her opportunity to have another child or even two if she lives until 60. It would seem possible that many women could produce a child in her mid forties and her late forties or early fifties. If a woman is to give up the opportunity to produce another child or two, then she needs an equivalent benefit to her inclusive fitness – such as her own children producing more grandchildren.

If a woman gave up her fertility to assist her daughters, then she would need to be the difference between life and death for two to four of her daughters' children, since two to four grandchildren are the genetic equivalent of one to two of her own children (25% relatedness versus 50% relatedness). While certainly a woman could assist her daughter, it would appear unlikely that she could be the difference between life and death for two, let alone four grandchildren.

Adding to this unlikelihood is exogamy. Exogamy rates for women can vary significantly: for the Walbiri of Australia the number of endogamic marriages is slightly greater than exogamic marriages; for the Tauade of Papua, eighty percent are exogamic; and while Napoleon Chagnon does not give a specific figure, it is evident that through village fissioning or alliance formation, numerous women are exogamic amongst the Yąnomamö.[102] In small communities, like the Murngin of north-east Arnhem land in Australia, all the women may marry exogamically:

> The forty or fifty-odd people who belong to a clan do not all live in the land of the clan. All sisters and daughters, since the rule of marriage is patrilocal after the first year or two, go to their husband's clan to live and rear their families. This means that fully half of the members of the clan live after the age of puberty with a group belonging to the opposite moiety. On the other hand, the wives of the clansmen have come from clans of the opposite moiety and make up the difference in population caused by the loss of the groups' females.[103]

Whatever the actual percentage of women that marry exogamically, a woman cannot expect her daughters to remain in her community. As well as marriage, there is also the possibility of abduction or the fissioning of the community that may separate a woman from her daughters. Napoleon Chagnon studied a village that had fissioned into three smaller communities, and investigated the relatedness of six people (four men and two women) in one village to 268 people in the three villages that had fissioned from the earlier village. He investigated this with respect to generations – relatedness amongst people of the same generation, which means it is not directly relevant to mothers and daughters, but gives some impression of relatedness after a village has fissioned. He found that the men had 44 demonstrable consanguineal kin (same surname in western terms) within the village at an average relatedness of 0.2900 and 52 demonstrable consanguineal kin in the other villages at an average relatedness of 0.1562 (a brother = 0.5000 relatedness). Males were thus able to keep close consanguineal kin of the same generation within their village (brothers, half-brothers, parallel-cousins), while the same category of males in the other villages were less related by about 50%. By contrast, females had 30 consanguineal kin of the same generation (sisters, half-sisters, parallel-cousins) within their village at a relatedness of 0.1672 and 37 consanguineal kin of the same generation in other villages at a relatedness of 0.1469. Females thus possessed similar levels of relatedness to consanguineal kin within their village and in the other villages. Thus, within the same generation, when a village fissioned, men and women ended up with similar numbers of consanguineal kin within each village, but the men's kin were twice as related to themselves by comparison to the women (0.2900 versus 0.1672).[104] That women appear to be less likely to remain with sisters and half-sisters would suggest that women would also be less likely to remain with their actual mothers. In another study, Chagnon found that amongst the sexes, females-to-females had a lower level of relatedness compared to males-to-males (82.9% compared to an average for all of 91.7%). In other words, nearly twenty percent of the women had no relatives at all in their village (they were most probably abducted).[105]

Female exogamy makes the likelihood that a woman could increase her inclusive fitness by undergoing menopause and assisting her daughters extremely unlikely. If a woman is likely to only produce two or three daughters, who would be likely to produce four to six grandchildren (eight to fifteen at the very most), then the number of co-resident grandchildren would quickly be reduced if one of the daughters married exogamically, was abducted or if the community fissioned.

Furthermore, if a woman's children's lives were dependent on the presence of a resident grandmother and children's death rates were significantly increased by the lack of a grandmother, then female exogamy would be expected to be rare. Men would attempt to ensure that their wife's mother was present to ensure the future of his children, but they do not. We should expect much higher death rates amongst babies whose mothers do not reside with their own mothers. But there does not appear to be any such evidence. Additional to this is the fact that very few females in the community live to the age of forty-five or older. Napoleon Chagnon noted that amongst the Yąnomamö, 20% of the individuals are girls aged 0-14, but only about 5% of the individuals are women aged 45 or older.[106] Thus the likelihood that a female reaches post-menopausal age is about 25%. If children were totally dependent on the existence of their post-menopausal grandmothers to care for them or face death, then death rates amongst infants would be enormous. There has been evidence amongst the Hadza, a group of only 300-400 hunter-gatherers in Africa, that post-menopausal women are capable of producing large amounts of nutrition;[107] however, others have interpreted the data differently suggesting that women produce much less nutrition than the men[108] and comparisons in other communities of the foraging ability of older women have not created similar observations.[109] In any event, I have already discounted the possibility that nutrition is an issue creating greater complexity, since nutritional shortages lead to either death or stunting, not to increased complexity, let alone increased life longevity. The argument that 50-60kg female hominids needed to live until fifty or sixty years in order for infants to avoid starvation is nonsensical, to say the least – if nutritional shortages were an issue, longevity would decrease not increase as would body size.

It also needs to be remembered that women in these communities are not isolated; they are not women living in houses separated from kin and friends. Women in cooperatively-simple societies are surrounded by other mothers, future mothers, and older women who may be able to assist a woman with her child. It would seem unlikely that a woman would need a resident nurse for her children when she was surrounded by twenty or more other women or girls who might be able to assist her. If women were isolated then the need for her mother to assist might be understandable, but, as reported above, the fact that women go to such lengths to form alliances with non-kin women suggests that she would have ample opportunities for assistance. We might also expect that women would assist one another when they knew they would, in turn, need assistance later. (It should also be noted that sexual selections would be easier when an ally could care for any children the woman might have. By comparison, it's unlikely a mother would assist in the same way, since sexual selections could undermine the cooperative arrangements of her sons.)

A fifth argument against this idea is the timing of women's menopause. Women will give birth to their first child when they are perhaps 16 or 17; this means a woman will become a grandmother in her mid-thirties if her first child is a girl. By the time a woman goes through menopause in her early forties, if her first child was a girl, her daughter could be pregnant with her third child or about to wean her second. If her second child was a daughter, then the second daughter could also be about to wean her first child by the time the mother is weaning her last child. If a woman was absolutely vital to her daughter's success as a mother we should expect that menopause would begin in a woman's mid-thirties, but instead it occurs in her early-to-mid forties.

Finally, one might also ask how difficult is it to be a human mother? While human children are certainly more difficult than the infants of other primates, it would seem amazing that a woman would need her own personal nurse, despite the presence of other women. By contrast female chimpanzees, gorillas, and orangutans, with, though still very dependent, less difficult infants are able to manage the feat with almost no assistance.

The idea of menopause to assist a daughter's children appears to be unfeasible. In terms of the timing, the social environment, women's behaviour in forming bonds with other women, and, most importantly, exogamy and community fissioning, none of these scenarios fit in with the idea of female menopause and the assistance of children.

If a woman is giving up her ability to produce more children, then she must be able to increase her inclusive fitness in another way. If she can gain little benefit from aiding her daughters, then there is only one other alternative – her sons. Assisting sisters is unlikely, for the same arguments apply to sisters as daughters: exogamy, fissioning, others present to lend assistance, alliances with non-kin, abduction, and so on.

Since women are exogamic, but men are not, firstly we can conclude that while women may separate from their daughters, they will remain with their sons. Even if all of a woman's daughters marry a man in a different community, she will remain with her husband-by-arrangement, and her sons.[110]

Consider the timing of menopause as with her son's marriages. In cooperatively-simple societies, men tend to marry when they are in their early-to-mid twenties, and this fits in much more closely with the timing of menopause. If a woman gives birth to a son when she is seventeen, she is likely to be in her forties when he marries. If her second child is a son, then she may have given birth to him at the age of 21 or 22 at the time, and he may marry when she is in her mid-forties or later. Thus a woman will begin to go through menopause and will be probably free of any infants at the time when her first son is likely to become a father.

Thirdly, sons are capable of producing far more children than their daughters. While a woman can only produce her own children, a man can produce children with four or five or even more wives. Indeed, if a woman were to remain with the children that were the most likely to increase her inclusive fitness, she would remain with her sons and assist their wives rather than assist her own daughters. If her two or three sons married five or six wives, then even if she had two or three daughters, her sons would out-produce her daughters by a factor of two.

For a woman, her daughters are a safe bet, while her sons are a gamble. A son may have no wives and be out-produced by his sister, or a woman's sons may have five or ten wives between them. Indeed, we should expect women to concentrate on their sons more than their daughters.

Fourthly, I argued earlier that if a woman went through menopause she might give up the ability to have one or two more children and therefore she would need to save the lives of two or four grandchildren to come to the same reproductive conclusion. Yet there is no evidence that resident grandmothers make a difference to survival rates amongst children in cooperatively-simple societies. However, by contrast, this chapter and previous chapters reveal that men are prone to suffering the equivalent of a child's death – cuckoldry. For a man, if his wife's child is not his own, then this is, genetically, the same as if the child died, or something close to it. Therefore, while it might seem implausible, unless the grandmother created a vaccine, for a woman to be the difference between life and death for two or four grandchildren, all she would need to do is prevent her sons from being cuckolded to be the difference between her grandchildren being related to her or another man's mother: in genetic terms, dead or alive.

If humans evolved with arranged marriages, then her sons' wives-by-arrangement did not select them. A significant amount of this book has considered cuckoldry, and this is not simply an issue for men, but also their mothers. If a man is cuckolded he loses a child or the equivalent of fifty percent of his own genes from the gene pool. His mother loses half that amount, the equivalent of twenty-five percent of her own genes from the gene pool. Because of primogeniture, with the eldest son being the most likely to benefit from the betrothal of his sisters, this further amplifies the loss. If a woman produces two or three sons and they have five wives between them, then it is her eldest son that is most likely to have the majority of the wives. This is equally true for her eldest grandson: while a woman is obviously not concerned about the paternity of her eldest son, she will be as concerned about the paternity of her eldest grandson as her son will be.

For instance, her eldest grandson will be the most likely to become polygynous. If her eldest grandson is the progeny of another woman's son, then she could lose an enormous amount of her inclusive fitness. If her grandson is very successful and has four wives and fifteen to twenty children, then those children are the equivalent of 1.875 to 2.5 children if she had given birth to them herself. If a woman's eldest, very successful grandson is the product of cuckoldry, then she could lose the equivalent of two children in the next generation, and all it takes is for her son's wife to copulate with another man once.

Since primogeniture is normal in cooperatively-simple societies, we should expect that as soon as her first son reaches a marriageable age, a woman should stop her own reproductive efforts and watch her eldest son's wives-by-arrangement.

This is especially true once we consider that a few men can be very successful. While some will, no doubt, suggest that most men marry monogamously in cooperatively-simple societies, and have only a few children, this is not of interest. Some men have ten wives, and can eventually leave hundreds of great, great, grandchildren. Evolution does not operate on averages, it works because some are more successful than others, and we are interested in the successes far more than the average. We are interested in those that have the biggest impact in defining the gene pool.

Unlike a daughter's ability to obtain many assistants in childcare, a man does not have as many assistants in watching his wives-by-arrangement. Men do not have access to the women's exchanges of information, and they may need to hunt, to raid other communities, to engage in lengthy political discussions, and so on. Men will not always be present in the community, and will be incapable of constantly watching their wife or wives. Even if the men have sisters in the community, their sisters may not be able to tell their brothers of their wives' sexual selections because they may also have sexually selected. If a sister has sexually selected a man other than her husband-by-arrangement then she compromises her ability to inform her brother of his wife-by-arrangement's infidelity. If she tells her brother, then the woman affected may reveal the sister's sexual selection to the sister's husband-by-arrangement. Fertile women are compromised by their potential need to sexually select. Furthermore, his sisters may have married exogamically and have left the community. A brother cannot expect that his sister will be able to inform him of his wife-by-arrangement's behaviour.

If a man and his brothers are out of the community, and if his sisters are severely compromised or have married exogamically, then who will watch his wives-by-arrangement? If his father is alive, he may well be out of the community as well, as part of the hunt, the attack on another community, or visiting another community. The fact that the marriages are arranged is fundamental to this argument. In these communities men do not love or trust their wives-by-arrangement.

As to evidence in the anthropological studies, most of the evidence is indirect. Meggitt found amongst the Walbiri of central Australia:

> After the menopause a woman is called *mil-gari*, "one who has an eye", a term whose use reflects the incidence of blindness (chiefly due to trachoma) among older people. The old women constitute cliques, generally centred on the widows' camps which are hotbeds of gossip.[111]

I wonder if there is a different reason for calling an older woman an 'eye'. In their summary of Australian Aboriginal material, Catherine and Ronald Berndt state that older women are incorporated into sacred men's rituals, such as the fertility cult, the *ubar*:

> Another feature of the *ubar* is that women with white hair, 'almost ready to die', are permitted to witness sections of the ritual on the men's sacred ground. Each woman who does so must be accompanied by a close male relative, such as a son, who formally introduces her to the *ubar*: and she herself has no active role in the proceedings. The explanation of this is that by the time she has reached such an advanced age she is reasonably safe: the power of the *ubar*, and of the sacred rites, cannot harm her. Younger women, and children, must be protected; men must mediate between them and the

potentially dangerous forces which are basic to human life, but too strong for any but initiated men to cope with directly. Some women say that the *ubar* Mother, like the Kunpipi Mother, prefers to have ritual dealings with men: that she is jealous of women, almost as a co-wife is jealous, but also in the same way that a mother may help and cherish her sons more consistently and for a longer period, if she can, than her daughters – who are expected to be able to fend for themselves at an earlier age. The *ubar* woman is a Mother, concerned for the wellbeing of her adherents: but she is a rival too. Men must be careful in approaching her, making use of the appropriate ritual channels: and this caution is even more imperative for women.[112]

I find it interesting that these Aboriginal people specifically state that the women help their sons and not their daughters, and that the *ubar* initiated women prefer to have dealings with men and become jealous of women. Warner gives the example of a Murngin woman's participation in her grandson's initiation:

> Sometimes an old woman, usually the initiate's mother's mother (mari), goes through the camp snatching the men's spears from them and threatening the younger women with the weapons. She is taking a threefold role in this ritual act: (1) expressing the antagonism of the women's division of the sex and age structure toward the men's division; (2) disciplining the younger women and thereby increasing the solidarity of the women's division; and (3) acting the part of a very old woman, which gives her special privileges, since old women are "more like a man".[113]

It is interesting that Australian Aborigines describe women becoming jealous and antagonistic towards younger women and that they become 'more like a man', reflecting the hypothesis.

Women and their daughters-in-law are in an antagonistic relationship, which is especially so in societies with arranged marriages. Since a woman will know the importance of sexually selecting, she will be especially antagonistic towards her daughters-in-law. Amongst the Saramaka of South America, Sally Price remarks upon the interaction between a wife and her husband's kin:

> The relationship between a woman and her husband's kin (especially his female kin, with whom she interacts most frequently) are, more often than not, distinctly strained.[114]

> The most common allegation against a husband's kin, however, is malicious gossip. Women frequently present themselves as the victims of false rumors spread by their husband's relatives.[115]

> A woman and her mother-in-law exercise great restraint with each other, and there is no expectation that feelings of real warmth will develop between them. Women see themselves as being under the thumb of their husband's mother when they are in the same village. Mothers-in-law are said to enjoy the prerogative to make specific demands on their daughters-in-law (e.g. to carry out or cut firewood) and even though such services are more often offered by the younger woman than demanded by the older, the

principle looms large in Saramakas' abstract image of this relationship. When, to cite
just one example, a woman said she intended to explain to a recently married younger
woman that she was her (distant classificatory) mother-in-law, the woman she was
talking to chimed in supportively, "yes, just bring her all the rice you need pounded
and she'll understand." The aspect of my life in the United States that Saramaka women
were most curious about was the nature of my relationship with my mother-in-law. Did
I ever say her name out loud? Did I ever walk behind her while she was bent over the
laundry at the river? If a photo was taken of men, would I show it to her? And so on.
Specific rules of etiquette reinforce this image. For example, a woman should not sit on
a stool that belongs to her mother-in-law, and a mother-in-law may not be the one to
help a new wife unload a special basket that she brings on her first visit to the husband's
village.[116]

In her experiences with the Yąnomamö, Helena Valero observed similar tensions:

> Fusiwe's mother talked with me a great deal: mothers-in-law don't like to talk to their
> sons' wives, but she told me so many things. She said that she was not afraid of me and
> that I must not be afraid of her, because she was very old.[117]

Valero also noted the manner in which sexual selections were divulged to the husband-by-
arrangement:

> Many women were always faithful, others not. It often happened that the mother, the
> father, or the brother-in-law said to the son of the betrayed man: 'Warn your father that
> your mother has gone with another man while he was out.'[118]

In modern Western nations with communities of people that use arranged marriages, these
issues are becoming apparent:

> Gina Singh, 26, told a court that she was bullied, isolated and became seriously ill after
> entering into an arranged Sikh marriage. Ms. Singh, from Nottingham, sued her former
> mother-in-law, Dalbir Kaur Bhakar, who imposed a 17-hour daily regime of housework,
> forced her to bleach her skin and cut her hair and severely restricted contact with her
> family. She was forbidden to leave the house alone and was not allowed to listen to the
> radio, read newspapers or watch television.
>
> Jasvinder Sanghera, who runs Karma Nirvana, a network of refuges for Asian women
> fleeing domestic violence, said: "It happens day in, day out. In the Asian tradition when
> a woman marries, she has to leave her family home to live in her matrimonial home. The
> mother-in-law has a very strong position within the family hierarchy. You are taught to
> obey your mother-in-law. The mother-in-law knows that and exercises that power.[119]

In the Oman, Unni Wikan noted that: '...women tend to spend most of the day at their
own homes in the company of children and such co-residents as there may be; for example,

a mother-in-law, mother, or sister(s)-in-law. About 65 percent of married women live in households of more than one nuclear family.'[120]

The evolution of menopause would have occurred as soon as male ancestors could achieve high rates of polygyny. The only time that menopause would become more successful is when a woman's son possessed both a chance of having numerous wives and possessed a high rate of potential cuckoldry. Those females that watched their sons' wives-by-arrangement were more successful at preventing cuckoldry and left more descendants than those females who kept reproducing. All it would take is for particularly successful males, with numerous wives, to suffer lower rates of cuckoldry if the males' mothers had passed through menopause, and the trait would become common.

It should be noted that once a woman has passed through menopause she is assumed to no longer be sexually active. Mervyn Meggitt notes amongst the Walbiri:

> A man should distribute his sexual favours without bias among his wives; but it is accepted that a comparatively old woman is no longer sexually active and, therefore, need not share her husband's bed.[121]

It is perhaps interesting, that the only other female mammal that passes through menopause are pilot whales and orcas. In resident orcas, a female matriarch stops reproducing and remains with her sons and daughters. Whether this is true in the less well studied transient and offshore orcas is yet to be seen. As to whether the mother and son are connected in the same manner, the North Gulf Oceanic Society website states:

> Older, post-reproductive females like AB10 ("grannies") seem to direct the activities of the entire pod. They are very possibly both leaders and repositories of knowledge for the pod. Often when an old female dies, her sons are not long to follow. The reason for that life-long tight bond between an adult male killer whale and his mother is a mystery. Perhaps these males lose status in the group when their mother dies.[122]

> The AB10 subpod is actually a single matrilineal group composed of a mother, AB10 and her two presumed adult sons (another probable son AB4 recently died). For reasons that are unclear, matrilineal groups that are primarily comprised of males tend to travel at a distance from the rest of the pod.[123]

As with the usual beliefs that all animals are desperate to survive in the harsh natural world, arguments for menopause in female orcas have mirrored those for humans: the passing on of knowledge and assisting other females with their young.[124] Considering that these animals can weigh 7.5 to 10 tonnes, one can assume that their diet must be adequate – otherwise they would be smaller. Arguing that a female goes into reproductive senescence in order to assist others in the group who might otherwise starve simply does not make sense when we are weighing the animal by the tonne. In any event, 'The details of mating relationships in killer whales remain obscure'.[125]

Kin

Kin bonds should be very strong, though not as important as they are for men since women are far more likely to live in an environment without close kin. Since a woman's children will be preferentially married to their cross-cousin, her brother's children, we should expect that women will have strong bonds with their brothers. Since brothers will also protect and avenge their sisters, a woman's health, safety, and peace may depend on the presence or strength of her brothers. If her community fissions into two groups, then whether she remains with her brothers may depend upon their strength. Whether she marries endogamically or exogamically may also depend upon the strength of her brothers and other male kin. These arguments will also be true for her father and her father's brothers.

The strength of such bonds are reflected in women preventing their brothers from being attacked by other women, as Meggitt states of the Walbiri:

> A woman is expected to defend her brother from insult and may use force to do so. Should two brothers quarrel over a woman, especially a mistress, their sisters are likely to belabour her, for she has caused the dissension.[126]

Alternatively there may be antagonism since a girl's husband-by-arrangement will be selected by her father and the marriage enforced by her brothers. It may be her brothers that force her to remain in a violent relationship. They may also prevent her from sexually selecting a man, and there are occasions where brothers are known to have killed a sister because of her refusal to end an extra-marital sexual selection.

Relationships between sisters and between mother and sister should also be strong. A mother and her daughters can be expected to assist each other in attacking opponents, in childcare, and in all aspects of life, assuming that the woman remains in her natal community. Mervyn Meggitt noted the women that acted as midwives for a pregnant woman and found that 17 of the 39 midwives were sisters, 12 were mothers, and 4 were the husband's sister.[127] In terms of violence, we should expect women to support their sisters in the same manner that men will support their brothers.

Alternatively, antagonism may occur due to competition between sisters, and jealousy between co-wives who are sisters (which is common due to sororal polygyny) is found regularly in the anthropological studies.

Caring for children is obviously vitally important. In cooperatively-simple societies, many children may die in infancy, usually due to disease. Considering their fragility, we should expect that women would be quick to prevent any harm to their children. Meggitt gives an example of this from the Walbiri:

> Della nagamara, the two-years-old daughter of Millie nabanangga, misbehaved in her mother's absence. Minnie nabanangga, a younger co-wife of Millie, slapped the child several times and berated her. Hearing the cries, Millie rushed to the camp. She took in the situation at a glance and struck Minnie on the head with a heavy club. The husband later upbraided Millie for being too free with her blows, but he confided to me that he could not really blame her for defending her child.[128]

Such incidents can escalate remarkably:

> The half-brothers Donny and Yarry djabangari usually camped together in friendly
> fashion. One afternoon, however, one of Donny's young sons hit Yarry's child. The
> mothers of both children intervened and were soon involved in a scuffle with digging
> sticks. Donny came to his wife's aid, striking her opponent several times. Yarry at once
> attacked Donny, and the two men, both tall and powerfully built, engaged in a vicious
> fight. Donny speared Yarry in the arm, but had his head split open with a club in return.
> Younger men, chiefly their brothers-in-law, separated them, but the quarrel quickly
> flared up again. Donny felled Yarry with a club, splitting his head in several places.
> Throughout the affray, the "M.M.G." of the combatants were quite at a loss. As they
> could not attack the assailants of the Sr.D.S. in the approved manner, they ran about
> the camp wailing, swearing and brandishing weapons. Everyone was thoroughly upset
> by the quarrel, which clearly revealed the difficulties created when close brothers fought
> each other.[129]

There has been some speculation on the idea that women evolved with great
intelligence to become better mothers. While women would have evolved to respond to the
needs of their infants, with the resolving of the difficulties of childbirth and bipedal movement
being an obvious example, it is arguable that, if anything, women might have evolved to be
too intelligent. It is not uncommon to read in anthropological reports that women will
commit abortion or infanticide because they simply cannot be bothered with another four
years of pregnancy and nursing. Amongst the Aboriginal women in the Kimberleys, Phyllis
Kaberry found:

> Apparently, however, children are not necessarily desired as was made clear in the
> statements of numerous women. They disliked the prospect of child-bearing; its pain
> and trouble, and the burden of carrying the baby about afterwards. There were few
> children and many of the women confessed to having committed abortion to avoid the
> contingency.[130]

The care of other people's children can also lead to difficulties, particularly in
stressful times. During a period where the Yąnomamö men had left their village for some
time, on a raiding mission:

> After several days the women were so frustrated and anxious that fights began to break
> out among them. One woman got angry because another one, her sister and cowife, left
> her to tend a small baby. When the mother returned, the angry one picked up a piece of
> firewood and bashed her on the side of the head with it, knocking her unconscious and
> causing her ear to bleed profusely.[131]

One must question any association between women's intellectual capacity and child
raising, because women's intelligence must surely be well beyond the needs of childcare.
Indeed, their level of intelligence would surely verge on being detrimental, and lead to

infanticide, abortion or contraception, something other female animals are less capable of performing. Equally, one might ask that if women achieved their level of intelligence for motherhood, why were other mammal mothers not similarly endowed with large brains?

The vast majority of women's morphological and behavioural peculiarities are aimed at selection and not motherhood. While infant chimpanzees and gorillas cling on to their mother's hair, which is an important part of their transport, women have lost all the hair to which a child could cling. This makes holding a child far more difficult to say the least, but evidently the loss of the ability of the child to cling to the mother is of less significance than the benefit of naked skin to attract a male that is unable to groom or smell her as part of sexual selection. Movement with an infant is more difficult for bipeds than quadrupeds: chimpanzees and gorillas can move swiftly with a clinging infant. Then there are the difficulties of birth created by biped morphology. The changes involved in becoming a female human have not assisted the ease of childcare or birth, indeed, instead they have made life more difficult and sometimes, as with childbirth, extremely dangerous.

If the brain is important in childcare, then the benefit must be astonishing considering the difficulties the large headed child creates for childbirth. Maternal death rates due to childbirth can be astonishing. Currently, in African nations, maternal mortality rates per 100,000 births are: Benin 880; Chad 1,500; Malawi 580; Mali 630; Mozambique 980; Niger 920; Nigeria 1,100; Uganda 1,100; and Zambia 870. This means that in each live birth, the woman has about a one percent chance of dying. Fertility rates (in the years 2000-2005) are between 5.42 (Nigeria) and 8.0 (Niger) and for each of these births a woman has a one percent chance of dying. Considering that the percentage of births with skilled attendants can be between 16% (Chad and Niger) and 60% (Benin), then one can suggest that in traditional societies with no skilled attendants the death rates would be higher.[132] This means that women in these societies will face a one in twenty chance of dying in childbirth, or higher. This does not include the number of women who experience injuries, such as obstetric fistula, where an obstructed labour leads to the child dying, and necrosis of the tissues connecting the bladder, vagina and rectum. Obstetric fistulas cause permanent incontinence that can affect urine or both urine and faeces. Statistics are not exact, but in East Africa it is believed that for every thousand births, there are two to five cases of fistula.[133] Again, considering the number of births that women will experience, the number of women rendered permanently incontinent from a fistula will be greater than one percent and could be as high as three percent. Such women are often abandoned or ostracised by their communities. One only need regard the number of people arguing against c-sections in our society to perhaps consider that many people in Western nations are somewhat ignorant of the true risks involved in childbirth. The evolution towards a riskier childbirth, with, first, bipedal movement and thus a reduced birth canal, and, secondly, an increased infant brain size, suggest that whatever the selection pressures towards these changes were, the safety of both mother and child were not nearly as important as the requirement for intelligent bipeds. Added to this is the relatively premature state of the newborn human infant and the high levels of attention infants require, which many have suggested is due to the requirement for the head to pass through a narrow passage during birth. Again, if the needs of the child and mother were paramount, then this would seem illogical, that in order to obtain an intelligent parent to care for the child, the infant needs to be born many months premature – compared to other mammals, many of which are able to walk or run soon after birth – and with a serious risk of death or injury to both mother and child. Added to these issues is the lack of other animals with high levels of intelligence. Faced with such a lack of logic, I argue the issue of parenting can be dismissed.

This is not to suggest that parenting is irrelevant or simple, but rather that it is simply not a significant selection pressure that has led to women's *peculiar* behaviours and morphologies. Almost any female can create an infant; the challenge for the human female is to raise a healthy child likely to become successful, and women's intelligence and body are surely aimed at attempting to create the best child she can in terms of physical and intellectual capabilities, in other words, in terms of sexual selection of genes. This is really a question of sexual selection and not nurture for no level of nurture can make a child with below average genes a more intelligent or physically brilliant child, as reflected in John Bruer's book *The Myth of the First Three Years*. No level of nurture will turn a *Homo erectus* into a *Homo sapiens*. No level of nurture can make a sickly baby survive the rigours of infancy in a cooperatively-simple society. Certainly, different levels of care and attention may have some impact on the result, but whether a baby survives a disease or whether a male child can compete in violent attacks with other men is something that will depend more on genetics.[134] The mother might only be able to do all she can, and hope for the best. In a Western society with enforced monogamy, differences in inclusive fitness and genetic quality will not be so obvious; however, in a cooperatively-simple society with polygyny, the link between differences in inclusive fitness and genetic quality may be strong. High rates of infant and juvenile mortality emphasise this in terms of the immune system. The contrary position is to argue that women in cooperatively-simple societies, whose children die one after another, in an environment where half of the children might be dead by the age of ten, are simply bad mothers. This is highly unlikely.

Concluding Remarks

I argue that women possess several alliances. Firstly, as they pass through puberty, they will require a female ally to increase their capacity to sexually select. I would argue that homosexual behaviour may also be used to create close alliances, though less commonly. Secondly, their brothers will be a tremendously important ally, providing her with protection from her husband-by-arrangement's violence and also as an attractive prospect for other women to sexually select, thus making her a desirable candidate as an intermediary. Their final alliance will be with their sons as she is the best person to prevent her sons' wives-by-arrangement from seeking to sexually select other men.

Compared to the male's manipulation of complex genealogies, some might wonder at female intelligence. Consider the bower bird and the male's complex bower, by comparison the female might appear quite dull. Yet generations of females have selected males based on their success at bower building. Is the male that performs the behaviours more intelligent, or the female that selects the male's ability to perform the behaviours the more intelligent? How much of human intelligence is due to careful female selection over generations?

Chapter 10

Communication

Verbal Communication

When asked to choose the one thing that separates humans from other animals most people will choose our capacity to use speech. Complex human language will often be described as our most important behaviour. This is not surprising, for whenever we wish to do anything of any importance we will invariably use language as part of the task.

However, I have argued throughout this book that our most important trait is the capacity to cooperate, and that we do this through forming arranged marriages. It is the capacity to form cooperative unions between genetically unrelated people through arranged marriages that allows us to form large cooperative groups. I have also argued that most of our physical and behavioural features stem from this fact. Yet it is obvious to any observer that we use speech constantly. Thus one may well ask, is the capacity to use speech essential in order to be able to form arranged marriages?

The answer to this is no. The reasons for this will be considered in greater detail in the following chapter, where I will give a step-by-step example of how complex societies using arranged marriages may have evolved. I will limit my argument in this chapter to say that in order for hominids to form arranged marriages there is no necessity for speech only an intuitive understanding that a sexual union will take place in the future. For instance, a male may point to his daughter and to another male's son. The other male may point to his son and to the other male's daughter. As long as the males have an intuitive understanding of the process, no language is needed. All we might need to include is one symbolic statement – such as two wedding rings or a wedding process – and both males, as well as all other adults and juveniles, will understand that an arranged marriage has been organised. I think most people could watch a silent film, with no written explanations, and easily understand that a marriage has been organised. Equally, I believe most people could imagine that if the people in an entire community were mute, then those people would still be able to organise arranged marriages with ease. Sexual unions do not require words, only an intuitive understanding. We can see that birds, mammals and many other animals are able to engage in a 'marriage' without needing to specifically use language. They are able to agree to build a nest together and protect and feed infants without needing to make verbal pledges of agreement. The step for parents or other relatives to make the decision for them is only a small step.

Cooperation itself is thus not dependent upon verbal language, which, again, will be explained in further detail in the following chapter. If this argument is correct, and it is true that our ancestors could use arranged marriages to build cooperative societies without the need for verbal language, then this begs the question: why do we need to talk?

Consider a group of ancestors that have the same capacity for complex verbal language as a chimpanzee or gorilla, that is, almost none at all. If these animals possessed the intuitive understanding that they could build cooperative societies by organising arranged marriages, then they would, without words, be able to form complex societies. If the Smiths and the Richardsons understand that by exchanging sisters and daughters they can bind their two families together and form a more robust and competitive community, then they will be able to do so without complex verbal language.

At this point, I expect most readers will be making various demonstrative outbursts. I suspect most readers will have difficulty imagining a society without written language let alone one without verbal language. There are many who will argue that 'consciousness' is not possible without verbal language and that humans alone are conscious. Humans use all kinds of methods in order to separate themselves from the rest of the animal kingdom. Verbal language is invariably used as the basis for why we are superior to other animals. The internal monologue, the voice of our 'consciousness', is perhaps the most commonly used excuse for this.

According to this type of argument, if we do not say the words in our head 'I am hungry' then we are not hungry. Thus if a human says the words 'I am hungry' and goes to the fridge, then this is an entirely different event to a dog going to its food bowl. Of course, in reality, both the human and the dog have exactly the same experience: there is an internal mechanism that triggers the desire to eat and both the human and the dog respond in exactly the same way. The fact that the human might use words to describe the response is neither here nor there. If a person contends that of course the dog knows it is hungry and that this is why the dog is going to its food bowl, then there will be the usual response of 'ah, but how do you know the dog knows it is hungry?'. These kinds of arguments are often used in philosophy to create an environment without genes, flesh and blood.

Perhaps the ultimate example of how important some people believe verbal language is are the arguments of Benjamin Whorf, who argues that human speech affects not merely how they talk, but how they think. In other words, if there is no word for sky, can humans think about the sky, let alone talk about it? Indeed, according to some, the tail does wag the dog, and quite vigorously, apparently. This is one of the pinnacles of the 'culture creates human' argument.

I am including this solely because it reflects the importance that humans place on language. It separates us from other animals, it makes us 'conscious', and it makes us superior. Indeed, it might even 'make' us. Language is the one trait we are least likely to be capable of understanding rationally.

Of course, these arguments are simply ridiculous. Human behaviour, like the behaviour of other mammals, is not based on words, but biological processes. Our desire to eat, to have sex, to win arguments or fights, to sleep, to fear pain, to experience joy and exultation, and so on, are based on exactly the same biological processes that occur in other mammals. Our euphoria or despair upon winning or losing a fight is not based on our capacity to use words, but rather due to a flood of hormones and possibly neurotransmitters. From an evolutionary point of view, whether we can describe the win or loss in Shakespearian tones is neither here nor there (most people describe the most important events in their life as leaving them 'speechless'); if the win or loss has affected our capacity to pass on our genes then this is all that matters and our euphoria or despair reflects this outcome. A mouse or a dog will have exactly the same experience. From a biological point of view, all that counts is the impact on the individual's capacity to increase their genetic influence in further generations.

All this was anticipated by Charles Darwin. In *The Descent of Man*, Charles Darwin considered the intelligence and behaviours of other animals at length. Darwin argued that humans were not the only users of tools more than 90 years before Jane Goodall's observations of chimpanzees ended the concept of 'man-the-toolmaker':

> It has often been said that no animal uses any tool; but the chimpanzee in a state of nature cracks a native fruit, somewhat like a walnut with a stone. Rengger easily taught an American monkey thus to break open hard palm-nuts; and afterwards of its own accord, it used stones to open other kinds of nuts, as well as boxes. ...and I have myself seen a young orang put a stick into a crevice, slip his hand to the other end, and use it in the proper manner as a lever. The tamed elephants in India are well known to break off branches of trees and use them to drive away the flies; and this same act has been observed in an elephant in a state of nature. ... In these several cases stones and sticks were employed as implements; but they are likewise used as weapons. ... Mr. Wallace on three occasions saw female orangs, accompanied by their young, "breaking off branches and the great spiny fruit of the Durian tree, with every appearance of rage; causing such a shower of missiles as effectively kept us from approaching too near the tree."[1]

Yet, for Darwin, the use of tools was only the beginning of non-human animal complexity:

> Most of the more complex emotions are common to the higher animals and ourselves. Every one has seen how jealous a dog is of his master's affection, if lavished on any other creature; and I have observed the same with monkeys. This shews that animals not only love, but have desire to be loved. Animals manifestly feel emulation. They love approbation or praise; and a dog carrying a basket for his master exhibits in a high degree self-complacency or pride.[2]

Animals manifestly enjoy excitement, and suffer from ennui, as may be seen with dogs, and, according to Rengger, with monkeys. All animals feel Wonder, and many exhibit Curiosity. They sometimes suffer from this latter quality, as when the hunter plays antics and thus attracts them; I have witnessed this with deer, and so it is with the wary chamois, and with some kinds of wild-ducks. Brehm gives a curious account of the instinctive dread, which his monkeys exhibited, for snakes; but their curiosity was so great that they could not desist from occasionally satiating their horror in a most human fashion, by lifting up the lid of the box in which they were kept. I was so much surprised by this account, that I took a stuffed and coiled-up snake into the monkey-house at the Zoological Gardens, and the excitement thus caused was one of the most curious spectacles which I ever beheld. ... I then placed the stuffed specimen on the ground in one of the larger compartments. After a time all the monkeys collected round it in a large circle, and staring intently, presented a ludicrous appearance. These monkeys behaved very differently when a dead fish, a mouse, a living turtle, and other new objects were placed in their cages; for though at first frightened, they soon approached, handled and examined them. I then placed a live snake in a paper bag, with the mouth loosely closed, in one of the larger compartments. One of the monkeys immediately approached, cautiously opened the bag a little, peeped in, and instantly dashed way. Then I witnessed what Brehm has described, for monkey after monkey, with head raised high and turned on one side, could not resist taking a momentary peep into the upright bag, at the dreadful object lying quietly at the bottom.[3]

Darwin concludes:

It has, I think, now been shewn that man and the higher animals, especially the Primates, have some few instincts in common. All have the same senses, intuitions, and sensations, - similar passions, affections, and emotions, even the more complex ones, such as jealousy, suspicion, emulation, gratitude, and magnanimity; they practise deceit and are revengeful; they are sometimes susceptible to ridicule, and even have a sense of humour; they feel wonder and curiosity; they possess the same faculties of imitation, attention, deliberation, choice, memory, imagination, the association of ideas, and reason, though in very different degrees.[4]

Experimental evidence is now supporting Darwin's conclusions. Robert Butler discovered that if he placed mirrors so that he could see the behaviour of the monkeys in an experiment, the monkeys stopped their activity and simply watched him. Harold Harlow later found that monkeys were capable of sustaining interest in a puzzle for 10 hours without food reward or expectation of a reward:

More important than the demonstration of the puzzle solution, however, was the finding that monkeys will perform on this problem for prolonged periods of time and still not become completely satiated with the task. For example, after the animals had learned the puzzle, the experimenter reset the puzzle every 6 minutes throughout the course of a 10-hour test session. Even under this condition, designed exclusively for satiating manipulative behavior, the monkeys continued to work on the problem for the entire period. Although there was a decrease in the actual number of devices manipulated as the session progressed, one or more devices were manipulated nearly every time the puzzle was reset. The persistence of manipulative behavior strongly suggests that manipulation is not a secondary motive conditioned upon some primary motive, but that it is in itself primary.[5]

In another experiment, a box was created with two doors, one of which was locked. The doors were colour coded with a coloured card, with yellow indicating open and the blue indicating locked. The monkey needed to push against the yellow card in order to open the door, and this was the monkey's only reward. 'The data on the persistence of visual exploratory behavior are of enormous theoretical interest. Rhesus monkeys, when tested until they failed to push against either door within a 10-minute period, performed for many continuous hours. In fact, one animal worked for nearly 20 hours before it finally refused to respond to visual incentives.[6] In a further experiment, where the monkey simply had to open the door, various objects were placed in a cage outside the door, and the time the monkey opened the door for each object was recorded. The results showed that the monkey opened the door for the longest when another monkey was in the chamber outside the door, second was an electric train in the chamber, third was desirable food, and fourth was nothing in the chamber.[7] These results were recorded in the 1950s when other animals were simply viewed as machines; the fact that a monkey was more interested in an electric train than desirable food went against all predictions.

While we may believe that a child playing a video game for endless hours is a reflection of unique human attributes, this has been shown in experiments not to be the case. Our experiences are thus a variation on the experiences of other animals, and language itself is simply one of these variations.

Other people, using a Western point of view, will point to libraries, statutes, academic papers and books, designs and patents, and so on, as to the absolute importance of language to our existence. However, people in cooperatively-simple environments live successfully without all these things. They need no books or patents to make everything they need. Napoleon Chagnon describes the Yąnomamö's possessions:

> They have hammocks, baskets, a few very crude poorly fired claypots which have now disappeared in the last twenty years, bows and arrows, and not much else. A whole village of Yanomamö can pack up in five minutes and go off into the forest, and carry everything they own. So their technology and the number of material items they have is very, very limited, almost as though they are nomadic hunters and gatherers, but they are not.[8]

It is doubtful that language is needed to show one another how to create these possessions, since comparative examples already exist and an individual is able to show another how to create the same item. I would argue that language, though no doubt used, is not necessary to be able to *show* another person how to make bows, arrows, claypots, hammocks, or the other requirements for such a life. Indeed, as any book with instructions will attest, language is an awkward method of instruction, and usually diagrams are used to best effect.

Other knowledge, such as which plants to eat, can also be passed on without language. A mute person would be equally as successful as a speaking person in being able to point out which food is good to eat and which is bad or poisonous. The same will be true for the hunting of animals, where, lest the animal be alarmed, language will be non-existent. The same will be true for warfare, which will be little different from a hunt. Chimpanzees are certainly able to engage in hunting and a form of warfare without complex verbal language.

While some people might quibble about the importance of language in each of the above scenarios, what I believe is definitely correct is that language is not essential for any of the activities. All of the above activities could occur without complex language.

My point is that people have a tendency to believe that language causes their behaviours. I would argue that the behaviours already exist; verbal language simply gives people the ability to describe the behaviours. People cooperate because they have the genetically-based instinct to cooperate. People reproduce because they have the genetically-based instinct to reproduce. Every major behaviour we possess will be based on built-in predispositions, and language simply allows us to communicate with others about our predisposed behaviours.

Perhaps the ultimate example of this difference between language and innate behaviours comes from people suffering from Williams syndrome, such as the following example about a girl named Crystal:

> In describing her future aspirations, Crystal, a 16-year-old adolescent, states: "You're looking at a professional book writer. My books will be filled with drama, action, and excitement. And everyone will want to read them. I'm going to write books, page after page, stack after stack... I'll start on Monday." Crystal describes a meal as "a scrumptious buffet," an older friend as "quite elegant", and her boyfriend as "my sweet petunia"; when asked if someone could borrow her watch, she replies, "My watch is always available for service." Crystal can spontaneously create original stories – she weaves a tale of a chocolate princess who changes the sun color to save the chocolate world from melting; she recounts with detail a dream in which an alien from a different planet emerges from a television. Her creativity extends to music; she has composed the lyrics to a love song.
>
> In view of her facility with language, proclivity for flowery, descriptive terms, and professed focus on drama and action, her aspiration may seem plausible; but in fact Crystal has an IQ of 49, with an IQ equivalent of 8 years. At the age of 16, she fails all Piagetian seriation and conservation tasks (milestones normally attained in the age range of 7 to 9 years); has reading, writing and math skills comparable to those of a first or second grader; demonstrates visuospatial abilities of a 5-year-old; and requires a babysitter for supervision.[9]

Here is an adolescent with remarkable language skills and yet a very low level of intellectual ability – people suffering from Williams syndrome are unable to draw and copy basic stick figures, scoring below even Down syndrome sufferers, when matched for age.[10] Despite not being able to draw an object so that it could be recognisable, these children are capable of correcting ungrammatical sentences with ease: Examiner: 'She slept the baby'; Crystal: 'She slept in bed. She slept *with* the baby' (Crystal's emphasis).[11] She demonstrates that there is no definitive connection between speech and intelligence. Children with Williams syndrome are capable of linguistic complexity, but are not able to rearrange rods in order of smallest to largest.[12] Researchers have thus concluded: 'Such findings suggest that linguistic functions can occur separate from cognitive functions'.[13] Speech is very much the icing on our intellectual cake, it allows us to communicate our complex innate behaviours: Crystal is a person with the icing, but not the cake – she lacks the innate behaviours to give a reason for the complex language.

Furthermore, why did humans not evolve using sign language? Certainly, gorillas, chimpanzees, and bonobos have been shown to be capable of using sign language; they have shown themselves to be able to use hundreds of signs.[14] As I was writing this book there was a newspaper article about the gorilla, Koko, using sign language to tell her handlers that she was in pain, point to her mouth, and describe her pain on a level of 1 to 10.[15] Once the pain level reached closer to ten, a dentist was called in.

The issue is thus not merely one of complex language, but why must the complex language be verbal?

Humans are naturally adept at using sign language even if they use verbal language. A.W. Howitt in his *The Native Tribes of South-East Australia* observes:

> It cannot be said that the use of signs by the Australian aborigines is in any way due to paucity of language, their being fully competent to provide for every mental or material necessity of their life. Those who have had the opportunity of becoming intimately acquainted with these savages in their social life will agree with me in this statement, and no one can feel the slightest doubt who has heard one of their orators addressing an assembly of the men, and with a flow of persuasive eloquence moulding opinion to his will.
>
> It is somewhat remarkable, and at the same time difficult to explain, that the use of gesture language varies so much in different tribes. Some have a very extensive code of signs, which admit of being so used as to almost amount to a medium of general communication. Other tribes have no more than those gestures which may be considered as the general property of mankind.
>
> The occurrence or absence of gestures as an aid or substitute for speech does not, as far as I can ascertain, depend on social status, or the locality in which a tribe lives. Yet, so far as I can venture to form an opinion from my own observations, and from the statements made to me by correspondents, the use of sign language is more common in Central and North-eastern Australia than in the South-eastern quarter of the Continent.
>
> The reason for this may perhaps be found in the vast extents of open country, plains, sandhills, and stony tracts which occur in the interior of Australia, as, for instance, in the Lake Eyre basin.
>
> A stranger is seen there from afar off, and can be interrogated at a safe distance by gesture language as to who he is, where he comes from, and his intentions. When I first saw some of the Cooper's Creek blacks, I was struck by their use of gestures, at a safe distance, and which I took to be either a defiance or a command to depart. In reality they were the sign for peace and the sign for interrogation as to our destination, or as to our reason for being there. Afterwards, when I became better acquainted with them, I came to see that these gestures were part of a complete system of hand signs, by which a person might be interrogated, informed, welcomed, or warned. In the coastal regions or in the forest-clad mountain ranges which lie alongside the Great Dividing Range, separating the coast lands from the interior, such would not be the case, and gesture language could not be made use of at a distance excepting in rare cases.
>
> I venture this supposition, but without laying much stress upon it.
>
> The different degree in which gesture language is made use of may be best seen

by taking a few illustrations from tribes within my knowledge.

The Dieri have a very full code of signs which suffice for ordinary needs of communication. A widow is not permitted to speak until the whole of the white clay which forms her "mourning" has come off without assistance. During this time (perhaps for months) she communicates by gestures alone.

As an instance of the value of such a means of communication between tribes speaking different languages, I give the following.

In 1853 seventeen of the Wirangu tribe were driven in from their country to the west of Lake Torrens by a water famine. They came across, and made for Elder's Range in hope of getting water. Here they fell in with the Arkaba blacks, who received them very kindly and hospitably for about a fortnight, when the appearance of rain induced the visitors to take their departure homewards. They did not understand a word of each other's language, and it was merely by gestures that they managed to communicate with each other.[16]

This is one of many examples of gesture language that Howitt gave:

Where? What? What is it? etc – Place right hand at left breast palm outwards, then move it up at an angle of 45° with the horizon, hold up for a moment, and let drop; when moving the hand, jerk up the chin (Wurunjerri). Hold the right hand opposite to and higher than the shoulder, gradually turning the hand so that at last the palm is upwards; or do this so that the movement of the hand upwards and forwards only brings it level with the face (Yantruwanta). Throw up the hand higher than the head, then let it fall palm upwards (Dieri).[17]

Thus, any suggestion that Howitt speaks of a very simplistic form of communication is wrong; these gestures are complex.

Susan Goldin-Meadow, who has been researching gestures and sign language in children for three decades, comments of the human capacity for sign language:

A priori we might have expected sign languages to be constructed differently from spoken languages. After all, sign languages are processed by eye and hand, whereas spoken languages are processed by ear and mouth. But, in many ways, the languages are not different. Sign languages all over the world are characterized by the same hierarchy of linguistic structures (syntax, morphology, phonology), and thus draw on the same human abilities as spoken languages. Furthermore, children exposed to sign language from birth acquire that language as naturally as hearing children acquire spoken language to which they are exposed, achieving major milestones at approximately the same ages.[18]

She explains how much simpler it is to create a sign language rather than a spoken language, since it is easier to create gestures that reflect an event, rather than a sound (and, I might add, few spoken English words sound similar to what they are attempting to signify, such as elephant and automobile – they are not signified with a trumpeting sound or the sound of an engine, respectively):

> However, the manual modality makes sign languages unique in at least one respect. It is relatively easy to use the manual modality to invent representational forms that can be immediately understood by naïve observers (e.g., indexical pointing gestures or iconic miming gestures). As a result, communication systems can be invented on the spot in the manual modality, which means that sign systems have the potential to provide a window onto the process of language creation. Indeed, deaf individuals have often found themselves in situations where they needed to create a language *de novo*.[19]

Such is the imperative of the need for humans to communicate with language, it has been found that where children do not possess an opportunity to communicate with others with a ready made language, they simply invent their own (this does not apply to isolated children), as was observed amongst deaf Nicaraguan children, who, gathered together, invented their own sign language, including the basic hierarchical structures.[20] Indeed, the children that created the original sign language were followed by a second generation of children who made changes – changes that were not fully acquired by the now adolescent and adult originators of the sign language. The researchers concluded: 'It appears that the processes of dissection, reanalysis, and recombination are among those that become less available beyond adolescence'.[21]

Yet, although humans are capable of complex communication with gestures, we have evolved with verbal language. Furthermore, the evolution of language must have occurred slowly. There must have been a time when it was an advantage to use a hundred words, and then, at a later time, to use five hundred words, then a thousand words, and so on.

Finally, all this must be closely linked to reproduction – there must be a distinct advantage in passing on genes. If so unusual a trait is to evolve, then we must expect that the basis for its existence must be very closely linked to benefits in reproduction. I have already argued that our unusual traits could not be a response to predators or food shortages. If our ancestors were starving, then they would fail to reproduce or the next generation would become stunted – they would not evolve to become more complex. In other words, our unusual traits reflect imperatives beyond survival, as Darwin predicted in *The Descent of Man*.

Thus we need an answer that gives us the following outcome: there must be a benefit in possessing a hundred words rather than none; it must be essential that the words are communicated vocally; and it must be closely linked to reproduction.

I have already argued that humans use arranged marriages to form cooperative societies, and language is not necessary to create these marriages. Thus, we have the complex society, it has provided for itself and can compete with other complex societies, and I am arguing that this is possible largely without verbal complex language. The complex society may be pre-human, being an earlier, less complex hominid society with fewer individuals.

So why would we need verbal language?

To answer this, it is first worth considering the difference between sign language and verbal language. If we watch two people using sign language and two people speaking, what is the difference between what is occurring? People can use sign language to communicate equally as effectively as people using verbal language. The information might well be the same information. Through facial expressions and body language, the people are able to convey the same level of emotion. Some might say that the two people speaking might be doing something else while they speak, but invariably if people have something important to say to another person and attempt to do something else they will be told to 'Stop what you're doing and talk to me'. The more important the message, the less likely the person will attempt to do anything simultaneously. One never sees a speaker at an important function attempting to perform a secondary act while giving an important speech. This applies equally to the listener; speakers become irritated if listeners are doing other things, such as using mobile phones, while they are speaking. It is often seen as a sign of not merely inattentiveness, but disrespect.

It would appear that people are capable of conveying the same information to the other person using both sign language and verbal language – so what is the difference? The difference lies with the observer: only verbal language is deceptive. Sign language is not deceptive since everyone can see what is being said if they have a clear view of the people's hands. Since people are using their hands to communicate they cannot even hide what they say behind the cover of their hands, something people commonly do when wishing to communicate verbally in a deceptive manner. People can also lower their tones to a level where a person sitting a metre away will not be able to discern what they are saying.

So, why would people need to communicate deceptively? Since the evolution of such an unusual trait needs to be closely linked with reproduction then we must consider how people reproduce. If all the marriages are arranged, then this is obviously not something for which deceptive communication is needed. However, if people wish to sexually select outside of the marriage – recognising how dangerous sexual selections can be, as has been well documented in previous chapters, – then here is a reason why people might need to reproduce in a manner necessitating deceptive communication.

In fact, the benefit of verbal speech has already been touched upon in previous chapters: by using verbal language women can sexually select men without anyone except herself, the man, and the intermediary knowing the information being communicated. A woman can speak to her ally, her best friend, and her ally can pass the words to her own brother. Or, alternatively, the man can speak to his sister, and she can speak to the woman that sexually selected her brother (through eye contact, for instance). If humans evolved with arranged marriages the only way to safely sexually select is deceptively, and the only way to select deceptively is through verbal communication.

This creates a fundamentally important link between the evolution of verbal communication and reproduction: verbal communication leads to copulation and the passing on of genes.

Consider humans in societies with arranged marriages attempting to sexually select without verbal communication. If they pointed, people could see to whom or where they were pointing. Even if they used sign language, their words could be seen. All it takes is for a woman to sign the name of the man she is attempting to sexually select, and if her husband-by-arrangement or his allies observed the sign then her husband would guard or attack her to prevent the possible selection. Alternatively, if she whispers the words to her messenger, the messenger whispers the words to the messenger's brother, and they meet deceptively, nobody in the community will know of the sexual selection unless they are actually caught meeting each other.

A few words would allow a female hominid to sexually select a male. If she told a female 'his name and a place' to the male's sister, and the male's sister told her brother 'her ally's name and a place' then they would be able to meet at the designated place. All it would take is a few basic words to signify the place where they might meet, and each other's names.

This is not a dramatic movement from the capacity of chimpanzees, gorillas and bonobos, who have been taught basic sign language and can use nouns. All it would take is for a chimpanzee, gorilla or bonobo to make a basic grunt to signify him or herself (name) and to label a place with a different grunt. The intellectual development between the hypothetical animal I am referring to and these three close relatives would be minimal. For instance, one chimpanzee, Lana used lexigrams:

> A computer-based keyboard system was created that provided the chimpanzee Lana ... with a large number of keys, each having a distinctive geometric symbol called a *lexigram* embossed on its surface. The lexigrams functioned as words, and the lexigram grammar, the rules by which the lexigrams were joined, was monitored for correctness.[22]

The fact that the chimpanzees were identifying the words linked to the lexigram was ensured by moving the symbols so that the chimpanzee could not simply memorise where the symbol was located on the keyboard.[23]

> She innovatively called a cucumber, "banana which-is green"; an overly ripe banana, "banana, which-is black"; and the citrus orange, "apple which-is orange (colored)." Comprehensive analyses clearly indicated that Lana's productions could not be satisfactorily attributed either to rote learning of sequences or to imitation...[24]

However, such suggestions have been criticised:

> A widely cited example of Washoe's ability to create new meanings through novel combinations of her signs is her utterance, *water bird*. Fouts reported that Washoe signed *water bird* in the presence of a swan when she was asked *what that?* Washoe's answer seems meaningful and creative in that it juxtaposes two appropriate signs in a manner consistent with English word order. Nevertheless, there is no basis for concluding that Washoe was characterizing the swan as a "bird that inhabits water." Washoe had a long history of being asked *what that?* in the presence of objects such as birds and bodies of water. In this instance, Washoe may have simply been answering the question, *what that?*, by identifying correctly a body of water and a bird, in that order.[25]

Jane Goodall has commented: 'But this argument cannot be applied to her insistent demands for a *rock berry*, which turned out to be a brazil nut'.[26] Laura-Ann Pettito has questioned whether chimpanzees can communicate an understanding of an abstract noun:

For one thing, chimps cannot, without great difficulty, *acquire* the word *fruit*. While apes seem to have some capacity to associate words with concrete things and events in the world they inhabit, unlike humans, they seem to have little capacity to acquire and readily apply words with an *abstract* sense. Thus, while chimps can associate a small set of labels with concrete objects in the world (*apple* for apples, *orange* for oranges), they have enormous difficulty acquiring a word like fruit, which is a classification of both apples and oranges. There is no tangible item in the world that is literally fruit, only instances or examples of this abstract kind-concept that seems to exist only in human heads.[27]

(With this in mind, the reader may wish to remember that in humans the entire classificatory kinship system links abstract concepts to reproduction.) Chimpanzees may even have difficulty understanding the concept of a noun:

Chimps, unlike humans, use such labels in a way that seems to rely heavily on some global notion of *association*. A chimp will use the same label *apple* to refer to the action of eating apples, the location where apples are kept, events and locations of objects other than apples that happened to be stored with an apple (the knife used to cut it), and so on and so forth – all simultaneously, and without apparent recognition of the relevant differences or the advantages of being able to distinguish among them. Even the first words of the young human baby are used in a kind-concept constrained way (a way that indicates that the child's usage adheres to "natural kind" boundaries – kinds of events, kinds of actions, kinds of objects, etc.). But the usage of chimps, even after years of training and communication with humans, never displays this sensitivity to differences among natural kinds. Surprisingly, then, chimps do not really have "names for things" at all. They have only a hodge-podge of loose associations with no Chomsky-type internal constraints or categories and rules that govern them. In effect, they do not ever acquire the *human* word *apple*.[28]

This perhaps is a little unfair. No doubt the chimpanzee desires an apple and may associate anything to do with the apple in an attempt to communicate the message. It would not be surprising if chimpanzees had difficulty using language since it is entirely foreign. Yet it reflects that there is a limited vocabulary. Even if it were accepted that chimpanzees were combining words to convey their demands, a child of the same age would know the word. It is perhaps more important to emphasise that they can learn and use some language.

Chimpanzees may not be able to communicate with language to any significant degree, but can they understand language? Sue Savage-Rumbaugh and her colleagues argued that they can with the research concerning the bonobo Kanzi.[29] Kanzi's foster mother, Matata, was trained on lexigrams, which Kanzi observed until he was two and a half years old, and Kanzi also appeared to understand some English. At the age of eight, Kanzi's understanding of verbal English was tested and compared with a child aged between 1-and-a-half and two, Alia. To test their understanding, instructions were issued from behind a one-way mirror so that they could not use body language to interpret the instructions:

Once Kanzi and Alia had become comfortable with the situation, the experimenter issued an instruction from behind the one-way mirror: "Put the ball on the pine needles." "Give the lighter to Rose." "Give Rose a hug." "Get Rose with the snake." "Knife the sweet potato." "The surprise is hiding in the dishwasher." "Take the [toy] snake outdoors." "Go to the refrigerator and get a banana." "Go get the carrot that's in the microwave." "Make the doggie bite the snake."[30]

Over the course of the experiment Kanzi was given 415 trials. He performed 246 of these, 59 percent, promptly and correctly. He carried out another 61 instructions correctly but only after some hesitation or after the instruction had been repeated or clarified. 100 were carried out partially but not fully. On 4 trials Kanzi did not respond at all, and 4 responses were totally incorrect. Alia was given 407 blind trials. As Table 1 shows, she made a few more flat-out mistakes than Kanzi did, but overall, her results were not much different.

For both Kanzi and Alia, the results are overwhelmingly better than could have been expected by chance. Dozens of responses might have been made to any of the instructions. Many objects might have been manipulated, many locations were available for the objects, and many different actions could have been performed with each of them. No one would hesitate to attribute Alia's success to her ability to understand English. Can there be any further reason to doubt that an eight-year-old bonobo can understand as much?

To be sure, Alia and Kanzi did not behave in identical ways. During the period of her testing, Alia was rapidly learning not only to understand, but to speak, and unlike Kanzi, she frequently commented verbally on what she was doing.[31]

He was able to respond correctly to three types of sentences where word order was significant. He responded correctly to 33 out of 42 examples (79 percent) with forms like "Put the ball on the rock" or "Put the rock on your ball" where word order was crucial. Even his mistakes were not usually simple reversals, but various other kinds of errors. With instructions such as "Take the umbrella outdoors" or "Go outdoors and get the umbrella," Kanzi responded correctly on 38 out of 46 trials. When he heard a sentence for the first kind, he would survey the objects around him, apparently searching, but with sentences of the second kind he would more often move directly to the place instructed without bothering to look at the things nearby. In such sentences, to be sure, the verbs are also different, so that word order is not the only clue to the meaning, but neither Kanzi nor Alia had trouble understanding a considerable number of different words and a variety of sentence types.

Kanzi's receptive skills give better evidence of linguistic ability than has ever been demonstrated by any other nonhuman primate who has been trained to use symbols, whether these were spoken words, deaf signs, plastic chips, or buttons that needed pressing. Indeed, Kanzi's comprehension demonstrates a degree of linguistic competence that linguists have often presumed to be exclusively human.[32]

Thus, in great apes there may be a very wide gap between being able to understand and being able to communicate.

In many respects, the above results should not be surprising. Chimpanzees have been shown to be capable of organising hunts of monkeys and 'warfare' against neighbouring communities, so the idea that a bonobo could understand basic commands is unsurprising. The major difference is that when the bonobo or chimpanzee is expected to communicate, they are only able to do so in the most rudimentary manner, far below their level of comprehension of verbal commands.

What this suggests is that chimpanzees and bonobos can link sounds to nouns and even verbs, but they simply lack any internal mechanism to be able to communicate these sounds, or to desire to communicate in an abstract sense. There is also a lack of any real capacity to engage in abstract concepts.[33] Importantly, for the issue at hand, however, is not whether chimpanzees or bonobos are capable of language, but rather, do they have the intellectual basis to be able to link sounds with items – and there appears to be little doubt that they are able to do this, with comparable success to a child aged between 1-and-a-half and two. Since the basic structure of linking sounds to nouns (and even verbs) appears to exist in chimpanzees and bonobos, then what is needed is the impetus to actually need to not merely understand the link, but that they must have been able to communicate the link. There must have been a selection pressure that any capacity to communicate these sounds was absolutely vital for reproductive success.

The link between extra-pair sexual selections and the need for verbal, deceptive language creates the selection pressure. Once the message has been passed between the female and male, then, at some time, the female might move out of the group in order to urinate, to collect food, or any other excuse and the selected male might, some time afterwards, move out of the group, in the opposite direction. He then joins her at the landmark, and they copulate. They then return to the group in different directions and at different times. If either the male or the female have a gene that allows for an increased capacity for verbal language, the genes would be passed on to the infant.

Once verbal communication was equated with the ability to sexually select, any trait for an improved verbal capacity would be selected. The ability to use a greater number of nouns would allow a male or female to designate their meeting place with greater accuracy; the use of words to denote times would allow for males and females to specify a time in the future; and basic grammar would allow for more complex messages. Thus 'Lee... stump' (or grunts that signified such a name and a tree stump, or other prominent land mark) might lead to 'meet Lee... stump', 'meet Lee... old stump', 'meet Lee at old stump... morning', 'meet Lee at the old stump tomorrow morning', 'meet Lee at the old stump by the river tomorrow morning', 'can you ask Lee to meet me at the old stump by the river tomorrow just after the sun rises'.

Even at the earliest stages grammar would be preferentially selected. The literal difference between 'meet Lee... stump' and 'Lee... meet stump' could lead to confusion, especially for a hominid of lesser intelligence that may have difficulty rearranging the words. The creation of prepositions – 'meet brother at stump', - and adjectives – 'meet brother at old stump', - would quickly follow where they increased the likelihood of a successful meeting. It might also be noted that giving someone directions, even with signs and numbers depicting streets and addresses, is a complicated task. In the environment in which cooperatively-simple societies exist, with no such signs or numbers definitively depicting places, it must be doubly difficult.

Grammar is a requirement for complex language. Chimpanzees that have been taught using signs have difficulty communicating with any refinement. One chimpanzee, Nim, was taught sign language and used 125 signs.

> ...Nim made utterances containing as many as 16 signs (give orange me give eat orange me eat orange give me eat orange give you me).[34]

Without even simple grammar, a simple message can become confusing or meaningless. Susan Goldin-Meadow comments:

> Chimpanzees and bonobos, our closest primate relatives, are able to learn the words of the system but not the underlying or surface structures that organize those words. Moreover, they use those words only to make requests (of humans), and not to make comments about the world around them.[35]

The capacity to not merely use nouns, but to arrange the nouns into organised sentences, is a basic requirement for the conveyance of messages.

If verbal language is a pre-requisite for sexual selection in a society with arranged marriages, males that possessed no verbal communication skills would not be selected, and neither would females that were incapable of acting as messengers. In order to become an intermediary, a female would need to posses verbal language to a limited degree, as would the male being selected, and the selecting female. Males, knowing that their verbal communication skills were paramount to sexual selections, would display their skills. The evolution of complex language would, however, be female oriented. Since I have argued that the female intermediary is the sister of the selected male, the intermediary and the selected male will possess very similar genes. Thus if the female selects a female as her intermediary and the intermediary has an excellent capacity for verbal language, then many of these genes for verbal language should also exist in her brother or half-brother. Thus the selection process for complex verbal language – despite male display – is almost entirely female directed. No matter how attractive the male, if his sister or half-sister cannot use verbal language, other females have almost no manner of communicating with him without it being obvious. The key to the entire process is the female intermediary; without her capacity to convey messages the male and female cannot sexually select one another.

I have placed three fundamental aspects to the evolution of verbal language: firstly, that it must be verbal and not signed; secondly, that an ancestor with 20 or 50 verbal words must be more reproductively successful than an ancestor with no words; and, thirdly, that there must be a definite link between reproduction and verbal language. I would argue that the idea of verbal language being used to convey deceptive messages for sexual selections would give a basis for all three fundamental aspects in the evolution of verbal language.

Since a female does not select her partner in a society with arranged marriages, the importance of being able to sexually select another male is important. Her husband-by-arrangement may suffer from disease, a physical handicap, infertility, or possess a lesser level of intellectual or physical capabilities compared to other males. For a female, the difference between selecting and not selecting may be between having a child or no children, or between possessing intelligent or strong children or less intelligent or physically weaker children. For the female this is not a minor issue, but may be the difference between producing a highly polygynous son, a monogamous son, a son with no wife-by-arrangement, or a son that dies in infancy. Considering that, in cooperatively-simple societies, around fifty percent of children do not live past the age of ten, the importance of sexual selection cannot be underestimated. It is vital to remember that all the marriages are arranged, and thus a female may be married to a male she never would have selected, to a weak male who simply has extensive social contacts, or to a male who was once healthy but has become affected by disease.

All the major human behavioural traits are being created by the arranged marriage system: cooperation, sexuality, fighting, personal friendship, and so on. Verbal language is simply a trait that emerges indirectly from the cooperative system of arranged marriages as it evolves. The evolutionary basis I have described shows how language emerges because it must emerge. Language does not emerge because it seems beneficial; if this were the case every animal would use complex verbal language. It emerges because it must emerge, because there is no other way that females can safely sexually select. And we can say that other animals do not use verbal language due to the fact that they did not evolve with a complex system of cooperation like the arranged marriage system that humans use which purports to nullify sexual selection. Other animals do not exist in a system of reproduction where female choice is systematically prevented.

Those that wish to argue that verbal language is beneficial in terms of food gathering or other mechanisms of survival need to explain why no other animal uses complex verbal language. The fact is that they cannot, nor will they ever be able to do so. Darwin predicted this and never used natural selection to explain the evolution of verbal language in humans – he argued that singing and display was the basis for the evolution of human language. Despite Darwin's position as the greatest thinker in biology, to this day biologists and others that use evolutionary arguments continue to ignore Darwin's arguments concerning human evolution. My argument is entirely Darwinian in that it extends his argument that sexual selection is the basis for the evolution of complex verbal language, but in a manner that Darwin could not have considered given that anthropology, as a discipline, barely existed.

The animal in question is not human; its pre-human form will be discussed in the following chapter. Once the arranged marriage system begins and females are prevented from selecting males, and have access to alternative males, then the selection pressure for the use of verbal language would begin. As males would be forming complex political and social bonds with other males and cementing these bonds with arranged marriages, so females would be selecting female allies and attempting to select other males using these female allies and verbal language. As the arranged marriage systems became more complicated, the brain of the animal became more complicated, the capacity to consider abstract concepts – such as the arranged marriage system – became more complicated, then verbal language became more complicated.

I suspect that the above argument will seem implausible to many people. Firstly, most people have an instinctive disgust for extra-marital sexual selections and a strong belief in the importance of marriage – without analysing why they feel this way. Secondly, many people believe that language is what makes them human:

> Speech is so essential to our concept of intelligence that its possession is virtually equated with being human. Animals who talk are human, because what sets us apart from other animals is the "gift" of speech.[36]

This is not an unusual statement. Shamatari villagers, upon seeing their first non-Yąnomamö, commented after hearing Napoleon Chagnon speak:

> Then I spoke, and again they marveled. I spoke a "crooked' version of Yanomamö, like the Bisaasi-teri do, but they understood me: "*Whaaa! A akahayuwo no modahawä* [He is capable of language]."[37]

Perhaps the ultimate example of the pre-eminence of language comes from the gospel of John: 'In the beginning was the Word, and the Word was with God, and the Word was God'. In religion, the non-physical nature of the supernatural understandably increases the importance of the non-physical form, thus magnifying the importance of language.

Thirdly, many people simply do not understand evolution as well as Charles Darwin. Darwin knew that any aspect of an animal that was not obviously aimed at survival could not be explained through natural selection; he thus devised the argument of sexual selection and applied it not merely to the peacock, but major human traits. Darwin's arguments in *The Descent of Man* have been mostly ignored by the vast majority of evolutionary theorists for 130 years. For this period of time, almost all of the people involved have attempted to explain human language in terms of survival, despite the fact that Darwin never thought this was remotely logical.

Fourthly, many people struggle with the emphatically biological aspects of evolutionary thought. People want to examine human behaviour and find something resembling complex linguistics, mathematics, morality or theology. Surgeons do not cut open the body, though, and discover exquisite moral arguments: they discover blood, muscle tissue, bones, mucus, faeces, urine, and so on. Every evolutionary argument must be based in the latter set of categories. Many academics feel most comfortable in discussing complex ideas rather than discussing sexual reproduction. Yet, sexual reproduction is at the very heart of all evolutionary arguments for sexual animals. The fact that arranged marriages rarely rate a mention in discussions on human evolution reflects how far from the core issue of reproduction most theorists prefer to be.

If one were to present to people (1) a film of a person giving a very complex lecture on an academic subject, and then (2) a film of a young woman speaking to her friend (in whispers so no-one could hear, including the audience), her friend (in whispers so no-one could hear, including the audience) speaking to her half-brother, and later her half-brother and the first young woman meeting and copulating, and then to ask the viewers which would be most important for the evolution of verbal language, I suspect many people, and probably almost all academics, would choose the first film. This is despite the fact that of the two films only the first could be remade using sign language without altering the intent of the film, and only the second film leads to reproduction. In fact, every lecture in every university in the world could be made in sign language – as long as students had a clear sight of the lecturer's body. The fact that the lecturers can give lectures on complex subjects is no different to people in cooperatively-simple societies discussing the complex aspects of the arranged marriage system that they utilise – and both could be achieved using sign language. In *Grand Valley Dani*, Karl Heider discusses the difficulties the Dani children have in learning their kinship system:

> But many Dani are not competent. They do not understand their own system and they make errors when talking about it. These people are, of course, the younger children, who were mentally quite normal, but had just not achieved full understanding. I did one study to discover the steps through which the children go in order to learn their own system... By 1970, when I was asking these questions, I had only months, not years to get an answer. So instead of following a few children as they were growing up, I talked to more than 150 children at different ages, almost 100 of whom turned out to be noncompetent. They were old enough to know some things about sibs and moieties and their own genealogies, but not far enough to avoid errors. The most common sorts of errors, which they made were claiming that their parents were of the same sib or moiety, or that they were of the same sib or moiety as their mothers, or different from their fathers.
>
> By arranging these noncompetent children in order, one can get a probable sequence for how they learn about moiety and sib exogamy and patrilineality. The most common route seems to be to learn the principles of sib exogamy first, then sib patrilineality, then moeity patrilineality, and finally moiety exogamy.[38]

Napoleon Chagnon found similar mistakes by sub-adult males concerning relatives in the generation above, though he speculated that the mistakes could be intentional:

> While adult male informants show a pronounced tendency to move female kin from non-wife categories into wife categories, sub-adult males do not. Sub-adult males tend to reclassify females into the mother category. It is tempting to suggest that having more 'mothers' is better for potential fitness of parentally dependent youngsters than having more 'wives'.[39]

Chagnon emphasises how important these systems are for the children's success:

> There is the more general question of why parents want their children to know and understand the rules as intimately and accurately as possible. If, as Alexander argues, rules generally function to thwart the reproductive strivings of males and foster in-group unity, then knowing the rules accurately makes it easier to break them

effectively. Thus Yanomamö fathers not only break the rules on behalf of their sons' reproductive success, they also encourage them to learn genealogies and the rules of classification as accurately as possible so they can do it for their sons. Knowing the rules makes it easier to demand that others follow them by quickly detecting and protesting the kin misclassifications of others that threaten your own reproductive interests and those of your sons.[40]

This is no different to children in modern schools learning algebra. People are able to consider complex concepts not because they can speak – in fact, the capacity to verbalise words has nothing to do with the capacity to understand complex concepts – but because they use a complex cooperative social system. This cooperative social system requires a brain able to understand complex concepts in order to utilise the system – a brain with far greater capacity for intellectual thought than a chimpanzee's. This cooperative social system is directly linked to reproduction. When the children above are discussing their social system, they are discussing who they can marry, who is their natural competitor for marriage, who is their natural ally, who they must avoid (such as male avoidance of the mother-in-law, whose daughters the males will marry), and so on. We possess the ability to understand complex concepts and children must have a desire to learn those complex concepts if they are to succeed in a socially complex environment. Yet, there is no reason why those complex concepts could not be conveyed purely in sign language.

The instigating reason for why we use verbal language is because between about five and twenty percent of all reproduction will occur outside of the arranged marriage. Females must be able to use deceptive communication in order to organise these events. This is so vital for females – who do not sexually select their partners, and who may be partnered with a sub-standard, infertile, or disease-ridden partner – that every gene supporting verbal language would have been utilised by females. Even before the cooperative system became extremely complex, these females would have created a selection pressure towards an increasingly complex verbal language that aimed at accuracy of exchanging information rather than the capacity to convey complex concepts. It is vital to appreciate this: the exchange of exact information is the basis for verbal language, not the conveyance of complex concepts.

Verbal Language and Female Alliances

If a female is to select a male using verbal communication, then she will need to know a close ally of the male in order to pass the information to him. She will need an intermediary – the best friend. Secondly, she will need to ensure that the information is kept a secret. I have already considered these aspects at length in the previous chapter, so I will merely consider these needs with respect to the evolution of verbal communication.

In the formation of a close ally, females form strong bonds based on the unlikelihood of the leaking of information. The more information a female discloses to her ally with no disclosure of the information to other females, then the stronger the bond between the two females. Thus females would evolve with the capacity to test her ally repeatedly by revealing information to see if that information is revealed to others. Once she has revealed important information and found that the information has not been disclosed, then she can be more confident that a sexual selection can proceed through her ally.

The choice of a good ally may be the difference between having a polygynous son or a son that dies in infancy; the choice of a poor ally may lead to a violent attack upon her by her husband-by-arrangement and his allies. We should expect that females would evolve increasingly complex verbal communication skills to reveal information to a potential ally and to see if that information is revealed to others. The greater the skills of verbal communication, the greater the ability to be able to test a potential ally: the better the ally, the better the capacity to sexually select.

When information is leaked, we invariably describe this as gossip. Karl Heider observed amongst the Dugum Dani in Papua New Guinea:

> The economic demands on the individual Dani are relatively slight, and even the women, who spend more time working than men or children, have ample opportunity to sit around gossiping. The men work hard for short periods when they are building a house or watchtower, or opening new gardens. But much of the time is spent sitting around a fire in a men's house or at a watchtower, perhaps smoking, perhaps weaving, but usually talking.[41]

Robin Dunbar in his work has suggested that human societies use gossip to knit a society together, an alternative to grooming.[42] I have already argued that arranged marriages – the exchange of the capacity to reproduce for cooperation – is the method that I believe humans use to knit their society together. Furthermore, people usually view gossip as a disruptive rather than a cohesive force. If arranged marriages are the method by which people knit their society together, if sexual selections outside of the arranged marriages can cause violence between paramours and arranged-husbands, and if sexual selections are instigated and undermined by verbal messages, then the view of gossip as being a negative force in a society is understandable. What knits the society together is not language dependent: what is language dependent undermines the society.

The very nature of gossip requires a complex language involving references of what a person did or said to another person at a particular time in a particular manner. While English has shed some of this precision, one only need look at the Latin languages to appreciate that languages are often extremely precise concerning what happened and when something happened, with numerous verb conjugations to define an occurrence as precisely as possible. Amongst cooperatively-simple societies, this precision also exists; Napoleon Chagnon observed of the Yąnomamö language:

> The Yanomamö language is very precise about what is known firsthand and what has come from secondhand, or hearsay, sources.[43]

The conception of primitive peoples having gruntlike languages whose poor vocabularies require lots of sign gestures is not only ignorantly wide of the mark, it is also ironic: the working vocabulary of most Yanomamö individuals is probably much larger than the working vocabulary of most people in our own culture. While it is true that the absolute content of our language's vocabulary greatly exceeds the content of Yanomamö vocabulary, it is also true that we know much less of ours than they do of theirs. When you do not have a written language you have to store more in your head.[44]

In the previous chapter I have considered how females try to control the lives of other females. If a female's sexual selection of a male is learned by other females, then she is at their mercy lest this information is leaked to her husband-by-arrangement. We should thus expect that not merely would females use verbal communication in order to form potential allies for sexual selections, but we should also expect females to use verbal communication to influence society. A female might also stave off a sexual selection of a brother, a husband-by-arrangement, or of an ally's husband-by-arrangement, or to instigate an attack on a rival female. By gathering knowledge, a female can prevent sexual selections of a desired male, protect males from sexual selections of other females (who might be married to extremely dangerous husbands-by-arrangement), and extend this influence to her ally or allies.

If verbal language evolved so females could sexually select deceptively outside of arranged marriages, then, rather like the rings on the stump of a tree, we should expect the first ring around the nucleus of sexual selection to involve the formation of an ally that is necessary for the deceptive sexual selection, and the second ring to be the exchange of information to attempt to prevent or control sexual selections by other females. Alternatively, the nucleus could be the formation of the ally, the first ring the actual sexual selection of the male, and the second ring the capacity to control sexual selections through the accumulation and revelation of information. The latter might be the more correct situation since, without the ally, the female has difficulty communicating with the male, thus the evolution of verbal language may be simply to form female allies at its first emergence.

If this were true we would expect female verbal language skills to be generally superior to males. In her book *Sex Differences in Cognitive Abilities*, Diane Halpern summarises the studies and finds that males are far more likely to suffer from speech and reading impediments and disabilities, such as stuttering and dyslexia. Studies also reveal that girls are superior users of language:

> Martin and Hoover (1987) conducted a large-scale longitudinal study in which they examined children's scores on the Iowa Test of Basic Skills in each grade from Grades 3 to 8 for 4,875 girls and 4,497 boys. They reported that girls scored higher on tests of spelling, capitalization, punctuation, language usage, reference materials and reading comprehension. It is important to note here that the between–sex differences were quite large. In Grade 8, for example, two thirds of the highest scoring students on the language tests were female. Results from the Differential Aptitude Test of Spelling also show a significant female advantage... As psychologists learn more about the nature of verbal abilities, new tests have been devised that show very large sex differences. Hines (1990) found very large differences on tests of associational fluency (which is a fancy term for generating synonyms). Her results showed a huge female advantage, with

an effect size of 1.2! Similarly high values have been reported on a consonant-vowel matching test, with a female advantage effect of 1.3 (Block, Arnott, Quigley, & Lynch, 1989). These are enormous effect sizes – so large that tests of statistical significance are not even needed.

Female superiority on verbal tasks may seem reminiscent of the stereotype that females talk more than males, but it is important to keep in mind that it is the quality of the speech produced and the ability to comprehend or decode language that is being assessed, not merely the quantity. Studies in naturalistic mixed-sex settings show that males talk more and interrupt more than females (Bilous & Krauss, 1988).[45]

Indeed, it might also be asked why females are stereotyped as more talkative than males, rather than to use it as an excuse. We would stereotype males as more likely to engage in physical play than females, but would we expect females to be superior if we artificially forced females to engage in more physical play than males? Studies have also found that sex differences between boys and girls show up in the earliest years; Diane Halpern summarises the evidence:

Of all the cognitive sex differences, differences in verbal ability are among the first to appear developmentally. Girls between 1 and 5 years of age are more proficient in language skills than their male counterparts (McGuiness, 1976; Smolak, 1986). There is also some evidence that girls may talk about 1 month earlier than boys and produce longer utterances than boys (Gazzaniga el al., 1998; Moore, 1967; Shucard, Shucard, & Thomas 1987). There are significant sex differences in the rate of vocabulary growth during the toddler years. On average, there is a 13-word difference in vocabulary size between girls and boys at 16 months of age, which grows to a 51-word difference at 20 months and a 115-word difference at 24 months (Huttenlocher, Haight, Bryk, Seltzer, & Lyons, 1991). These researchers found that the differential rate in vocabulary growth was unrelated to how much mothers spoke to their children – mothers spoke as much to their infant sons as to their infant daughters. They concluded that "gender differences in early vocabulary growth seem to reflect early capacity differences" (Huttenlocher et al., 1991, p. 245).[46]

Studies have also shown that women's verbal abilities improve when they are mid-cycle in their menstrual cycle when they are likely to be ovulating. Hampson and Kimura found that 'women performed significantly better on speech articulation (speed of reciting a tongue twister), a test of manual dexterity, and verbal fluency in the high-hormone or midleutal phase than during the menstruation portion when hormones are low. By contrast, performance on spatial tasks (e.g. size of errors on the Rod and Frame Test) was better during the low-hormone phase of the cycle than the midleutal phase'.[47] There have been some findings that increased estrogen and testosterone have positively and negatively, respectively, affected verbal ability in various subjects (healthy men, post-menopausal women, transgender), while other studies have found no impact.[48] It would not be surprising if the relationship between immediate hormone levels and traits such as verbal ability were complex.

If language were more important for one particular sex then we should expect that the members of that sex would be superior users of language. That girls, in large studies, have been found to be consistently superior users of language than boys would suggest that girls have a greater need for using language than men. If women evolved with a complete dependence on verbal language for sexual selection, then these findings would not be surprising. Indeed, any casual observer of the sexes would readily observe that women's social contacts are entirely dependent upon language; by contrast, men and boys may spend hours together with minimal conversation. Any person going into a school would observe that girls in their early teens would be talking and boys in their early teens would be engaged in physical play. No doubt, what would be heard between these girls would involve something like the following:

> We spend hours talking. Just talking. Sometimes it's about just one thing, like what happened last night, or yesterday at school. I tell her what happened, and then she'll ask a hundred questions. "What did she do then?" or "How did she say it?" and "Who else was there?" And it goes on and on...[49]

It is difficult to imagine a boy saying these words. Boys rarely engage in verbal communication of such precision in a social setting.

Men and Language

The major difference in the way that males and females use language is that females use language for a specific purpose, while for males, language is an expansion of their normal behaviours. Girls and women specifically exchange information in a manner that has no equivalent in other species. If one walked into a school yard, then one would see boys engaging in physical play, which would probably be expected if one observed males in numerous other species of primate. By contrast, only in the human species would juvenile females be in small groups verbally exchanging pieces of information.

Deborah Tannen in *You Just Don't Understand* explains the basic difference between men and women's use of verbal language.

> For most women, the language of conversation is primarily a language of rapport: a way of establishing connections and negotiating relationships. Emphasis is placed on displaying similarities and matching experiences. From childhood, girls criticize peers who try to stand out or appear better than others. People feel their closest connections at home, or in settings where they *feel* at home – with one or a few people they feel close to and comfortable with – in other words, during private speaking. But even the most public situations can be approached like private speaking.
>
> For most men, talk is primarily a means to preserve independence and negotiate and maintain status in a hierarchical social order. This is done by exhibiting knowledge and skill, and by holding center stage through verbal performance such as storytelling, joking, or imparting information. From childhood, men learn to use

talking as a way to get and keep attention.[50]

The difference in the use of language is most obvious in private and public environments. Women are more likely to talk in private environments, while men are more likely to speak in public environments:

> For everyone, home is a place to be offstage. But the comfort of home can have opposite and incompatible meanings for women and men. For many men, the comfort of home means freedom from having to prove themselves and impress through verbal display. At last, they are in a situation where talk is not required. They are free to remain silent. But for women, home is a place where they are free to talk, and where they feel the greatest need for talk, with those they are closest to. For them, the comfort of home means the freedom to talk without worrying about how their talk will be judged.[51]

Women are most comfortable speaking intimately, while men tend to be silent when in an intimate situation and are more verbose in public moments, mostly when there are other men or competitors present. For women conversation is about creating possible alliances: for men it is about the struggle for dominance. Tannen describes this from personal experience:

> What happens when I talk to people at parties and social events, not fellow researchers? My experience is that if I mention the kind of work I do to women, they usually ask me about it. When I tell them about conversational style or gender differences, they offer their own experiences to support the patterns I describe. This is very pleasant for me. It puts me at center stage without my having to grab the spotlight for myself, and I frequently gather anecdotes I can use in the future. But when I announce my line of work to men, many give a lecture on language – for example, about how people, especially teenagers, misuse language nowadays. Others challenge me, for example, questioning me about my research methods. Many others change the subject to something they know more about.[52]

Such a response from men is entirely understandable; for a man to ask another man in an open way about his work would be to defer to him and to place himself in an inferior position. A woman would do this instinctively, but if the woman she deferred to did not reciprocate and defer back, it would annoy the first woman and, no doubt, end the conversation.

Men use language as an extension of themselves, to be aggressive, to portray themselves in a certain manner. Language is like clothing: it allows men to create a certain image of themselves, to advertise themselves, to spoil other men's advertisements, to accentuate their emotions, to accentuate their attacks on others or to defend themselves against others, to portray confidence, to portray trustworthiness, to portray cleverness, and so on. Language allows men to accentuate their behaviour. To that extent, women's use of language is a unique behaviour amongst primates, while men's use of language is simply an expansion of behaviours exhibited by other primates, a tool to further their competitive cause.

This is not to suggest that men's use of language is simple; like any complex behaviour it demands practise. Cicero, in his work *On the Orator*, defines what is important:

For, first, one has to acquire knowledge about a formidable quantity of different matters. To hold forth without this information will just mean a silly flow of windy verbiage. And then one has to be able to choose one's words well, and arrange them cleverly. It is also essential to have an intimate understanding of every emotion which nature has given to mankind: it is in the processes of calming or exciting the feelings of an audience that both the theory and practice of oratory find their fullest expression. Other requirements include a certain sparkle and wit, and the culture appropriate to an educated man, and a terse promptitude both in repartee and in attack. A sensitive, civilized lightness of touch is also desirable. One's memory, too, must be capable of retaining a host of precedents, indeed the complete history of past times. Nor is it by any means advisable to be ignorant of the law and existing statutes.

As regards delivery, I am sure I need not go into a great deal of detail. The principal relevant factors include physical deportment, gestures of the arms, facial expression, voice production, and the avoidance of monotony. ...

To go back to the memory, this serves as a universal treasure house, and it has to be given the safe keeping of every single aspect of the speech one is going to make, all its substance and all its words. Without this precaution, however remarkable all these items were when the orator originally conceived them, they will one and all be totally wasted.

So there is no longer any need to wonder why orators are rare. They have to display a whole host of different qualifications: and to possess even a single one of them is a considerable achievement. So let us urge our sons, and everyone else whose glory and reputation are dear to us, to appreciate the gigantic scope and range of this activity.[53]

Some might view that this method of speaking is essential in order to be dominant in a society, but this is not true. Alliances between men are formed through marriage coalitions and genetic similarity, and men's opinions can rarely be won over since they must follow the decision to whomever they owe allegiance. This is no different to modern politics; as much as politicians may argue their case in parliament, they almost always vote in accordance with the party to which they have pledged allegiance, and not to would be seen as an act of betrayal. What we observe are normal attempts to gain an edge over an opponent for other politicians or the general public to observe – but they rarely make any difference when it comes to the outcome.

One of the most important methods of men using verbal language is that men can compete without resorting to violence. In language the idea of fighting with swords and with language is often equated, with the word 'fencing' meaning to both fight with swords and debate. The word 'debate' comes from the Latin *de*, from, and *batuere*, to strike or beat. In recent years, sex differences in verbal aggression have been examined in children and adolescents, and it has been confirmed that males have a higher level of use of not merely physical aggression, but direct verbal aggression.[54]

Such is the belief that words do hurt, in modern societies defamation is actionable in court and can lead to damages being paid. As the history of duels in Europe attests, verbal aggression is dangerous. Insults can lead to serious violence, as this example from Hallpike of the Tauade in Papua shows:

> Aima Kamo also killed A— K— (of Kilamalumi hamlet) because he insulted
> Aima's sons Peeri and Maia.
> He said 'cut off your sons' penises and eat them'. He said this because they had
> defecated in his stream and he had come and drunk from it. So he was angry when he
> saw the faeces in it and insulted them. A— K— had no children. He was speared in the
> throat by Aima. Tauru Ivei and Lume Ivei came and collected his body. They took him
> back to Oropo and killed pigs on his body and buried him. He was a *malavi*. After some
> time Aima killed a pig and gave it with dog's teeth to Tauru and Lume.[55]
> [*Malavi* = rubbish man]

Killings due to insults appear to be rare, but often lead to fights, as Warner observed amongst
the Murngin: 'The use of obscenity and profanity against a man always results in a camp
fight'.[56] The victor in a fight can continue to use verbal aggression to present their superiority.
Chagnon gives an example from the Yąnomamö:

> Rerebawä had already been given one of the daughters in marriage and was
> promised her younger sister as his second wife. He was enraged when the younger sister,
> then about 16 years old, began having an affair with another young man in the village,
> Bäkotawä, making no attempt to conceal it. Rerebawä challenged Bäkotawä to a club
> fight. He swaggered boisterously out to the duel with his 10 foot-long club, a roof-pole
> he had cut from the house on the spur of the moment, as is the usual procedure. He
> hurled insult after insult at both Bäkotawä and his father, trying to goad them into a
> fight. His insults were bitter and nasty. They tolerated them for a few moments, but
> Rerebawä's biting insults provoked them to rage. Finally, they stormed angrily out of
> their hammocks and ripped out roof-poles, now returning the insults verbally, and
> rushed to the village clearing. Rerebawä continued to insult them, goading them into
> striking him on the head with their equally long clubs. Had either of them struck his
> head – which he held out conspicuously for them to swing at – he would then have
> the right to take his turn on their heads with his club. His opponents were intimidated
> by his fury, and simply backed down, refusing to strike him, and the argument ended.
> He had intimidated them into submission. ... Rerebawä had won the showdown and
> thereafter swaggered around the village, insulting the two men behind their backs at
> every opportunity.[57]

Oratory becomes an important part of responding to insults in a public manner, as this
example from Hallpike shows in relation to a dance held by the Tauade:

> Before the dance, the chief had been challenged.
> 'Hey!' Look at these fellows. Amo Kau, who wants to be a great man. He
> thinks he over-tops everyone, when he doesn't even come up to our ankles. He's inviting
> a crowd of big men to his Gabe. He's sure they'll come. We're going to have a good laugh
> - there'll be no one!'

These disagreeable comments had been relayed to him [Amo Kau] and very probably exaggerated. That is quite normal.

If one provokes a person, one does it thoroughly.

No one forgets an injury. It is one of the great pleasures of the Gabe to be able to repay someone in his own coin: injury for injury, humiliation for humiliation.

On the night of the dance itself, the *Gab'u babi* [hosts] congratulate one another on the tremendous success of the occasion. One might suppose oneself transported into the middle of a Mutual Admiration Society.

There comes a point when the chief can stand it no longer. He must declaim his joy, his good fortune, his pride, and also his revenge. He takes his revenge there and then.

'Listen, all of you. I have something to say to you. You, So-and-So, you have challenged us to get these dancers to come. And to make your words come true, you have visited them, and exhorted them, and even bribed them not to come. Well! Have they come? I will take them by the hand and introduce them to you, so that you can get a good look at them. Do you recognize them? What is this person's name? And what is his village? And what is this one's name, and from what village does he come? And this one, and this one, and again, this one? So take a good look, and see how fine they are. I am a great man, the son and grandson of great men, an An'Uta, and everyone respects me. You can go and find other dancers like these, and we will come and see them when it is your turn to give a Gabe.'

In corners, people are whispering 'It is well done. He [the challenger] can say nothing.'[58]

The complexity of the human brain derives from the arranged marriage system. The arranged marriage system presents a reproductive system of unique complexity, and the human brain evolved to be able to manipulate this environment. If men use verbal language to manipulate this environment, it does not mean that the verbal language creates the social environment. People are born into a social environment of arranged marriages, they instinctively respond to this social environment and attempt to manipulate the environment, and language is simply one of the tools used to manipulate the environment. Girls and women use language to manipulate the environment from their own perspective, by creating or limiting the sexual selections that may enhance their own or their allies reproductive success (inclusive fitness) and prevent it being undermined by sexual selections by other females. Males use language to enhance their capacity to improve their own social status within the existing hierarchy. Heider notes the overwhelming importance of the exchange system as against other factors amongst the Dani of Papua (Um'ue is a Big Man):

Men achieve importance by success in many fields. If one asks the Dani how to become a Big Man, they always say first, by killing many enemies and second, by having lots of wives and pigs. ... In any case, straight-out killings may be less important than being clever enough to carry off a successful raid. In a battle, brave but relatively unimportant men like Wejak were often very visible on the front lines; men like Um'ue seemed to prefer action in the thickets off to the side.[59]

What seems to characterize the important men more than anything else is their skill in manipulating the exchange system; in other words, the degree to which they have established ties with many others through the exchanges.[60]

Wealth per se does not bring influence, but a large household, several wives, and lots of pigs and shell goods, all go together as marks of the clever competence of the leader.[61]

Chagnon observed that individual aggression from a leader can play an important role in both village fissioning and village cohesiveness. He observed this with regard to one leader in particular, Möawä, in his text in studying Möawä's village, the Mishimishimaböwei-teri.

Möawä, more than any other Yąnomamö I have known, was the archetypical expression of his culture's values and political ideals. He was fierce, a man of great personal courage and renown, strong, an excellent marksman, and forceful to the point of awesomeness. He took what he wanted when he wanted it, and scorned those who would pretend to impose their will on him. People feared him because he was capable of great violence and expressed it unhesitatingly. But people looked up to him, if they were not his competitors, and did what he demanded of them. He was a man of his word, and when he spoke or commanded people listened and obeyed. When he threatened, everyone knew he meant what he said. Because of his forceful bearing and abilities, he was able to attract a large following and hold it together. His enemies despised him, but they feared him as well. Life for the obedient in his village was secure and relatively happy, but for the competitor or politically ambitious, it was dangerous and strained. Competitors did not last long as coresidents, for Möawä's overbearing manner led to disputes and fights, and his competitors – mostly agnates – sooner or later packed up and left. At lest three of his agnates – Sibarariwä, Nanokawä, and Reirowä – were very much like him, but not nearly so able or successful. There was a long history of their joining and leaving Möawä's village in recent years, a process of fusion and fission that hinged on how well they got along with Möawä.[62]

Perhaps the most difficult parameter to measure is the role of leadership in holding villages together. I am convinced that villages in this area of the tribe grow to be large because headmanship is strongly developed. There is a correlation between size of village and, for lack of a more precise term, "strength" of headmen. Strong leaders like Möawä add a great deal to the stability of the village and manage to keep order and organization a jump ahead of the smouldering conflicts that threaten to tear larger villages apart. ... Möawä is a very young man, and it has been only in the past few years that he has managed to acquire a large following. ... He is, as the Yąnomamö refer to headmen, "... the one who *really* lives here." Like strong leaders in other large villages, Möawä depends a great deal on his aggressiveness to keep the village intact. He, like other strong leaders in Patanowa-teri and Hasubowa-teri, has had to, from time to time, settle disputes by killing the individual who caused them.

> These are very drastic measures, but they might represent the additional leadership element required to allow a village to grow from 150 people to 200 or more people. Quantifying this element is, perhaps, impossible, and it may be that only subjective evaluations are possible.[63]

As can be seen from these examples, verbal capacity in itself is of little value if it cannot be backed up by the person making the statements. Demands for cooperativeness within the group are of little value if the cooperativeness cannot be enforced, with perhaps, as in the example above, the actual killing of the uncooperative person being the most extreme example. It needs to be remembered that there are no jails in these societies, and people in Western nations who might be armed bank-robbers or murderers are as free as anyone else. So while Möawä's killing of a coresident may seem extreme, the man may have been someone who in a Western society may be viewed as uncontrollable and have spent much of his life in jail. In any event, the above accentuates the lesser importance of verbal capacity amongst males as against females. Amongst males, the capacity to link verbal capacity – particularly in regard to demands – and the ability to enforce those demands, even with lethal force, are inseparable. Chagnon noted that this is a daily issue for headmen, though the following two headmen went about achieving this end in different ways:

> Like all headmen, Möawä carefully weighed his requests and rarely made one that did not turn out the way he wanted it to. I was at first startled and then very intrigued by this consistent pattern among headmen, since it took consummate skill and ability to size up each situation, day after day, and consistently appear to have ones' orders followed and requests satisfied. Kąobawä was very accomplished at this and, in his dealings with me was very gracious and polite. Should there be any doubt in his mind that I might deny a request, it was never made in such a way that he asked for something and I refused. He would indicate, for example, that one of his friends from a distant village arrived to visit and needed a machete, but he had none to give the man. Then he would pause, look at me and say something like: "Dearest nephew, do you not have one lying around that you are not using?" It worked every time. Möawä, on the other hand, would exercise a different style, one that was harsher. He would say: "Older brother! Give that man your machete! Do not be stingy with your possessions!"[64]

The long-term implications for extreme aggression, however, may be destructive. The Monou-teri were dominated by a very aggressive man:

> The headman of the village, Matowä, was a particularly aggressive man. According to Rerebawä and Kąobawä, Matowä was the only fierce man in the entire village, the true *waiteri* (fierce one) of the group.[65]

> Matowä frequently seduced the wives of other men, a factor that led to regular

feuding in the village, and resulted in a number of club fights. Of the numerous affairs he had, two in particular illustrate the nature of possible consequences. His youngest brother was married to an abducted Shamatari girl. Matowä seduced her, thereby enraging his brother. The young man was afraid to vent his anger on the real culprit, his brother, so, instead, he shot the wife with an arrow. He intended only to wound her, but the arrow struck her in a vital spot and she died.

Manasinawä, a man of some 55 years at the time I learned of this incident, joined Matowä's group with his wife and young daughter. He fled from his own village in order to take refuge in a group that was raiding his own village, as he wanted to get revenge against them for a wrong they had committed. Matowä, who already had several wives, decided to take Manasinawä's wife from him and add her to his own household. This resulted in the final club fight that led to the separation of Kąobawä's group from the Monou-teri.[66]

Matowä was eventually killed by men who might normally be considered allies:

The man who fired the fatal arrow into Matowä's neck was a son of a man the Monou-teri shot in their raid. Two of the men who shot Matowä were his classificatory brothers (members of the same lineage), three were brothers-in-law (including the man who shot the fatal arrow), and one was a man who had been adopted into the Patanowä-teri village as a child, after he and his mother were abducted from a distant Shamatari village.[67]

The implication of this for the village, with their only 'fierce' man dead, was that the village became extremely weak and leaderless, and were forced to live with their protectors (this is documented in chapter 4, above). Thus aggression can be both productive and counterproductive, and if a particularly aggressive man dominates a village and other potential leaders leave the village, then, upon his death, there can be a vacuum leaving the remaining villagers' future in jeopardy.

The above gives some idea of the manner in which language is used for males, but it is obvious that language is simply an aspect, and a lesser one, of other attributes of importance for males, namely: exchanges of marriage and individual prowess with both good decision making and the ability to use violence to enforce cooperation from others.

No doubt many will argue that information gathering is important to men, and this is true. Consider the following extract from Chagnon's Yąnomamö. Chagnon had gone to a village where the people had never met a non-Yąnomamö, yet:

I was flabbergasted at the detail and accuracy of what they knew about me. They knew I had a wife and two children, and the sexes and approximate ages of my children. They could repeat with incredible accuracy conversations I had had with Yąnomamö in many different villages. One of them even wanted to see a scar on my left elbow. When I asked what he meant, he described in intimate detail a bad fall I had taken several years earlier on a trip to Reyaboböwei-teri when I had slipped on a wet rock and landed on my elbow, which bled profusely. He even quite accurately repeated the string of Yąnomamö vulgarities I uttered at the time, and my complaint to my guides that their goddamned trails foolishly went up and down steep hills when they could more efficiently go around them! For people who had never before seen a non-Yąnomamö, they certainly knew a great deal about at least one of them![68]

It is obvious from the above, that people in one community can learn information about people in other communities that they have never met, and, possibly, will never meet, which is not so different from television, newspapers and gossip magazines.

The gathering of information also may be of interest to women: they can learn about the reputation of other men and their exploits, or the trustworthiness, beauty or intelligence of other women, which may be useful information if they encounter the individual's community.

There is no doubt that men can gain quite a lot through learning important pieces of information to enhance their own capacity to compete against other men. Yet such pieces of information are rather like paintings to the walls and roof of a house. A young man will exist in a structure of marriage coalitions and kinship, which no little snippet of information will affect. However, it might be beneficial to know that the man, with the attractive wife, is unwell or away hunting; or to know that another man in a similar position in the hierarchy is injured and that now might be a good time to insult him.

Between communities, the gathering of information is also important, particularly in relation to the injuries or deaths of adult males in other communities or, as in chapter 4, whether a particular community is on the receiving end of a large number of attacks and thus more vulnerable. There is, however, no reason why such information could not be exchanged purely through sign language, and thus, while this is a benefit, it is not a reason for the evolution of verbal language. The studies with the chimpanzees of Gombe suggest that chimpanzees are equally adept at taking advantage of the sudden weakness of other groups:

> Some dramatic encounters between Kasakela and Kalande groups were observed. Once, for example, Figan and four other males were routed by a larger group of Kalandeites and fled, in silence, back towards the north and safety. Two Kasakela males disappeared: first the strong young male Sherry and, the following year, old Humphrey. And, while we shall never know for sure, we thought it more than likely that they were victims of intercommunity aggression. After this, with only five adult males remaining, the Kasakela community not only continued to lose ground in the south, but also in the north where the large Mitumba community, seizing the opportunity, began to extend its territory southward. By the end of 1981, four years after Sniff's death, the Kasakela range was only about five and a half square miles – scarcely big enough to support the eighteen adult females and their families. I even feared that we might lose the whole community.[69]

Chimpanzees, like many animals, call in a chorus to rival groups, giving adversarial groups a signal of strength or an implication of weakness. Jane Goodall noted that this possibly played a role when the community was reduced to four adult males and a few juvenile males:

> These youngsters – Mustard and Atlas, Beethoven and Freud – lacked the strength
> and social experience to be of much use during an actual attack, but the sound of their
> calling and noise displays, added to that of the four remaining senior males, may have
> given their neighbours the illusion that the Kasakela community was more powerful
> than it really was.[70]

The similarities between chimpanzee behaviour and human behaviour in intercommunity conflict are remarkable. It is unlikely that for a chimpanzee-like ancestor, the requirement for verbal communication would have been necessary to expand on these behaviours.

I argue that the most important aspect of verbal language amongst males is as a tool to increase their influence within their society. Yet, the influence of language amongst men is small, since men are always placed within the rigid structure of the arranged marriage system that defines their cooperative place within the community. Men simply don't gain the same benefits from verbal language as women do. And, invariably, amongst the males, any significant changes in the community will be enforced with force rather than words.

Storytelling, Music and Visual Art

Some people view the evolution of art as being the equivalent of the evolution of humanity itself. In this sense, art is invariably described as 'culture'. Whenever an archaeologist discovers a new set of cave paintings or a necklace, then some people back-date human evolution, equating the finding of these discoveries as being the difference between being pre-human and human. This suggests that hominids before humans were incapable of making jewellery or being able to consider self-decoration. Considering that, as was stated in the second chapter, 400,000-year-old spears have been found, this assumption seems highly questionable.

I would like to suggest that a necklace compared to the complexity of the arranged marriage system is the equivalent to the artwork in the foyer of a skyscraper compared to the skyscraper itself: the complex web of human cooperation tends to make an object used for self-decoration seem rather unimportant. We cannot place too much importance on a piece of jewellery used for self-decoration that may simply have been thrown away and replaced with another. Furthermore, we cannot be presumptuous, and all it would take is for a 500,000-year-old necklace to be found for many books to need to be rewritten with a new definition of humanity. Since this has already occurred for 'man-the-toolmaker', there is no reason why it could not occur to 'culture'. While many view 'culture' as the essence of freedom of the individual from biological constraints, the belief that biology is a constraint stems from a lack of understanding of biology: the biological organism can adapt to its environment. The idea of the biological machine mindlessly operating rarely exists in nature: the animal, particularly the more complex animal, is highly adaptive and innovative, searching for novel interactions with its environment to improve its capacity to compete. As Richard Dawkins argued in *The Extended Phenotype* some organisms, such as beavers and humans, have a profound effect on their environment, and can shape their environment to suit their purposes, whether it is a dam and lodge or a clearing and shelter.[71] It is not evolutionists who attempt to reduce nature to a series of machines, but those with the restricted view of the natural world, who merely view a deer or wolf as a mechanised organism incapable of innovation. Humans, rather than the only organism capable of innovation, simply possess a greater ability for innovation. Human culture, rather than unique, is simply an extension of the innovation shared by most, and possibly all, mammals. Though a doe attempting to eat a novel leaf or berry in an area beyond her normal territory may seem small compared to human ingenuity, they are both attempts to extend the individual's influence on their environment to their own advantage. It may not even be relevant to nutrition: a bird may discover a berry with a specific cartenoid that increases the colour in its feathers, or, as in the bower bird's case, uses the berry to external decorative purpose in his bower. In these cases, the animals are instigating their influence on the environment, rather than acting like a virus, and simply adapting due to the interaction of the species with the host over many generations. This is not Lamarckism, the genes are the same, but the genes create behaviours, such as curiosity, that enable the animal to adapt to its environment in innovative ways. We should not view other animals as simply dullards operating according to set behaviours with no capacity or intent to alter their behaviours. One example comes from recent human interactions with giant squid that made a rare appearance near the surface. They were far from reticent:

> Roger Uzun, a veteran scuba diver and amateur underwater videographer, swam with a swarm of the creatures for about 20 minutes and said they appeared more curious than aggressive. The animals taste with their tentacles, he said, and seemed to be touching him and his wet suit to determine if he was edible.
>
> "As soon as we went underwater and turned on the video lights, there they were. They would ram into you, they kept hitting the back of my head," he said.
>
> "One got ahold of the video light head and yanked on it for two or three seconds and he was actually trying to take the video light with him," said Uzun... "It almost knocked the video camera out of my hands."[72]

The attempts by biologists to quantify behaviour should not be viewed merely as an attempt to appreciate the biologically driven processes, but also how those biological processes attempt to innovatively interact with the environment. Other than perhaps very short-lived animals, all animals are innovative and not machines, they are biological and adaptive to their environment. For humans, this adaptiveness is invariably aimed at cooperation: horses, roads, trains, automobiles, and planes have all been utilised or developed not as attempts to simply be cultural, but to expand the individual's ability to cooperate with larger networks of people. They are simply variations of the pre-existing desire to cooperate with larger groups of people to increase one's individual ability to compete. Culture, as it is known, may be viewed as the unique behaviours of the individual aimed at increasing its capacity to compete within its environment. Jewellery, for instance, as reflected in the amount of status afforded to it (gold or diamonds), is competitive.

Some of the most interesting decorations might be considered as an extension of the web of human cooperation. Napoleon Chagnon observed that the Yąnomamö men engage in fights in which they offer their head to their opponent to strike with a pole so that they may respond in kind. The results are that:

> Needless to say, the tops of most men's heads are covered with deep, ugly scars of which their bearers are immensely proud. Some men, in fact, keep their heads cleanly shaved on top to display these scars, rubbing red pigment on their bare scalps to define them more precisely.[73]

Of the same people, Helena Valero recalled of the men:

> When they have killed, they become *unocai*. Then they put those little sticks in their ears and tie them also on the inside of their wrists, quite long. From this you can tell they have killed.[74]

It is interesting to note that evidently some men did not want to be identified as possessing this status:

> One of those who had felt his body to be really soft [indicating the death of the enemy he had shot with an arrow], said two days later: 'Today I woke up feeling strong. I think the man I shot is not dead; I want to throw away the little sticks.' He took them off his wrists and out of his ears and threw them away.[75]

Or this could be simply because he was confident that the man he shot really would survive, which might prove somewhat embarrassing. Valero, herself, was also decorated:

> Meanwhile the woman began to say: 'Now you are almost a woman. How can you adorn yourself, if you have no holes? Come, we will make holes in the middle of your lip and at the corners of your mouth.' I already had holes in my ears. I answered: 'You make your holes, for you are of this race; I am not.' But the woman continued: 'Come, it doesn't hurt at all.'
> The lip is much harder; with *pashuba* palm thorns it is impossible to make the holes. It was a man, the husband of the woman with whom I was living, who prepared a fine point with very hard *bacaba* wood and pierced my lip. With *inajá* wood it becomes inflamed, and so they don't use it. Before pushing it through, he burnt the end a little, so that it would not become inflamed. The lip was hard; he pushed and hurt me. It remained all very painful and half paralysed until it healed. With these same sharp points of *bacaba* wood they make the holes through the nose and the lobes of the ears. In the case of males they always make a hole in their lower lip and their ears.[76]

In Australia, piercing of the nose, the removal of teeth, circumcision, incisions in the penis, and scarring of the body are common as part of male initiation rituals.[77] Similar rituals for girls can include the cutting of genitals with stone or wood (such as the breaking of the hymen), the burning of pubic hair, and the painting of the girl's body.[78] People may also scar themselves when a relative is scarred:

> Or again, and this practice is widespread, they may commemorate a boy's first initiation, his formal transition from childhood to adulthood, and show their sympathy for him at the same time, by scarring their own bodies. For each category of relative a specific part of the body is allocated: and this is not identical for all areas. Among the Maung, for example, a boy's mother cuts three horizontal scars on her belly (this is fairly widespread); his sister, the same on the calf of each leg; his betrothed wife, or any girl who calls him close 'husband', the same on her buttocks, just above the thighs. And so on.[79]

It is interesting how often decoration or ritual scarring involves pain and mutilation of some part of the body. This is true for men and for women, and continues to our own day with the practice of tattoos and piercing, which are becoming more common. It appears that the willingness to accept pain and exhibit pain is an important part of becoming an adult or respected individual within the community, though this is less true in Western society. It may be that the exhibition of one's willingness to endure and accept pain may be a desirable trait in both females and males, since sexual selections and arranged marriages can often lead to extreme violence, the presentation of the capacity to endure pain may be an important emblem. In societies where violence is common, the presentation of signs of previous violence, even if inflicted ritualistically, seems to be more common. A good example of this is tattooing amongst the Maori in New Zealand.

Tattoos and piercings are invariably used to reflect an individual's desire to engage in risk-taking behaviour, and are often aimed at the same sex rather than the opposite sex. In Western societies the suggestion of the willingness to endure pain appears to be a method of intimidation. In cooperatively-simple societies, the willingness to endure pain appears also to concern community cohesion in a more violent social environment. In the description above, the method by which Australian Aborigines were scarring themselves according to their relationship with another person is an example where self-decoration or mutilation is linked with kinship and marriage.

Besides self-ornamentation, there are numerous possibilities for display in story-telling, dancing, and singing. It is more likely in these that the ability to exhibit one's personal qualities comes to the fore.

Myths are usually described as a spiritual experience; in reality, in cooperatively-simple societies they appear to be used for entertainment, while also stating the rules of the society or to explain the existence of something. Chagnon explains of the Yąnomamö:

> Sex is the big thing in Yąnomamö myths – the general relationships between men and women at the level of comparative status on the one hand and their relative biological attributes on the other. Sex is also a big thing in everyday life among the Yąnomamö, as it is elsewhere, and much of their humour, insulting, fighting, storytelling, and conceptions revolves around sexual themes. If I were to illustrate the dictionary I have

been patiently collecting on my field trips, it would be, as one of my puckish graduate students once commented, very good pornography![80]

Some of the characters in Yąnomamö myths are downright hilarious, and some of the things they did are funny, ribald, and extremely entertaining to the Yąnomamö, who listen to men telling mythical stories or chanting episodes of mythical sagas as they prance around the village, tripping out on hallucinogens, adding comical twists and nuances to the side-splitting delight of their audiences. Everybody knows what Iwariwä (Caiman Ancestor) did, and that part cannot be changed. But how he did it, what minor gestures and comments he made, or how much it hurt or pleased him as he did it is subject to some considerable poetic license, and it is this that is entertaining and amusing to the listener. Occasionally the inspired narrator will go a bit beyond what is acceptable, a violation for which his own peers might good-naturedly forgive him, but a violation that people in other villages might object to, as they did when I taped the narrative and had people in other villages comment on it.[81]

The same importance on entertainment exists for Australian Aborigines:

The time and place for stories is, most often, around the camp fires at night, or at any period when people can spare an hour or two for relaxation. And they are told, primarily, for pleasure. Occasionally they serve other purposes too. They may be admonitory or instructive, pointing a moral, or imparting information in an agreeably easy way. ... Some people, are, of course, more adept at this than others. They can hold an audience spellbound even when the tales they tell are familiar in every detail.[82]

Invariably, such is the nature of attempting to present either Western or non-Western societies as 'serious' societies, the true nature of the mythological stories is often forgotten. Perhaps the Old Testament, for instance, is full of so many remarkable stories simply because they were more entertaining than others which were forgotten and not recorded. Certainly the Old Testament is far more entertaining than the New Testament, and the former is far closer to a cooperatively-simple society in its view of life and the complexity of the society in which it was written. I have read comments from missionaries that the stories and presentation of Jesus Christ are laughed at by people in cooperatively-simple societies, and that the missionaries decide to start with the Old Testament first. Jesus Christ's meekness and martyrdom reflect weakness rather than the strength that impresses people in these societies, where 'turning the other cheek' is viewed as something that will merely incite further insults or attacks. The concept that the meek shall inherit the Earth could equally be described as the cooperative shall inherit the Earth, that the general message of the New Testament is that long-term cooperativeness is more successful than short-term aggression. Judging by the success of Christian societies in terms of wealth and health, this may well be correct.

My point is that mythological stories are invariably believed to be a reflection of men and women's desperate desire to make sense of their world, when, perhaps in reality, they are designed foremostly as entertainment, and secondly as a way of presenting rules or an understanding of the social world.

The telling of such stories, and any singing or dancing, allows individuals to present their intelligence, wit, compassion, physical prowess, sexual behaviours, and so on, to members of their own community and, if present, other communities. In societies where men and women may have little capacity to exhibit their feelings, the telling of such stories allows individuals to exhibit qualities that were previously unnoticed by members of the opposite sex. Females may carefully take note of male behaviour exhibited during the telling of a story. The licence available to individuals in telling stories allows males to convey self-control or lasciviousness, dependability or impulsiveness, empathy or ruthlessness, confidence or over-confidence, and so on. Females watching males conveying stories may thus be able to gather information, and males may be able to deceive females by misrepresenting themselves. Story-telling allows sexes to communicate indirectly in a society where direct communication is extremely limited. If a story includes sexual behaviour, then a male may be able to present how he may behave with a female, conveying tenderness or dash, which may attract the interest of differing females.

Individuals may also exhibit qualities to their own sex that they are a desirable and entertaining companion. These qualities may be especially important when a community has members of a visiting community. Females may be more interested in making a close ally of an entertaining female, or a female who, through her story-telling, exhibits behaviours that are attractive (such as dependability). In the chapter on female sexual selection of males, I have quoted Kaberry's documentation of women's corroborees or ritual meetings, where women, away from the men, are able to define their interactions with other women through songs and dances. The same preferences may be true for males, who may much prefer to spend time with an entertaining male than a dull male, regardless of political requirements.

Exhibitionism is most evident in Hallpike's descriptions of large dances held by Papuans; it is remarkable how little has changed from Papuan dance ceremonies to modern dance halls, including the use of smoke for decoration and the importance of endurance:

> Great clouds of smoke from these brands swept across the dance yard, blown by the strong wind, and in the lurid and fitful yellow glare could be seen the incredible figures of the dancers themselves. Like moving idols, they were no longer the stunted little men of the daylight – they had donned their immense feather head-dresses made of magnificent plumes on cane frames strapped to their heads and torso and were transformed into superhuman monsters 12 feet high. These extraordinary impressive figures were gathered into a pack of sixty or so fighting men, each man – for only men form the body of night dancers – carrying two raw bamboo poles; these they thumped rhythmically on the ground in unison, producing a sonorous and powerful accompaniment to their songs. As I watched them dancing and singing up and down the yard with tireless energy, they seemed to have taken on the qualities of the *agotevaun*, the culture heroes before the dawn of time, beautiful as birds, with the endurance and virility of giants, melodiously exulting in their strength, and roaring their triumph across the ranges. All through the night they danced, for about ten hours, the weaker dropping out in the course of the night, but the strongest unflagging in the dance until

the light of dawn broke over the great valley of the Aibala.[83]

Hallpike underlines the sexually competitive aspect of the dance and dismisses the group selection ideas of Wynne-Edwards (that undernourished males are ignored by females for the benefit of the species, thus allowing women to select available, single males):

> But undernourished and depressed individuals would have trouble attracting women in any case, whether or not they danced badly. While sexual display, expressed in the endurance and exuberance of their dancing and the finery of their plumage, is an essential element in Tauade dancing, the fact remains that the dancers are usually married, and excite reciprocal passion in their hosts' wives, as well as in the unmarried girls. So one of the features of the dance has always been the sexual jealousy caused by married women who are tempted to be unfaithful by the sexual allurements of the dancers, and much violence has traditionally been the result, from men who beat their women for admiring the dancers, and from battles between tribes caused by women later leaving their husbands and going to live with men to whom they have first been attracted at a dance. Fastre records that when he was first at Popole, one dance in particular led to twenty-five killings in this way. Thus the sexual element in the male display during dancing is not primarily meant to serve as a means of pairing off the unmarried couples; it is essentially narcissistic, and the women are there basically as mirrors for male self-esteem.[84]

Napoleon Chagnon observed similar features in Yąnomamö feasts:

> For the Yąnomamö participant in a feast, the feast itself has its significance in the marvelous quantities of food, the excitement of the dance, and the satisfaction of having others admire and covet the fine decorations he wears – and hopefully an opportunity to have a clandestine sexual affair with one of the host women. The enchantment of the dance issues from the dancer's awareness that, for a brief moment, he is a glorious peacock that commands the admiration of his fellows, and it is his responsibility and desire to present a spectacular display of his dance steps and gaudy accoutrements. In this brief, ego-building moment, each man has an opportunity to display himself, spinning and prancing about the village periphery, chest puffed out, while all watch, admire, and cheer wildly, as shown in the film *The Feast*.[85]

Helena Valero gives an example of people singing in a Yąnomamö society, when the Mahekototeri visited her community (*epena* is a hallucinogen):

> In the evening, they performed their dances with palm leaves and arrows. Before they began their songs, the *tushaua* of the Mahekototeri said once more: 'Let no man take *epena*. At night, while we are singing, it is possible that the enemy may listen and come near. Nobody notices while singing. You go and sit down on the path at a distance,' and he sent some men to guard the *shapuno*. No one took *epena*. When night fell, the *tushaua* Kashihewe shouted: 'Sing, sing, Namoeterignuma; I want to listen to

the song of Namoeterignuma.' But the young women were afraid to sing. After a little while the *tushaua's* sister stood up and went to sing: she called other friends too. I did not want to sing, but she said: 'Come; otherwise they will say that you did not want to come because you were afraid.' She took me by one arm and said to me: 'You will always stay with me; I will not leave you with other women.' She went into the middle of the square; I was on one side and the eldest daughter on the other. First she sang a long song, a beautiful one, which I do not remember.

The women sing women' songs and the men sing men's songs. Every tribe has its favourite songs, and the men are never too shy to sing; but the women are often too shy. Generally the woman who sings is alone; sometimes she has under her arm two other young girls and, while she is singing, she goes to and fro with the two companions. When the woman has finished her song, all the others answer in chorus. When a man sings, he often stands alone in the middle of the square, leaning with both hands held high on his pointed bow upright on the ground; sometimes he places his hand on the shoulder of another man. Around him, the others listen in silence, standing still; when he finishes his song, they intone the chorus, dancing with steps forwards and backwards.[86]

Story-telling, dancing, and self-decoration certainly allows an individual to magnify their personality and enhance their prestige and desirability as a companion or sexual partner. It also provides entertainment. Yet no wives are obtained via such methods, and it is certainly possible that the headman of the village may have no acting ability, no sense of humour, and little magnetism, but through careful political manipulations may have organised for himself, his sons, or brothers, numerous marriages and allies. Equally, since the community is bonded through these marriages, if few stories were told and little body painting or other decoration were performed then the society would not collapse. These behaviours exist for people to increase their capacity to compete within the larger fabric of society, in either directly attracting a potential sexual partner, or gaining a new ally.

Such behaviour is also used to bond the group, as is often seen in Western nations through the unified singing of supporters of football teams, the choreographed dancing during the opening of a major event (such as the Olympics), or the presentation of the soldiers and their weaponry in a parade. Napoleon Chagnon documented the Yąnomamö preparation for battle:

The procession to the line-up took about 20 minutes, as about 50 or so men participated. When the last one was in line, the murmurs among the children and women died down and all was quiet in the village once again. ... Then the silence was broken when a single man began singing in a deep baritone voice: 'I am meat hungry! I am meat hungry! Like the carrion-eating buzzard I hunger for flesh!' It was Torokoiwä, one of Matowä's brothers. When he completed the last line, the rest of the raiders repeated his song, ending in an ear-piercing, high-pitched scream that raised goose bumps all over my arms and scalp. A second chorus, again led by Torokoiwä, followed the scream. ... They listened as the jungle echoed back their last shout, identified by them as the spirit of the enemy. They noted the direction from which the echo came. On hearing it, they pranced about frantically, hissing and groaning, waving their weapons, until Kąobawä calmed them down, and the shouting was repeated two more times. At the end of the third shout for the third repetition, the formation broke... When they reached their hammocks, they all simulated vomiting, passing out of their mouths and bodies the "rotten flesh" of the enemy they had symbolically devoured in the line-up.[87]

Karl Heider noted that the Dani of Papua perform ritual dancing and singing after the battle, as if to taunt their opponents even further, which appears to be a part of the ritual:

> The communication with the ghosts is most evident after one side succeeded in killing an enemy. Then for two days they hold a ceremony called *edai*. On a special *edai* field in plain view of the enemy, the entire confederation comes to dance and sing. The dance is simple: groups of people move back and forth, or around in circles, shouting, whooping the *jokoik* call, and singing. Men, women, and children join in, wearing as much finery as they can and carrying weapons. This sense of spectacle is heightened by the appearance of the women, for this is the only time when they wear shells, feathers, and furs and carry spears and bows and arrows. For two days they dance, sing, and display trophies taken from the enemy, weapons as well as ornaments. The noise probably reaches across to where the enemy are holding the funeral. When the Widaia had cause to *edai*, they came to the ridge of the Wakawaka battlefield, where they could be seen as well as heard from the Dugum Neighborhood. But the *edai* has a double significance. The people are aware that in killing an enemy they have placated their own ghost, but they also know that now the enemy has a fresh ghost of its own which demands placation with blood.[88]

In Western societies people are so accustomed to such behaviour that they barely take note of it: fly-overs by jet fighters at major events, synchronised flying by jet fighter pilots, or the declaration of the end of the Iraq war by the then President George W. Bush aboard an aircraft carrier in a highly orchestrated presentation for the media. Modern societies have tended to symbolise the presentation of military might through the most expensive piece of modern military technology available for the individual, the jet fighter, while ritualistic singing by the group is left to the sporting fan, as they attempt to urge on their local sporting group to defeat the sportsmen of a foreign town or city. The New Zealand rugby union team is one of the few that have incorporated an indigenous people's ritualistic war-dance into a pre-game ritual (a 'haka').

Concluding Remarks

The evolution of verbal language fits into the larger evolutionary framework of arranged marriages. It would evolve hand-in-hand with the evolution of women's alliances, particularly best friendships, which is based on the exchange of secrets. It would also assist in the evolution of the other distinctly human traits that enable for sexual selection at a distance, such as: relative nakedness, visual communication of emotions, female breasts, and the dispersal of scent at a distance. All these morphological and behavioural traits are only beneficial if the necessary method of selecting is at a distance and involves meeting deceptively. Verbal language thus forms the basis for the evolution of these other traits, for without the capacity to meet, there is no point in attracting a potential sexual partner from a distance. This suggests that verbal language evolved before these other traits, albeit, most probably in a more restricted form. The arranged marriage system, in which male and female are married without selecting one another forces the need to select deceptively.

Males use verbal language as a generalised method of increasing their influence, rather than the specific method of increasing their influence, as females do. Bodily adornment, singing, and dancing – whether by the group or individual – are logical implications of individual sexual selection at a distance and competitive cooperation within the arranged marriage system. Cave painting is probably a reflection that the people in question possessed a lot of free time.

The reader should appreciate that what I am proposing is an entire system, based on the cooperative method of exchanging cooperation for the ability to reproduce. The other traits, including human language, flow on from this method of cooperative alliance.

Finally, many people assume thinking and using language are the same event. To a human, who uses speech as part of their thought processes, such a conclusion appears to be self-evident. Yet, to a chimpanzee, such an assumption would seem nonsensical.

An animal should be able to think with ease without the use of words. One example of this is the silent movie; people can watch an entire silent movie and are able to understand the entire narrative assisted with a few written explanations.

In her work in Gombe, Jane Goodall and her colleagues noted the rise of one chimpanzee, Mike, to alpha male position in the hierarchy. To Jane Goodall, Mike appeared to be thinking and planning his rise to alpha male. If this is correct, then it would appear that intelligence might be as important to chimpanzees as it is to humans in their individual success.

Mike's rise to the number one or top-ranking position in the chimpanzee community was both interesting and spectacular. In 1963 Mike had ranked almost bottom in the adult male dominance hierarchy. He had been the last to gain access to bananas, and had been threatened and actually attacked by almost every other adult male. At one time he even had appeared almost bald from losing so many handfuls of hair during aggressive incidents with his fellow apes.

When Hugo and I had left the Gombe Stream at the end of that year preparatory to getting married, Mike's position had not changed, yet when we returned four months later we found a very different Mike. Kris and Dominic told us the beginning of his story – how he had started to use empty four-gallon kerosene cans more and more often during his charging displays. We did not have to wait many days before we witnessed Mike's techniques for ourselves.

There was one incident I remember particularly vividly. A group of five adult males, including top-ranking Goliath, David Graybeard, and the huge Rodolf, were grooming each other. The session had been going on for some twenty minutes. Mike was sitting about thirty yards apart from them, frequently staring toward the group, occasionally idly grooming himself.

All at once Mike calmly walked over to our tent and took hold of an empty kerosene can by the handle. Then he picked up a second can and, walking upright, returned to the place where he had been sitting. Armed with his two cans Mike continued to stare toward the other males. After a few minutes he began to rock from side to side. At first the movement was almost imperceptible, but Hugo and I were watching him closely. Gradually he rocked more vigorously, his hair slowly began to stand erect, and then, softly at first, he started a series of pant-hoots. As he called, Mike got to his feet and suddenly he was off, charging toward the group of males, hitting the two cans ahead of him. The cans, together with Mike's crescendo of hooting, made the most appalling racket: no wonder the erstwhile peaceful males rushed out of the way. Mike and his cans vanished down a track, and after a few moments there was silence. Some of the males reassembled and resumed their interrupted grooming session, but the others stood around somewhat apprehensively.

After a short interval that low-pitched hooting began again, followed almost immediately by the appearance of the two rackety cans with Mike close behind them. Straight for the other males he charged, and once more they fled. This time, even before the group could reassemble, Mike set off again, but he made straight for Goliath – and even he hastened out of his way like all the others. Then Mike stopped and sat, all his hair on end, breathing hard. His eyes glared ahead and his lower lip was hanging slightly down so that the pink inside showed brightly and gave him a wild appearance.

Rodolf was the first of the males to approach Mike, uttering soft pant-grunts of submission, crouching low and pressing his lips to Mike's thigh. Next he began to groom Mike, and two other males approached, pant-grunting, and also began to groom him. Finally David Graybeard went over to Mike, laid one hand on his groin, and joined in the grooming. Only Goliath kept away, sitting alone and staring toward Mike. It was obvious that Mike constituted a serious threat to Goliath's hitherto unchallenged supremacy.

Mike's deliberate use of man-made objects was probably an indication of superior intelligence. Many of the adult males had at some time or another dragged a kerosene can to enhance their charging displays in place of the more normal branches or rocks, but only Mike apparently had been able to profit from the chance experience and learn to seek out the cans deliberately to his own advantage. ... [e]ventually Mike was able to keep three cans ahead of him at once for about sixty yards as he ran flat-out across the camp clearing.[89]

Even after a demonstration, the other males were not able to copy Mike's success with the kerosene cans. Any suggestion of Mike copying a human demonstration would be incorrect, since, firstly, humans do not perform charging displays, and furthermore, even with a chimpanzee's demonstration, the other chimpanzees could not successfully mimic the behaviour.

Charging displays usually occur at a time of emotional excitement – when a chimpanzee arrives at a food source, joins up with another group, or when he is frustrated. But it seemed that Mike actually *planned* his charging displays; almost, one might say, in cold blood. Often when he got up to fetch his cans he showed no visible signs of frustration or excitement; that came afterward when, armed with his display props, he began to rock from side to side, raise his hair, and hoot.

Eventually Mike's use of kerosene cans became dangerous – he learned to hurl them ahead of him at the close of a charge. Once he got me on the back of my head, and once he hit Hugo's precious movie camera. We decided to remove all the cans, and went through a nightmare period while Mike tried to drag about all manner of other objects. Once he got hold of Hugo's tripod – luckily when the camera was not mounted – and once he managed to grab and pull down the large cupboard in which we kept a good deal of food and all our crockery and cutlery. The noise and the trail of destruction were unbelievable. Finally, however, we managed to dig things into the ground or hide them away, and like his companions Mike had to resort to branches and rocks.

By that time, however, his top-ranking status was assured, although it was fully another year before Mike himself seemed to feel quite secure in his position.[90]

Is Mike planning his displays with the kerosene cans? Did Mike consider what to do if the other chimpanzees decided not to be impressed by his kerosene cans? Did Mike imagine how he might have used Hugo's tripod as part of his display? Did Mike plan to aim at the lower ranked males first and then head towards the alpha male, Goliath? He certainly was specific as to which chimpanzees he headed for first with the kerosene cans, and only headed for the alpha male (Goliath) once he was convinced of the cans' intimidatory factor. When the kerosene cans were removed, his attempts to find alternatives suggests that he certainly knew what he was doing and why. If this was a human being acting in this manner, we wouldn't question that Mike knew what he was doing, planning and practising his behaviour. In fact, I suggest that only those deeply obsessed with human superiority would question that Mike thought all this through.

Returning to the example of the silent movie, why couldn't Mike plan his displays and *think* them through? He might see himself in his mind hurling the cans towards the other males, and their reaction. He might play the entire event through in his mind like a silent movie, without requiring a single word. He cannot, of course, communicate his intentions though. Thought does not require words, it requires imagination, and the capacity to see, in the brain, the events unfolding. Perhaps other animals may even be able to imagine the smell or the sound as much as visual images. The imagination is simply the capacity to bring forth variations of prior experiences and play them out without actually experiencing them.

Perhaps as an exercise, the reader can consider how many of their thoughts actually require words, and how many can simply be played through with images.

Steven Pinker and Ray Jackendoff suggest several 'domains of human concepts which are probably unlearnable without language... For example, the notion of a "week" depends on counting time periods that cannot all be perceived at once'[91] and 'numbers themselves'[92] are two suggestions. The problem here is that people in cooperatively-simple societies do not have days of the week, nor do they have a counting system beyond one, two and many. Does this matter? It does if we are considering the evolution of human language, because it did not evolve to count days of the week or to count to a hundred, though it can be adapted to these ends. Even so, I can suggest days merely by pointing to the sun and revolving my arm going down, going up, going down again, and then having an accepted concept for past or future (clockwise or anti-clockwise, for instance). Another suggestion is 'the distinction between cross- and parallel cousins'.[93] Yet I can simply point to my mother, my mother's brother, and their son or daughter – my cross-cousin! I can use the same method to represent a parallel cousin. I can convey it without a single word being used. Remember, we are considering whether the understanding is dependent on language. I can create a distant cross-cousin marriage, by pointing to a man, his sister, her daughter and then to the man's brother's father's brother's son's son – and then miming an action that suggests marriage. With a stick and dirt, I can map out the genealogy of an entire kin group, their relationships (horizontal lines for marriage, downward lines for birth), and who has died (cross them out). Another example given is formal social roles 'such as "justice of the peace" and "treasurer"'.[94] Again, these roles do not exist in cooperatively-simple societies. If a chimpanzee can understand which male is alpha male without words, there seems no reason why a human cannot understand which male is a headman or a shaman without requiring words. Entire hierarchies can be demonstrated using diagrams: the entire court system can be reduced to a diagram, and all I need to do is point to where that official is on the diagram and represent which cases their court is responsible for by using mimes of events (theft, murder, compensation). The term 'justice of the peace' will mean nothing to a person, unless the justice's duties are explained, all of which could be mimed. The notion that we only understand something because it is given a word is incorrect; we understand something because of its context, which we must be able to imagine.

If we cannot imagine the meaning of a word, then how can we understand it? How much language can be demonstrated through mime as the vehicle of communication (as against sign language, which is not mime)? If the word cannot be demonstrated in this manner, then does it even have a definite meaning, or, like some academic terminology, is it ambiguous? Conscious or unconscious, for instance, can be mimed, but consciousness is a highly ambiguous term, with no specific definition; we have only a vague appreciation of its meaning. Is there any non-ambiguous word that cannot be understood and represented through mime or in a diagram? Is there any such word that could also exist in a cooperatively-simple environment?

The point of this argument is to demonstrate that: (1) thoughts do not require words, only the internal consideration of images, smells, sounds, touch and other experiences (such as pain); (2) humans can imagine numerous ideas without the need for words, including very complex ones; and (3) humans can think of complex events occurring, but only need language (verbal or signed) when there arises the need to communicate those events. For instance, when Einstein explained his theory of relativity, he used a thought experiment, which the reader must be able to see in their brain.[95]

Some people argue that language is so complicated that its evolution is almost impossible. If I can imagine an orange in my brain (including the odour), then all I need to do is to affix a few sounds to these complex images and odours and I have a word. One can get a chimpanzee's brain, with all the pre-existing neurons for images, odours, and so-on, and label these with sounds and then link up the neurons. If I see an orange, in my mind I recognise the orange, imagine the taste and texture of the orange, and the word for orange is on the tip of my tongue.

If I can see a person or animal running, then I can have this image in my brain as well, and I can affix an appropriate set of sounds to that event (run or gallop). I might see a horse moving quickly and my brain recognises the image and the word is simultaneously in my mouth, all at once. Or I might see an image of a horse (a photograph or painting), and the word pops into my mouth (gallop), because I can see that the image fits with the one in my brain. In order to be able to communicate with greater precision, I then need to explain to the other person when this is occurring – future, past, present, and so on, tense. When I have the image of the horse galloping in my brain, I link it to several different sounds reflecting the tense. Neurons are so tiny that we can have a huge number of different neurons, even in different languages, to define the same image in our brain. If we speak four different languages, our brain follows different pathways depending on the language.

The neurons to define other senses, such as smell, may actually require more neurons, and this would be particularly true if we happened to be a dog. While a dog has no verbalised noun for a horse, general or individual, it would be able to tell exactly which horse it was by the smell, as they can with humans when they are used as trackers – even if they haven't met the human, only smelt a piece of clothing worn by the human.

People underestimate the complexity of other aspects of the brain – tactile, olfactory and auditory experience. An orange may have a large number of neurons dedicated to it, with the sounds, even in multiple languages, being a fraction. If we picked up an orange and smelt it, and it smelt like a banana, we would be surprised. If we picked up an orange and it felt like a pineapple, we would be surprised. We have neurons in our brain dedicated to all these events, including the sounds for language. The number of neurons in our brain dedicated to something as simple as an orange may be surprisingly large. Adding neurons for sounds to the pre-existing neurons dedicated to an object may not be as complex as we might think, but it might be expensive in energy.

Does this understate the complexity of language? No, instead it is others that have understated the complexity of the brain itself. We think language is complex, but we do not think that a dog's brain is complex, despite the fact that the dog can link a stranger to a piece of clothing merely by a unique smell. How is a common verbalised name for a dog (such as Rover) more impressive than a unique identifying smell? For other animals scent can be an entire method of communication, marking territories and presence as we use fences and business cards. If we underestimate the brains of other animals, how can we discover how the ability to think evolved?

Our capacity to communicate with words is limited. We know the tactile or olfactory experiences of many objects yet have no specific words for them. We know many smells or colours, but have few specific words for them, unless we are a specialist in the area. Our language is actually very limited and dedicated to those words that we require for regular communication. In reality, our imagination (and our experience) is far more complicated than the words we use to describe it. Language is a poor tool compared to the power of our internal imagination. Many celebrate language as the greatest human ability, but what they really mean is the capacity for thought and imagination, which we use language to attempt to convey.

Thinking and language are not inextricable. A non-human animal should be able to think merely by playing the events out in its brain, like a silent movie. Anyone who has seen animals asleep, such as a cat, would observe that they appear to be responding to internal stimuli, in the same manner that humans do when they dream. The fact that we can have an entire experience when we sleep presents the power of the internal imagination. Do other animals possess such an internal imagination as well? Do cats chase mice or other cats in their dreams? If they do, then a non-human animal should be able to consider many possible eventualities by playing the events out in its brain, with alternative endings, as it might in a dream. (Indeed, this might be the purpose of dreams, to enable the capacity to think.) We do not need to use language for our thoughts, we can achieve the same outcome by simply playing the events out in our brain.

The mammalian brain is itself remarkably complicated. People who argue that they cannot imagine what life would be like without language lack imagination. Humans do not need language to experience a large number of remarkable events, as giving birth, sexuality, sport and silent movies attest. The silent movie represents the fact that thought does not require words, only images and an understanding, even if the movie is as complex as *Metropolis*. Finally, the human brain is large not to communicate thoughts, but to understand very complex events involving hundreds of known, unborn or dead individuals over large periods of time. Language does not give us the capacity to think about these events, but merely to communicate them.

Chapter 11

The Evolution of Social Organisation

The arranged marriage system using cross-cousin marriages allows us to imagine the social organisational structures of the hominids up to and including humans. We have the fundamental building block of society, the cross-cousin marriage:

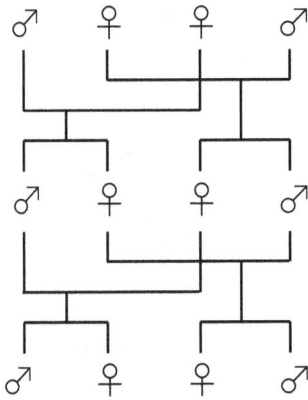

Diagram 1.

If humans evolved with cross-cousin marriage systems then the above is the fundamental building block of human societies. Anthropology has shown that human societies use this basic method to construct societies with up to 100 or 150 adults in a single community, which can be interwoven with neighbouring communities of a similar or smaller size. If we consider that complex hominids use this system to create complex societies, then we can also consider that simpler hominids used simpler versions of the same system.

Using the cross-cousin marriage system, I will attempt to infer the social structures used by human ancestors, and also their behaviours and morphologies, since, as I have argued, behaviour and appearance are linked to the social structure.

This will be a theoretical argument; I will not attempt to draw too many comparisons between my theory and archaeology and paleoanthropology. The main reason for this is simple: if the theory is correct it should predict the discoveries made by archaeologists and paleoanthropologists.

My theoretical explanation of how we evolved from an animal of no greater intelligence than a chimpanzee will be simple and straightforward, and will occur in three steps.

In this chapter, each diagram representing the four hominids is an ideal representation of what occurs over several generations. Each diagram represents four generations, and living individuals may exist in two or three of the four generations, depending on longevity. Deaths due to disease or war are not depicted, so the actual realistic size of a group will vary from what is represented. What is represented is an ideal, without considering factors that will occur in normal life. Yet it does not take much imagination to *reconstruct* the ideal, with the possibility of larger numbers of females, adolescents and infants; high infant mortality; deaths due to competition (warfare); female mortality due to childbirth; fissioning into smaller groups; and the many other possible contingencies that occur in normal life. These normal events would affect the representation of a diagram, but would not affect the basic structure depicted in each diagram. Since taking into account such occurrences would complicate the diagram, I have preferred to represent each hominid in its ideal form to ensure that what is most evident is the structure of the social group.

The First Hominid

The first stage of human evolution is the formation presented above. We should expect that the first in the hominid line was a two male harem with two or more females. The feature of this animal is that two males should exist in a community with their respective sisters, and copulate with the sisters of the opposite male.

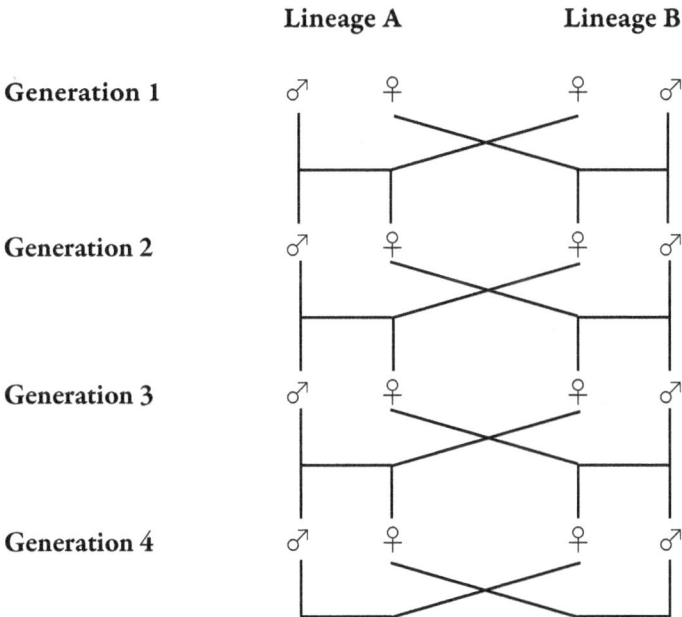

Diagram 2: the first hominid.

First hominid: detailed

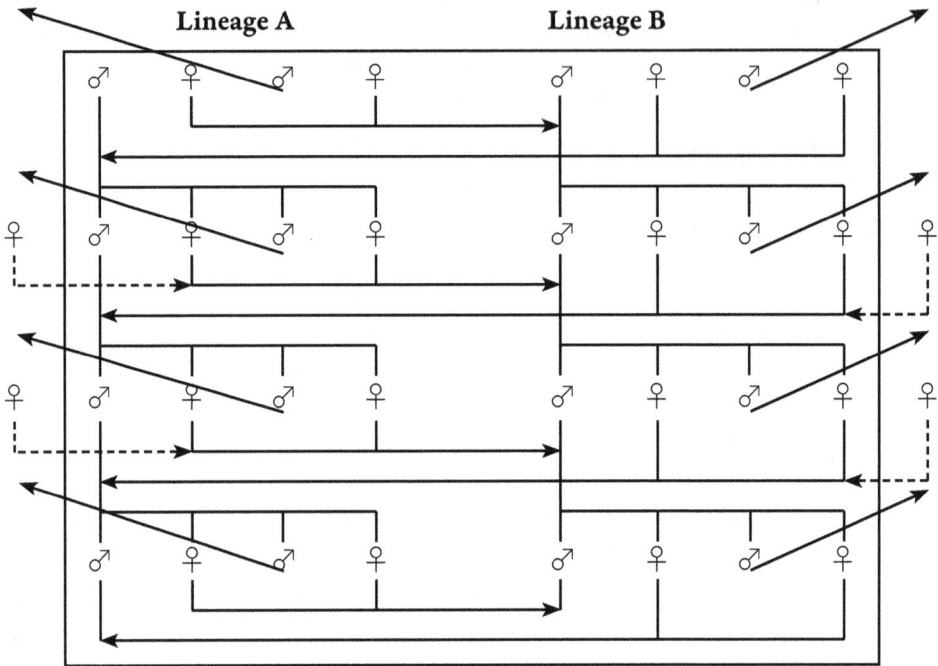

Diagram 3: a more detailed view. Four generations are depicted, and living members may exist in two or three generations. There are two lineages, with the males in one lineage mating with the females in the other lineage (horizontal lines with arrows depict 'marriage'). Extra males leave the group (as depicted in arrows leading outside of the group). Dominant males and females remain in the group. Females can enter the group, by being abducted from other groups or leaving a natal group. The depiction is ideal and does not take into account deaths. For instance, a dominant male may die and be replaced by a younger brother (if present) or eldest son, with the former remaining in the group rather than leaving the group. Males that leave the group will attempt to forge their own groups.

What is depicted in Diagram 2 and Diagram 3 is a simple structure: two males and their sisters; the males mate with the females in the other lineage; the males are replaced by their eldest son; the new male mates with the respective sisters of the other eldest son; and thus the chain is passed down from generation to generation. In reality, this system could become far more complicated, but initially it is best to consider it in its ideal situation.

Importantly, neither the males nor the females need high intelligence. There is no reason for competition between the two males because the females are either sisters or sexual partners, and whichever is a potential sexual partner for one male is a sister to the other. There is no reason for competition for mating rights: all the females fall into the two categories of sister or sexual partner, and no female falls into the category of sexual partner for both. The only manner in which competition can arise is between the two adult males if a female enters from outside the group or between brothers fighting to take the father's position. This assumes that daughters will mate with the next generation of males rather than a male in their father's generation. Again, there is some possibility for competition between males in the other lineage if a female enters puberty and there are two generations of males (father and son) in the other lineage. This could occur through avoidance. In zoos, female chimpanzees refuse to mate with older male chimpanzees. Females are normally exogamic, but obviously in zoos this is not possible, and the response by females is refusal to copulate (normally females copulate with all resident males).[1]

Thus the system works without any need for contracts or intelligence: the males simply mate with the available females – their cross-cousins.

One requirement is that two males are required to protect the group, rather than only one male, as is the case for gorillas. The basis for such a requirement will be dealt with later in this chapter, which considers the importance of the biped body structure (see The emergence of numerical advantage, below).

The reason why such a group structure – cross-cousin marriage – would evolve rather than two full brothers copulating with females is that only the former is stable. This is one of the tremendous benefits of the cross-cousin marriage organisational structure: the group is stable over many generations. Males and females do not need to leave the group in order to find a sexual partner, since they are born into a group in which they will have a sexual partner. By contrast, if the males mated with all available females, then all the progeny would potentially be half-brothers or -sisters. In this case, as with chimpanzees, the females would be exogamic.

This creates an important distinction between chimpanzees and humans. Our social organisational structure is stable and self-perpetuating. Humans, because of the specific nature of marriage, know how they are related to other humans.

By contrast, chimpanzees do not have a stable or self-perpetuating social organisational structure. The males mate with all available females; therefore, all progeny are potentially half-brothers and half-sisters. Individuals could only guess at their relatedness, except maternally. This genetic similarity no doubt enhances male cooperativeness (all males are potentially highly related), but it also leads to female exogamy. Males require females to migrate into their social group for their lineage to be perpetuated. That chimpanzees evolved in a different manner might reflect differing social and physical environments, perhaps greater population density and faster evolution to larger group sizes. Chimpanzees may have evolved to communities with eight adult males relatively quickly compared to hominids, requiring a faster resolution to rewarding cooperative behaviour, this being females mating with all resident males, ensuring all have a chance to pass on their genes.

The self-perpetuating factor of the cross-cousin marriage gives those that possess that structure a tremendous advantage in competition. If self-perpetuating groups existed and groups reliant on exogamy existed, then the former will be more successful over time. They will attract females from the exogamic groups, but not yield any females since the females have no reason to leave. Slowly, the exogamic groups will become less common, since they will always need to attract females from other groups, yet always be yielding females to the self-perpetuating groups.

In order for humans to evolve, the human structure of cross-cousin marriage must exist from the very beginning. In other words, individuals must know their genetic relationship to other individuals in the group; as soon as paternity is blurred, as with chimpanzees, then this is a dead end in terms of human evolution. Paternity must always be apparent and specific for humans to evolve.

The basic structure of the cross-cousin marriage system can begin and operate effectively without any higher intelligence required. This animal will probably be similar to a gorilla, with little more intelligence, nor any other human-like traits. It is basically a two male harem, with each male's harem consisting of the other male's sisters. Males that are pushed out of the community by elder or dominant brothers will need to form pairs. These pairs will then cooperate to attack and evict males from other communities or abduct females from other communities.

In order for evicted males to create successful communities, males need the 'understanding' that they need to mate with certain females and not others. Males that do not possess this understanding and mate with all available females will not produce a self-perpetuating group over the generations, and will simply lose their females to the self-perpetuating groups of males where the males have created the cross-cousin marriage system.

Alternatively, it may be the case that where there are several brothers in both lineages, the group may split into two, with one set of cross-cousin males taking a sister or more with them and creating an entirely new group that does not need to attract new females. This may occur where there is a particularly successful group, with numerous brothers and sisters in both lineages (see Diagram 4, below).

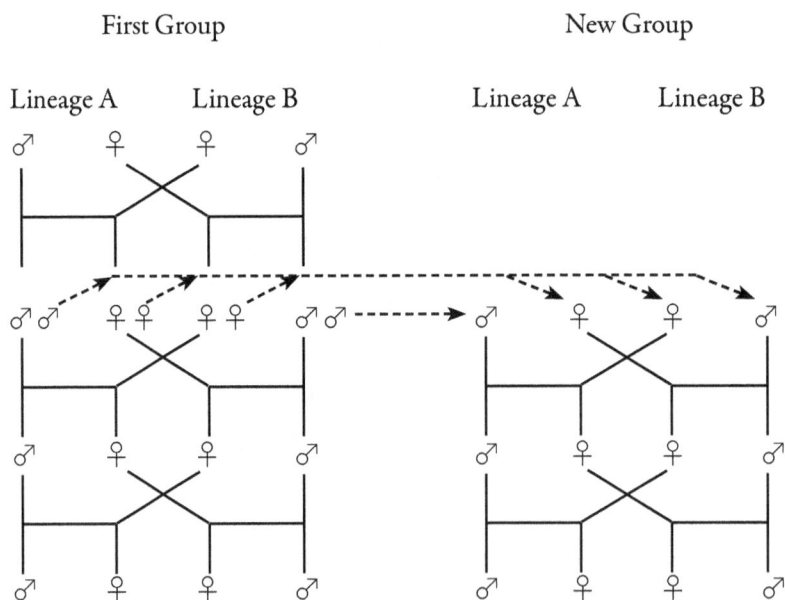

Diagram 4: formation of a second group, with fissioning of two cross-cousin males with sisters to form a new group.

The Second Hominid

Lineage A **Lineage B**

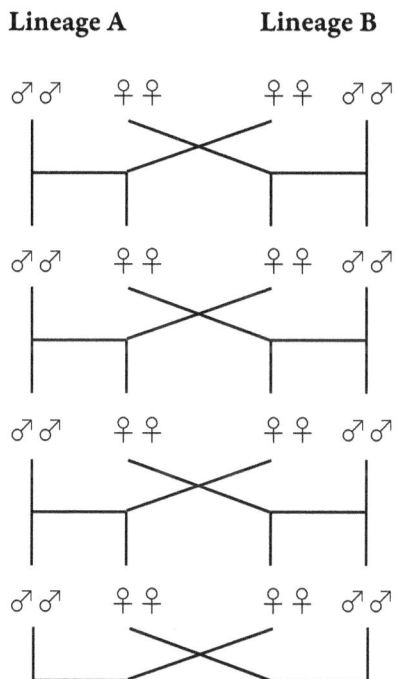

Diagram 5: two males in each lineage.

Second hominid: detailed

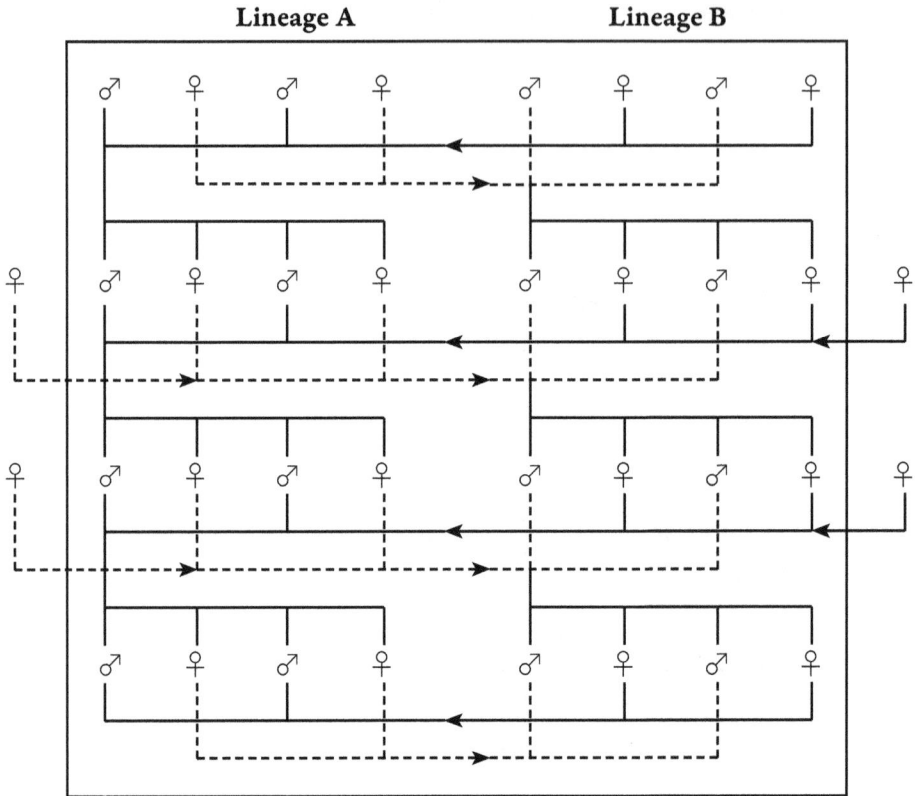

Lineage A   **Lineage B**

Diagram 6: more detailed picture. The basic alteration from the first hominid is that a brother of the males in each lineage remains in the group. Females enter the group from other groups through either exogamy or abduction. Lines with arrows depict the direction of marriage, with females moving to the other lineage. Note that deaths are not detailed; at any particular time, living members may exist in two or three generations. The diagram represents those that have lived in each generation rather than those presently living.

The next step occurs due to a larger level of available nutrition in the environment. If there is more nutrition available, there will be a higher density of individuals, and thus more competitors. In order to remain competitive, the group needs to increase its size. The individuals increase in number rather than increase in body size, for reasons that will be considered below (see The emergence of numerical advantage).

The increase in group size will be small. The dominant male in each lineage will allow one or two younger male siblings to remain in the group and to mate with a female of the other lineage. Thus there is one dominant male in each lineage, and he will have a younger brother, or two, who will mate with the females in the other lineage. The basic structure remains the same: one dominant male in each lineage, with a minority of offspring in each lineage produced by younger brothers (or half-brothers or parallel-cousins). Instead of only two adult males in the group, there are now 4 or 6.

While this does not seem like a major change, it is.

In the first hominid only two adult males existed in the group, each with their own sisters, and there was little possibility for competition between the males. In the second hominid, there will be more than one adult male in each lineage. This means there can be competition between brothers in each lineage to reproduce with cross-cousin females in the other lineage. It also means that offspring in each lineage will be more genetically diverse. In a single lineage there can be brothers, sisters, half-brothers, half-sisters, and female and male parallel-cousins. Obviously, if two brothers produce male offspring, the infants will be parallel-cousins (of the same patrilineage, or, in a Western sense, with the same surname). There is no reason, therefore, why a male may not mate with his female parallel-cousin, thus undermining the cross-cousin marriage system.

The outcome of this is that there is a greater possibility for conflict because of the greater genetic diversity in each lineage. In order for the group to be stable, this potential for conflict needs to be resolved.

There are two pressures. One is external, an increased competitiveness in the surrounding social environment, leading to the formation of larger groups in order to compete. The other pressure is internal, where greater genetic diversity creates greater internal conflict. This conflict, where external pressures force groups to enlarge but the enlargement creates internal conflict, is a consistent factor for the increase in intelligence required to maintain the cooperative nature of the group (and continues to be a factor for humans, as has been documented in previous chapters).

At this stage, these animals do not possess intelligence much beyond what a chimpanzee possesses, so they do not yet have the capacity to make 'deals' for cooperation. Thus, for the moment, I suggest that the movement towards greater group size is based on primogeniture. One male in each lineage dominates the matings. Perhaps one or two brothers, half-brothers or parallel-cousins remain in each lineage and are allowed to mate with one or two females in the other lineage. Genetic diversity is kept at a minimum if most of the females are mating with only one male cross-cousin.

Those hominids able to maintain a large group size with minimal internal competition were thus able to form larger, cooperative, competitive groups. Other groups that were incapable of forming large, stable groups had their females removed by these larger groups.

The group size is increased, genetic variation in each lineage is minimised, and the basic pattern of cross-cousin marriage can be maintained, since most of the infants born will be fathered by only one male in each lineage. In the first hominid, cross-cousin marriage evolved, while, in the second hominid, primogeniture evolved – both factors I have emphasised as normal parts of human reproductive behaviour in earlier chapters.

The lesser males in each lineage remain in the group because if they leave, they will need to obtain their own territory and females. Secondly, they remain because if the head male in a lineage dies, then a lower ranked male will take his place allowing patient males to increase their reproductive potential in time. Thirdly, they benefit by assisting a close genetic relative to reproduce. The levirate (the passing of a male's wives to his brother) thus develops at a very early stage in human evolution.

The cross-cousin marriage system must remain intact. If all the males mate with all the females then paternity is obscured. If paternity is obscured then the females become exogamic, and this leads to chimpanzees. Only those groups that maintain the cross-cousin marriage system possess a self-perpetuating group where sexual partners are already present – a distinct advantage to those that must continuously seek out or attract females. Those males that have genetically diverse females in the group have immediate sexual partners without needing to leave the group, and they can also seek out or attract females from other groups, increasing group size and genetic diversity.

A small but significant increase in intelligence is necessary for such a group to exist. There is the possibility for greater levels of competition, since brothers within the same lineage can compete to copulate with the females of the other lineage, and the females can try to select between the brothers. Thus both males and females must 'know' that they can only copulate with their designated partner in the other lineage. Impulses must be controlled to ensure that the group remains cooperative and competitive: the short-term outlook of chimpanzees and gorillas gives way to the longer-term outlook. Cooperation would deteriorate in groups where males and females did not 'know' that they were meant to only copulate with their designated partner and attempted to mate with other available partners. These groups would be uncooperative, uncompetitive, and disappear. The females of these groups would be removed by the more cooperative groups.

This increase in intelligence comes about due to a selection process. Over tens or hundreds of thousands of years, genes for the long-term view rather than impulsive behaviour were selected. This occurs in both females and males. Younger brothers, half-brothers or male parallel-cousins desist from mating with females even though they are present. Females desist from mating with these males, despite the opportunities. This is not to suggest that these matings may not occur, but rather that the vast majority of infants are produced through the two dominant males. The dominant males also enforce their mating rights over subordinate males.

For such a selection process to occur, the competition between groups needs to be intense. Groups with only two males quickly become uncompetitive. It becomes essential for groups to possess four to six males. Originally, this process is extremely difficult. Females do mate with the male they prefer, and fights between males occur regularly. Amongst this milieu, genes for cooperative behaviour emerge, and those groups that possess these cooperative genes are more likely to maintain the expected mating patterns, and have reduced levels of internal tension. The brains of the animals increase in size, allowing them to 'think' before they act impulsively, and thus to restrict impulsive behaviour that causes internal dissension. Groups without genes for increased cooperation are unable to maintain the larger group size and are more likely to fission into two or more groups. Groups able to maintain the large group size due to individuals restricting their behaviour are less likely to fission and are more competitive. The cooperative groups that possess the self-perpetuating group structure of the cross-cousin marriage system are more successful. There may be, even at this early stage, an element of female selection, where females occasionally select males that are more cooperative and less impulsive over their 'designated male'.

Instead of a group of chimpanzees where all males act impulsively and attempt to mate with all available females, what exists is a more cooperative, more highly structured social group. The groups are dominated by a single male in each lineage. Younger males wait patiently to supplant the dominant male, or are occasionally allowed to mate with a single female. Females accept to mate with the dominant male, or with the younger male they have been designated. They possess the cooperative genes that enable them to control their behaviour, genes that have been selected over many generations in a very competitive environment. Whenever groups fission into smaller groups with only two males, these groups are uncompetitive – males are defeated and females removed by the larger groups. Only those groups with the larger numbers of males, with the controlled behaviour of the males and females, continue to exist.

If the group becomes quite large with more than six males present, then the group may fission into two self-perpetuating groups. This is similar to Diagram 4, above, but with more males and females present. Thus whenever a group becomes so large that there are eight or more adult males, it may fission into two groups, both with a dominant male in each lineage and supporting males. As the number of males increases, the limits of cooperation are reached and increased dissension occurs. Two male cross-cousins may leave the group, taking a sister or more each (with whom the other male cross-cousin will mate) and also a younger male or two.

At this stage, it is unlikely that if males are ejected without females that they will be able to create a self-perpetuating group since they would need to remove females from groups containing four or six males in order to create a group with females. Thus, if two males are ejected without females, they may have little possibility of reproducing. Males must remain in the group, or wait until they can leave with other males and females to create their own competitive group. This is another aspect of the selection process: uncooperative males that do not abide by 'the rules' are ejected and have no capacity to reproduce, leaving only the more cooperative, patient males. Since all the males are highly genetically related (most will be brothers, half-brothers or cross-cousins) then levels of dissension remain small compared to that possible between humans in larger communities.

Though this seems a small change, it has at least doubled the number of males in the group from two to four or six.

The Third Hominid

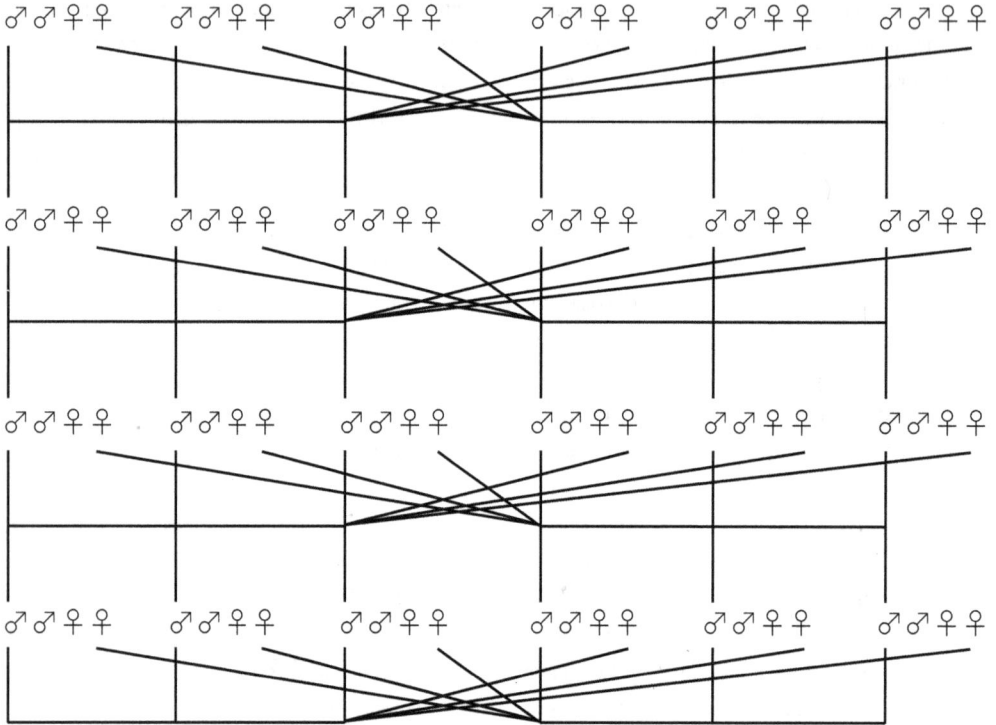

Diagram 7: the third hominid.

Third hominid: detailed

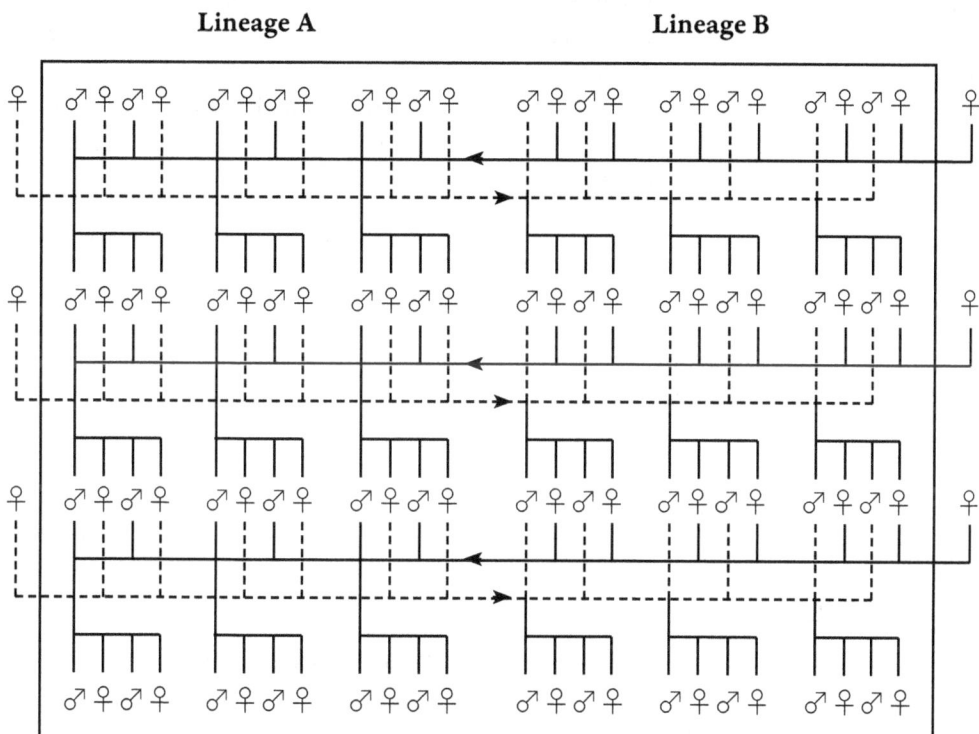

Diagram 8. Extra dominant males and their younger brothers (brother-alliances) are included in each lineage. Females are abducted, or enter the group, from other communities. Lines with arrows depict the direction of marriage, with females moving to the other lineage. Note that deaths are not detailed. Thus, at any particular time, living members may exist in two or three generations, and many in each generation, including an entire generation, may no longer be living. The diagram represents those that have lived in each generation rather than those presently living.

The change from the second to the third hominid is a significant change. It is initiated by the same basis for all changes towards greater complexity: a greater amount of nutrition, which leads to a higher density of individuals, which leads to greater competition, which leads to intense pressure to form larger groups.

The movement towards greater complexity could occur in the following fashion:

	Lineage A	Lineage B
First hominid:	♂	♂
Second hominid:	♂ (♂)	♂ (♂)
Third hominid:	♂(♂) p-c ♂(♂) p-c ♂(♂)	♂(♂) p-c ♂(♂) p-c ♂(♂)

(Single male in each lineage, followed by a dominant and subordinate male (in brackets), then three groups of parallel-cousins (p-c) each with a dominant and subordinate males.)

In the first hominid, I argued that there was a single male in each lineage. In the second hominid, I argued that there was a dominant male in each lineage, supported by one or two lesser males (such as brothers, half-brothers or parallel-cousins). In the third hominid, I argue that there is an increase in the number of dominant males in each lineage, and each dominant male will be supported by several lesser males. The movement from the second hominid to the third hominid requires an intermediary step, as presented below (see Diagram 9). I will now call the dominant male plus subordinate males the brother-alliance, as defined in the fourth chapter.

Transformation from hominid 1 to hominid 3 represented in a single lineage within a community

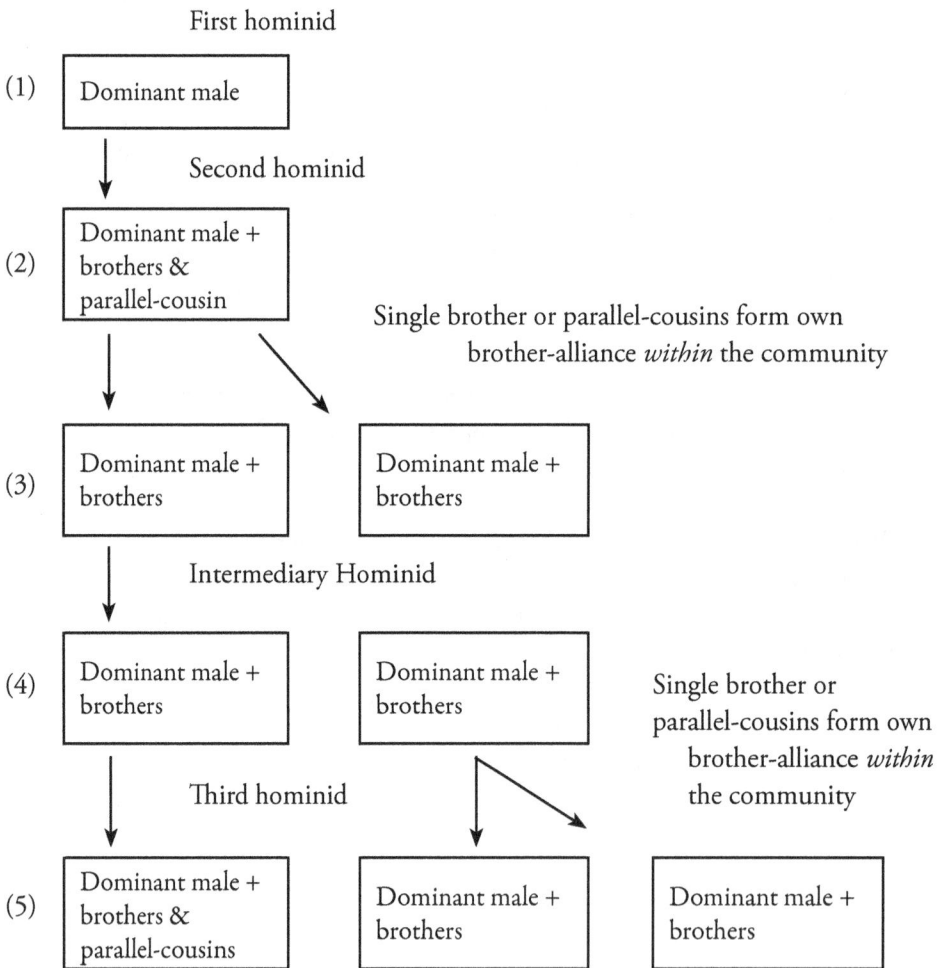

First hominid

(1) | Dominant male

Second hominid

(2) | Dominant male + brothers & parallel-cousin

Single brother or parallel-cousins form own brother-alliance *within* the community

(3) | Dominant male + brothers   Dominant male + brothers

Intermediary Hominid

(4) | Dominant male + brothers   Dominant male + brothers

Single brother or parallel-cousins form own brother-alliance *within* the community

Third hominid

(5) | Dominant male + brothers & parallel-cousins   Dominant male + brothers   Dominant male + brothers

Diagram 9: representing the movement from a single male in each lineage to three dominant males, each with subordinate males within the community. Only one lineage is represented, with the mirror image occurring in the other lineage. This represents the change from the first hominid (line 1), to the second hominid (line 2), to the intermediary hominid (lines 3 and 4), and the third hominid (line 5).

Diagram 9, above, represents the change from the first hominid to the third hominid in a single lineage. In the first line (1) the first hominid is represented as the solitary male. In the second line (2) the solitary male is now the dominant male supported by one or more lesser males – it has evolved into the second hominid. In the third line (3) perhaps two half-brothers or parallel-cousins have formed their own brother-alliance *within* the lineage, with one male becoming the dominant male amongst the supporting brothers, half-brothers and male parallel-cousins. They remain and form their own brother-alliance within the community, rather than leaving to form a separate community. In the fourth line (4) this is consolidated. In the fifth line (5), a similar expansion occurs, as in the third line (3), so that there are now three brother-alliances each with a dominant male and supporting lesser males within the lineage. In lines (1) to (5) we have thus gone from a single male to three dominant males with their supporting males in each lineage.

Only one lineage is depicted in Diagram 9. The same events are mirrored in the opposite lineage. Thus a community of two males (first hominid) has evolved into a community of four to six males (second hominid), which evolves into a community of eight to twelve males (intermediary hominid), which then evolves into a community of twelve to eighteen males (third hominid).

Natural growth is the reason for the evolution of extra dominant males in each lineage. There are larger levels of nutrition that allow for the formation of larger communities. Larger communities entail greater genetic diversity of males and females within the group, which means greater competition and conflict, as well as the potential for greater female sexual selection.

In the second hominid, increased genetic diversity amongst males would have led to the community splitting into two communities. In the second hominid, if there were several parallel-cousins or half-brothers in each lineage, then these males may have taken their sisters to form their own community. Thus there would have been two communities each with two lineages, each with a dominant male and his younger brothers. For example, if the community increased in size then this may have led to competition for the dominant male position between two males in the same lineage. Since the dominant male has the most 'wives' then this leads to competition over this position. One of the competing males pairs with a male in the other lineage and takes several females and subordinate males with them to create a new community where they are now the dominant males. The argument being that the second hominids do not possess the capacity to form larger cooperative communities, and increasing size leads to fissioning into smaller groups.

In the third hominid, instead of forming a distinct community, they remain within the community, but form a separate competitive group within the community – a brother-alliance. Thus, within the community, the sons of the older and younger brothers remain in the community and compete for females in the other lineage.

Initially, perhaps two dominant males and their younger male siblings form factions in each lineage. With time, greater genetic diversity and larger groups emerge, with perhaps three dominant males and their younger siblings forming factions in each lineage. This occurs because increased levels of nutrition leads to larger population densities and increased competition. Fissioning into smaller, more stable communities becomes less competitive, and these intermediary and third hominids are forced to maintain the less stable, larger communities.

The consistent pattern, however, is that of a dominant male in each lineage, exchanging females with a dominant male in the other lineage. This remains the same from the first hominid to human beings. With greater nutrition there are more males, more dominant males supported by subordinate males, exchanging more females, forming an increasingly complex group – but the basic structure remains the same.

There are now many males that can compete for females. Females can choose amongst many males creating more reasons for conflict within the group. Thus, there is the difficult balancing act that leads to the evolution of greater cooperative instincts in a social environment where there are greater reasons for conflict.

In the third hominid, human-like intelligence begins to evolve as a response to the increased complexity of the group. The intelligence is necessary to maintain the group and to manipulate the group to the individual's advantage (or close genetic relative of the individual's advantage – inclusive fitness).

Greater group size and complexity emerges because it must emerge. As nutrition increases, so there are more individuals per unit area, and more small groups competing within each unit area. Any group that can form and sustain a larger group size thus possesses a tremendous advantage over these smaller groups in competition. Any genetic changes that allow for greater cooperative behaviour and a greater group size will be favoured.

Selection for greater cooperative behaviour will occur in two ways. Males with cooperative genes will be able to form larger groups and possess an advantage over those that exist in smaller groups. Selection will also occur by females preferentially selecting more cooperative males, in recognition that uncooperative males exist in smaller, uncompetitive groups. Females will thus observe that their infants, particularly male infants, must possess the cooperative genes otherwise their male infants will be more likely to be uncooperative adult males and exist in smaller, less competitive groups.

The most obvious change would be a larger brain that is able to consider future actions. These males would be able to control their impulsive behaviour by thinking through their actions first. In these communities the males are surrounded by genetically diverse females, and the males' immediate instinct will be to copulate with these females. If the males act in this manner, then the stable system of cross-cousin exchange would disappear and end all possibility of cooperation. Mutations that allow for restraint are essential if genetically diverse males and females are to co-exist in relative harmony. The restraint occurs because the male sees the attractive female, but 'thinks' through his behaviour first, and desists – the opposite of chimpanzee behaviour.

In order for males to accept restrictions on their impulsive behaviours, they require long-term benefits. These benefits are created through the exchange of female cross-cousins. In the first and second hominids, I argued that such an exchange demanded little intelligence since females are largely the progeny of a single dominant male in each lineage; thus each dominant male simply mates with the sisters and half-sisters of the other male. In the third hominid, there are now up to three dominant males in each lineage, and this is a much more complex scenario.

It is in this scenario that males begin to create future contracts. This would be a pivotal event in the evolution of human intelligence.

Evolution of Contracts: Intermediary Hominid

In the *intermediary* hominid (between the second and third hominid) I argued that there are two dominant males in each lineage. Here, a dominant male can potentially exchange his sisters and half-sisters with those sisters of three other dominant males, two cross-cousin dominant males in the other lineage, and also with the parallel-cousin dominant male in his own lineage.

If the dominant male exchanged his sisters and half-sisters with *all* other dominant males, then this leads to a few sisters being exchanged amongst many males. If he exchanges sisters and half-sisters with dominant males *only* in the other lineage, the females are exchanged amongst only two dominant males. If the male has three sisters and half-sisters, and he exchanges them only amongst two males in the other lineage, then each dominant male can expect to receive either one or two females; if he exchanges them amongst three dominant males, then each male might receive one, two or none.

Since the males are limiting their impulsive behaviours for short-term gain in the expectation of a long-term benefit, they require an excellent chance of obtaining that long-term gain. In other words, the males need a good chance of 'getting married' in order to control their impulsive desires. They agree not to try to take other males' 'wives' from them, or to copulate with as many females as possible, in exchange for gaining their own 'wife'. If each male is exchanging with three other males, the possibility of one dominant male ending up with many 'wives' and another dominant male of having one or no 'wives' is high. Conversely, if two dominant males exchange females with two other dominant males then the chance of a dominant male receiving no 'wives' is low. When there are few females to exchange, if 2 exchange with 2 then the possibility of inequity is lower than if each 1 is exchanging with 3 others.

These 'marriages' begin to be planned before the female passes through puberty to increase the expectation of a future benefit in exchange for limiting impulsive, short-term gains. Therefore, the fathers must have begun to actually plan their daughters' 'marriage'.

The movement towards a group with two dominant males in one lineage exchanging females with two dominant males in the second lineage is a small step towards increased complexity. Instead of exchanging females with one dominant male, it is now two dominant males.

Evolution of Contracts: Third Hominid

In the third hominid, the expectation of brothers organising these 'marriages' will be completely replaced by the father or, if not present, an uncle organising most of the marriages. Fathers would begin planning the 'marriage' of their daughters and the process of the father 'giving away' his virgin daughter will have begun hundreds of thousands of years ago.

In the third hominid, where there are three dominant males in each lineage, then the possibility of inequity in exchange is even greater. A dominant male can exchange daughters with five other dominant males: three cross-cousin males in the other lineage, and two parallel-cousin males in his own lineage. Each dominant male can exchange daughters with up to 5 other dominant males and there is a greater chance that one or more dominant males will not gain the long-term benefits they require to control impulsive, short-term desires. Alternatively, if the rules of cross-cousin exchange are maintained, then 3 are exchanging with 3, and the chance that each dominant male will gain a long-term benefit is greater, than each exchanging with 5.

Thus, for the intermediary and third hominid, 2 exchanging with 2 and 3 exchanging with 3 should lead to more stable groups than each 1 exchanging with 3 or each 1 exchanging with 5, respectively. Those that maintain the cross-cousin marriage system should be more stable than those that allow for marriage between cross- and parallel-cousins. Thus the idea of the 'incestuous' marriage, a marriage between parallel-cousins, should have developed before humans evolved.

The process towards three dominant males in each lineage is one of trial and error, on a mass scale. There may be, over a period of time, thousands of such groups, all competing with neighbouring communities. Eventually, the progeny that emerge as the most successful are those that are able to balance cooperation and competition between members of the community. These groups are the most stable and most competitive, and, I argue, maintain the concept of exchange between lineages.

As there is cooperation, there is also competition between members of the same group. Dominant males that are better able to manipulate the system of exchange within the group ensure that their sons have more 'wives' and their daughters have more stable unions. Those that simply organise a 'marriage' for their daughter to his 'wife's-brothers'-sons will be less successful than those that specifically select the most trustworthy and successful dominant males. Since there are now numerous potential partners for their daughters, the males that carefully select a trustworthy and successful male ally will be more successful than those that are unable to 'see' into the future. Equally, those that manipulate the system successfully by receiving a daughter for their son, while not giving a daughter in return, and do so without upsetting the cooperative group as a whole, will be more successful than those that cause the cooperative group to fission because of their manipulation. Genes for cooperation and manipulation will benefit the possessor. Genes for foresight, to improve the capacity to 'see' potential cooperation or manipulation in the future will benefit the possessor.

The implication of this is that genes for intelligence are selected. What emerges are males that are not merely capable of cooperating with first, second or more distant cousins, but males that are able to contemplate the future, and to be able to manipulate the cooperative system to their advantage. If some males are 'thinking' years ahead – or however many years until their daughters pass through puberty – and competing males are incapable of doing this, and being able to see into the future gives a distinct advantage, then genes that enable such forward thinking will be selected. Long-term memories that can recall who cooperated, who reciprocated, and who cheated over long periods of time become beneficial. Over generations, males that *cannot* remember who reciprocated and who cheated will be more likely to be cheated again. A male might thus prefer one male over another due to something that occurred ten or twenty years before, and the males need brains to be able to store this level of information. It might even be a father's resentment toward another male that a young male has observed, a resentment that the young male maintains toward the resented male's sons, even if the male did not observe the act that caused the resentment.

The genes for intelligence are not merely those that allow for cooperation, but those that allow the individual to manipulate the cooperative system. The male that preferentially selects an influential dominant male with five sons as his daughters' 'husbands', may be more successful than the male that preferentially selects the weaker dominant male and his two sons simply because their ancestors have exchanged females in previous generations. His daughter's children will have more males to protect them. Those that are able to see into the future and observe who will be weak and who will be strong in the future will increase their capacity to place their genes (daughters) into more successful brother-alliances; those that are unable to see into the future will only be able to place their genes (daughters) into more successful brother-alliances through luck.

All the very human characteristics of human exchange thus emerge before human beings exist. Males are aware which females are possible candidates for marriage and which are not – remembering that the concept of 'incest' must exist, since marriage must not be within lineages. The male must be able to categorise females as 'marriageable' or 'unmarriage-able' (something that will develop in humans into 'husband', 'wife', 'sister', 'brother', 'father', 'mother', 'father-in-law', and 'mother-in-law'). Capacity to see into the future, self-awareness, knowledge of rules, categorisation and knowledge of genealogy must exist before human beings evolved. In order for the male to know which females are marriageable and which are not he must have some knowledge of genealogy – who married whom in previous genera-tions, and thus whether the female is in his own lineage or the other lineage. He must thus have evolved the brain power to be capable of such knowledge.

While this appears to be a large increase in intelligence in only one generation, the reader must remember that there is an intermediary step. In the intermediary male, these increases in intelligence would be emerging, and only in the third hominid would they exist in a more complete form.

Thus what we have is an animal with a level of intelligence and behaviours that are, for the first time, obviously human-like. The males can organise marriages, see into the future, categorise others in the community, know complex rules, and behave in a more sophisticated manner compared to all other life forms. They are able to control their impulsive behaviours in exchange for reproductive benefits that might only be accrued ten years into the future. The benefit they and their progeny receive is a stable, competitive group in an extremely com-petitive environment where small groups can no longer compete. Males that could organise more wives for themselves and their younger siblings – without undermining the coopera-tion in the group – would have left more descendants; males that were either unable to orga-nise more wives or caused the community to fission into smaller groups, were less likely to leave many descendants.

Some may question why males would organise marriages in advance rather than sim-ply marrying the female to the male when she passes through puberty. Organising marriages in advance increases the expectation that males in the community can gain a reproductive benefit in the future and that they should remain in the community and be cooperative. If females are only 'married' off to their future 'husbands' when she passes through puberty, then males have no expectation that they will be rewarded for cooperating. In the first and second hominids there was no issue for the dominant males as to whether they would be rewarded, because each 'married' all or almost all of the available females – they had no competitors, since females were either future 'wives' or sisters. Now that there are competitors in each lineage, males have no such confidence. Groups where males could 'see' into the future, and organise 'marriages' years in advance would be able to give the males in their group the confidence that they will be rewarded for cooperative behaviour and controlling impulsive behaviour. Groups with males that were unable to do this would possess less confident males since they would not know, until the female reaches puberty, whether they will get 'married' or not. Males, unsure of their capacity to reproduce, would be more likely to act impulsively compared to those that were given the confidence of future reproductive success because the 'marriages' are organised before the female reaches puberty. The less confident males are of gaining a reproductive benefit, the less cooperative they will behave – and may take females and cause serious fights that will lead to the community fissioning, leading to two smaller communities that can be exploited by larger communities. Concepts of 'honour', the importance of reputation, will be expected, not merely for males looking to be reliable partners, but for females who are expected to mate with their designated male 'husband'. The daughters of reproductively 'unreliable' females may be less desirable as future 'wives'. Reproductive rewards for cooperation emerge simultaneously with the evolution of long-term, strategic politics and 'family' reputation.

Meanwhile, the females are deceptively selecting increasingly cooperative, larger brained males within these groups, forming a second process of this evolution.

Females are beginning to use vocal exchanges of information to organise sexual selections with preferred males (though these make up only a small proportion of actual births). The information is highly volatile, involving short-term gains for both the female and the selected male, but undermining the long-term benefits that bind the group. To that extent, enforcement of an arranged marriage, or jealousy, should also have emerged. Females are beginning to form alliances with preferred females to gain access to the brothers of those females. Thus any female ally must possess language skills to be able to act as an intermediary between her brother and her female ally. The females begin to exchange information through increasingly complex vocalisations, testing one another for their capacity for trustworthy behaviour. Female alliance formation and sexual selection leads to the evolution of verbal language as females focus on language as the method of alliance.

Verbal language should thus evolve before humans have evolved. It should have evolved at this point because, for the first time, females possess both a restricted capacity to select a long-term partner, as their partner is chosen for them, and possess numerous opportunities for alternative reproductive partners. In fact, it should have begun to evolve with the intermediary hominid.

Vocal language also evolves in males, but this is primarily a derivative of its evolution within females. Females form allies with other females that possess a superior capacity for complex vocalisations – an essential factor for unobserved, deceptive sexual selections. Females only select verbal skills in males through selecting their verbally complex sisters, with whom the males share similar genes. Even if a male possesses tremendous vocal skills, a female cannot sexually select him without possessing his sister as a trustworthy intimate or she must otherwise take tremendous risks.

Females have accepted a male being chosen for them, because it increases the cooperativeness of the group. If females were able to select their preferred male, then the majority might select the same male, undermining cooperation within the group. Females accept the male selected for them, enabling males to organise cooperative communities in which the females can reproduce in safety.

Since those males who are better at political networking get more 'wives', then the female aim is to produce the polygynous son who is polygynous because of his capacity to both cooperate and successfully manipulate the cooperative system. Females must be sexually selecting those males whom they perceive to be better at forming cooperative networks than other males. In other words, the females select 'upwards' in the social system, choosing high status males where status is determined by how many supporters a male has created. However, female sexual selection both aids and undermines cooperation within the community: it undermines the bonds of the arranged marriages by creating children outside of those 'marriages', but since the females select males with greater capacity to cooperate than their 'husbands' they do increase the cooperative capacity of future generations. A female might, for instance, be 'married' to a particularly impulsive and aggressive male who creates disunity within the community, but she deceptively sexually selects a more cooperative male as the father for one or more of her children.

Females hide their ovulation from the male chosen for them, and form female allies with the hope of selecting alternative genes, if necessary (though this might have evolved in the second hominid). In response, males begin to control the behaviour of the females chosen for them. Males guard their females and attack them at any sign of 'unfaithfulness'. Females control their own behaviour in response.

Menopause in females would be unlikely since the number of infants any male could produce with several females may not be of such a significant number as to benefit the mother of such a male to undergo infertility in later life. The potential for polygyny of males in these communities may not yet be high enough to warrant females leaving more descendants by guarding their sons' wives than by continuing to reproduce themselves.

The overall change from the second to the third hominid seems large; however, it is simply an enlargement of a pre-existing structure. A dominant male, after all, is simply expanding from exchanging females with only one dominant male to choosing between two (the intermediary hominid) and then three dominant males (the third hominid). Previously, arranged marriages existed merely due to the limitation of a male being in the presence of females that were either mostly sisters or cross-cousins, but now, males are specifically organising the marriages of their sisters, daughters or nieces with selected males (for nieces, males may assume duties on behalf of deceased male relatives).

The evolution of various human traits emerges in a step-by-step manner: an extensive memory and careful observation, enabling an individual to be able to 'see' into the future

in order to choose trustworthy males over cheating males; greater cooperativeness; significantly greater brain size; and verbal communication, enabling females to sexually select more
competitive and cooperative males as fathers for their offspring. Hairlessness may also have
emerged, allowing for greater accuracy in female sexual selection of healthy males without
spots or rashes (depending on the existence of fire for warmth). Increased odour may have
evolved for selection against unhealthy bacteria that may suggest disease. It is at this point
that all the major human characteristics would begin to emerge. However, these changes are
all based on the pre-existence of other major human characteristics: known paternity, primogeniture, cross-cousin marriage, and the co-residence of males whose paternity is known
– these all evolved in the first and second hominids.

The idea that males would need to know the rules in the community, know the genealogy of other members of the community, begin to betroth females at an early age to gain the
political support of other males for long periods of time, to see into the future to imagine the
position of other males in the community, and to observe the behaviour of other males and
whether they are honourable – these are simply the obvious implications of the maintenance
of an increasingly complex and stable cross-cousin marriage system. The idea that the female,
being in a community with many genetically diverse males to herself, and yet married to a
male she never chose, would respond by hiding the timing of her ovulation and use deceptive
methods of selection is also an obvious implication of her scenario.

We should expect that if, on another planet, another life-form existed that also had
evolved a cross-cousin marriage system, then they would exhibit exactly the same traits, without exception – in the same way that anthropologists have found that human beings using the
cross-cousin marriage system possess very similar societies in different physical environments.

It is important to note that the hominids must form somewhat larger groups, and,
afterwards, evolve the social behaviours that are capable of knitting the community together.
This argument may seem paradoxical. In order for these animals to evolve the capacity to
exist in larger communities, there must be a selection pressure preferentially selecting genes
for the formation of larger communities. The preferential selection for genes that enables
individuals to exist in larger groups *cannot* occur if the animals exist in small groups. This
would be the equivalent of an animal evolving traits to exist at high altitude, while actually
living at low altitude – which is impossible. In order for there to be the preferential selection
of genes for the animal to become successful in a particular environment, in this case a social
environment, then the animal must exist in that environment already. Thus the formation of
larger groups comes first; the evolution to be able to successfully knit together such a group
and manipulate such a group occurs afterwards. Such an evolution, it must be emphasised,
is gradual, as the existence of an intermediary stage between the second and third hominid
suggests.

Increased levels of nutrition would have allowed for the continual arising of large
groups, which fission into smaller groups. With time, some individuals would have existed
with genes for greater levels of cooperation and the descendants of these individuals would
have been able to maintain larger, stable groups.

In one's mind, one can envisage an environment where there is more nutrition available; larger groups form and fission, form and fission. Eventually individuals emerge that

possess cooperative traits and the groups containing the descendants of these individuals *maintain* the larger, stable groups that are highly competitive in this rich environment. These groups dominate their environments, while others that are unable to sustain larger groups because of a lack of the cooperative mechanisms are, in turn, dominated. There are then even larger levels of nutrition. Larger groups form and fission, form and fission, until individuals emerge that possess even more cooperative traits that allow for their descendants to *maintain* even larger, stable groups. These larger groups then dominate those previously large groups, which are now relatively small. (Thus the intermediary hominid dominates the second hominid, and, in turn, the third hominid dominates the intermediary hominid.)

The evolution of these hominids is due to abundance rather than scarcity. The formation of complex cooperative behaviours comes from the need to form large, formidable, stable groups in a highly competitive, abundant environment. Due to the abundant environment, large groups emerge because of higher levels of fertility, and, over time, some of these large groups are able to stabilise because the members of the groups possess unusual genetic traits that allow them to maintain large group sizes.

It would *not* be the case that simply because there is more food available that the individuals are able to form large groups. As the complexity of the cross-cousin marriage system attests, and, indeed, our own remarkable intelligence attests, *cooperation is not simple*, and the evolution of these large groups would have slowly evolved with larger brains, verbal communication, and many other complex traits. The evolutionary path from the second hominid to the third hominid may have taken many hundreds of thousands of years.

This third hominid would, if we were able to meet it, be eminently human in many ways, yet obviously lacking that level of sophistication that marks the human being as something quite different. The difference between the third hominid and the fourth hominid is thus a matter of degree rather than major changes. The last movement towards humanity is simply a step towards greater complexity.

Fourth Hominid

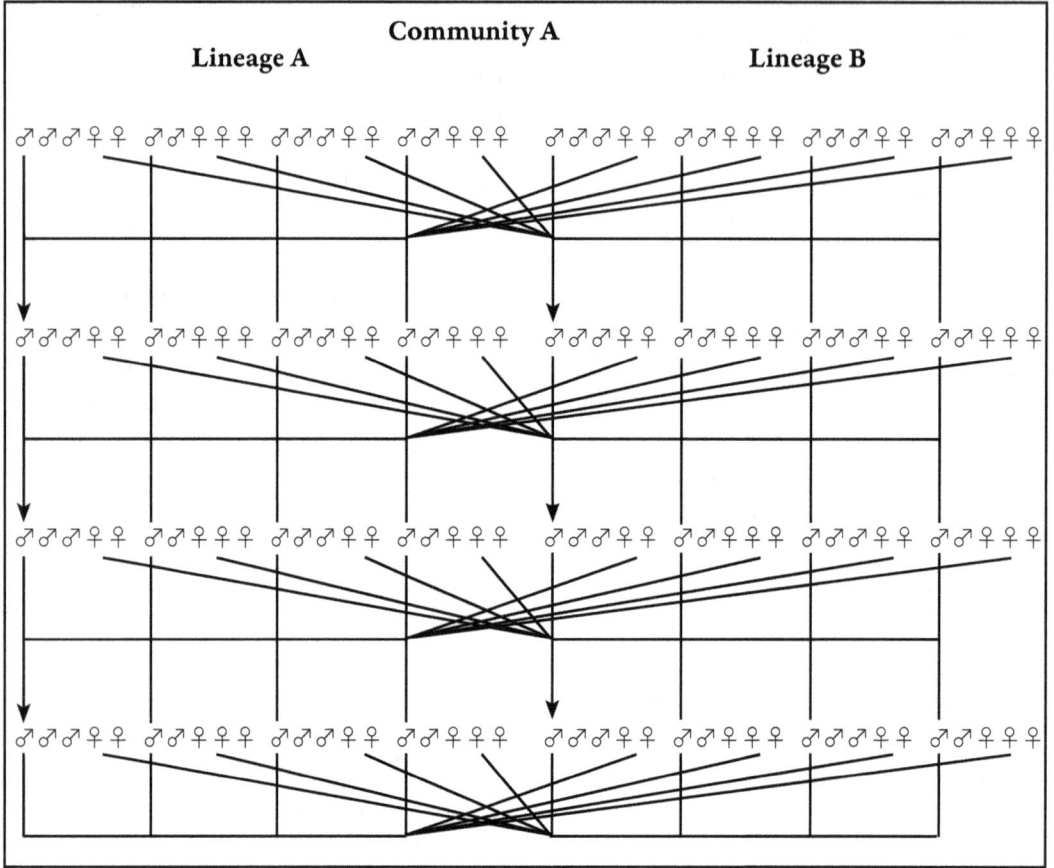

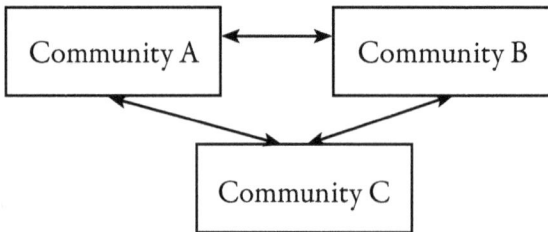

Diagram 10: the fourth hominid community includes more males and intermarrying between communities. In the upper diagram, four generations are presented, with two lineages exchanging females for marriage. In the lower diagram, the exchange between communities is presented.

Fourth Hominid: detailed

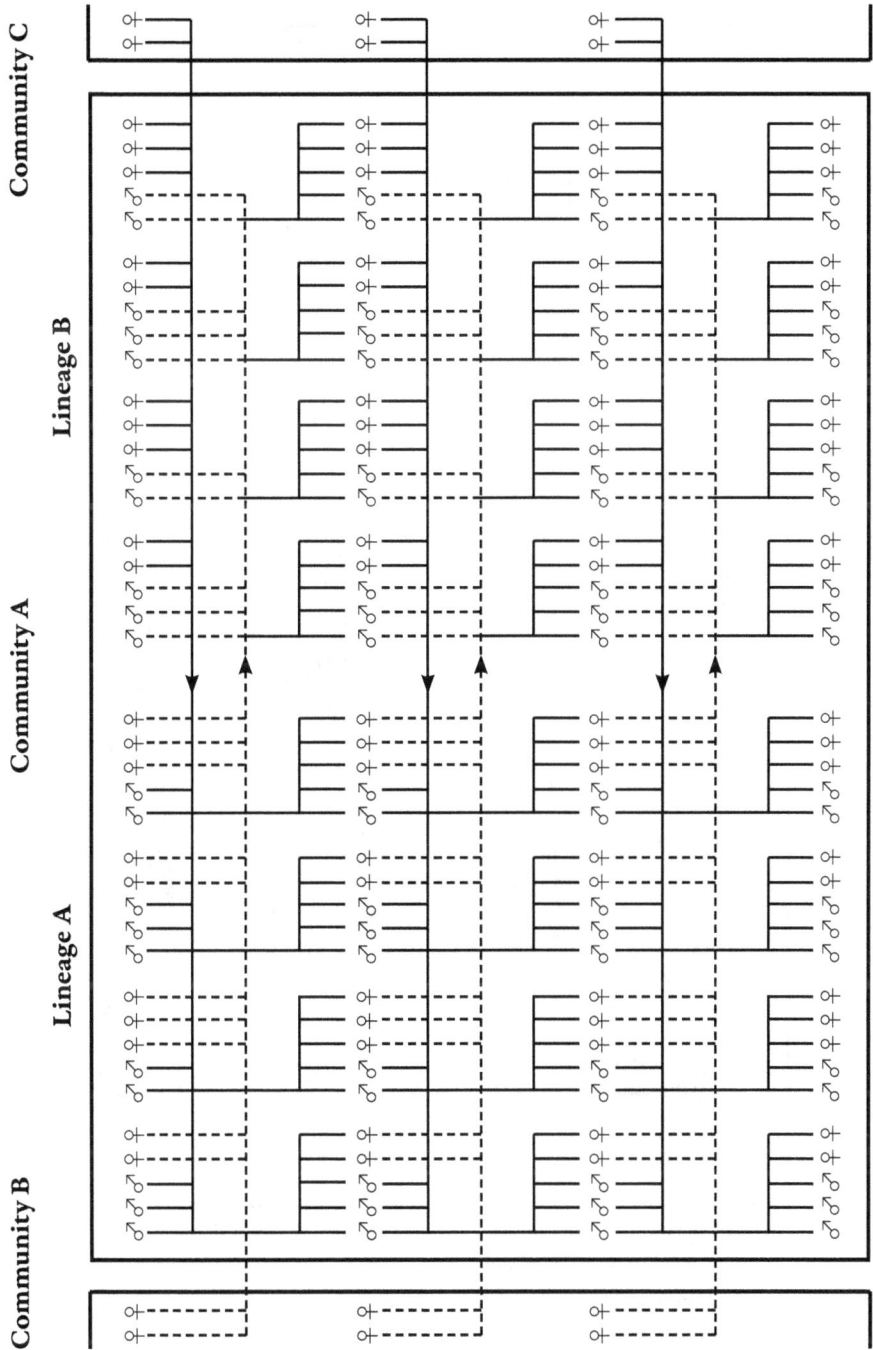

Diagram 11: community, A, is intermarried with communities B and C. There is an extra dominant male, with younger brothers (brother-alliance) in each lineage. At any particular time, living members may exist in two or three generations, and many in each generation, including an entire generation, may no longer be living. The diagram represents those that have lived in each generation rather than those presently living.

Fourth Hominid with three lineages

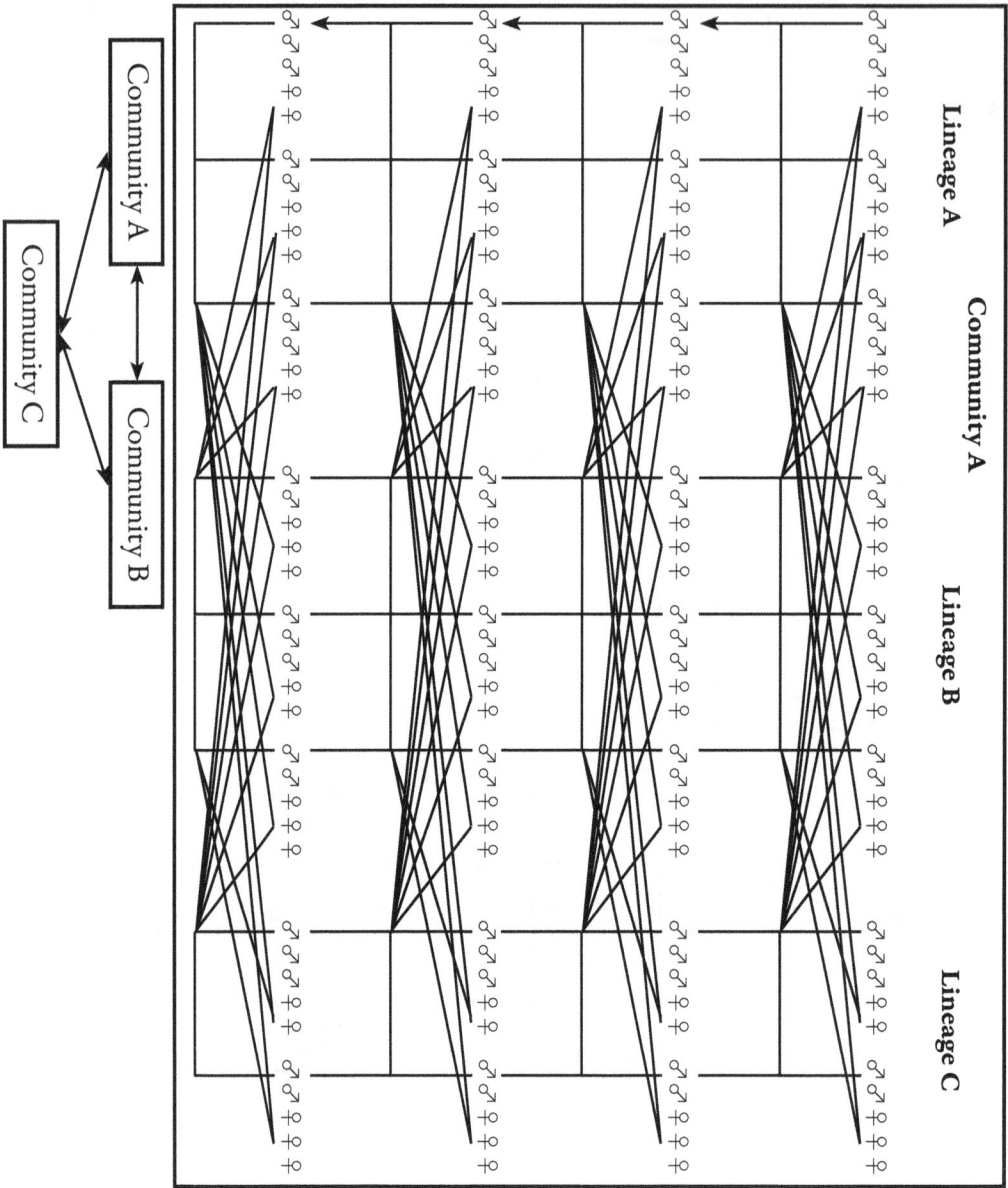

Diagram 12: the community consists of three rather than two lineages, with two larger lineages, each with three sets of brother-alliances (parallel-cousins), allied with a smaller lineage with only two sets of brother-alliances (parallel-cousins). Community A (top) is then allied with two other communities – B and C (see bottom of diagram).

Fourth hominid with three lineages: detailed

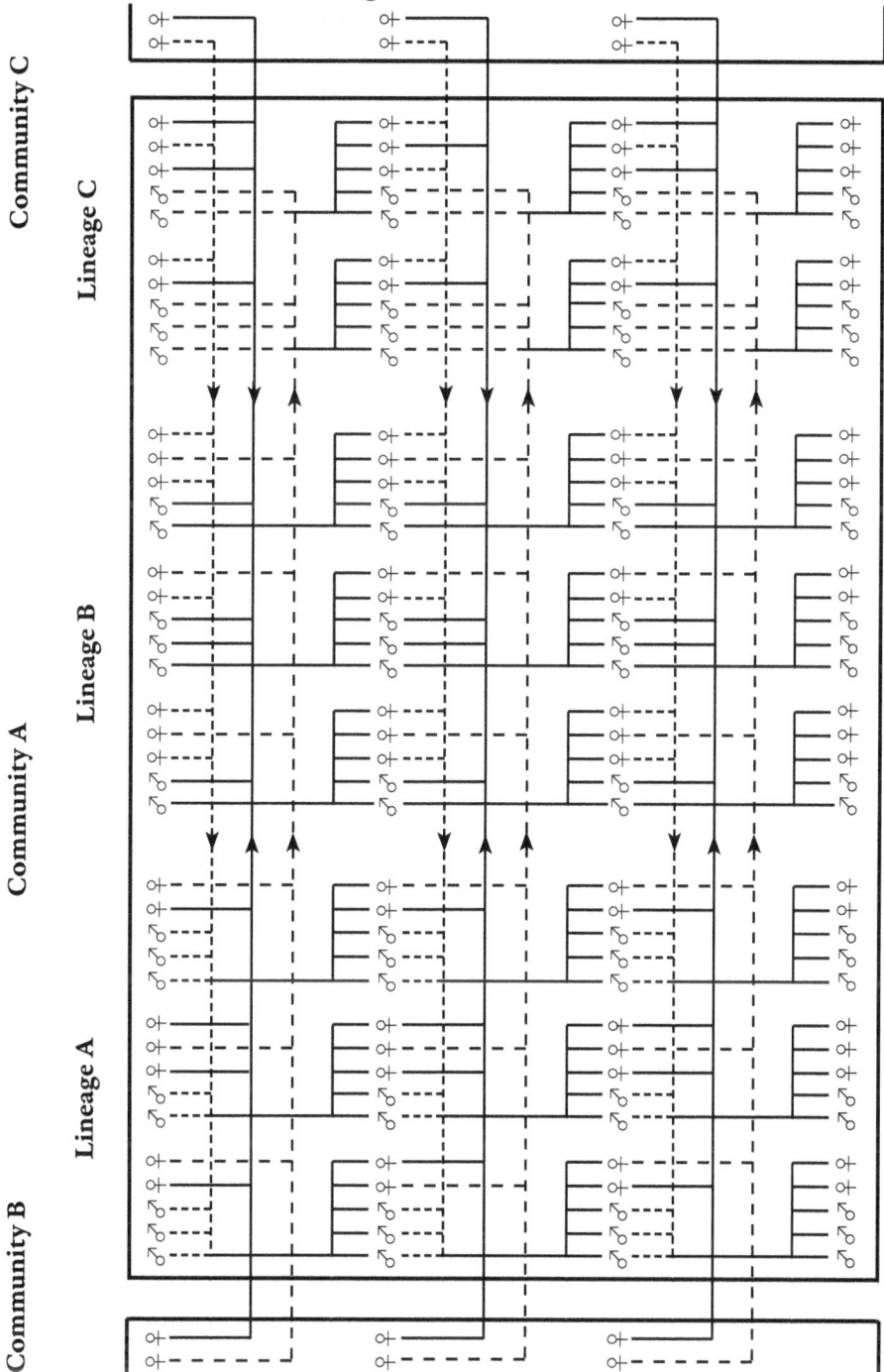

Diagram 13: Community A possesses three lineages that exchange marriage for cooperation. Community A exchange marriages with two allied communities, B and C. In each
generation, the first line represents the flow of women from lineages B and C into lineage A;
the second line, lineages A and C into lineage B; and the third line, the flow of women from
lineages A and B into lineage C.

The fourth hominid is a human being. Thus there is little point in explaining the appearance or behaviour of the fourth hominid; however, there is the issue of the fourth hominid's evolution from the third.

The reader can see from the diagrams that I have given two examples of the fourth hominid, one with two lineages (Diagrams 10 and 11) and an alternative group with three lineages (Diagrams 12 and 13). Both versions possess cooperative links with external communities. The diagrams suggest about ninety people (half are children) in each generation. If we assume a mortality rate of forty percent cumulative for each generation, then if three generations co-exist, this would mean a community of 105 individuals, half being children. This mortality rate reflects Chagnon's research showing 46% to be aged 0-14, 27% aged 15-29, 19% aged 30-44 and 9% over 45[2] (remembering, females define generations and can be grandmothers in their mid-thirties). Communities may be smaller or larger than the ideal presented, since this example should possess only 26 adult males.

The first version with two lineages is a simple alteration of the third hominid: basically, there are four dominant males and their younger siblings in each lineage rather than three dominant males in each lineage. However, a major change is that the community now needs not merely more political alliances within the community, but political alliances between communities. In other words, the level of competitive cooperation has exceeded the limits of the number of individuals that can cooperatively exist within one community using the cross-cousin marriage system.

The size of the community is limited by the dilution of alliances through marriage between members of the group. If brother-alliances (groups of brothers, half-brothers and possibly close male parallel-cousins) have only a small number of daughters to exchange, there will be a limit on their capacity to form links with other brother-alliances. As the number of individuals in the community increases, the method by which one group of brothers can form links with other lineages in the community diminishes. Equally, the number of brother-alliances within the lineage competing for marriages with other lineages increases.

Due to even greater levels of available nutrition, the environment has become increasingly competitive. The number of males required for a community to be competitive has outstretched the number that can cooperatively coexist in a single community, due to the limitations of the arranged marriage system as a method of forming alliances. Alliances now stretch beyond the community, and separate communities now engage in alliance formation using arranged marriages. This is the greatest change from the third hominid to the fourth hominid: the fourth hominid male has cooperative bonds that he cannot see and that he does not encounter on a daily basis. They exist in his mind and he only engages with them on occasion. The alliances between communities are weaker than those within a community, but they must exist.

The number of available males for females to sexually select has increased dramatically, not merely amongst the larger number of males within the community, but also amongst the visiting males from other communities. Females are also aware of the precariousness of selection and its potential to undermine cooperation. The importance of political alliances is as important to the females as it is to the males.

Hopefully, by this stage, the reader is aware of how truly remarkable human beings are and why they need such enormous brains. What is finding food compared to the precarious and enormous difficulty of having to organise such cooperative arrangements merely to be able to exist and reproduce? Most people that are involved in human evolution work with the idea that humans in their 'natural' state resemble something closer to my representations of the second hominid than the fourth hominid. The complexity of the human brain can be directly related to the requirements for the formation of these cooperative arrangements.

Increased nutritional levels lead to increased competition and the pressure to form larger communities. Those communities most able to form the larger communities are most competitive and successful. The evolution of the more cooperative fourth hominid is the same as outlined above for the first-to-third hominids: as communities increase in size they become more likely to fission into smaller groups; however, as the traits for greater cooperativeness evolve, some communities will be able to maintain their increased size. Those communities able to maintain their increased size become more successful, removing females from the communities that are unable to maintain a large size and which have fissioned into smaller groups.

Such cooperative traits have expanded to include external communities. Previously, when a single community of third hominids fissioned into two groups they became two politically distinct groups; for the fourth hominid, when a single community fissions into two communities, the political ties may remain. Genes are selected for cooperating with members of a larger group, and for the maintenance of ties with other communities. Those with genes for creating alliances with other communities were able to gain a much greater advantage over those that were only able to cooperate with individuals in their own community.

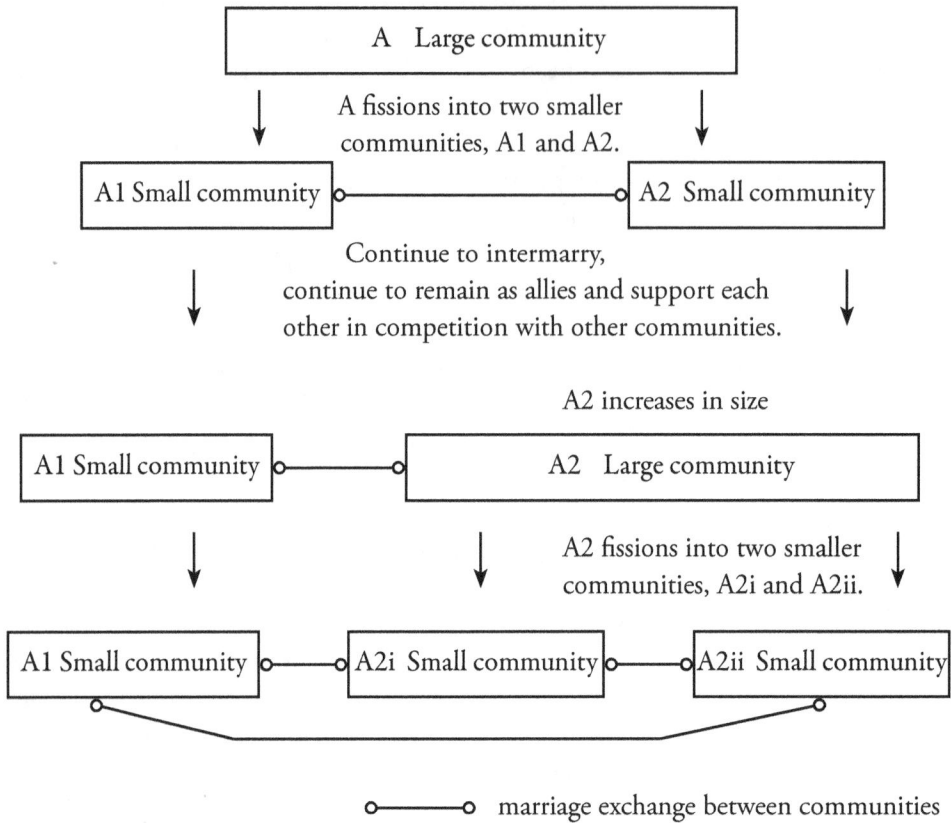

Diagram 14: A large community fissioning into smaller communities over time, while main-taining close proximity and intermarriage. The communities remain allies and support each other in competition with other groups.

In Diagram 14, if the animals were the third hominid, then A1, A2i and A2ii would all be politically distinct communities; instead, ties are maintained and the three groups become allies – able to support each other in conflict with other communities. In many ways, the great difference between the third hominid and fourth hominid is not necessarily greatly increased complexity within the community, but rather, greatly increased complexity between communities.

In the above example, I suggest that this would have evolved with fissioned com-munities supporting each other; however, this would be expanded, as in Diagram 15, below, to include neighbouring communities that may be more genetically diverse. Thus the com-munity B1 forms an alliance with another neighbouring community (A2ii), having fissioned from an ancestral community of both communities several generations previously.

This level of complexity obviously requires a remarkable brain to maintain such alliances. Initially, with the first hominid, males possessed no choices as to whom they would exchange sisters with other than the other resident male, and required limited intelligence.

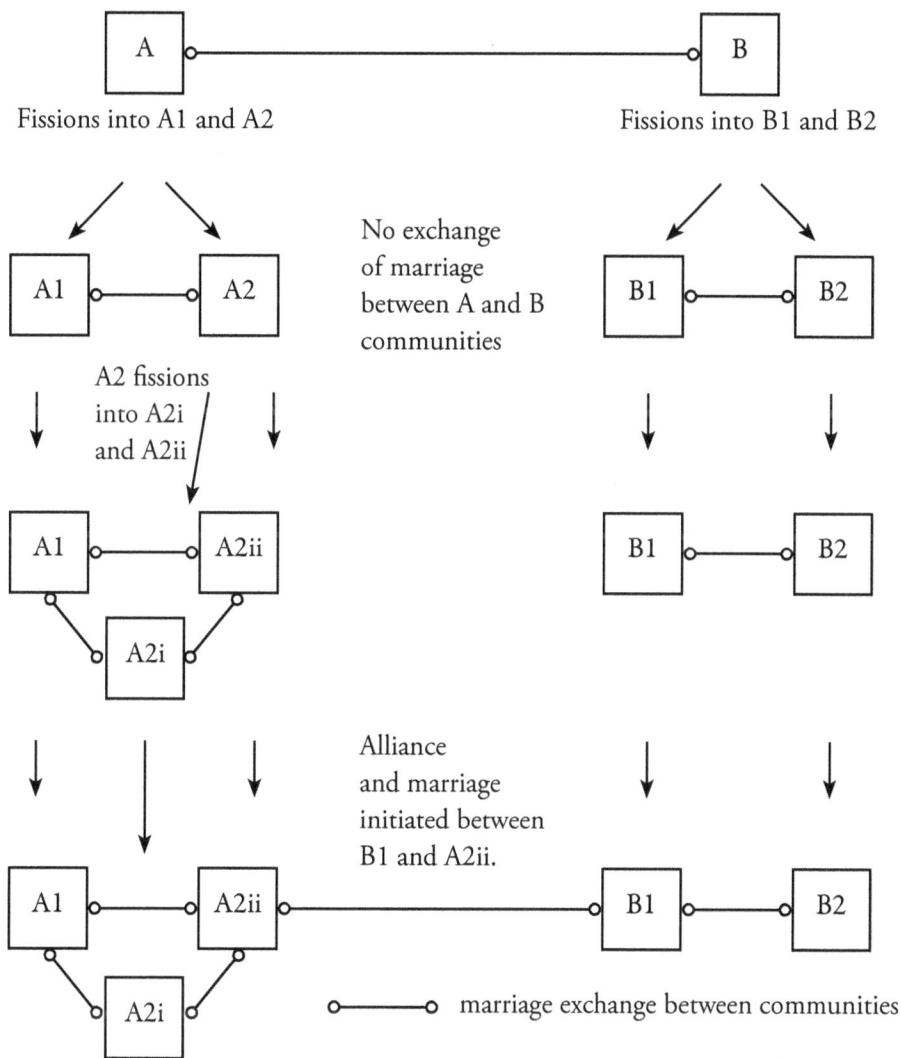

Diagram 15: movement of intermarriage between several communities derived from the same historical communities, eventually including intermarriage between two previously related groups.

In the second and third hominid, choices for allies became more complicated as the number of resident males increased. For the fourth hominid, males are now required to exchange females with numerous resident and non-resident males in order to maintain their competitiveness. Males that were able to increase their competitiveness two- or three-fold with such alliances would quickly usurp males that were incapable of forming such alliances. The benefits of such behaviour are: the capacity to maintain better nutritional environments, the ability to protect members of the community from others of the same species; the ability to respond to any hostility with vigorous retaliation; and the ability to remove females from other communities, either directly or through intimidation (protection). It is not suggested that the importance is the capacity to engage in modern warfare involving strategic attempts for annihilation of large groups of people – all the competition is between people of similar genetic heritage. It is highly unlikely that, until in the past few tens of thousands of years, any such events will have occurred.

The expansion to include external communities as cooperative partners requires not merely that the fourth hominid is capable of such levels of cooperation, but also to be able to manipulate this scenario. The fourth hominid needs to not merely envisage the future of its own community, as the third hominid did, but to envisage the future of surrounding communities as part of its strategic, long-term alliance formation and competitive behaviour. The fissioning of the community into separate, yet allied groups reflects how this development can occur in a realistic manner: the links are not being made with groups of completely foreign individuals, but with individuals whose ancestors were once part of the same group. As the cooperative capacity of the fourth hominid develops, it becomes increasingly capable of not merely maintaining links with recently fissioned groups, but also more genetically diverse groups. The fact that this occurs was observed by Napoleon Chagnon, when the village he was studying, the Upper Bisaasi-teri, made allies with genetically distinct Shamatari villages (see page 104).

This is a tremendous change. The third hominid possessed a brain capable of maintaining and manipulating a community of perhaps thirty adults, or thereabouts. Now, this third hominid has evolved into a fourth hominid that is capable of maintaining and manipulating a community of fifty adults or more, and also possesses the capacity to form cooperative links for further support in terms of protection and aggression with several communities of a similar size. The brain has moved from being able to consider a group size of perhaps sixty (half being children), to a maximum group size of perhaps four hundred (four communities of 100 persons each, half being children) or more. This tremendous increase in the number of cooperative individuals incorporates mostly external communities rather than a massive increase in the size of the resident community.

The morphological and behavioural changes from the third to the fourth hominid will mostly be a matter of degree. Intelligence will have improved in order to consider more complex genealogies and with it the capacity to appreciate even more complex abstract concepts: the brain must be capable of knowing or deducing hundreds of interwoven kinship links. The individuals must be capable of foreseeing political changes not merely within their own community, but between communities, and to be able to manipulate those changes. With a longer period of adolescence for the brain to acquire maturity, marriage contracts have extended to up to fifteen years, and the capacity of individuals to 'see' into the future will reflect such time dimensions (including their own mortality). With the increase in the technology of penetrative weapons (such as spears), the requirements for physical strength have diminished. Communities aim at creating excess levels of nutrition for feasts with which to present their success to other communities as part of the alliance formation system, and mere subsistence is no longer the nutritional requirement. Females' selection for vocal abilities and trustworthy behaviours in female allies increases to include both resident and non-resident females. Females select males on the basis for even greater levels of cooperative behaviour, by selecting high status (politically) males. Females would now possess the ability to increase their inclusive fitness by undergoing menopause since males now have a much greater capacity to be polygynous. Fire and now clothing enable for the selection of higher levels of nakedness. Females attempt to sexually attract males from not merely their own community but other communities who may not know their capacity to reproduce a first male child (males select females with breasts that demonstrate their history of breast-feeding).

The changes from the third hominid to the fourth hominid reflect increased pressure to form alliances with other communities. Males need to know all the individuals within those communities and their ancestors (genealogies) so that they can manipulate the arranged marriage system. Males need to select political allies not merely within their own community, but between communities, so that males are able to increase both their capacity for competitiveness but also protection. For females, the ability to select at a distance becomes imperative to increase the available gene pool beyond their own community, requiring not merely the physical nakedness and breasts, but also an alliance with the male's sister or half-sister so that the selection can occur unnoticed.

From the third hominid, humans have massively expanded their capacity to cooperate, compete and sexually select. The dimensions of the brain now require not the knowledge of 30 adults, but potentially 200 adults, and their ancestors. If competitors are included, then these dimensions are expanded even further since a community with three allied communities may have several enemy communities, and they would need to know the individuals in those communities as well. As has been demonstrated, alliances and enemies are momentary, and allies can become enemies and enemies can become allies, thus enforcing the requirement to know all possible individuals that an individual may confront. The competitiveness that yielded such changes must have been immense. The available nutrition must have been extensive. Communities that would normally have fissioned into entirely separate communities, must now fission into communities that remain close by and continue to interact, assuming that the fissioning wasn't caused by something so catastrophic that the communities became enemies.

The flip-side to this is that males became increasingly capable of forming alliances with males of increasingly greater genetic diversity. Males were thus able to form partnerships with perhaps a parallel-cousin, perhaps six times removed from their own parentage or greater, something which no other mammalian male can even consider, yet alone perform. The increased competition yields increased cooperation.

Nutrition

The competitive nature of the environment in which the fourth hominid evolved must have been truly astonishing in evolutionary terms. It is remarkable that animals with the complexity of the third hominid – astonishingly intelligent in their own right – were now placed in an environment where they were obsolete and incapable of forming the cooperative alliances needed to remain competitive. This environment must have been extremely plentiful in terms of food, and, it is plausible, that the only way such an animal could have evolved is if it possessed a supply of food that was fairly unlimited.

Humans possess the ability to form alliances with very large numbers of people and do this instinctively. People using horticulture are able to form large cooperative communities simply using arranged marriages. This suggests that the earliest humans were capable of doing this, since no-one is suggesting that humans have altered radically with the use of horticulture. While some might suggest that horticulture allows for overly large community sizes and levels of cooperation, if humans had not already evolved the capacity to possess such levels of cooperation, then large communities would not be possible, even with horticulture. If humans evolved with maximum community sizes of thirty adults, then even with the nutritional increases from horticulture, thirty adults would remain the maximum community size. The size of the human brain reflects the social complexity of the environment in which humans evolved, a brain capable of forming partnerships well beyond thirty adults.

One of the difficulties in envisaging how humans could have evolved these remarkable cooperative abilities is that our assumption is that people used a hunter-gatherer manner of obtaining food, and much of the documentation of this comes from nutritionally poor environments such as Australia. Aborigines in Australia exist in very small groups and simply are not capable of forming very large groups because of nutritional restrictions.

This raises two possibilities: that the environment in which humans evolved was very rich using a hunter-gatherer method of obtaining nutrition; or, that horticulture existed before humans. In order for humans to evolve the capacity to create cooperation between large numbers of people, they must have possessed large amounts of nutrition, and have evolved within this environment of large amounts of nutrition. As previously argued, animals or plants that have evolved to exist at high altitude must be living in that environment to evolve those adaptations, so too humans must have evolved in an extremely rich nutritional environment in order to develop the cooperative capacity to be able to form large networks of cooperation. The environment must have been so rich that it also enabled the evolution of the tool for such cooperation, the human brain, which requires very large amounts of energy, supposedly 20 percent of total energy requirements despite weighing less than 3 percent of total body weight. These nutritional levels must have existed continuously for a long period of time for the evolution of the increased brain size.

Humans did not enter a new environment and evolve; they evolved in an environment surrounded by similar hominids who were competing for exactly the same nutritional resources. Pre-human hominids did not move to a novel environment, such as North America, and, in an environment without competitors, evolve as humans with enormous nutritional resources. Furthermore, the evolution occurs because of competition between similar hominids, forcing hominids to form larger and larger groups in order to compete. The competitive social environment is the evolutionary environment: it is this competition that causes the evolution to occur.

There is evidence that fire existed well before the first evidence of human fossils,[3] and hunting and gathering had also previously existed. There appears to be evidence that *Homo neanderthalensis* hunted woolly mammoths, which, if true, suggests that hominids with less intelligence than humans had perfected hunting animals the size of an Asian elephant. If *Homo neanderthalensis* had pushed hunting to its natural limits, hunting some of the largest land animals, then this would suggest that the nutritional resources from hunter-gathering simply were not enough for the nutritional requirements to exist for the evolution of humans. If hunter-gathering could supply the nutritional requirements for such high levels of competition, then one would expect that the Neanderthals would have attained human-like intelligence, or that the evolutionary process would have halted at this point.

This leaves only one alternative: that horticulture existed before humans. Horticulture created an environment in which nutrition was so large that it pushed the boundaries of group size and meant that humans evolved to possess the ability to use cooperative arranged marriages with external communities. An environment existed in which very high levels of nutrition were continuous and so the human brain could evolve.

There is no evidence that horticulture existed before humans. This is not surprising, since there is evidence of bone needles and even buttons 26,000 years ago; however, there is no evidence of the accompanying clothing.[4] If clothing does not survive for large periods of time, then the likelihood of plant material surviving is extremely low.

With horticulture people are able to maintain very high density levels. In the Papuan highlands, the Dani of Papua use sweet potatoes as their basic crop[5] and with this are able to support a population density of a remarkable 175 people per square kilometre.[6] The Tauade of Papua using yams and pandanus nuts possessed a population density of between 19 and 62 persons per square mile.[7] One of the problems with these comparisons is that the people are often using imported crop plants that are not indigenous to the region.

One of the most over-looked aspects of the behaviour of Western people is their desire to control their environment, including their floral environment. People seem to have an almost inbuilt desire to enjoy manipulating the floral environment, even if they live in apartments or flats with no available soil. The rapidity in which humans have been able to breed new crop types, as in the Americas, where people for several thousand years, bred maize and other plants for their own purposes, makes one wonder whether similar events had occurred in Africa and Europe. Anyone who looks through the anthropological literature would see that people using a hunter-gatherer method of food acquisition are actually uncommon, while those using hunting and horticulture are common. Horticulture appears to be the standard method of food acquisition, while gathering appears to be used only by those in environments that cannot support horticulture. Again, this is complicated by the use of non-indigenous crops.

While it is more than understandable that people are hesitant to make suggestions outside the evidence, there is certainly a history in science of making suggestions based on logic before the evidence becomes available. I suggest that the major increases in nutrition were hunting and gathering, then the use of fire, and finally horticulture. That horticulture was in use before humans existed appears to me to be an inescapable possibility. The density levels of the hominids that became humans must have been very high, and it does not seem plausible that a hunting and gathering method of acquiring food could sustain such density levels. The human brain is capable of forming very large communities and to form cooperative alliances with external communities, and the brain's size suggests that pre-human ancestors must have existed in large communities for such a brain capacity to evolve. Without a fairly unlimited food supply, I believe human evolution might be inconceivable. Considering that ants can farm fungus and that beavers can build dams, the possibility of pre-humans taking plants, re-planting them near their community, and selecting the better plants, is not implausible. This is especially true when we consider that they were already expert manufacturers of spears.

The Emergence of Numerical Advantage

There is one major question left to be considered: why did our ancestors become more numerous rather than larger? Most mammals, such as the giraffe or the elephant or the whale, tend to increase in size rather than form larger, cooperative societies. These animals become physically larger and their intelligence remains roughly the same. So why did we not follow in the same direction? Why are there many humans with remarkable intelligence rather than hominids that are larger than gorillas but with a gorilla's intelligence? This issue underpins human evolution, and, in many ways, might be described as the most important issue of all.

For instance, in the first hominid, I argued that there were two males in the community exchanging females. But why are two males required rather than only one? Why not simply one large male gorilla-like animal, and why wouldn't the single male have simply increased body size to protect his females and infants rather than increasing in number?

The only explanation is that something must have made numerical advantage more important than physical size in our ancestors.

Consider gorillas and chimpanzees. A major difference between gorillas and chimpanzees is where they find their food. Chimpanzees tend to find their food in the trees (fruit and meat), while gorillas find their food on the ground. Even in areas where chimpanzees and lowland gorillas occupy the same territories, gorillas eat more herbaceous plants than chimpanzees, as reflected by higher fibre contents in their faeces.[8] Caroline Tutin observes:

> The diets of chimpanzees and mountain gorillas differ greatly. The chimpanzees are mainly frugivorous (fruit-feeders), supplementing their diet with leaves, insects, and mammalian prey... while the mountain gorillas are folivorous (foliage-feeders), with 95% of their food being leaves, stems, and pith... An important implication of this dietary difference is the extent to which food availability varies over time. The amount of food available to mountain gorillas changes little as, apart from bamboo shoots, their herbaceous food plants are abundant in all seasons. Chimpanzees, like all frugivores, are faced with both regular seasonal fluctuations in available fruits and unpredictable fluctuations throughout the year.[9]

Tutin also notes that though gorillas are much larger than chimpanzees, the females reproduce at a faster rate. It might seem that this is logical due to their less sporadic source of nutrition.

Being ground dwelling animals that can find their nutritional source on the ground, gorillas are unrestricted in their body size and can grow to whatever weight the food resources will provide. Chimpanzees, by contrast, can only grow to a body size that the trees will allow. Anybody who has seen the photos of chimpanzees in Gombe – where large chimpanzees sit on tiny branches eating fruit – will appreciate that chimpanzees must be at the limit of how large they can grow. Male and female chimpanzees have similar body weights. Male gorillas generally weigh double that of females. In her overview of primate injury and illness, Nancy Lovett observed: 'Among the great apes, gorillas seem to exhibit the fewest traumatic lesions,'[10] which may be linked to their ground-dwelling life. Studying chimpanzees, Jane Goodall and her colleagues noted 51 falls, two that led to the deaths of one adult male and one infant. 'Of the 51 falls, 31.4 percent occurred when chimpanzees trod or jumped on dead branches. Adult males fell most often during aggressive incidents, often when two were fighting each other. Female aggression accounted for three of the female falls. Infants and juveniles fell most frequently during play.'[11] Those listed are only those when the chimpanzees fell and hit the ground, and Goodall notes: 'Other falls not listed caused serious injury', or included a fall where the chimpanzee managed to stop before falling over a 15 metre cliff.[12] Considering that falls occur, and can, rarely, be lethal, could chimpanzees grow to 70kg or larger?

If we moved a smaller, chimpanzee-sized animal to a gorilla's environment and used the gorilla's source of nutrition, would we get several times as many chimpanzee-like animals? No. They should evolve and increase in size until they reached gorilla-sized proportions now that there are few limits on their size and weight.

A further example of this can be found in the orangutan. The orangutan male comes in two sizes: a female-sized male and a fully grown male. The fully grown male is upwards of 100kg and is restricted to moving on the ground or large limbs on trees. The female and the female-sized male are half the grown male's size. They weigh about the same as chimpanzees. These females and female-sized males can move about in the trees. The two body sizes for males are important as male orangutans compete vigorously and dominate sub-adult males.[13] Sub-adult males can remain in the lower body size for long periods of time and it is speculated that they remain small due to the presence of fully grown adult males.[14] This could be because they are waiting for a territory to become available. The smaller body-size allows sub-adult males to remain arboreal.

The point of these examples is that animals are restricted in size by their social and physical environments. Gorillas and adult male orangutans are the least limited and, in fact, evidently grow to whatever proportions the environment will allow. Chimpanzees, however, are limited by their food source and can only grow to whatever dimensions the trees allow. If chimpanzees found a new food source on the ground that was larger than the food source they find in the trees, then they could grow much larger.[15]

Chimpanzees live in a rich environment, yet the nature of their food source restricts how large they can grow. The outcome, I argue, is that rather than large chimpanzees we get numerous chimpanzees. Chimpanzees live in relatively high population densities and males must cooperate and form groups to compete with other groups that have multiple males. By contrast, male orangutans and gorillas live in food rich environments and can grow as large as the environment will allow; they live with single male territories and harems, respectively.

There is thus the possibility that the first hominid simply lives in an environment where it too, like the chimpanzee, is restricted by its food source. That, like the chimpanzee, it lives in an environment where the animals can only grow to a certain size, and instead of being a solitary, large 150kg male gorilla-like animal, there are three times as many 50kg male chimpanzee-like animals.

This, of course, cannot be the case. We are bipeds and utilise the ground. Like gorillas and male orangutans, we should have been unlimited in our physical size and could have grown to gorilla-like proportions.

To explain why numeracy is important in humans, we are faced with a serious paradox: we use the ground for movement and to obtain our food so there should have been no limit as to our physical size. So why did we evolve with large groups rather than large individuals when we were unlimited in body size? Why did our ancestors not follow the path of gorillas, whales, elephants, moose, elephant seals, and seemingly all other ground or sea dwelling mammals and grow to a maximum size?

At this stage, some might argue that numeracy is important for hunting, and that numerical advantage gives a considerable advantage for providing food in this manner. Yet, this cannot be the answer. This argument assumes that our ancestors cooperated in the need to find food. Yet no animal responds to starvation by forming larger groups and larger brains; instead, they form smaller groups and have smaller brains in a response to the food shortage. Food shortages lead to stunted progeny, not more intelligent, cooperative and complex progeny. If limitations of food are not the answer, then what is the imperative to evolve the capacity to hunt cooperatively? There is no *must* argument to explain such an evolutionary process since there is no selection pressure to hunt. In order to reproduce one must be healthy; if there is a sudden shortage of nutrition, the infants will become stunted. Animals simply do not evolve complex behaviours and morphologies as a response to sudden decreases in nutrition. Complex behaviours and morphologies require generations in order to evolve.

I suggest the following explanation. Our ancestors must have competed, as we do, using projectiles. It is through the competitive use of projectiles that numerical advantage overcomes physical advantage. The first and most important step in human evolution was the change to biped structure. It is the use of two feet rather than four that allows for effective and accurate use of projectiles and provides the basis for the importance of numerical advantage in competition. At a later date, it also provided the basis for increased levels of nutrition through hunting.

A biped body is superior to a quadruped body for throwing projectiles. If the reader takes a bow-legged, chimpanzee stance, they will appreciate how difficult it is to throw accurately with a chimpanzee-like physique. Chimpanzees do throw rocks, but they select large rocks and underarm the rock at an opponent or throw small rocks overarm at close range.[16] Jane Goodall gives one example, in an attempt by one community to intimidate a lone chimpanzee:

> Another young Kasakela male, Sniff, once taunted a large group of Kalande chimpanzees, including at least three fully adult males, quite by himself – his two companions had fled. The Kalande group was in a shallow, steep-sided ravine, calling loudly and charging about in the undergrowth. Sniff, uttering deep roar-like hoots, performed a spectacular display along a trail near the top of the ravine. As he charged he hurled at least thirteen huge rocks down onto the strangers. An occasional missile – a stone or a stick – flew up from the undergrowth below, but they fell far short of Sniff. Only when two Kalande

males raced towards him did Sniff retreat.[17]

Goodall notes that chimpanzees also threw rocks at baboons and humans, the first being Mr Worzle:

> Also, he was the first chimpanzee we ever saw throwing stones and other objects at baboons that approached and threatened him. True, Worzle sometimes hurled leaves if no more suitable missile was close at hand, and once threw a whole handful of bananas at an aggressive male baboon, to the obvious satisfaction of his opponent! As time went on, Worzle became more selective, usually picking up and throwing rather large rocks.
>
> We were not surprised when first we saw Worzle throwing a stone at a baboon, for we had seen chimps throwing things aggressively before. Rodolf had hurled a large rock straight at Hugo and me the first time he ever visited camp. He had, it seemed, followed Goliath along without really looking where he was going. When he suddenly awoke to the fact that there was a tent ahead, with us inside, he uttered a strangled call, stood upright, and threw his missile before bolting back to the bushes.
>
> Soon after we first saw Mr Worzle throwing at baboons, a number of the other males started doing likewise. These days nearly all the adult males use objects as weapons in this way, but though they usually choose fairly large stones they seldom hit the baboons unless they are at very close quarters.[18]

Goodall concludes on the similarities of the behaviours of chimpanzees and humans:

> Some of the chimpanzees' aggressive patterns are not dissimilar to some of ours. Like a human, an angry chimpanzee may stare fixedly at an opponent, raise his forearm rapidly, jerk back his head a little, run toward the adversary upright and waving his arms, throw stones, wield sticks, hit, kick, bite, scratch, and pull the hair of a victim.
>
> In fact, if we survey the whole range of the postural and gestural communication signals of chimpanzees on the one hand and humans on the other, we find striking similarities in many instances. It would appear, then, that man and chimp either evolved gestures and postures along a most remarkable parallel or that we share with the chimpanzees an ancestor in the dim and very distant past; an ancestor, moreover, who communicated with his kind by means of kissing and embracing, touching and patting and holding hands.[19]

Chimpanzees use rocks mostly to intimidate, rather than harm. By contrast, anyone who observes either a baseball pitcher or a cricket bowler will see that both use a similar technique: upright stance, with one straight leg firmly planted on the ground, the arm swings around the trunk and the other leg counterbalances to provide momentum. A fast bowler in cricket uses a ball with the likeness of a rock, and, though he must bounce the ball before it reaches the batsman above waist height, he can break a man's jaw or skull at a distance of twenty metres. Batsmen in cricket wear protection on almost every part of their body that faces the bowler. Baseball pitchers can throw the ball at similar speeds to bowlers in cricket, but use a much lighter ball. One can only imagine what would happen if pitchers used a cricket ball instead of a light baseball. The batters would certainly need far more than a helmet. I suspect that a baseball pitcher could easily fracture an unprotected batter's skull if he used a cricket ball.[20]

It is worth imagining what a biped chimpanzee, with many times the strength of a man, could do with a cricket ball! Jane Goodall and her colleagues used bananas to attract the chimpanzees at Gombe, and had enormous difficulties in ensuring some kind of equitable distribution with their limited supply since: 'For one thing, an adult male, if he has the chance, will eat fifty or more bananas at a sitting; for another, we were having more and more trouble with the baboon troop'.[21] The difficulties encountered in attempts to control the process of rewarding individual chimpanzees give some indication of the strength of chimpanzees:

> During most of that year the chimps' banana supply was our worst headache. For one thing, there never seemed to be enough boxes. Hassan was making more almost nonstop, but every day two or three – or even more – were put out of commission by one of the chimps. Even when everything had been sunk into cement, a few of the adult males still managed to break something. J.B. was the worst offender. He kept breaking off the steel handles of the levers, so we could not close them. And he managed to snap even strong cables, though the only part showing was a length of about seven inches between the cemented end of the pipe and where it was attached to the lever. It was a rather terrifying indication of the tremendous strength of these chimpanzees.[22]

A biped chimpanzee-like animal, I argue, could kill an unprotected man with a well aimed cricket ball. There are those who wonder how our biped ancestors survived on the plains of Africa considering the possibility of predators, yet an ancestor that could throw a rock with the manner and accuracy of a baseball pitcher and possessed the strength of a chimpanzee would have no difficulty with lions. Predators rarely attack animals that can fight back, since it is simply not worth the risk, and a biped chimpanzee that could accurately launch a heavy rock at more than 160km/h off a single step, like a baseball pitcher, would probably be able to fracture a lion's skull before it was within twenty metres.

If our ancestors, from the earliest stages, could fracture an opponent's skull from a distance, then having numerous individuals in the group becomes more important than becoming larger. In competition, having three smaller individuals that can kill an opponent twenty metres away will become more successful than a single individual three or four times larger who only possesses an advantage in hand-to-hand combat.

It has been observed that the canine teeth – used to inflict serious wounds, particularly by gorillas – are minimal in our earliest ancestors. This would suggest that hand-to-hand combat was rarely engaged in by our earliest ancestors and that competitors so rarely bit an opponent that the canine teeth diminished in size to be barely larger than the other teeth. (The human smile, presenting a lack of knife-like canine teeth, may be several billion years old.) It has also been observed that the skulls of *Homo sapiens* and the skulls of their ancestors differ in the shape of the skulls, as well as their size. The older skulls have a thick formation of bone above the eyes, and the eyes are less prominent in comparison to the skulls of *Homo sapiens*. One might deduce that the formation of bone over the eyes is protective – it is difficult to conceive of any other reason of why there is thick bone above the eyes. The brain of the animal barely emerges over the top of this brow ridge, which can be contrasted to human skulls with a significant amount of our forehead above the brow. One might conclude that the brow ridge protects the eyes and brain against blows, but it is difficult to conceive of what kind of blows might occur in a front-on manner. Certainly if the animals were fighting with clubs, one would expect the entire skull to be of thick bone. If the blows came from projectiles, however, it is certainly more likely that the eyes and brain would need to be protected, particularly if the manner of combat involves two groups of individuals (with the strength of chimpanzees) throwing stones at one another at a distance. Once penetrative objects, such as spears, were used in combat then we would expect that the brow ridge would no longer be necessary and would disappear, which it has. In the second chapter, I have already observed that our ancestors have possessed spears for at least 400,000 years.

The importance of biped movement is thus linked to the more important issue of why our ancestors increased their group size and needed to form large, cooperative groups with many members. If our ancestors did use projectiles in combat, then the importance of numeracy would be a lasting trait.

How could such a trait have evolved? The movement to a biped body structure is difficult to envision, simply because it is difficult to imagine the intervening stages as being beneficial. Chris Stringer and Peter Andrews observe:

> The adaptations for bipedalism are many and extend to all parts of the body. The head has to be balanced on top of the backbone instead of being slung in front; the backbone has developed curves to withstand stresses and to function as a spring; the hip has broadened and wrapped around the sides of the body to give better leverage to the muscles that maintain us in an upright position; the legs become longer and angled inwards to keep the centre of gravity along the midline of the body; and the feet have developed arches and the big toe has rotated in line with the rest of the foot to provide additional leverage.[23]

In comparing chimpanzees and humans, they observe:

> Humans have a broader pelvis and curved lower spine so that their centre of gravity is set back, enabling them to stand upright. Chimps by contrast lean forwards when they stand upright and cannot maintain that position for long. Humans also stand with their knees together so that their weight passes down through the centre of their body, and they have their toes lined up, lacking the opposable toes of chimps.[24]

In photographs of chimpanzees attempting to stand upright, this difference in how the body is weighted is quite obvious, with chimpanzees needing to bend their legs in order to correct for their forward weighted body. The most interesting issue is that of the hips and legs, as observed in *Australopithecine afarensis*:

> The knee joint is another significant joint surface in bipedal locomotion. The knee was

also the first fossil found that showed that *A. afarensis* was unequivocally bipedal. For a human to stand with feet and knees together, the upper leg has to angle in strongly because the tops of the legs are widely separated at the hip. The upper leg therefore makes a sharp angle with the lower leg at the knee joint, and this is quite different from the ape knee. Apes walk and stand with feet and knees apart, so that upper and lower legs are in a straight line, but *A. afarensis* has the human condition in its knee angle.[25]

An interesting question is whether the upper leg does need to angle in to allow for biped locomotion. I suspect that the answer to this might be no. It would appear feasible that a biped primate could have legs that, like a chimpanzee's, simply angle downwards from the outside of the body. In locomotion, however, such an animal could only stand still (assuming the spine, head, and trunk were re-weighted to allow for a biped stance) or move at a jog or fast walk; if it attempted to walk slowly, it would possess an awkward, falling gait. The reader can appreciate this by assuming a gait where the legs are well-spaced rather than close together. Those who argue that the importance of bipedalism is for locomotion need to explain why such an animal did not evolve, that is, one with legs that descend down from the hip rather than inwards. Such an animal would not have faced the difficulty of giving birth through a narrow hip and may have been able to run at higher speeds. Indeed, while at slow speeds such a body design might be unstable, at higher speeds it might be an advantage with powerful legs able to push from the outside of the hip directly to the ground, rather than, as with humans, through the bent leg joined at the knee. The bicycle is ample evidence that an unstable design at slow speeds is more than capable of making up for such instability at higher speeds; it has the odd design of two feet directly in line. Yet it works because momentum creates balance. A human leg is the equivalent of attempting to use a spanner with a bend in the middle: a great deal of energy is wasted. It is easy to imagine that a sprinter with legs that move directly from the hip to the ground in a straight line and with powerful leg muscles might easily out-sprint any human since the force would be directed down, rather than directed inwards and down. The design of the knee would also be simpler, and such an animal would not be hampered by the knee problems that plague humans due to the competing forces that are redirected through the knee. Issues of balance could be made up in other ways, such as wide feet. Indeed, the human foot must be considered one of the most paradoxical aspects of any primate body. Since we walk on two feet rather than four, one might expect that our feet would be very large and wide to increase the ability to balance, as they are in ground-dwelling birds. Instead, they are very narrow. Toes that might be used for extra grip and that might be designed to be large and strong, instead are small and weak. If one were designing feet for a biped primate, one might expect ours to be the more implausible and that a chimpanzee's foot would be a more likely alternative, since they are wide and spread out.

Yet, as the above details concerning *Australopithecine afarensis* present, our ancestors possessed inward sloping legs from the earliest stages. This has also been found to be the case in the Laetoli footprints, the four million year old footprints in volcanic ash, that present two animals walking in a biped manner, with one foot in front of the other, rather than with two feet at a width, as would be expected of a chimpanzee if it attempted to walk on two legs. The Laetoli footprints also present the feet to be narrow with small toes, similar to our own.

Alternatively, in order to allow for the capacity to throw, the legs must slope inwards so the animal can balance on one leg. This must be the case. The animal needs a leg placed close to the middle of its centre of balance to allow the rest of the body to pivot and provide momentum. If any animal without an inward sloping leg attempted to throw like a baseball pitcher it would fall over. The foot is also exactly as we would imagine it would be if throwing was the major issue. While wide feet and large, strong toes would create balance, they also get in the way if the animal is attempting to pivot with the throw. A body plan for throwing needs a small ball on which to pivot, and that is exactly what our foot possesses. There are no large toes or great width that would grip and undermine the speed with which the body could pivot.

Longevity of knees, relative ease of childbirth, the ability to balance on wide feet and well-spread toes, and the capacity to sprint with a force direct in a straight line from the hips to the ground: all this (and probably more) is sacrificed for a very specific body plan. The body plan is created, it appears, for one specific purpose: to allow the animal to balance on one leg and rotate on a small pivot.

This also suggests the manner of the changes. The first change in the animal's posture would be the leg, bent at the knee, towards the centre of the body. This would allow the body to be able to balance on one leg. A chimpanzee that could stand on one leg, even if it was only for a short time, could use its enormous strength to throw a projectile at great speed. The issues are balance, and its close relative, accuracy. In order to throw accurately, the primate must be able to maintain its balance. As long as the primate can maintain its balance, any shortcomings in body design, at this stage, can be made up for with brute strength.

Such a body plan change, however, is, even if easy to consider, difficult to imagine in reality. For instance, chimpanzees can move very quickly, with their legs and arms powering them along with tremendous speed. It is difficult to imagine why a chimpanzee or gorilla would evolve with legs with a kink directed towards the centre. While it might increase the capacity to throw with accuracy, at this stage, the animal is a quadruped and moves like a quadruped. If mutations that increased balance on one leg were so successful, then, we might ask, why do chimpanzees and gorillas continue with knuckle-walking? (Knuckle-walking is where chimpanzees and gorillas place their weight on their knuckles rather than their hands as they walk and run. This allows them to create a direct line of force from their shoulder to their knuckles, rather than placing all the force on their wrist by moving with their palm on the ground, as other primates do. This would seem feasible for very heavy animals that need to support a tremendous upper bodyweight allowing the weight to be placed directly on the ground via the knuckles, rather than being redirected through a joint – the wrist – that might cause serious wear and tear issues. Indeed the evolution of knuckle-walking reflects how animals evolve in seemingly unusual ways to reflect their circumstances. It again makes one wonder at humans with their legs bent inwards, with our entire weight being redirected through the knee joints, why there was no minimisation of knee wear and tear.)

One might expect forest-dwelling animals not to gain much of an advantage from the capacity to throw, even accurately. The presence of trees, branches, bushes and grasses increases the capacity for animals to engage in ambush attacks and to avoid projectiles. Throwing is only beneficial where there is an environment with a clear line of sight. The continued existence of chimpanzees, orangutans and gorillas as against many extinct hominids and the fact that they dwell in heavily forested areas may not be a coincidence. If projectiles were assumed to be the major factor in the hominid domination of its environments then these would be the only non-desolate environments where hominids would not be able to dominate. If projectiles are not deemed to be the major factor in hominid domination, then it is difficult to envision how chimpanzees, gorillas and orangutans have managed to continue to exist, considering all non-human hominids have become extinct.

In a more arid environment though, where the animal can gain a clear shot, throwing gives a clear advantage. In such an environment, a chimpanzee-like animal that was hampered in movement by a leg bent inwards might, by contrast, be enormously benefited by being able to balance on that leg and deliver a lethal projectile. Even if other chimpanzee-like animals were faster and more agile, they wouldn't get within five metres. Neither would lions or other predators.

The strange, deformed legs, bent inwards, would not be an impediment, they would be a benefit. This animal would dominate food sources, such as fruiting trees, merely by threatening to throw missiles at others in the fruiting tree. It would monopolise food sources, and be able to dominate its territory, leaving many descendants. Those descendants would quickly out-compete all those unable to balance on a bent leg and begin to compete amongst themselves. Evolution would favour those born with advantages to throwing. A foot unhampered by large toes and that could allow the body to easily pivot. A head that could remain still on top of the neck, rather than move with the shoulders due to it being slung forward for quadruped movement, allowing for greater accuracy and hand-eye coordination. Hips that acted as a platform for the upper body, rather than long and fitted into the piston-like legs as they are in chimpanzees. The body would become more balanced, with the legs lengthening and the arms shortening, minimising a tendency to over-balance because of a heavier upper body. The more accurate the animal was able to throw, the more likely it would be to hit its target, and defeat competitors, subsequently acquiring females and passing on genes for throwing proficiency. Such an evolution would occur at a much faster rate than any argument for natural selection, because those favoured in competition would defeat rivals and increase their capacity to pass on genes. Arguments concerning efficiency of energy rely on insufficient nutritional arguments that are incorrect. Furthermore, hamadryas baboons can exist in arid environments without issue.

The animal would have taken on a biped movement from the earliest stages, even if it were strange and somewhat uncoordinated in its gait, simply because, no matter how slow or awkward, it would be unbeatable as long as it could get a throw off before any predator or opponent could get close. Such an animal, that could throw with relative accuracy, and possessed three or four times the strength of a strong male human, would be potentially lethal to all opponents. It could roam non-forested areas without little fear of predators, and – as is the case for chimpanzees and gorillas – its only real danger would be its own species.

Bipedalism is, however, the lesser issue. The use of projectiles meant that numeracy was more important than body size and our species followed the path of ants rather than whales or gorillas. Even at the earliest stages our ancestors formed male groups: no solitary male was able to defend his females alone, no matter how large he was. He needed other males to defend themselves, the females and the infants; plus, he possessed an advantage over other lone males or smaller groups. By forming an alliance with another male, and swapping sisters, the males were not only able to create a group of males, they were also able to create a social environment into which their children could find a mate, without needing to leave the group. Thus the groups were self-sustaining. As more nutrition became available, rather than becoming physically larger, the males formed larger groups that allowed for larger numbers of males able to use projectiles. With the use of weapons that could pierce the body, the need for sheer strength was reduced and this allowed for a smaller, less-muscular body plan. As the level of competition increased with novel forms of obtaining food, so too the brain increased in size to be able to maintain and manipulate these groups to the individual's advantage (including genetic relatives). Females preferentially selected genes for intelligence and cooperative behaviour from amongst the males, in the same manner that they had previously preferentially selected the genes for the improved ability to use projectiles, in the hope of producing the ultimate reproductive achievement – the highly polygynous son.

I argue that a biped body structure was the first step towards human evolution, and all further steps were dependent upon this first step.

Concluding Remarks

The first major point to be taken from this chapter is that the arranged marriage system, using a dual organisation and cross-cousin marriages gives us a formal structure for human cooperation. It is possible to create a step-by-step evolution of humans using such a structure.

The second major point to be taken from this chapter is that, though the entire process of increasing group size must be underpinned by an increased availability of nutrition, this in itself does not lead to the evolution of complex cooperation. As I have argued, even if food quantities doubled, it is not logical that the group size would double; indeed, it is more logical that the size of the animals would double. If we doubled the food available for gorillas, the number of gorillas would not necessarily increase in the long-term, but the size of the animals would: two gorillas – particularly males – simply cannot compete with another that is much larger, perhaps twice their size, in hand-to-hand combat. The larger gorilla would not necessarily be any different in intelligence and behaviour than the current gorilla.

Human evolution, though underpinned by increased amounts of nutrition, is based upon two factors: firstly, that, in competition, numerical advantage rather than size is important; and, secondly, the tendency for larger groups to be more successful, with sexual selection within the group for more highly cooperative traits.

One final point that must be addressed is the issue of progress. Some may view this chapter as suggesting that the progress towards *Homo sapiens* was inevitable. This argument would be incorrect. Firstly, human evolution may be dependent upon one initial, major physical feature: the capacity to throw with force and accuracy, which is based upon the biped form. This highly unusual feature is paramount because of its tendency to favour numerical superiority over physical superiority, and thus our ancestors, when competing, formed larger groups rather than simply becoming physically larger, as is the case with most mammals. Human evolution may be dependent upon an unusual factor, which, as the above social organisational models suggest, required millions of years to evolve sufficient complexity for the specific traits of human-like intelligence and other behaviours to emerge.

Our ancestors may have been refugees forced from the fruit-rich environments of the forests. Forced from the forests where trees produce huge quantities of fruits for primates to gorge themselves, our ancestors appear to have been digging for tubers or cracking nuts with heavy rocks – time and energy consuming livelihoods compared to the chimpanzee's lifestyle. What we view as 'progress' and evidence of 'intelligence' can be equally viewed as energy-sapping methods to obtain nutrition in a difficult environment.

Even once the biped morphology took shape, the slow, awkward biped animal would have been no competitor for a chimpanzee-like animal, and projectiles would offer little benefit in thick forests. Once actually in the fruit-bearing trees, our ancestors would have been entirely deficient, with the chimpanzee's ancestors being perfectly adapted to the arboreal environment, and projectiles of little benefit once one is gripping a branch or trunk. Any assistance that may have been gained from a few colleagues remaining on the ground, attempting to ward off the arboreal competitors with rocks, may have been of little use in dense forest, with the rocks more likely to hit branches than their targets. Slow-maturing infants would have been particularly at risk to a chimpanzee's ancestor's incisors and attacks on infants alone may have been enough to ward off the biped competitors. Our ancestors may have existed in less nutritious environments for millions of years, chased from fruit-bearing trees by the stronger, more agile, faster chimpanzee-like animals, until slowly the importance of numerical superiority produced the increased levels of cooperation and intelligence that, in the past hundred thousand years, have made human beings the dominant animal that they are today. The unusual gait and mannerisms of our ancestors have evolved in unimaginable ways and yielded a remarkable animal. But anyone that viewed those early, hard-working hominids, existing in the less nutritious environments of Africa, may never have predicted their future dominance.

For the first few million years, our ancestors may have been little more than a slow-maturing baboon, whose young fell prey to chimpanzee-like animals, whenever they attempted to enter the richest of environments. If they had possessed sufficient intelligence to be aware of the fact, our chimpanzee-like cousins may well have viewed themselves as the kings and queens of the primate world.

Human evolution may very well have been a rags-to-riches transformation, as paradoxical as any evolutionary event.

Chapter 12

Conclusion

Overview

The argument in this work is that a human ancestor, not very different to a gorilla, entered an environment where using projectiles was more successful in competition than hand-to-hand combat; this instigated a rare situation where numerical superiority was more important than individual strength; that this animal began to use a cooperative system – the bilateral cross-cousin marriage system – in order to allow for greater numbers of cooperative apes; that nutritional increases (hunting, fire, etc.) led to increased competition and larger group sizes evolving; and that these larger group sizes required more complex forms of cooperation and more complex traits to maximise genetic success within those environments.

Each of these events is vital, but at the heart of the argument is that the cross-cousin marriage system explains the vast majority of our traits. If one could go back in time and study the ancestors of humans, then once the animal possessed a method of competition involving numerical superiority rather than sheer physical strength, then the cross-cousin marriage system would be a consistent factor for every animal. Fire or hunting may not exist, and various human traits such as high levels of intelligence or menopause may not exist, but the cross-cousin marriage system would exist in each animal after the attainment of a biped body structure. As the mechanism of cooperation for humans, it is at the very centre of increased cooperation over the millions of years of human evolution. Without it, there is no obvious reason for how our ancestors became more cooperative, more intelligent, or why they evolved such unique traits. It also fits the evidence of arranged marriages outside of highly industrialised environments.

A theory of human evolution needs to confirm six major requirements before it can be taken seriously. These consist of:

(1) uniqueness – the explanation should possess something vital that is unique to the animal compared to other animals;

(2) evidence – there should be evidence to support the argument;

(3) complexity – the argument should encompass a high level of complexity, befitting the animal in question;

(4) increasing complexity – the argument should explain why the animal needed to become increasingly complex;

(5) explanation of paradoxes – the paradoxes that currently confound should make sense with the argument; and,

(6) a good fit – the argument should be a good fit for the behaviour and morphology of the animal in question.

For instance, let's consider the argument that cooperative hunting played the major role in human evolution. This fails the uniqueness argument (wolves cooperatively hunt), the complexity and increasing complexity arguments (wolves are not as complex as humans), the explanation of paradoxes (wolves do not possess a brain capable of calculus and females do not have hidden ovulation), the good fit (there is no apparent reason why wolves should possess verbal communication), though it does succeed on the evidence test (humans do hunt). It would appear to fail in five of the six requirements.

In fact, cooperative hunting does not appear to be a good explanation for the traits of various cooperative hunters. Female orcas, for instance, enter a late stage of reproductive senescence (menopause), while female hyenas are dominant to males, neither of which is found in wolves. If the method of nutritional attainment is vitally important then these animals should be similar in their behaviours and morphologies, but they aren't. It is the fact that these animals cooperate that makes them different to lone hunters, and in order to understand their differing traits, it is necessary to understand how they effect this cooperation, to understand their social organisation.

In humans I have made a similar argument, that it is the manner in which humans cooperate that defines many of their unique behaviours and morphologies. Remove the complex cooperative mechanisms from humans, and we are left with either gorillas or orangutans, apes acting in small nuclear families or in complete isolation. Cooperation between members of the same species does not simply happen, it must evolve, in the same manner that legs and fins evolve.

Let's apply the six requirements to the main argument of this book: that humans evolved with the cross-cousin marriage system. This is a unique feature that makes humans different from all other animals. It enables complex societies to function using a method of exchanging long-term cooperation for a future reproductive benefit. There is no choice in long-term reproductive relationships. Such a cooperative or reproductive method is not found in any other animal. The concept of a long-term pair-bond between a male and female with no actual choice of the partner by either male or female is unique.

The evidence to support this argument is found throughout the anthropological material, summarised by Claude Lévi-Strauss in *The Elementary Structures of Kinship*. The very existence of anthropology is based on the study of the bilateral cross-cousin marriage system, beginning with the identification of classificatory terminology by Lewis Morgan in the 19th century. The desire to understand why people living in a natural environment labelled their mother's sister 'mother' rather than 'aunt', as an example, led to the formation of an entire discipline and over one hundred years of study of the cross-cousin marriage system as a method of cooperation and reproduction. Those who wish to contradict the evidence for the cross-cousin marriage system as the basic method of human cooperation and reproduction need to ask themselves: why does anthropology exist, why do arranged marriages exist, why did people in Western nations once use arranged marriages, why do remnants of this process still exist in Western culture (the father 'giving away' the bride) and why is this method of marriage so common amongst people living in humanity's natural state? The cross-cousin marriage system gives both a logical and evidence-based reason as to how humans cooperate without a market economy or central government.

As demonstrated in this book, the cross-cousin marriage system is complex, both to maintain and manipulate. It requires knowing hundreds if not thousands of living and dead relatives. It requires the individual to be able to imagine one's society ten or fifteen years into the future, in order to make a successful betrothal and marriage for one's daughters and sons. It requires the individual to be able to predict and prevent manipulations that may be planned years into the future, such as the relabelling of a woman from 'mother-in-law' to 'sister' so that her daughter might be marriageable for one's son many years into the future. The fact that this complexity is linked to reproduction further enhances the plausibility of this argument, since there are reproductive benefits from being able to engage in a system of such complexity.

It has also been demonstrated that the system is capable of developing increased levels of complexity from a simple cross-cousin marriage based society. It is possible to create such a society with two males and several females, and then to expand the complexity until there is more than twenty males and twenty females in each generation, utilising the same cooperative and reproductive system. I have argued that this depends upon the existence of throwing rather than hand-to-hand combat, as a pre-requisite for the evolution of the cross-cousin marriage system. Once numerical superiority is a requirement in combat, then an increasingly complex cooperative system will evolve as numbers increase.

Once the cross-cousin marriage system is used as the basis for human cooperation, paradoxes can be explained in a logical manner.

Why does marriage exist and why do married couples fight so much? Why do women hide their ovulation from their long-term partner? Why are all human sexual signals designed to operate at a distance, whether it is blushing, naked skin or eye-contact? Why is sexual emotion (love) so intense and short-term? Males and females evolved without selecting their long-term partner in a formal marriage system, and possess behaviours and morphologies to control the partner in the marriage. Females hide their timing of ovulation to increase the time taken for fertilisation by the non-selected partner. Selection outside of the arranged marriage is dangerous since the husband-by-arrangment uses violence to prevent selections. Humans possess numerous behaviours and morphologies to attract the opposite sex from a distance in a discreet manner. The positive emotions involved in selection are high due to the potential for a strong negative response from the husband-by-arrangement and his supporters. These emotions are short-term since they will rarely lead to a long-term relationship: the marriages are created through alliances between kin groups, not personal choice.

How do people cooperate with genetically unrelated individuals without a market economy? Why do we need such large brains? Kin groups exchange daughters in long-term contracts: an exchange of reproduction for cooperation. The cross-cousin marriage system yields cooperation in a complex manner, requiring a high level of intelligence to maximise success within this environment. Contracts for marriage and cooperation are made ten or more years into the future. This system is manipulated by knowing the ancestry of each individual and altering the classificatory terminology in order to maximise the number of wives for their close male kin.

Why would females go through menopause? To guard their sons' wives-by-arrangment: each act of cuckoldry is the equivalent of the death of an infant for her sons.

Since her sons' wives-by-arrangment did not select her sons, the risk of extra-pair selection is increased.

Why do humans have complex verbal language? Why do girls and women spend years creating highly emotional friendship networks? Why are their friendships based on the ability to withhold secretive information? Why are females superior users of verbal language than males? Females, if they decide to select a male outside of the arranged marriage, cannot communicate with the male directly, and must do so without others becoming aware of the sexual selection. They use complex verbal language, communicating with the desired male through a trusted female intermediary. The relationship with the female is based on the ability to keep secretive information hidden. Female sexual selection necessitates the existence of complex, verbal language and strong alliances between females.

For each paradox an original answer can be deduced from a logical consideration of the implications of the cross-cousin marriage system. An issue such as the evolution of human language is seemingly an insurmountable challenge to any theory, because it seems so implausible. Yet secrecy linked to reproduction within a complex social environment, allows for a plausible hypothesis to be suggested. The hypothesis also reflects female behaviour in adolescents and adults.

Finally, is the cross-cousin marriage system, as the basis for human evolution, a 'good fit'? With biped body structure enabling the ability to throw with lethal force and directing an evolutionary path of numerical superiority rather than individual physical superiority, the cross-cousin marriage system emerges as a stable method of creating cooperation between large numbers of individuals. It is a unique method of cooperation and reproduction creating a unique, unusual species. The cross-cousin marriage system creates biologically unique situations such as the male and female in a long-term relationship despite neither selecting the other, and all sexual selection being outside of the long-term relationship. It creates a complex cooperative system that requires intelligence to manipulate. While people can argue about the logic or evidence of the conclusions I have made, they cannot argue that there is a complexity to this system, a complexity that is not inherent in other arguments, and that it is plausibly a 'good fit'.

Consider instead that humans evolved with Western-style societies with choice of the reproductive partner and tit-for-tat cooperation. This argument fails the uniqueness test: other animals use 'tit-for-tat' cooperation (vampire bats) and choose their long-term partner (gorillas or gibbons), yet none have attained human intelligence. The argument fails the evidence test: there is little evidence that humans in their natural state select their long-term partner. The argument fails the complexity test and increased complexity test: there is no reason why such an animal needs to be exceedingly complex or to become increasingly complex; neither gorillas nor gibbons have become as intelligent as humans, despite using Western-style reproductive systems to varying degrees. It fails the explanation of paradoxes test: all the major questions concerning human behaviour and morphology remain paradoxes; for instance, there are at least five explanations as to why females hide their ovulation from their long-term partner when one assumes she selected the male.[1] The argument fails the good fit test: when one attempts to use assumptions for the social behaviour of a gibbon (selected monogamy) or gorillas (selected polygyny), none of the major questions concerning human evolution are easily answered. Why would these gibbon-like animals talk? Why would they cooperate with genetically diverse individuals? Nobody has answered these questions while assuming a Western-style social environment for human evolution to the level of a 'good fit'.

For people in Western societies, the cross-cousin marriage system is a new way of thinking and behaving, which is why anthropologists have studied it for an extended period of time. Humans in their natural environment do not simply find food, fight predators, enjoy culture, and live in small groups. The natural human environment is a complex environment, which appears to reflect human complexity despite the lack of technology.

The manner of human evolution is mostly due to sexual selection, the competition between members of the same species, activated by higher levels of nutrition and group size. It occurs in the following manner:

Evolution of biped body structure → necessity of multiple male groups rather than single male groups, despite ground-dwelling behaviour → occurrence of cooperation via cross-cousin marriage to create a stable group → more nutrition (due to fire etc.) → more individuals → more similarly sized groups → greater competition → formation of larger groups → more complex social groups → greater difficulty in manipulating the larger, more complex social group → selection by females of males able to manipulate the social group → success of males more able to manipulate the social group → evolution.

The evolution of major hominid traits occurs after increases in nutrition and not before. The reason why increased nutrition does not lead to increased social complexity can be observed in whales, giraffes, elephants, and dinosaurs: most species compete in a direct contact style, for which greater physical strength and size is an advantage. I argue that humans replaced contact-based competition with throwing at a distance and numerical superiority, which requires higher levels of cooperation. That our ancestors appeared to have lost their canine teeth – a formidable weapon for other apes – at a very early stage gives some evidence that hand-to-hand combat was less relevant at an early stage in human evolution, and also in defence against predators.

I have argued that genes for cooperation will be selected in order to give a simplistic reasoned argument, out of necessity. Is our higher level of cooperation based solely on a larger brain, or is it affected by hormones or other factors? Currently, these issues are too complex to consider. It should be noted that my argument is not that humans have a gene for arranged marriages, since, once personal freedoms were enabled with a central government and relative personal security, Western-style marriages were preferred. This reflects the plasticity of human traits, but the underlying behaviours remain the same. Arranged marriage systems and Western-style marriages with jobs and taxes both reflect the capacity to be part of a larger cooperative system.

This plasticity of behaviour is reflected in the ability of humans to think through their behaviours. While a male elk might attack a statue of an elk or even a clothesline, or a bird might attack its own reflection, humans tend not to respond in such a simplistic manner, though they have the underlying framework. A man may be aroused by a woman or intimidated by another male, but he possesses a high level of self-restraint, whether it be in sexual interest towards the female or aggression to other males. Yet, this capacity to think through one's behaviour is also genetically-based. The fact that people with the genetic disorder William's Syndrome have high verbal ability, an increased desire to socialise and connect with people, and yet appear to lack the capacity to understand the complexity of human interaction[2], gives some indication that genetic mutations can affect specific aspects of cooperative behaviours but not others.

If Western civilisation collapsed would people resort to the cross-cousin marriage system? If there were no police force, then people would return to living with their most trusted individuals: their close kin. If competition with other people were intense, they would then need to devise a method of cooperating with other kinship groups to increase their numbers for protection. While some might utilise short-term methods at first (such as sharing of nutritional resources), it would not be long before a stronger foundation was needed, and the cross-cousin marriage system gives a genetic benefit in exchange for cooperation. If society returned to a situation where there was no way to store wealth safely, then the only wealth the people would possess for exchange would be themselves. Humans would return to their social origins. It would not be long before men were promising their daughters to other men in exchange for cooperation and protection.

Regarding the Evidence Used

In presenting my arguments I have used both data and anecdotal evidence. One of the tremendous difficulties in understanding human behaviour in the natural human environment is that only a small amount of numerical data has been collected by anthropologists. In part this is due to the scarcity of long-term studies; that many of these studies were performed before large research grants existed; and that the anthropologist may need a good overview of the people before he or she will be able to determine which areas to focus on for data-based research, which could take a year in the field, a period of time longer than most anthropological field studies.

This lack of data and the extensive use of anecdotes, in my opinion, is unavoidable if one is to use the anthropological field research. Consider Jane Goodall's experience from her three decades with the chimpanzees of Gombe, particularly the difference between field and laboratory research:

> People sometimes ask why chimpanzees have evolved such complex intellectual powers when their lives in the wild are so simple. The answer is, of course, that their lives in the wild are not so simple! They use – and need – all their mental skills during normal day-to-day life in their complex society. They are always having to make choices – where to go, or with whom to travel. They need highly developed social skills – particularly those males who are ambitious to attain high positions in the dominance hierarchy.[3]

It is easier to study intellectual prowess in the lab where, through carefully devised tests and judicious use of rewards, the chimpanzees can be encouraged to exert themselves, to stretch their minds to the limit. It is more meaningful to study the subject in the wild, but much harder. It is more meaningful because we can better understand the environmental pressures that led to the evolution of intellectual skills in chimpanzee societies. It is harder because, in the wild, almost all behaviours are confounded by countless variables; years of observing, recording and analysing take the place of contrived testing; sample size can often be counted on the fingers of one hand; the only experiments are nature's own, and only time – eventually – may replicate them.

In the wild a single observation may prove of utmost significance, providing a clue to some hitherto puzzling aspect of behaviour, a key to the understanding of, for example, a changed relationship.[4]

These comments are equally applicable to human beings. If it weren't for Jane Goodall and other researchers, we might well be wondering why on earth chimpanzees need a large brain when all they do is find food and reproduce. Even today, many people seem more impressed by the fact that chimpanzees use twigs to eat termites and thus use tools, than the complex, competitive struggles between individuals within the same community or the lethal competition between communities. Many people make the same arguments for human beings, that all human beings do in the 'wild' is to run away from predators, find food, and use tools, and make the statements with no regard to fieldwork.

Unfortunately, the complexity of obtaining information in the 'wild', particularly for humans, appears to be poorly understood. I heard, for instance, one researcher demanding 'evidence' in response to arguments by a scientist that Australian Aborigines were, historically, well fed, and that their modern problems of diabetes and obesity were probably due to a bad diet rather than genes for a 'feast and famine' existence.[5] The researcher was talking about data evidence based on the testing of the null hypothesis, while the scientist was presenting only basic data or anecdotal evidence from the field. However, the data evidence that the researcher then claimed was the clincher were studies of gerbils found in Israeli deserts. Gerbils are nocturnal and have specific adaptations to survive desert life, including specialised kidneys that allow for days without water – somewhat unlike a 60kg human being, whose brain is ten times heavier than the gerbil. Apparently the research on a gerbil was more important than evidence collected on cooperatively-simple people themselves over the last hundred years in anthropology, which, because anthropologists rarely use the null hypothesis test, can apparently simply be dismissed out of hand. Jane Goodall relates a similar example from her own experience:

I shall never forget the response of a group of ethologists to some remarks I made at an erudite seminar. I described how Figan, as an adolescent, had learned to stay behind in camp after senior males had left, so that we could give him a few bananas for himself. On the first occasion he had, upon seeing the fruits, uttered loud, delighted food calls: whereupon a couple of the older males had charged back, chased after Figan, and taken his bananas. And then, coming to the point of the story, I explained how, on the next occasion, Figan had actually suppressed his calls. We could hear little sounds, in his throat, but so quiet that none of the others could have heard them. Other young chimps, to whom we tried to smuggle fruit without the knowledge of their elders, never learned such self-control. With shrieks of glee they would fall to, only to be robbed of their booty when the big male charged back. I had expected my audience to be as fascinated and impressed as I was. I had hoped for an exchange of views about the chimpanzee's undoubted intelligence. Instead there was a chill silence after which the chairman hastily changed the subject. Needless to say, after being thus snubbed, I was very reluctant to contribute any comments at any scientific gathering, for a very long time. Looking back, I suspect that everyone was interested, but it was, of course, not permissible to present a mere 'anecdote' as evidence for anything.[6]

It would appear that those researchers who use the laboratory believe that their work is superior, yet, compared to fieldwork, the work done in laboratories is highly artificial. How can behavioural studies of a chimpanzee in a cage, no matter how impressive the amount of data, be compared to studies of chimpanzees in their natural habitat?

The differences are amplified by human behaviour. How, for instance, could one study extra-marital sexual selections by human beings in their natural environment, when the sexual selections are intended to be secret? How does one study something that is meant to occur unobserved, and which, if successful, is known by no-one except the participants, even amongst other people in the society, particularly when one considers the ethical issues of paternity testing? Other problems are that most cooperatively-simple peoples are significantly affected by external cultural influences before they are studied. As Napoleon Chagnon noted in one interview:

> It became very clear to me after years of university-training, reading lots and lots of monographs about tribal peoples, that I had stumbled accidentally upon an extraordinarily unusual and short-lived opportunity. Because very few people were as remote and isolated as the Yanomamö were. And I realised from knowing how quickly acculturation can happen, that if I did not decide for an intense and long term commitment to learning about these people while they were still the way they were, that valuable opportunities to learn many important things about them would disappear.[7]

> One of the reasons that I felt it was urgent to study the Yanomamö was that I was one of the few anthropologists who had an opportunity to study a tribal society while warfare was still going on, and not being interdicted by the political state. Even though anthropology has a lot of literature about warfare and violence, the number of anthropologists who studied tribesmen while still at war you can count on the fingers of one hand.[8]

I found only one other study that could begin to compare to Chagnon's and that was Lloyd Warner's study of the Murngin in the 1930s. The paucity of such studies reflects the difficulties of not merely finding untouched peoples, but studying them. As one might expect, studying human beings that are at war is difficult for an isolated anthropologist whom might be considered as little more than an unimportant freak by the people being studied. If Western warfare was the issue in question, people would laugh at the idea that gerbil studies or studies of university students were superior to the data and anecdotal evidence gathered from the front. Yet, when it comes to indigenous peoples, this does not appear to be the case.

Scientists often take a dim view of anthropology, possibly due to anthropology rarely taking a data-based approach and its politics. However, if anthropologists had taken a reductionist approach, based on pre-conceived notions, then all the evidence we might possess is a large number of data-based studies concerning nutrition, tool use, and predator avoidance. The fact that the people engage in an entirely different social system would have been ignored in order to obtain data based on Western views of how these people live, rather than taking the naive view, and asking 'how do these people live?' before attempting a reductionist approach. There is little benefit in taking a reductionist approach if it is based on pre-conceived ideas, and the subject is poorly understood.

In this book I have placed the work of those in the field as being the best available evidence, even if it is anecdotal. By contrast, evidence from other human social environments, even if it is data-based, is treated as secondary and supportive. I have done this after recognising the tremendous differences between Western societies and cooperatively-simple societies, with selected versus arranged marriages and the form of cooperation – taxes and central government versus arranged marriages – being examples. While it is easy to be impressed by fancy graphs and findings of statistical significance, it is rather unfortunate if the data concerns an entirely different social environment or animal. Geneticists, for instance, might claim that they have found the genes for people to fall in love and marry, despite the fact that there is little evidence that people actually do this in non-Western environments. The controlled laboratory experiment can be excellent for understanding 'what', but without context it can be the equivalent of a plant in a pot instead of a forest when it comes to 'why'.

What has been attempted in this book is to utilise what data is available and to use the studies of the two best studied peoples, Australian Aborigines and the Yąnomamö. I have made various arguments, supported or elaborated by further data and anecdote. I am not using the anecdotes to prove my arguments, but often merely to give examples from the field of what I am arguing. Exact proof would require further fieldwork to test these hypotheses; however, one can expect that very little anthropological or scientific fieldwork of unacculturated peoples will be likely to occur. In my opinion, the information available, no matter how limited, cannot be usurped by studies of university students or other animals.

The Expected Response to My Work

Evolution is one of the most political branches in science, and there are numerous participants with personal preferences or ideological viewpoints, some of whom might be described as activists. It is unlikely that, no matter how much evidence is produced, there will ever be a single theory that everyone involved in evolution will ever agree upon. Some people are more interested in what human beings could be, rather than what they are. The most obvious issue of contention is violence. It is possible to read numerous textbooks concerning human evolution that do not possess a single reference to warfare.

Many people who work in the area of human evolution appear to deny the possibility of violence between our ancestors being a major evolutionary selection pressure. By contrast, Darwin argued that humans have competed vigorously over reproduction, as embodied in his theory of sexual selection. Those academics who wish to argue that there was no human violence until civilisation are closer to Marxism than to Darwin's ideas.

An example of the opposition to violence is the book *African Exodus*, in which Chris Stringer and Robin McKie consider the debate over the discovery of the front half of a skull and other fossils, at a quarry at Taung, which made their way to Raymond Dart, the professor of anatomy at Witwatersrand University, Johannesburg:

> He gave it the name *Australopithecus africanus* (Southern Ape of Africa), the first time the *Australopithecus* nomenclature was used. Although Dart recognized that adult features had still to develop on the Taung fossil... he was not restrained in claiming the species was a predecessor of modern humans, as intelligent, and made tools. Dart's affirmations were ignored by the British scientific establishment. African australopithecines were only apelike remnants, left behind as human evolution unfolded in Europe and Asia, it was said.
>
> We now know that Dart was right in some ways. *Africanus* has been shown to be much more ancient than the human fossils of Europe and Asia and could perhaps have been an ancestor of ours. However some of his other assertions went far beyond his meagre evidence. He concluded the species was made up of confirmed killers, "carnivorous creatures that seized living quarries by violence, battered them to death, tore apart their broken bodies, dismembered them limb from limb, slaking their ravenous thirst with the hot blood of victims, and greedily devouring living writhing flesh," he wrote in an essay on "The predatory transition from ape to man." This astonishing, almost pornographic outpouring was based on Dart's interpretation of the damaged skulls and bones of *africanus*, and other animal species, found at Taung, and later at Makapansgat and Sterkfontein. He argued that the injuries to both animal and hominid were those inflicted by crude weapons of bone and stone, produced as these ancient killers indulged in orgies of slaughter, and sometimes cannibalism.
>
> Dart's hypothesis was seized upon by Robert Ardrey, an American play-wright, who transformed his ideas into a series of sensational bestsellers, starting with *African Genesis* ... which promoted the same notion: that mankind's origins were bloody and violent. Far from evolving big brains and then tools, "the weapon fathered the man," Ardrey claimed. It was our development of stone axes and spears that triggered our evolution to our present status, a process that has been driven by the engines of war.[9]

It was a convenient idea, for it suggests that war is in our genes, and is beyond the control of even the most reasonable men and women. We should feel no guilt or responsibility for our bloody actions, the theory implies. Killing is instinctive and natural.[10]

Yet this whole extraordinary edifice was based on flimsy, and as we now know, misinterpreted evidence. Poor, maligned *africanus* probably did not use tools at all, never mind weapons, and far from being a hunter, was in fact the hunted. Those jumbles of skulls and bones found at Makapansgat, Sterkfontein, and elsewhere had been left by leopards and other predators who had brought their prey, including *africanus* males, females, and children, to their secluded lairs to eat them without being disturbed. The Taung child, for example, is now thought to have been the victim of an eagle which carried off its severed head to a nest. As for the damage marks on the other skulls and bones, these were caused not by human weapons but by predators' teeth, or by the sustained pounding they received as other bits of cadaver or rocks dropped on them. There is no justification for believing that humanity is innately depraved. We extrapolate about our nature from limited data at our peril.[11]

If intra-species violence played a role in human evolution, does this mean that 'humanity is innately depraved'? If so, would we then suggest that spiders that engage in cannibalism are immoral? Since chimpanzees have been shown to engage in lethal violence, are they an immoral animal? Ants are highly combative, does this mean they are innately depraved? What then of the morality of a female insect, that eats the male as he copulates with her? Can single-celled organisms be immoral and depraved?

The marks on fossils can be questioned and their cause debated. For instance, forty percent of the fractures found on Neanderthal fossilised bones are skull fractures and a more recent study confirmed a similar percentage for human fossils from the same period.[12] Are we really to believe that these were hunting injuries? What, however, cannot be debated is evidence of what a human being actually is. If there were few wars in the written history, and little evidence of violence in the anthropological texts, then it would be possible to infer a peaceful evolution, but this would be a very different animal to the one whose evolution we are attempting to explain. We are attempting to explain the evolution of an animal that has created thousands of nuclear weapons (as well as universal education and health care).

If the ancestors of humans engaged in lethal violence, then this is a fact, not a moral conclusion. There is no innate morality in the process of evolution nor in any animal. An animal is no more moral than a rock: evolution is no more moral than gravity. In a social group of a species, however, animals may form a negative or positive view of another animal's behaviour, which is the defining concept of morality.

An example of how facts can be interpreted is found in Karl Heider's study of the Dani of Papua, *Grand Valley Dani: Peaceful Warriors*. Note the oxymoron in the sub-heading. Heider found war deaths were numerous:

Another way to figure war deaths is to look at genealogies. Here I found that 100 of 350 (or 28.5 percent) of deceased males and five of 201 deceased females (2.4 percent) were killed in warfare.[13]

Yet, in interpreting this, he concludes:

A distinction must be made between violence and aggression, for it will help us to sort out Dani behavior. ...I would like to suggest here that violence is the use of force against

another, while aggression, although usually violent, implies an inner state, and intention
to overcome or dominate an enemy. Using the terms in this way, we can say that killing a
pig in a ceremony is certainly an act of violence, but it is not aggression. So although the
Dani are violent in warfare, even in war they are not often aggressive in any useful sense
of that word. ... The data presented in this book, however, show the violence of warfare
in the total context of Dani life, a context which turns out to be quite nonaggressive.[14]

Thus we get the subtitle of the book: even though more than twenty-five percent of males
died in warfare, these people are non-aggressive and peaceful (warriors), because what is
important is not the actual percentage of deaths, but their inner state, which is peaceful.
These conclusions come from one of the handful of anthropologists who actually recorded
the numbers killed in warfare, one of only a few such statistical records. Yet, so strong is
the demand that all people are innately peaceful, that even he uses semantics to argue the
seemingly unarguable.

It is easy to take up the moral agenda and to view oneself as moral and superior.
Creationists, for instance, feel that their agenda is entirely moral and use moralistic arguments
to attack anything – including evolution – that might undermine their moral agenda. What
is the difference between arguing that a violent past or an ape-like beginning is evidence of
innate depravity? Both are opinions.

It is not the lack of morality that we should fear, but the moral agenda. It is moral
reasoning that humans use to absolve themselves of guilt, that those they are killing are
inferior or subhuman. Whether it is racial purity; the wrongs of the past; or differences in
class status, racial status, sexuality, religion, or political manifestos; humans create their own
moral agenda to support their actions. The moral agenda is used to unify and concentrate
group behaviour. The moral agenda needs no evidence either; the moral superiority of those
that argue the agenda is evidence enough of its rightness. This is the difference between a
moral agenda and an evidence-based agenda.

We do not enjoy our current freedom because of papers published in academic
journals, but rather due to those who fought and died for our current freedoms. As violence
has annihilated and enslaved, it has also liberated. Those that equate violence with depravity
undermine the efforts of the millions that have died fighting for our current freedom and
prosperity so they might live in an ethereal dream. A society that automatically allows all its
members to live peacefully has never existed, it is something that has been fought for and
obtained.

Does an amoral evolution suggest that there is no morality or meaning to life? The
current way of life – democracy, modern medicine, modern engineering, and so on – are
all the product of people questioning their circumstances or the limits of their existence.
Our current way of life is the reflection of millions of people who worked, argued, fought
and died in order to question their circumstances. The capacity of the individual to question
their circumstances, to develop their own moral position and purpose in life, is possibly the
greatest freedom of all. It is vastly superior to the capacity to simply accept what is written in a
book and behave like an automaton or that one's moral position should not have to be fought
for, but should come ready-made in an ideal world and accepted unquestioned.

By the end of 1941, Ernst Chain had isolated, purified and proved the curative effect of penicillin, with Howard Florey and Norman Heatley, after it was first discovered by Alexander Fleming. Chain was also of Jewish ancestry and had been forced to leave Germany after the Nazi party banned Jews from holding positions in universities in 1933. In 1970, he commented:

> Science itself has no ethical quality in so far as it throws light on the laws of nature. Responsibility for the use or abuse of its achievements must rest with society.[15]

This lack of ethical quality is science's greatest asset. There is no moral or ethical agenda, there is no political agenda, no religious agenda, and no prejudice. Science remains a cold, objective argument that cuts through others' moral agendas. It is one rare moment when we may know the truth. It is an asset that should never be lost.

Violence will not be the only issue in this book. The arguments regarding the selective mechanisms for language and menopause, for instance, may also be viewed poorly. This book is intended to explain human behaviour and morphology, not to moralise human behaviour. The work is intended to explain human cooperation, whether that is used to create hospitals or warfare.

A Cross-disciplinary Approach

This book brings together various disciplines involved in the study of human beings into a coherent, single picture. Specialisation can be a problem in the study of human beings, which has been broken up into many fields: archaeology, law, history, psychology, medicine, anthropology, paleoanthropology, and so on. The educational system only gives people enough information to know part of the entire human picture; nobody is taught or given an introduction to the entire picture.

Edward Wilson has already addressed this issue in detail using William Whewell's concept of consilience: 'as literally a "jumping together" across the scientific disciplines',[16] and gap analysis: 'the systematic attempt to identify domains of phenomena in which important discoveries are most likely to be made'.[17]

> The social sciences, I believe, will advance more rapidly if they adopt a consilient worldview and the gap analysis suggested by it leading to reductionist analysis. They have failed to give this approach a try, except in a few sectors such as biological anthropology, largely because of their aversion to biology. The reasons for the aversion are complex, stemming partly from the effort of the social science disciplines – anthropology, economics, political science, and sociology – to maintain intellectual independence, partly from the daunting complexity of the subject, and partly from fear of the misuse of biology to support racist ideology.[18]

Even where there has been a joining of two disciplines, this may not be enough. Evolutionary psychology has emerged as an area to study human behaviour from an evolutionary viewpoint, yet it has failed to take into account the full range of human behaviour. While it is interesting to study how men and women at university want to select a long-term partner on a 1 to 7 scale, in terms of understanding innate behaviour, it is unfortunate if we then discover that in cooperatively-simple societies marriages are dictated by arrangement rather than selection, the people are illiterate, and they do not have a counting system beyond two.[19]

If architecture was divided into ten disciplines and each architect's status was reflected in how well they knew ten percent of architecture, regardless of their knowledge of the other ninety percent, then we could have a situation where no architect could design a building: this is how the study of humanity currently stands. People who wish to understand human evolution simply have no choice but to understand the cross-cousin marriage system and the manner of life of humans in their natural state. Currently, arranged marriages rarely rate a mention in almost all summaries and studies concerning human behaviour.

The double helix is viewed as central to understanding genetics. Without the double helix, DNA would simply be a chaotic, unstructured group of molecules. I argue that the cross-cousin marriage system is equivalent to the double helix. Without the cross-cousin marriage system, explanations of our cooperativeness and unusual intelligence, behaviours and morphologies, will simply be based on unrelated facts and observations.

Fortunately, due to the existence of anthropology, we do not have to wait years for the cross-cousin marriage system to be studied. Many studies have been completed and the observation that arranged marriages may well be the social structure on which human cooperation is based can be made. It might be argued that Claude Lévi-Strauss's *Elementary Structures on Kinship* is anthropology's equivalent to the Crick and Watson paper.[20] Lévi-Strauss's text precedes the double helix paper by a few years and to this day it has been ignored outside of anthropology.

Concluding Remarks

This book is the reflection of a personal desire to better understand human beings. The argument is based on explaining how an animal could evolve with highly developed cooperative abilities and a remarkable level of intelligence to facilitate this cooperation. This is a book about cooperation, how it evolved, and its implications for behaviour and morphology.

Charles Darwin first argued that human beings evolved primarily through sexual selection in *The Descent of Man, and Selection in Relation to Sex*. His arguments have generally been ignored by academics that attempt to explain human evolution. They almost always prefer to argue that it is survival rather than selection that directed human evolution, despite the fact that Darwin possessed the opposite belief in 1871. This work may give a firmer basis for Darwin's argument that sexual selection is the evolutionary force behind human evolution and extend it into a working model.

Endnotes

Chapter 1 Introduction

[1] Chagnon, N.A. (1992a). Yąnomamö (4th edn.). Forth Worth: Harcourt Brace Jovanovich, p. 1.

[2] Radcliffe-Brown, A.R., (1975 [1950]). Introduction to A.R. Radcliffe-Brown and D. Forde, (eds), *African Systems of Kinship and Marriage*. London: Oxford University Press, p. 1.

[3] Mackay, J. (2000) *The Penguin Atlas of Sexual Behaviour*. New York: Penguin Reference. (Including eighty percent in China at the time.)

[4] Connolly, B. (2005). *Making 'Black Harvest': Warfare, Film-making and Living Dangerously in the Highlands of Papua New Guinea*. Sydney: ABC Books for the Australian Broadcasting Corporation, p. 256.

[5] For popular overviews: Diamond *Why is Sex Fun?*; Buss *The Dangerous Passion* and *The Evolution of Desire*; Ridley *The Red Queen*; Miller *The Mating Mind*; Pinker *The Blank Slate*; Stringer and Andrews *The Complete World of Human Evolution*; Johanson and Edgar *From Lucy to Language*; Sawyer and Dean *The Last Human*. See also more academic texts: Daly and Wilson *Sex, Evolution and Behavior*; Symons *The Evolution of Sexuality*; and Darwin's major texts.

[6] Stiles, D. (2001). Hunter-gatherer studies: the importance of context. *African Study Monographs*, Suppl.26: 41-65, p. 55-56.

[7] Stiles, D. (2001). Fn. 7, p. 55.

[8] Hill, K. and Kaplan, H. (1988). Tradeoffs in male and female reproductive strategies among the Aché. Parts 1 and 2. In L. Betzig, M. Borgerhoff Mulder and P. Turke (eds), *Human Reproductive Behavior: A Darwinian Perspective*. Cambridge: Cambridge University Press. Hill, K. and Hurtado, M. (1996). *Aché Life History: the Ecology and Demography of a Foraging People*. New York: Aldine de Gruyter.

[9] Chagnon, N.A. (1974). *Studying the Yąnomamö*. New York: Hold, Rinehart and Winston, page 92.

Chapter 2 Competitive Cooperation

[1] Darwin, C. (2004 [1871]). *The Descent of Man, and Selection in Relation to Sex*. London: Penguin, p. 684.

[2] Thieme, H. (1997). Lower palaeolithic hunting spears from Germany. *Nature*, **385**, 807-810.

[3] Dennell, R. (1997). The world's oldest spears. *Nature*, **385**, p. 767-768.

[4] See, for instance, Jeffreys, M.D.W. (1965). The Hand Bolt. *Man*, **65**, 153-154. Darlington, P.J. Jr. (1979). Group Selection, altruism, reinforcement, and throwing in human evolution. *Proceedings of the National Academy of Sciences of the United States of America*. **72, No. 9**, 3748-3752. O'Brien, E. M. (1981). The Projectile capabilities of an Acheulian handaxe from Olorgesailie. *Current Anthropology*, **22: 1**, 76-79. Calvin, W. H. (1983). A stone's throw and its launch window: timing precision and its implications for language and hominid brains. *Journal of Theoretical Biology*. **104**, 121-35. For an overview of the importance of projectiles, see Crosby, A.W. (2002) *Throwing Fire: Projectile Technology through History*. Cambridge: Cambridge University Press. It should be noted that the use of a handaxe as a projectile weapon has been found to be unlikely after tests; however, it is doubtful that humans have the strength of a chimpanzee-like animal, which early hominids probably would have possessed.

[5] Chagnon, N.A. (1992a). *Yąnomamö* (4th edn.). Forth Worth: Harcourt Brace Jovanovich, p. 106.

[6] Chagnon, N.A. (1992a). Fn. 5, p. 106.

[7] Chagnon, N.A. (1992a). Fn. 5, p. 105. This has some interesting implications that reflect how little difference exists between people in cooperatively-modern and cooperatively-simple environments:

> Be that as it may, the Yąnomamö pit Culture against Nature, as we often do, and see the cosmos in many contexts as an 'either/or' moeity. Thus, an animal, captured in the wild, is 'of the forest', but once brought into the village, it is 'of the village' and somehow different, for it is then part of Culture. For this reason, they do not eat their otherwise edible pets – such as monkeys, birds, and rodents – for to them, it is similar to cannibalism: eating something 'cultural' and therefore 'humanlike'. Nothing disgusted the Yąnomamö more than my matter-of-fact comments that we ate our domestic animals, such as cattle and sheep, and many a missionary gave up in frustration after having attempted to introduce chickens at mission posts. (Page 105.)

[8] Chagnon, N.A. (1992a). Fn. 5, p. 106.

[9] Kummer, H. (1995). *In Quest of the Sacred Baboon: A Scientist's Journey*. Translated by M. Ann Biederman-Thorson. Princeton, N.J.: Princeton University Press.

[10] Chagnon, N.A. (1988a). Life histories, blood revenge, and warfare in a tribal population. *Science*, **239**, 985-992, p. 987.

[11] Chagnon, N.A. (1992a). Fn. 5, p. 205.

[12] Sinclair, D. and Dangerfield, P. (1998). *Human Growth After Birth* (6th edn.). Oxford: Oxford Medical Publications, p. 156.

[13] Sinclair, D. and Dangerfield, P. (1998). Fn. 12, p. 156-7.

[14] Sinclair, D. and Dangerfield, P. (1998). Fn. 12, p. 157.

[15] Sahlins, M. (1972). *Stone-age Economics*. Chicago: Aldine-Atherton, p. 5.

[16] Sahlins, M. (1972). Fn. 15, p. 6-7.

[17] Sahlins, M. (1972). Fn. 15, p. 7-8.

[18] Sahlins, M. (1972). Fn. 15, p. 14.

[19] Sahlins, M. (1972). Fn. 15, p. 23. Quoting Richard, L. (1968). What Hunters Do for a Living, or, How to Make Out on Scarce Resources. In *Man the Hunter*. (eds R. Lee and I. DeVore) Chicago: Aldine. See also Lee, R. (1969). Kung Bushmen Subsistence: An Input-Output Analysis. In A. Vayda (ed.), *Environment and Cultural Behaviour*. New York: Natural History Press.

[20] See Chagnon, N.A. (1992a). Fn. 5, p. 91-96 for a longer explanation.

[21] Chagnon, N.A. (1992a). Fn. 5, p. 96.

[22] Spencer, B. (1914). *Native Tribes of the Northern Territory of Australia*. London: Macmillan and Co., p. 9-10.

[23] Spencer, B. and Gillen, F.J. (1968 [1899]). *The Native Tribes of Central Australia*. New York: Dover Publications, p. 37.

[24] Eyre, E. (1964 [1845]). *Journals of expeditions of discovery into Central Australia, and overland from Adelaide to King George's Sound, in the years 1840-1*, volume 2. Adelaide: Libraries Board of South Australia, p. 206-207.

[25] Eyre, E. (1964 [1845]). Fn. 24, p. 250-1.

[26] Eyre, E. (1964 [1845]). Fn. 24, p. 276.

[27] Eyre, E. (1964 [1845]). Fn. 24, p. 276-278.

[28] Eyre, E. (1964 [1845]). Fn. 24, p. 278-279.

[29] Eyre, E. (1964 [1845]). Fn. 24, p. 253.

[30] Eyre, E. (1964 [1845]). Fn. 24, p. 254-255.

[31] Neel, J.V. (1977). Health and disease in unacculturated Amerindian populations. *Ciba Foundation Symposium*, **49**, 155-68.

[32] Neel, J.V. (1977). Fn. 31, p. 155.

[33] Neel, J.V. (1977). Fn. 31, p. 155

[34] Neel, J.V. (1977). Fn. 31, p. 156.

[35] Neel, J.V. (1977). Fn. 31, p. 157.

[36] Neel, J.V. (1977). Fn. 31, p. 157.

[37] Neel, J.V. (1962) Diabetes mellitus: A "thrifty" genotype rendered detrimental by "progress"? *American Journal of Human Genetics*, **14**, 353-62.

[38] Neel, J.V., Weder, A. B. and Julius, S. (1998). Type II diabetes, essential hypertension, and obesity as 'syndromes of impaired genetic homeostasis': the 'thrifty genotype' hypothesis enters the 21st century. *Perspectives in Biology and Medicine*, **42**, 44-74, p. 48.

[39] Neel, J.V. and others (1998). Fn. 38, p. 48.

[40] Darwin, C. (2004 [1871]). Fn. 1, p. 684.

[41] Myles, S., Lea, R.A., Ohashi, J., Chambers, G.K., Weiss, J.G., Hardouin, E., Engelken, J., Macartney-Coxson, D.P., Eccles, D.A., Naka, I., Kimura, R., Inaoka, T., Matsumura, Y. and Stoneking, M. (2011). Testing the thrifty gene hypothesis: the Gly482Ser variant in PPARGC1A is associated with BMI in Tongans. *BMC Medical Genetics*, **12**: 10. The paper states: 'Polynesians likely experienced long periods of cold stress and starvation during their settlement of the Pacific and today have high rates of obesity and type 2 diabetes (T2DM), possibly due to past positive selection for thrifty alleles.' No evidence is provided for this assumption, nor any theoretical argument. In what time frame does this occur? How long would it take for such a change in metabolism to occur, and, if so, why has it not reverted to a normal metabolism? Considering that these people entered lands where the local animal population had not encountered humans, one must suspect that hunting would have been relatively easy.

The heavy physique of Polynesians has been one basis for the suggestion that BMI (body mass index) standards may need to be reconsidered for Polynesians due to their different body size and composition. Polynesians were also found to prefer BMI measurements that were above the healthy weight standards of the WHO (18.5-25kg/m^2) for both men (28 kg/m^2) and women (26 kg/m^2); in other words, what Polynesians considered healthy, the WHO would consider to be overweight. The preference for males was slightly under the WHO standard for obesity (30 kg/m^2). This is in part due to normal standards being based on people of European descent. It is also in part due to Tongans possessing higher levels of fat-free mass (muscle, bone, water etc.) at the same BMI compared to Caucasians. Craig, P., Halavatau, V., Comino, E. and Caterson, I. (2001). Differences in body composition between Tongans and Australians: time to rethink the healthy weight ranges? *International Journal of Obesity*, **25**, 1806-1814.

That Tongans possess higher levels of fat-free mass wouldn't surprise anybody who has watched rugby: Polynesians are well built. Tonga has a population of merely 100,000, but is currently rated 11th in rugby union, while Samoa, population 183,000, is rated 7th, slightly behind England and Ireland and ahead of Wales, whose national sport is rugby (population 3 million). In physically demanding, robust sport, these people can compete with much larger nations. How can this be based on 'thrifty-genes' and a feast and famine diet? How do people survive with a feast and famine diet, and end up with greater percentages of bone and muscle?

[42] Trivers, R.L. (1972). Parental investment and sexual selection. In B. Campbell (ed.), *Sexual Selection and the Descent of Man*. Chicago: Aldine-Atherton, p. 136-179.

The relative parental investment of the sexes in their young is the key variable controlling the operation of sexual selection. Where one sex invests considerably more than the other, members of the latter will compete among themselves to mate with members of the former. Where investment is equal, sexual selection should operate similarly on the two sexes. The pattern of relative parental investment in species today seems strongly influenced by the earlier evolutionary differentiation into mobile sex cells fertilizing immobile ones, and sexual selection acts to mold the pattern of relative parental investment. (Page 174.)

[43] Hesiod (1914). *Works And Days*, translated by H.G. Evelyn-White, lines 69-82. <http://www.sacred-texts.com/cla/hesiod/works.htm> (accessed April 5, 2010).

[44] Hesiod (1914). Fn. 43, lines 90-96.

[45] Hays, H.R. (1964). *The Dangerous Sex: the Myth of Feminine Evil*. New York: G.P. Putnam's Sons, p. 81.

[46] The Electronic Text Corpus of Sumerian Literature. (2006). *A hymn to Inana (Inana C)*. <http://etcsl.orinst.ox.ac.uk/cgi-bin/etcsl.cgi?text=t.4.07.3> (accessed March, 29, 2010).

[47] For an overview see Hays, H.R. (1964). Fn. 45.

[48] Trivers, R.L. (1971). The evolution of reciprocal altruism. *Quarterly Review of Biology*, **46**, 35-57.

[49] Trivers, R.L. (1971). Fn. 48, p. 42.

[50] Axelrod, R. and Hamilton, W.D. (1981). The evolution of cooperation. *Science*, 211, 1390-6. Axelrod, R. (1984). *The Evolution of Cooperation*. New York: Basic Books. For a summary see Dawkins, R. (1989). *The Selfish Gene* (rev. edn.). Oxford: Oxford University Press, p. 202-235.

[51] Clutton-Brock, T.H. and Parker, G.A. (1995). Sexual coercion in animal societies. *Animal Behaviour*, **49**, 1345-65. Boyd, R. and Richerson, P.J. (1992). Punishment allows the evolution of cooperation (or anything else) in sizeable groups. *Ethology and Sociobiology*, **13**, 171-195. Clutton-Brock, T.H. and Parker, G.A. (1995). Punishment in animal societies. *Nature*, **373**, 209-216. For a summary, see Pusey, A.E. and Packer, C. (1997). The ecology of relationships. In J.R. Krebs and N.B. Davies (eds.), *Behavioural Ecology: An Evolutionary Approach*, 4th edn., (pp. 254-283). Oxford: Blackwell Science. It is also briefly mentioned as 'aid in combat' by Robert Trivers in Trivers, R.L. (1971). The evolution of reciprocal altruism. *Quarterly Review of Biology* **46**, 35-57, p. 40.

[52] Hallpike, C.R. (1977). *Bloodshed and Vengeance in the Papuan Mountains: the Generation of Conflict in Tauade Society*. Oxford: Clarendon Press, p. 11.

[53] Goodall, J. (1999 [1988]). *In the Shadow of Man*. London: Phoenix, p. 118.

[54] Goodall, J. (2000 [1990]). *Through a Window: Thirty Years with the Chimpanzees of Gombe*. London: Phoenix, p. 47.

[55] For an overview of the benefits of dominance in chimpanzees, see Muller, M.N. and Wrangham, R.W. (2001) The reproductive ecology of male hominoids. In P.T. Ellison (ed.) *Reproductive ecology and human evolution* (pp. 397-427). New York: Aldine, p. 415-417. Mitani, J.C., Watts, D.P. and Muller, M.N. (2002). Recent developments in the study of wild chimpanzee behavior. *Evolutionary Anthropology*, **11**, 9-25, p. 12-14. In one study using genetic testing, half of the infants were sired by high-ranking males, including five by the alpha male: Constable, J., Ashley, M., Goodall, J. and Pusey, A. (2001). Noninvasive paternity assignment in Gombe chimpanzees. *Molecular Ecology*, **10**, 1279–1300.

[56] Goodall, J. (2000 [1990]). Fn. 54, p. 83-94.

[57] Goodall, J. (2000 [1990]). Fn. 54, p. 87. For further examples, and an overview of research see Watts, D.P., Muller, M., Amsler, S.J., Mbabazi, G. and Mitani, J.C. (2006). Lethal intergroup aggression by chimpanzees in Kibale National Park, Uganda. *American Journal of Primatology*, **68**, 161–180. Death rates due to violence appear to be half those for humans, at 277 per 100,000 (0.27% per year as against 0.5% per year, as calculated by Keeley); however, chimpanzee rates rise above human conflict to 810 per 100,000 if 'suspected' killings are included, fifty percent higher than humans: Wrangham, R.W., Wilson, M.L. and Muller, M.N. (2006). Comparative rates of violence in chimpanzees and humans. *Primates*, **47**, 14-26. Reasons for chimpanzee violence could be that infant survival and female reproduction are linked to habitat quality, with females in poor quality habitats with lower levels of preferred fruits take longer to reproduce and have higher infant death rates: Thompson, M.E., Kahlenberg, S.M., Gilby, I.C. and Wrangham, R.W. (2007). Core area quality is associated with variance in reproductive. *Animal Behaviour*, **73**, 510-512. Other reasons are that chimpanzees appropriate females (usually juvenile) from defeated neighbours and may attract females (which are exogamic) if they have a larger territory: Goodall, J. (1986). *The Chimpanzees of Gombe: Patterns of Behaviour*. Cambridge: Cambridge University Press. Goodall, J., Bandora, A., Bergmann, E., Busse, C. Matama, H., Mponga, E., Pierce, A. and Riss, D. (1979). Inter-community interactions in the chimpanzee population of the Gombe National Park. In D.A. Hamburg and E.R. McCown (eds.) *The Great Apes* (pp. 13-54). Menolo Park: Benjamin/Cummings. Nishida, T. (1979). The social

structure of chimpanzees of the Mahale Mountains. In D.A. Hamburg and E.R. McCown (eds.) *The Great Apes* (pp. 73-122). Menolo Park: Benjamin/Cummings.

[58] Darwin, C. (2004 [1871]). Fn. 1, p. 211-212.

[59] Darwin, C. (2004 [1871]). Fn. 1, p. 211.

[60] Cialdini, R.B. (1993). *Influence: The Psychology of Persuasion* (rev. edn.). New York: Quill, p. 229. Doob, A.N. and Gross, A.E. (1968). Status of frustrator as an inhibitor of horn-honking responses. *Journal of Social Psychology*, **76**, 213-18.

[61] This might then provoke the question: why does the government simply not deceive everyone? If the government is so powerful, why does the government not simply deceive the members of the public? This sometimes occurs, as with dictatorships. However, the reason governments do not continually deceive the general public is because Western societies have evolved a method of preventing such deception. The power is split between two bodies: the legislature and the executive or government. The legislature can make laws, but not enforce the laws. The executive can enforce the laws, but not make the laws. Then there is a third body, the courts, who make decisions as to the laws – but can neither create the laws nor enforce the laws. The executive is normally headed by the president; the legislature is normally headed by the prime minister; and the courts are headed by a chief justice. By splitting the power in this method, no one individual can ever take control. Furthermore, how the government spends the available taxes must pass through the legislature; thus there must be agreement between the two organisations as to how the taxes are spent. The formation of political parties mollifies this concept somewhat, since the same group may control both the executive and the legislature. This is the separation of powers. While it may not work perfectly, it reflects the attempts of complex societies to prevent deception by those in power. A further aspect may be the relative status (wealth) of the government to the non-government population.

[62] Cialdini, R.B. (1993). Fn. 60, p. 213.

[63] Milgram, S. (1963). Behavioural study of obedience. *Journal of Abnormal and Social Psychology* **67**, 371-78, p. 376. See also Milgram, S. (1974). *Obedience to Authority: an Experimental View*. London : Tavistock Publications.

[64] Cialdini, R.B. (1993). Fn. 60, p. 290. See also Meeus, W.H.J. and Raaijmakers, Q.A.W. (1986). Administrative obedience: carrying out orders to use psychological-administrative violence. *European Journal of Social Psychology*, **16**, 311-24.

[65] Cialdini, R.B. (1993). Fn. 60, p. 208-229.

[66] Sereny, G. (2001). *The German Trauma: Experiences and Reflections 1938-2001*. London: Penguin, p. 76.

[67] Sereny, G. (2001). Fn. 66, p. 76.

[68] Sereny, G. (2001). Fn. 66, p. 81-82.

[69] Sereny, G. (2001). Fn. 66, p. 80-81.

[70] Cialdini, R.B. (1993). Fn. 60, p. 229.

[71] 'Although outranked, Thompson ordered the lieutenant and his men to stand down. "Thompson put his guns on the Americans and said he would shoot them if they shot another Vietnamese," William Eckhardt, the chief prosecutor at the My Lai court martial, said. A furious Thompson reported the massacre in progress to army superiors who ordered a ceasefire. Thompson also ordered two other helicopters to evacuate about a dozen wounded villagers to hospital for treatment.' 'As a combat pilot Thompson was shot down four times in Vietnam. He suffered a broken back in his last crash and received a Purple Heart and Distinguished Flying Cross.' Johnson, A. (2006). My Lai massacre hero buried. *The Australian*, January 12.

[72] Cicero (1971). *On the Good Life*. Translated by Grant, M. Harmondsworth: Penguin, p. 191. Similar sentiments are found in the Old Testament, 'A faithful friend is a sure shelter; whoever finds one has found a rare treasure' (Ecclesiasticus, 6:14).

[73] Goodman, M. (1997). *The Roman World, 44BC - AD 180*. London: Routledge, p. 22.

[74] Goodman, M. (1997). Fn. 73, p. 193.

[75] Lévi-Strauss, C. (1969). *The Elementary Structures of Kinship* (rev. edn.). Translated by Bell, J.H. and von Sturmer, J. R. and edited by Needham, R. Boston: Beacon, p. 52.

[76] Lévi-Strauss, C. (1969). Fn. 75, p. 52-53.

[77] Lévi-Strauss, C. (1969). Fn. 75, p. 53.

[78] Chagnon, N.A. (1992a). Fn. 5, p. 164.

[79] Keeley, L.H. (1996). *War Before Civilisation*. New York: Oxford University Press, p. 93. Author's italics.

[80] Keeley, L.H. (1996). Fn. 79, p. 90.

[81] Keeley, L.H. (1996). Fn. 79, p. 91.

[82] Keeley, L.H. (1996). Fn. 79, p. 91.

Chapter 3   Arranged Marriages

[1] Hiatt, L.R. (1996). *Arguments about Aborigines: Australia and the Evolution of Social Anthropology*. Cambridge, England: Cambridge University Press, p. 36.

[2] Hiatt, L.R. (1996). Fn.1, p. 36.

[3] Parkin, R. (1997). *Kinship: an Introduction to Basic Concepts*. Oxford, UK: Blackwell, p. 156-157.

[4] Parkin, R. (1997). Fn. 3, p. 158.

[5] Tylor, E. (1880). On the methods of investigating the development of institutions applied to the laws of marriage and descent. *Journal of the Anthropological Institute*, **18**, 245-272.

[6] Leach, E. (1970). *Lévi-Strauss*. London: Fontana/Collins, p. 102.

[7] Hiatt, L.R. (1996). Fn. 1, p. 42.

[8] Alexander, R.D. (1979). *Darwinism and Human Affairs*. Seattle: University of Washington Press.

[9] Lévi-Strauss, C. (1969). *The Elementary Structures of Kinship* (rev. edn.). Translated by Bell, J.H. and von Sturmer, J. R. and edited by Needham, R. Boston: Beacon, p. 70-71.

[10] Chagnon, N.A. (1992b). *Yanomamö: the Last Days of Eden*. San Diego: Harcourt Brace Jovanovich, p. 173.

[11] Genesis: 034: 016. The entire quote is worth citing. In fact, the arranged marriages are intended to create peace between two kin groups, after one man (Shechem) raped the daughter (Dinah) of a man (Jacob) in the second kin group. Jacob's family accept the pact for peace (GEN:034:016), but do so deceitfully. They enter Shechem's community – who are expecting peace – and slay them, taking their wives, children and livestock (GEN:034:029). Being in the land of Canaan, they then understandably fear attack from Shechem's relations – the Canaanites and the Perizzites – but God intervenes and states that if Jacob creates an altar to God, then God will protect them. Thus were created the chosen people, the Israelites, with God informing them: 'kings shall come out of thy loins'. It is a classic example of alliance formation and deceit. Note the explicit cross-cousin marriage agreement at GEN:034:009.

> GEN:033:018 And Jacob came to Shalem, a city of Shechem, which is in the land of Canaan, when he came from Padanaram; and pitched his tent before the city.
>
> GEN:033:019 And he bought a parcel of a field, where he had spread his tent, at the hand of the children of Hamor, Shechem's father, for an hundred pieces of money.
>
> GEN:033:020 And he erected there an altar, and called it EleloheIsrael.
>
> GEN:034:001 And Dinah the daughter of Leah, which she bare unto Jacob, went out to see the daughters of the land.
>
> GEN:034:002 And when Shechem the son of Hamor the Hivite, prince of the country, saw her, he took her, and lay with her, and defiled her.
>
> GEN:034:003 And his soul clave unto Dinah the daughter of Jacob, and he loved the damsel, and spake kindly unto the damsel.
>
> GEN:034:004 And Shechem spake unto his father Hamor, saying, Get me this damsel to wife.
>
> GEN:034:005 And Jacob heard that he had defiled Dinah his daughter: now his sons were with

his cattle in the field: and Jacob held his peace until they were come.

GEN:034:006 And Hamor the father of Shechem went out unto Jacob to commune with him.

GEN:034:007 And the sons of Jacob came out of the field when they heard it: and the men were grieved, and they were very wroth, because he had wrought folly in Israel in lying with Jacob's daughter: which thing ought not to be done.

GEN:034:008 And Hamor communed with them, saying, The soul of my son Shechem longeth for your daughter: I pray you give her him to wife.

GEN:034:009 And make ye marriages with us, and give your daughters unto us, and take our daughters unto you.

GEN:034:010 And ye shall dwell with us: and the land shall be before you; dwell and trade ye therein, and get you possessions therein.

GEN:034:011 And Shechem said unto her father and unto her brethren, Let me find grace in your eyes, and what ye shall say unto me I will give.

GEN:034:012 Ask me never so much dowry and gift, and I will give according as ye shall say unto me: but give me the damsel to wife.

GEN:034:013 And the sons of Jacob answered Shechem and Hamor his father deceitfully, and said, because he had defiled Dinah their sister:

GEN:034:014 And they said unto them, We cannot do this thing, to give our sister to one that is uncircumcised; for that were a reproach unto us:

GEN:034:015 But in this will we consent unto you: If ye will be as we be, that every male of you be circumcised;

GEN:034:016 Then will we give our daughters unto you, and we will take your daughters to us, and we will dwell with you, and we will become one people.

GEN:034:017 But if ye will not hearken unto us, to be circumcised; then will we take our daughter, and we will be gone.

GEN:034:018 And their words pleased Hamor, and Shechem Hamor's son.

GEN:034:019 And the young man deferred not to do the thing, because he had delight in Jacob's daughter: and he was more honourable than all the house of his father.

GEN:034:020 And Hamor and Shechem his son came unto the gate of their city, and communed with the men of their city, saying,

GEN:034:021 These men are peaceable with us; therefore let them dwell in the land, and trade therein; for the land, behold, it is large enough for them; let us take their daughters to us for wives, and let us give them our daughters.

GEN:034:022 Only herein will the men consent unto us for to dwell with us, to be one people, if every male among us be circumcised, as they are circumcised.

GEN:034:023 Shall not their cattle and their substance and every beast of their's be our's only let us consent unto them, and they will dwell with us.

GEN:034:024 And unto Hamor and unto Shechem his son hearkened all that went out of the gate of his city; and every male was circumcised, all that went out of the gate of his city.

GEN:034:025 And it came to pass on the third day, when they were sore, that two of the sons of Jacob, Simeon and Levi, Dinah's brethren, took each man his sword, and came upon the city boldly, and slew all the males.

GEN:034:026 And they slew Hamor and Shechem his son with the edge of the sword, and took Dinah out of Shechem's house, and went out.

GEN:034:027 The sons of Jacob came upon the slain, and spoiled the city, because they had defiled their sister.

GEN:034:028 They took their sheep, and their oxen, and their asses, and that which was in the city, and that which was in the field,

GEN:034:029 And all their wealth, and all their little ones, and their wives took they captive, and spoiled even all that was in the house.

GEN:034:030 And Jacob said to Simeon and Levi, Ye have troubled me to make me to stink among the inhabitants of the land, among the Canaanites and the Perizzites: and I being few in number, they shall gather themselves together against me, and slay me; and I shall be destroyed, I and my house.

GEN:034:031 And they said, Should he deal with our sister as with an harlot?

GEN:035:001 And God said unto Jacob, Arise, go up to Bethel, and dwell there: and make there an altar unto God, that appeared unto thee when thou fleddest from the face of Esau thy brother.

GEN:035:002 Then Jacob said unto his household, and to all that were with him, Put away the strange gods that are among you, and be clean, and change your garments:

GEN:035:003 And let us arise, and go up to Bethel; and I will make there an altar unto God, who answered me in the day of my distress, and was with me in the way which I went.

GEN:035:004 And they gave unto Jacob all the strange gods which were in their hand, and all their earrings which were in their ears; and Jacob hid them under the oak which was by Shechem.

GEN:035:005 And they journeyed: and the terror of God was upon the cities that were round about them, and they did not pursue after the sons of Jacob.

GEN:035:006 So Jacob came to Luz, which is in the land of Canaan, that is, Bethel, he and all the people that were with him.

GEN:035:007 And he built there an altar, and called the place Elbethel: because there God appeared unto him, when he fled from the face of his brother.

GEN:035:008 But Deborah Rebekah's nurse died, and she was buried beneath Bethel under an oak: and the name of it was called Allonbachuth.

GEN:035:009 And God appeared unto Jacob again, when he came out of Padanaram, and blessed him.

GEN:035:010 And God said unto him, Thy name is Jacob: thy name shall not be called any more Jacob, but Israel shall be thy name: and he called his name Israel.

GEN:035:011 And God said unto him, I am God Almighty: be fruitful and multiply; a nation and a company of nations shall be of thee, and kings shall come out of thy loins;

GEN:035:012 And the land which I gave Abraham and Isaac, to thee I will give it, and to thy seed after thee will I give the land.

[12] Valero, H. (1970). *Yanoámo: The Narrative of a White Girl Kidnapped by Amazonian Indians, as told to Ettore Biocca*. Translated by Dennis Rhodes. New York: Dutton, p. 31. Valero is living with the Kohoroshiwetari. The Kohoroshiwetari are soon to be raided by the Karawetari.

[13] Valero, H. (1970). Fn. 12, p. 33.

[14] Valero, H. (1970). Fn. 12, p. 35.

[15] Valero, H. (1970). Fn. 12, p. 34.

[16] Valero, H. (1970). Fn. 12, p. 271-272.

[17] Valero, H. (1970). Fn. 12, p. 273.

[18] Chagnon, N.A. (1992b). Fn. 10, p. 150.

[19] Valero, H. (1970). Fn. 12, p. 157.

[20] Warner, W.L. (1969 [1958]). *A Black Civilization: a Study of an Australian Tribe* (rev. edn.). Gloucester, Mass.: P. Smith, p. 144. Note that most deaths are due to revenge for an earlier killing:

Of seventy-two recorded battles of the last twenty years in which members of Murngin factions were killed, fifty were for blood revenge – the desire to avenge the killing of a relative, usually a clansman, by members of another clan. Of these, fifteen were deliberate killings, against the tradition of what is fair cause for a war, and because it was felt that the enemies had killed the wrong people. Ten killings were due stealing or obtaining by illegal means a woman who belonged to another clan. Five supposedly guilty magicians were killed by the clan members of victims of black magic. Five were slain for looking at totemic emblem under improper circumstances and thereby insulting the owning clan and endangering the clan's spiritual strength.

The idea underlying most Murngin warfare is that the same injury should be inflicted upon the enemy group that one's own group suffered. This accomplished, a clan feels satisfied; otherwise, there is a constant compulsion toward vengeance, causing continuous restlessness among those who are out to "buy back" the killing of one of their clansmen. The stealing of women provokes the same spirit, since the group feels itself injured; and only the return of the woman and a ceremonial fight, or the stealing of another woman, will satisfy the hurt to its self-esteem, unless the clan has retaliated by killing or wounding one of the enemy clansmen. The same feeling is instigated by the improper viewing of the totem – an insult and an injury to the entire clan. (Page 148.)

[21] Warner, W.L. (1969 [1958]). Fn. 20, p. 66.

[22] Howitt, A.W. (1996 [1904]). *Native Tribes of South-East Australia*. Canberra: Aboriginal Studies Press, p. 195-280.

[23] Keeley, L.H. (1996). *War Before Civilisation*. New York: Oxford University Press, p. 84.

[24] Bourke, J. (2007). *Rape: Sex, Violence, History*. Emeryville, CA: Shoemaker & Hoard, p. 386.

[25] Chagnon, N.A. (1996). Chronic problems in understanding tribal violence and warfare. In G.R. Bock and J.A. Goode (eds.), *Genetics of Criminal and Antisocial Behavior. Ciba Foundation Symposium*, **194** (pp. 202-232). Chichester, New York: Wiley, p. 222.

[26] Warner, W.L. (1969 [1958]). Fn. 20, p. 77.

[27] Chagnon, N.A. (1992a). *Yąnomamö*, (4th edn.). Forth Worth: Harcourt Brace Jovanovich, p. 190.

[28] Some societies have variations on this basic theme. For instance, some Muslim societies have a preference for parallel-cousin marriage rather than cross-cousin marriage. Several commentators argue that this undermines the argument for universal cross-cousin marriage; however, it should be noted that the focus is on cooperatively-simple societies. If more complex societies vary the basic pattern, this is a reflection of the complexity of the cooperative system of their society. Once societies gain the capacity to create status as an abstraction (such as money), then they are no longer a cooperatively-simple society, and cannot be viewed as evidence of an exception to a universal feature in cooperatively-simple societies.

[29] Berndt, R.M. and Berndt C.H. (1977). *The World of the First Australians* (2nd edn.) Sydney: Ure Smith, p. 88.

Chapter 4 Male Alliances

[1] Cicero, M.T. (1971). *On the Good Life*. Translated by Michael Grant. Harmondsworth: Penguin, p. 140.

[2] Hallpike, C.R. (1977). *Bloodshed and Vengeance in the Papuan Mountains: the Generation of Conflict in Tauade Society*. Oxford: Clarendon Press, p. 201. Hallpike notes a third killing of a man who it was believed to have a disease and concludes:

It seems, therefore, that the killing of the defenceless stranger, and killing by stealth, perpetrated by members of one tribe against another, are patterns of violence that are distinguishable from warfare, which is usually generated by some provocative act, usually theft, witnessed by many and leading quickly to an armed confrontation. If blood was spilt because of this, raiding often followed, together with considerable destruction of gardens, killing of pigs, and sometimes loss of life. (Page 196.)

[3] Hallpike, C.R. (1977). Fn. 2, p. 201.

[4] Cicero, M.T. (1971). Fn. 1, p. 187.

[5] Cicero, M.T. (1971). Fn. 1, p. 188-189.

[6] Chatwin, B. (1987). *Songlines*. London : Jonathon Cape, p. 201.

[7] Williams, G.C. (1966). *Adaptation and natural selection; a critique of some current evolutionary thought*. Princeton, N.J.: Princeton University Press. Hamilton W.D. (1964). The genetical evolution of social behaviour (I and II). *Journal of Theoretical Biology*. **7**, 1-16; 17-52. Note that actual rates of

relatedness tend to vary slightly, but I have kept the theoretical level for simplicity.

[8] Hamilton W.D. (1964). Fn. 7.

[9] Saran, A. B. (1974). *Murder and Suicide among the Munda and the Oraon*. Delhi: National Publishing House. Varma, S. C. (1978). *The Bhil Kills*. Delhi: Kunj Publishing House. Daly, M. and Wilson, M. (1982). Homicide and Kinship. *American Anthropologist*, **84**, 372-378.

[10] Warner, W.L. (1969 [1958]). *A Black Civilization: a Study of an Australian Tribe* (rev. edn.). Gloucester, Mass.: P. Smith, p. 51-52.

[11] Warner, W.L. (1969 [1958]). Fn. 10, p. 150.

[12] Wilson, M. and Daly, M. (1985). Competitiveness, Risk-Taking, and Violence: The Young Male Syndrome. *Ethology and Sociobiology*, **6**, 59-73.

[13] Warner, W.L. (1969 [1958]). Fn. 10, p. 74-75.

[14] Warner, W.L. (1969 [1958]). Fn. 10, p. 49-50. In her account of her stay with the Yąnomamö, Helena Valero recounted a situation when her husband, the tushaua, Fusiwe, found one of his wives consorting with his brother. He then ordered his wife to stay with his brother by sending her hammock to him. The example suggests that there is an expectation of generosity on the part of the older brother toward the younger.

> I went back to the tushaua; behind me came also his mother who said to him 'Why have you sent the hammock? I don't' want that woman for my other son.' Fusiwe replied: 'Mother, let her say with him; let her live with him as he has no other woman. In the meantime let him take this one; when she is older still, he will look for another, younger one.' The mother answered: 'But I think you are jealous of her and that's why you have sent to give her the hammock.' 'If I had been jealous, I would not have sent the hammock, but I would have struck your son and the woman with my nabrushi. I am certainly not jealous of my brother; I think, poor brother, he has no wife and I have so many, I'd better give him one of mine. For days I have been thinking about it and now today I have my opportunity.' Then his other went up to him and said quietly: 'Then, if that's how it is, you should give him another one, the youngest; otherwise everybody will say that you gave given him the oldest.' Then that woman came back with us.
> (Valero, H. (1970). *Yanoámo: The Narrative of a White Girl Kidnapped by Amazonian Indians, as told to Ettore Biocca*. Translated by Dennis Rhodes. New York: Dutton, p. 141.)

[15] Warner, W.L. (1969 [1958]). Fn. 10, p. 52.

[16] Chagnon, N.A. (1992a). *Yąnomamö* (4th edn.). Fort Worth: Harcourt Brace Jovanovich, p. 27-28.

[17] Chagnon, N.A. (1992a). Fn. 16, p. 155. The number of people studied was 1327, a sizeable number of the total Yąnomamö population, estimated at 20,000.

[18] Chagnon, N.A. (1982). Sociodemographic attributes of nepotism in tribal populations: man the rule-Breaker. In King's College Sociobiology Group (eds.), *Current Problems in Sociobiology* (pp. 291-318). Cambridge: Cambridge University Press, p. 294.

[19] Chagnon, N.A. (1982). Fn. 18, p. 294.

[20] Warner, W.L. (1969 [1958]). Fn. 10, p. 43-44.

[21] Daly, M. and Wilson, M. (1998). *The Truth about Cinderella: a Darwinian View of Parental Love*. London: Weidenfelt & Nicolson, p. 4. 'Fairy tales about malevolent stepfathers are scarcer than those about stepmothers, but no cheerier.'

[22] Chagnon, N.A. (1988b). Male Yąnomamö manipulations of kinship classifications of female kin for reproductive advantage. In L. Betzig, M. Borgerhoff Mulder and P. Turke (eds.), *Human Reproductive Behaviour: A Darwinian Perspective* (pp. 23-48). Cambridge: Cambridge University Press, p. 29.

[23] Chagnon, N.A. (1988b). Fn. 22, p. 29.

[24] Chagnon, N.A. (1992b). *Yanomamö: the Last Days of Eden*. San Diego: Harcourt Brace Jovanovich, p. 36.

[25] Chagnon, N.A. (1992a). Fn. 16, p. 155.

[26] Chagnon, N.A. (1979a). Mate competition, favoring close kin, and village fissioning among the

Yanomamö Indians. In N.A. Chagnon and W. Irons (eds), *Evolutionary Biology and Human Social Behavior* (pp. 86-131). North Scituate, M.A.: Duxbury Press, p. 96.

[27] Warner, W.L. (1969 [1958]). Fn. 10, p. 81. Warner explains the importance of the cross-cousin system in further detail:

> The marriage of a male ego to his wife is something more than the modern marriage contract to any woman out of a general group. He marries into his wife's family of orientation, a group of four personalities who are his mother's brother, his mother's brother's wife (who is also part of another family), his wife's brother, and his wife's sister, who is also mother's brother's daughter. Ego, then, is constantly looking to his mother's people for his wives while he is a member of his own family of orientation and while forming his family of procreation, and when he has a son he still seeks the father of his mother's group to obtain wives for his male offspring. On the other hand, he has nothing to obtain during childhood and early manhood from his father's side of the family. When he has a daughter old enough for marriage, he wants a son-in-law; but he is then not a seeker but the one sought after by his sister's son, which puts him in a very strong position. (Warner, W.L. (1969 [1958]). Fn. 10, p. 43.)

[28] Chagnon, N.A. (1979a). Fn. 26, p. 106.

[29] Cicero, M.T. (1971). Fn. 1, p. 160.

[30] Chagnon, N.A. (1979a). Fn. 26, p. 106.

[31] Chagnon, N.A. (1979a). Fn. 26, p. 107.

[32] Chagnon, N.A. (1982). Fn. 18, p. 306.

[33] Warner, W.L. (1969 [1958]). Fn. 10, p. 32-33.

[34] One example of the deceit than can occur between parallel-cousins is the following extract from Chagnon's Yąnomamö: 'Two of the raiders were Matowä's parallel cousins, men he would call 'brother.' The raiders caught Matowä outside the new garden searching for honey. ... He was looking up a tree when the raiders shot a volley of arrows into his body, at least five of which struck him in the abdomen.' (Chagnon, N.A. (1992a). Fn. 16, p. 193.)

[35] Chagnon, N.A. (1974). *Studying the Yąnomamö*. New York: Hold, Rinehart and Winston, p. 166-67.

[36] Warner, W.L. (1969 [1958]). Fn. 10, p. 51.

[37] Warner, W.L. (1969 [1958]). Fn. 10, p. 66-67.

[38] Valero, H. (1970). Fn. 14, p. 66.

[39] Chagnon, N.A. (1988a). Life histories, blood revenge, and warfare in a tribal population. *Science*, **239**, 985-92, p. 987.

[40] Chagnon, N.A. (1988a). Fn. 39, p. 989.

[41] This is perhaps most obvious from the following example from a more modern society, the Punjabi of Pakistan and Northern India, where the giving and receiving of women is viewed specifically in terms of weakness and strength, respectively (Hershman, P. (1981). *Punjabi Kinship and Marriage*. Delhi: Hindustan Pub. Corp. (India)).

> A brother gives his sister in marriage to another man and in so doing not only makes a gift of that which he most jealously guarded but also exposes himself to dishonour. ... To give a woman in marriage is to place oneself in a position of inferiority to the taker; to take a woman is to assume a position of superiority to the giver; and to exchange women is to assume a position of equality. Punjabis resolve the problem of having to give their sisters in marriage and yet at the same time of pressuring their honor, in two quite distinct ways: Punjabi Muslims maintain and exchange their women within closed groups thus preserving their honor within the group by arranging the marriages of their sisters to one another; while Punjabi Hindus and Sikhs solve the problem by accepting the inferiority of the wife-giver role and by creating from this premise a system of exogamy based upon the principle of non-exchange. (Page 191.)

[42] Meggitt, M.J. (1974 [1962]). *Desert People: A Study of the Walbiri Aborigines of Central Australia.*

Sydney: Angus & Robertson, p. 122.
[43] Chagnon, N.A. (1992a). Fn. 16, p.188-189.
[44] Chagnon, N.A. (1992a). Fn. 16, p. 154.
[45] Chagnon, N.A. (1992a). Fn. 16, p. 158.
[46] Warner, W.L. (1969 [1958]). Fn. 10, p. 94.
[47] Warner, W.L. (1969 [1958]). Fn. 10, p. 27-28.
[48] Chagnon, N.A. (1982). Fn. 18, p. 313.
[49] Phyllis Kaberry, in a society more greatly affected by Western intervention, observed that the vast majority of unions were within the correct relationships; however, amongst the Lunga, 16.8 were 'wrong' (40 of 238), amongst the Wolmeri 4.5 percent, amongst the Djaru 19.3 percent, and amongst the Punaba 26.0 percent. (Kaberry, P.M. (1939). *Aboriginal Woman: Sacred and Profane*. London: George Routledge, p. 126.) Like Chagnon, she comments:

> In fact those who are the guardians of tribal law frequently break it, though it is doubtful if they are using their position as a cloak for this. (Page 130)

> We have seen that the system is essentially elastic, and it is adapted to daily life – a fact which is not altogether astounding if we accept the proposition that in the first place it was evolved in response to a complex set of needs. It cannot, therefore, be regarded as a nemesis that the natives have brought upon themselves: as a mechanism which, designed primarily to control and satisfy desires within the limits of environment and society, has ended by dominating existence and destroying the means of satisfaction. In one sense it is more rigid than our own system, but we must remember that the less complex character of aboriginal society enables us to see more clearly the restrictions imposed on the individual, whereas in our own society they tend to be camouflaged to a certain extent by an overt cult of individuality, and such tags as liberty, freedom and equality. (Pages 125-126.)

[50] The same preference for polygyny can be found in the Old Testament, for instance, in Book 17, Esther, where King Solomon of Israel 'delighted' in having numerous concubines:

> EST:002:003 And let the king appoint officers in all the provinces of his kingdom, that they may gather together all the fair young virgins unto Shushan the palace, to the house of the women, unto the custody of Hege the king's chamberlain, keeper of the women; and let their things for purification be given them:
> EST:002:013 Then thus came every maiden unto the king; whatsoever she desired was given her to go with her out of the house of the women unto the king's house.
> EST:002:014 In the evening she went, and on the morrow she returned into the second house of the women, to the custody of Shaashgaz, the king's chamberlain, which kept the concubines: she came in unto the king no more, except the king delighted in her, and that she were called by name.

As did his son, King Rehoboam in Book 14 2 Chronicles:

> 2CH:011:021 And Rehoboam loved Maachah the daughter of Absalom above all his wives and his concubines: (for he took eighteen wives, and threescore concubines; and begat twenty and eight sons, and threescore daughters.)

Who, in turn, was defeated by Abijah:

> 2CH:013:021 But Abijah waxed mighty, and married fourteen wives, and begat twenty and two sons, and sixteen daughters.

By contrast, the New Testament is suggestive in its support of monogamy, in Mark, Luke and Corinthians (to 'put away' is to leave):

> MRK:010:011 And he saith unto them, Whosoever shall put away his wife, and marry another, committeth adultery against her.
> MRK:010:012 And if a woman shall put away her husband, and be married to another, she committeth adultery.

> LUK:016:018 Whosoever putteth away his wife, and marrieth another, committeth adultery: and whosoever marrieth her that is put away from her husband committeth adultery.

> 1CR:007:001 Now concerning the things whereof ye wrote unto me: It is good for a man not to touch a woman.
> 1CR:007:002 Nevertheless, to avoid fornication, let every man have his own wife, and let every woman have her own husband.
> 1CR:007:003 Let the husband render unto the wife due benevolence: and likewise also the wife unto the husband.
> 1CR:007:004 The wife hath not power of her own body, but the husband: and likewise also the husband hath not power of his own body, but the wife.

[51] Chagnon, N.A. (1992a). Fn. 16, p. 191.

[52] That men are more likely to fight for their status rather than over women has also been observed by Anne Campbell in her studies of male and female gang members. Campbell, A. (2002). *A Mind of Her Own: the Evolutionary Psychology of Women.* New York: Oxford University Press.

[53] Heider, K. (1979). *Grand Valley Dani: Peaceful Warriors.* New York: Holt, Rinehart and Winston, p. 70.

[54] Chagnon, N.A. (1988a). Fn. 39, p. 990.

[55] Berndt, R.M. and Berndt, C.H. (1951). *Sexual Behavior in Western Arnhem Land.* New York: Viking Fund, p. 204-205.

[56] Chagnon, N.A. (1988a). Fn. 39, p. 990.

[57] Valero, H. (1970). Fn. 14, p. 121-122.

[58] Chagnon, N.A. (1992a). Fn. 16, p. 3.

[59] Chagnon, N.A. (1992a). Fn. 16, p. 166-168.

[60] Chagnon, N.A. (1992a). Fn. 16, p. 159-160.

[61] Chagnon, N.A. (1992a). Fn. 16, p. 161.

[62] Chagnon, N.A. (1992a). Fn. 16, p. 160-161.

[63] Valero, H. (1970). Fn. 14, p. 144-145.

[64] Chagnon, N.A. (1992a). Fn. 16, p. 46-7.

[65] Hallpike, C.R. (1977). Fn. 2, p. 149.

[66] Hallpike, C.R. (1977). Fn. 2, p. 142.

[67] Chagnon, N.A. (1992a). Fn. 16, p. 193.

[68] Chagnon, N.A. (1992a). Fn. 16, p. 193-194.

[69] Chagnon, N.A. (1979a). Fn. 26, p. 103.

[70] Chagnon, N.A. (1980). Kin-selection theory and the Yąnomamö Indians. In G.W. Barlow and J. Silverberg (eds.) *Sociobiology: Beyond Nature/Nurture? Special Symposium Publication No. 35, American Association for the Advancement of Science* (pp. 545-574). Boulder: Westview Press, p. 553.

[71] Warner, W.L. (1969 [1958]). Fn. 10, p. 50.

[72] Chagnon, N.A. (1977). *Yąnomamö: The Fierce People* (2nd edn.). New York: Harcourt Brace Jovanovich, p. 78-79.

[73] Chagnon, N.A. (1977). Fn. 72, p. 79.

[74] Chagnon, N.A. (1977). Fn. 72, p. 79.

[75] Chagnon, N.A. (1977). Fn. 72, p. 80.

[76] Chagnon, N.A. (1977). Fn. 72, p. 42.

[77] Chagnon, N.A. (1977). Fn. 72, p. 79.

[78] Chagnon, N.A. (1996). Chronic problems in understanding tribal violence and warfare. In G.R. Bock and J.A. Goode (eds.), *Genetics of Criminal and Antisocial Behavior. Ciba Foundation Symposium 194* (pp. 202-232). Chichester, New York: Wiley, p. 223-224.

[79] Chagnon, N.A. (1992a). Fn. 16, p. 193.

[80] Chagnon, N.A. (1992a). Fn. 16, p. 194-195.

[81] Chagnon, N.A. (1996). Fn. 78, p. 225.

[82] Chagnon, N.A. (1996). Fn. 78, p. 224-225.

[83] Chagnon, N.A. (1996). Fn. 78, p. 226-227.

[84] Chagnon, N.A. (1996). Fn. 78, p. 223.

[85] Chagnon, N.A. (1996). Fn. 78, p. 223.

[86] Chagnon, N.A. (1992a). Fn. 16, p. 26.

[87] Chagnon, N.A. (1992a). Fn. 16, p. 27.

[88] The implications of the need for a complex genealogical understanding for the evolution of the human brain cannot be underestimated. The following comes from A.R. Radcliffe-Brown concerning Australian Aborigines:

> When a stranger comes to a camp that he has never visited before, he does not enter the camp, but remains at some distance. A few of the older men, after a while, approach him, and the first thing they proceed to do is to find out who the stranger is. The commonest question that is put to him is "Who is your maeli (father's father)?" The discussion proceeds on genealogical lines until all parties are satisfied of the exact relation of the stranger to each of the natives present in the camp. When this point is reached, the stranger can be admitted to the camp, and the different men and women are pointed out to him and their relation to him defined... If I am a blackfellow and meet another blackfellow that other must be either my relative or my enemy. If he is my enemy I shall take the first opportunity of killing him, for fear he will kill me. This, before the white man came, was the aboriginal view of one's duty towards one's neighbour.
>
> (Radcliffe-Brown, A.R. (1913). Three Tribes of Western Australia. Journal of the Royal Anthropological Institute, 43, 143-170, p. 151, quoted in Levi-Strauss, C., The Elementary Structures of Kinship, page 482.)

The system of marriage thus defines a person in relation to all others.

This is as abstract a level of reasoning as algebra. Another observer, A.B. Deacon, commented:

> It is perfectly clear that the natives (the intelligent ones) do conceive of the system as a connected mechanism which they can represent by diagrams... The way they could reason about relationships from their diagrams was absolutely on a par with a good scientific exposition in a lecture-room.
>
> (Deacon A.B. (1927) The regulation of Marriage in Ambrym. Journal of the Royal Anthropological Institute, 57, 325-42. Quoted in Levi-Strauss, C., The Elementary Structures of Kinship, page 126.)

> The old men explained the system [of marriage] to me perfectly lucidly; I could not explain it to anyone better myself... It is extraordinary that a native should be able to represent completely by a diagram a complex system of matrimonial classes... I have collected in Malekula, too, some cases of a remarkable mathematical ability. I hope... to be able to prove that the native is capable of pretty advanced abstract thought.
>
> (Deacon, A.B. (1934) Malekula: A Vanishing People in the New Hebrides. London, p. xxii-xxiii.

Quoted in Levi-Strauss, C., The Elementary Structures of Kinship, page 126-7.)

The view that people in cooperatively-modern societies are the only ones capable of complex thought is highly questionable. One must then question whether there is a similarity between the capacity to think in terms of genealogies involving potentially hundreds of people (living, dead and even unborn) and the capacity to use algebra and think mathematically or logically. It is interesting to note that Napoleon Chagnon found that Yąnomamö men were slightly better at understanding genealogical relationships than Yąnomamö women – unsurprisingly, since for men this is the difference between bachelorhood, monogamy or polygyny, while women had no control over whom they married. Chagnon decided to test his ideas:

> In three different villages I had informants of both sexes and all ages tell me what kinship term they used for everyone in the village. To make sure we both knew who I was talking about, I simultaneously showed each informant a polaroid picture of each person as I whispered the name of that person in to his/her ear and asked 'What do you call so-and-so?' I timed their responses.
>
> I also knew from many previous years of fieldwork in these villages how everyone was genealogically related back for or five generations, and knew what they should have called the kinsmen that were related to them in the specifiable genealogical ways if they were following their 'rules' of classification.
>
> I did this with 100 informants of both sexes and all ages and got nearly 12,000 'kinship classification responses, as well as a measure of the time it took each informant to classify all members of the village.
>
> The results were as I predicted they would be. Males were faster than females at classifying their kin, suggesting that they knew more about genealogy and kinship than the females. They had the information 'at the tip of the tongue', but women and girls frequently had to do some genealogical algebra of the sort, 'Well, I call his father so and so and his sister such and such, so I suppose I'd call him "husband". This information also demonstrated that significant numbers of females in the village actually did not use specific terms for many co-residents, but used 'vague' terms that did not require genealogical knowledge of any depth. (Chagnon, N.A. (1992a). Fn. 16, p. 157)

Chagnon also found that males were reclassifying women into the 'wife' category – attempting to commit 'incest':

> When the kinship classification so of 'adult' males (17 and older) were examined, it was clear that they showed a statistically significant pattern of reclassifiying women in the 'wife' category more than into any other category for female relatives: The bias was to move females from reproductively useless categories and put them into the only reproductively useful category, 'wife'. A detailed analysis of the women they moved into this category showed that most of these females were young and had high reproductive value, i.e., were not old women past their reproductive prime. (Chagnon, N.A. (1992a). Fn. 16, p. 158)

By thus possessing, in their minds, a systematic understanding of the genealogical relationships of potentially hundreds of people, men are able to potentially increase the number of wives they or their close kin can marry. By knowing how to categorise people in an abstract manner, men are able to increase the number of women they can marry. Abstract thought and reproduction are thus intrinsically linked. Is this the basis of complex mathematical and logical thought? Considering that these people invariably do not possess a counting system beyond 'one, two and many', it would appear to be the only possible basis for complex, abstract thought.

Stanner, who studied the Marinbata of Australia, observed:

Those who doubt the aborigines' power of such abstract reasoning can never have heard them expounding to their tribe's fellows how ŋinipun (subsections) should work, by inference from the theory to the case under attention. In this way an abstraction becomes a flesh-and-blood reality. (Stanner, W.E.H. (1936). Murinbata kinship and totemism. Oceania, 7(2), p. 202. Quoted in Levi-Strauss, C., The Elementary Structures of Kinship, p. 125.)

Very recent research appears to support these arguments. It has been shown that Australian Aboriginal children – whose only counting words are one, two and many – are as capable at counting as children with more words. They were found to be able to count precisely, despite having no actual words to denote the numbers.
To test the children, they could not, for example, ask questions such as "How many?" or "Do these two sets have the same number of objects?"

They had to develop special tasks, for instance by showing how to put out counters that matched the number of sounds made by banging two sticks together. "We found that Warlpiri and Anindilyakwa children performed as well as or better than the English-speaking children on a range of tasks, and on numerosities up to nine, even though they lacked number words." (Highfield, R. (2008). We are natural born mathematicians. Telegraph, August 18. <http://www. telegraph.co.uk/science/science-news/3349815/We-are-natural-born-mathematicians.html> (accessed Nov. 13, 2010.)

This suggests that we are able to perform mathematical functions quite naturally. (see Butterworth, B., Reeve, R., Reynolds, F. and Lloyd D. (2008). Numerical thought with and without words: Evidence from Indigenous Australian children. Proceedings of the National Academy of Sciences of the USA, 105, 13179-13184.)

[89] Roes, F. (1998). An interview of Napoleon Chagnon. Human Ethology Bulletin, 13 (4), 6-12, p. 8.
[90] Roes, F. (1998). Fn. 89, p. 8.

Chapter 5 Marriage

[1] King, M.L. (1991). Women of the Renaissance. Chicago: The University of Chicago Press, p. 34.
[2] Duby, G. (1994 [1988]). Love and Marriage in the Middle Ages, translated by Jane Dunnett. Cambridge: Polity, p. 26.
[3] Coontz, S. (2005). Marriage, A History: from Obedience to Intimacy or How Love Conquered Marriage. New York: Viking, p. 15.
[4] Coontz, S. (2005). Fn. 3, p. 123.
[5] Fraser, A. (1984). The Weaker Vessel: Woman's Lot in Seventeenth-Century England. London: Weidenfeld and Nicolson, p. 27.
[6] Fraser, A. (1984). Fn 5, p. 41.
[7] Fraser, A. (1984). Fn 5, p. 43.
[8] Chagnon, N.A. (1992a). Yąnomamö (4th edn.). Forth Worth: Harcourt Brace Jovanovich, p. 122-123.
[9] Valero, H. (1970). Yanoámo: The Narrative of a White Girl Kidnapped by Amazonian Indians, as told to Ettore Biocca. Translated by Dennis Rhodes. New York: Dutton, p. 133-134. Epená is a hallucinogenic drug.
[10] Chagnon, N.A. (1992a). Fn. 8, p. 155. For a larger set of data involving many societies see: Marlowe, F.W. (2005). Hunter-Gatherers and Human Evolution. Evolutionary Anthropology, 14, 54-67, in particular infant mortality and juvenile mortality rates of 24.83% and 41.36% for Old World hunter-gatherers (12 studies each), and 18.15% and 47.83% for New World hunter-gatherers (4 and 7 studies) respectively. These rates compare with Chagnon's findings, that infants have barely a better than 55% chance of reaching adulthood (page 58). Marlowe counted infant and juvenile mortality as

being within 1 year and 15 years, respectively.

11 Chagnon, N.A. (1992a). Fn. 8, p. 155.

12 Trivers, R. (1985). *Social Evolution*. Menlo Park, Calif.: Benjamin/Cummings Pub. Co.

13 Burbank, V.K. (1994). *Fighting Women: Anger and Aggression in Aboriginal Australia*. Berkeley: University of California Press, p. 59.

14 Hallpike, C.R. (1977). *Bloodshed and Vengeance in the Papuan Mountains: the Generation of Conflict in Tauade Society*. Oxford: Clarendon Press, p. 128.

15 Warner, W.L. (1969 [1958]). *A Black Civilization: a Study of an Australian Tribe* (revised edn.). Gloucester, Mass.: P. Smith, p. 70.

16 Chagnon, N.A. (1992a). Fn. 8, p. 125.

17 Leblanc, W. (1999). *Naked Motherhood: Shattering Illusions and Sharing Truths*. Milsons Point, N.S.W.: Random House Australia, p. 209-210.

18 Johnson, H. (1995). Risk Factors Associated With Non-Lethal Violence Against Women by Marital Partners. In C. Block and R. Block, *Trends, Risks, and Interventions in Lethal Violence: Proceedings of the Third Annual Spring Symposium of the Homicide Research Working Group*. Georgia, Atlanta: U.S. Department of Justice, 151-168, p. 151.

19 In New South Wales, Australia (1968-1986) and Canada (1974-1990) men were at least five times more likely to kill their wife if they separated and were no longer co-habiting, while in Chicago (1965-1990) they were twice as likely; however, Chicago had over twice as many killings. Women in the 16-19 year age group were twice as likely to be killed as any other age group in Canada and England (1977-1990). Wilson, M. and Daly, M. (1995). Uxoricide. In C. Block and R. Block, *Trends, Risks, and Interventions in Lethal Violence: Proceedings of the Third Annual Spring Symposium of the Homicide Research Working Group*. Georgia, Atlanta: U.S. Department of Justice, 169-178, p. 171. Wilson, M. and Daly, M. (1993). Spousal Homicide Risk and Estrangement. *Violence and Victims*, **8**, 3-16.

20 Shamu, S., Abrahams, N., Temmerman, M., Musekiwa, A. and Zarowsky, C. (2011). Intimate Partner Violence in Pregnancy in Africa. *PLoS ONE*, **6**, e17591.

21 Ballard, T.J., Saltzman, L.E., Gazmararian, J.A., Spitz, A.M., Lazorick, S. and Marks, J.S. (1998). Violence during Pregnancy: Measurement Issues. *American Journal of Public Health*, **88**, 274-276. The larger study quoted is: Amaro, H., Fried, L.E., Cabral, H. and Zuckerman, B. (1990). Violence during pregnancy and substance abuse. *American Journal of Public Health*, **80**, 575-579.

22 Johri, M., Morales, R.E, Boivin, J., Samayoa, B.E., Hoch, J.S., Grazioso, C.F., Barrios Matta, I.J. Sommen, C., Baide Diaz, E.L., Fong, H.R. and Arathoon, E.G. (2011). Increased risk of miscarriage among women experiencing physical or sexual intimate partner violence during pregnancy in Guatemala City, Guatemala: cross-sectional study. *BMC Pregnancy and Childbirth*, **11**, 49. For an overview of other studies see: Sarkar, N.N. (2008). The impact of intimate partner violence on women's reproductive health and pregnancy outcome. *Journal of Obstetrics and Gynaecology*, **28**, 266-271.

23 Meggitt, M.J. (1974 [1962]). *Desert People: a Study of the Walbiri Aborigines of Central Australia*. Sydney: Angus & Robertson, p. 90.

24 See, for instance, Russell, D.E.H. (1990). *Rape in Marriage*, (rev. edn.). Bloomington: Indiana University Press. Russell found that in July 1980 only three U.S. states had abolished the marital rape exemption, while five had partially restricted the exemption (page 21).

In Scotland, the first case where a man was convicted of raping has wife was in 1982 (Duffy) and the first case involving cohabitation was 1989. In England it was not until 1990 that the 'marital rape immunity rule' was recommended to be abandoned, and it was not until 1994 that the statutes excluded the term 'unlawful' – which had always been interpreted to mean outside of marriage – to define rape. In England, the first case where a husband was found to have raped his non-cohabiting wife was in 1990. (Painter, K. and Farrington, D.P. (1999). Wife Rape in Great Britain. In R. Muraskin (ed.) *Women and Justice: development of international policy*. Amsterdam: Gordon and Breach.)

       In Australia, rape was abolished in New South Wales in 1981 after the publication of a survey by the *Australian Woman's Weekly* which found that 13% of women who said they had been raped said that the perpetrator was their husband (one percent of all women). In 1985 the state of Victoria passed similar laws, and it was followed by all states, with South Australia the last in 1992. Canada made marital rape illegal in 1982, New Zealand in 1985, and Ireland in 1990. See Temkin, J. (2002). *Rape and the Legal Process* (2nd edn.). Oxford: Oxford. By the time Temkin's book was published, only 17 U.S. states had similar changes to the law, while 33 states continued to have exemptions in certain circumstances. Sweden altered its laws in 1965 so that marital rape was illegal, though the charge carried a lower maximum sentence than stranger rape; this was altered in 1984. (Elman, R.A. (1996). *Sexual subordination and state intervention: comparing Sweden and the United States*. Oxford: Berghahn Books, p. 90-91.) The Netherlands made marital rape illegal in 1995 and Germany did so in 1997. (McNamee, C. (2001). Rape. In R.J. Simon (ed.) *A Comparative Perspective on Major Social Problems*. Lanham, Maryland: Lexington Books.) Marital rape continues to be legal in otherwise wealthy nations such as Singapore and many poorer nations.

[25] Russell, D.E.H. (1990). Fn. 24, p. 57 and p. 64.

[26] Finkelhor, D. and Yllo, K. (1985). *License to Rape: Sexual Abuse of Wives*. New York: Holt, Rinehart & Winston, p. 6 –7.

[27] See Russell, D.E.H. (1990). Fn. 24, p. 68-71 citing Seger-Coulborn, I. (1976). *Vergewaltigung in der Ehe* (Rape in Marriage), Unpublished report (West Germany: Institut für Demoskopie Allensbach).

[28] *The Australian Women's Weekly* (1980), February 13, p. 27.

[29] For instance, William Blackstone, commenting in *Commentaries on the Laws of England in Four Books*:

> The husband also (by the old law) might give his wife moderate correction. For, as he is to answer for her misbehavior, the law thought it reasonable to intrust him with this power of restraining her, by domestic chastisement, in the same moderation that a man is allowed to correct his apprentices or children; for whom the master or parent is also liable in some cases to answer. But this power of correction was confined within reasonable bounds, and the husband was prohibited from using any violence to his wife, *aliter quam ad virum, ex causa regiminis et castigationis uxoris saue, licite et rationabiliter pertinet* [otherwise than lawfully and reasonably belongs to the husband for the due government and correction of his wife. Translation in Steele v. Steele, 65 F. Supp. 329 (D.D.C. 1946)]. The civil law gave the husband the same, or a larger, authority over his wife: allowing him, for some misdemeanours, *flagellis et fustibus acriter verberare uxorem;* [to beat his wife severely with scourges and sticks] for others, only *modicum castigationem adhibere* [to use moderate chastisement]. But, with us, in the politer reign of Charles the Second, this power of correction began to be doubted; and a wife may now have security of the peace against her husband; or, in return, a husband against his wife. Yet, the lower rank of people, who were always fond of the old common law, still claim and exert their ancient privilege: and the courts of law will still permit a husband to restrain a wife of her liberty, in case of any gross misbehaviour.
>
> (Blackstone, W. (1765) *Commentaries on the Laws of England in Four Books* (p. 444-445) Quoted in, with translations, Myers, J.E.B. (2005). *Myers on Evidence in Child, Domestic and Elder Abuse Cases*, volume 1, fourth edition. New York, Aspen Publishers, p. 736.)

[30] Chagnon, N.A. (1992a). Fn. 8, p. 125.

[31] Kaberry, P.M. (1939). *Aboriginal Woman: Sacred and Profane*. London: George Routledge, p. 153-54. (On the copy of her book that I was using one reader had written 'why would it?')

[32] Berndt, R.M. and Berndt C.H. (1977). *The World of the First Australians* (2nd edn.) Sydney: Ure Smith, p. 199.

[33] See Glass, D. (1995). *All My Fault: Why Women Don't Leave Abusive Men*. London: Virago.

[34] Meggitt, M.J. (1974 [1962]). Fn. 23, p. 106.

[35] Berndt, R.M. and Berndt, C.H. (1977). Fn. 32, p. 205.

[36] Goetz, A.T., Shackelford T.K. and Schipper, L.D. (2006). Adding Insult to Injury: Development and Initial Validation of the Partner-Directed Insults Scale. *Violence and Victims*, **21 (6)**, 691-706. These are, apparently, not abnormal levels compared to other studies (page 704). See also, McKibben, W.F., Goetz, A.T., Shackelford, T.K., Schipper, L.D., Starratt, V.G. and Stewart-Williams, S. (2007). Why do men insult their intimate partners? *Personality and Individual Differences*, **43**, 231-241, which found that insults were directly related to mate guarding and 'intersexual negative inducements' and 'intrasexual negative inducements'.

[37] Berndt, R.M. and Berndt, C.H. (1977). Fn. 32, p. 204-205.

[38] Berndt, R.M. and Berndt, C.H. (1977). Fn. 32, p. 205.

[39] Meggitt, M.J. (1974 [1962]). Fn. 23, p. 94.

[40] Wikan, U. (1982). *Behind the Veil in Arabia: Women in Oman*. Baltimore, Maryland: John Hopkins University Press, p. 36.

[41] Wikan, U. (1982). Fn. 40, p. 36.

[42] Wikan, U. (1982). Fn. 40, p. 37.

[43] Stanford, J.B., White, G.L. and Hatasaka, H. (2002). Timing intercourse to achieve pregnancy: current evidence. *Obstetrics and Gynecology*, **100**, 1333-41. Billings, E. and Westmore, A. (2000). *The Billings Method: Using the Body's Natural Signal of Fertility to Achieve or Avoid Pregnancy* (rev. edn.). Richmond, Vic.: A. O'Donovan.

[44] Warner, W.L. (1969 [1958]). Fn. 15, p. 78.

[45] Warner, W.L. (1969 [1958]). Fn. 15, p. 165.

[46] Warner, W.L. (1969 [1958]). Fn. 15, p. 70.

[47] In *Aboriginal Woman* Phyllis Kaberry found relative harmony: 'As elsewhere, a woman often welcomes an additional helper, and in the unions I encountered, the co-wives seemed to be on excellent terms, particularly where they were sisters. On the other hand the taboo on sexual intercourse during menstruation and pregnancy, the precarious nature of the food supplies, made another wife an asset to the man.' (p. 154.). However, Kaberry makes it clear that a co-wife is not essential and that a single wife is competent to perform all necessary tasks by herself (p. 113). Kaberry also noted that polygyny was rare in the people she was studying, representing only 12.6 percent and 7.4 percent of the two tribes she was studying (numbering 174 and 130 people, respectively). (Kaberry, P.M. (1939). Fn. 31, p. 114.)

    By contrast, Mervyn Meggitt found much higher levels of polygyny amongst the Walbiri: 42.5 and 34.3 percent in two samples (numbering 192 and 440 people respectively). Of 136 co-wives, only 17.6 percent were either sisters or half-sisters (the preferred relationship), though another 12.5 percent were doubtful sisters or half-sisters, suggesting their actual relationship couldn't be exactly determined. (Meggitt, M.J. (1974 [1962]). Fn. 23, p. 77-79.)

[48] Warner, W.L. (1969 [1958]). Fn. 15, p. 79-80.

[49] Meggitt, M.J. (1974 [1962]). Fn. 23, p. 109.

[50] Meggitt, M.J. (1974 [1962]). Fn. 23, p. 111.

[51] Batten, M. (1994 [1992]). *Sexual Strategies: How Females Choose Their Mates*. New York: G.P. Putnam's Sons, p. 74.

[52] Chagnon, N.A. (1992a). Fn. 8, p. 28.

[53] See, for instance, Buss, D.M. (2000). *The Dangerous Passion: Why Jealousy is as Necessary as Love and Sex*. New York: The Free Press, p. 210-215.

[54] Trivers, R.L. (1972). Parental investment and sexual selection. In B. Campbell (ed.), *Sexual Selection and the Descent of Man*. Chicago: Aldine-Atherton, p. 136-179.

[55] Malinowski, B. (1963 [1913]). *The Family among the Australian Aborigines: a Sociological Study*. New York: Schocken Books.

[56] Kaberry, P.M. (1939). Fn. 31, p. 86-87.

[57] Berndt, R.M. and Berndt, C.H. (1977). Fn. 32, p. 120.

[58] Chagnon, N.A. (1979a). Mate competition, favoring close kin, and village fissioning among the

Yanomamö Indians. In N.A. Chagnon and W. Irons (eds.), *Evolutionary Biology and Human Social Behavior* (pp. 86-131). North Scituate, M.A.: Duxbury Press, p. 103. Also, Stephanie Coontz observes:

> Many simple hunting and gathering societies place so much emphasis on sharing that a person who kills an animal gets no more of its meat than do his companions. A review of twenty-five hunting and gathering societies found that only in three did the hunter get the largest share of his kill. In most, the hunter was obliged to share the meat equally with other camp members, and in a few he got less than he distributed to others. Anthropologist Polly Wiessner observes that these customs create total interdependence among families "[T]he hunter spends his life hunting for others, and others spend their lives hunting for him."
> (Coontz, S. (2005). Fn. 3, p. 40.)

[59] Kaberry, P.M. (1939). Fn. 31, p. 155.
[60] Chagnon, N.A. (1979a). Fn. 58, p. 105.
[61] Meggitt, M.J. (1974 [1962]). Fn. 23, p. 95.
[62] See, for instance, Anonymous, (1966 [1890]). *My Secret Life*. Secaucus, N.J.: Castle Books.

Chapter 6   Female Sexual Selection of Males

[1] Valero, H. (1970). *Yanoámo: The Narrative of a White Girl Kidnapped by Amazonian Indians, as told to Ettore Biocca*. Translated by Dennis Rhodes. New York: Dutton, p. 139.
[2] Valero, H. (1970). Fn. 1, p. 141.
[3] Valero, H. (1970). Fn. 1, p. 141.
[4] Hallpike, C.R. (1977). *Bloodshed and Vengeance in the Papuan Mountains: the Generation of Conflict in Tauade Society*. Oxford: Clarendon Press, p. 149.
[5] Hallpike, C.R. (1977). Fn. 4, p. 216.
[6] Chagnon, N.A. (1992b). *Yanomamö: the Last Days of Eden*. San Diego: Harcourt Brace Jovanovich, p. 34.
[7] Valero, H. (1970). Fn. 1, p. 249.
[8] Meggitt, M.J. (1974 [1962]). *Desert People: A Study of the Walbiri Aborigines of Central Australia*. Sydney: Angus & Robertson, p. 98-99.
[9] Chagnon, N.A. (1992b). Fn. 6, p. 38-39.
[10] Chagnon, N.A. (1979a). Mate competition, favoring close kin, and village fissioning among the Yanomamö Indians. In N.A. Chagnon and W. Irons (eds.), *Evolutionary Biology and Human Social Behavior* (pp. 86-131). North Scituate, M.A.: Duxbury Press, p. 101.
[11] Hallpike, C.R. (1977). Fn. 4, p. 149.
[12] Hallpike, C.R. (1977). Fn. 4, p. 129-130.
[13] Meggitt, M.J. (1974 [1962]). Fn. 8, p. 180.
[14] Chagnon, N.A. (1992a). *Yąnomamö* (4th edn.). Fort Worth: Harcourt Brace Jovanovich, p. 187.
[15] Chagnon, N.A. (1996). Chronic problems in understanding tribal violence and warfare. In G.R. Bock and J.A. Goode (eds.), *Genetics of Criminal and Antisocial Behavior*. Chichester, New York: Wiley, p. 221.
[16] Valero, H. (1970). Fn. 1, p. 172.
[17] Meggitt, M.J. (1974 [1962]). Fn. 8, p. 176.
[18] Meggitt, M.J. (1974 [1962]). Fn. 8, p. 177.
[19] Buss, D.M. (2000). *The Dangerous Passion: Why jealousy is as Necessary as Love and Sex*. New York: The Free Press, p. 166.
[20] Buss, D.M. (2000). Fn. 19, p. 165.
[21] Greiling, H. and Buss, D.M. (2000). Women's sexual strategies: the hidden dimension of extra- pair mating. *Personality and Individual Differences*, **28**, 929-963, p. 943.

[22] Ellis, B.J. (1992). The evolution of sexual attraction: evaluative mechanisms in women. In J.H. Barkow, L. Cosmides, and J. Tooby (eds.), *The Adapted Mind*. New York: Oxford University Press, p 267-288, p. 277. Sadalla, E.K., Kenrick, D.T. and Vershure, B. (1987). Dominance and heterosexual attraction. *Journal of Personality and Social Psychology*, **52**, 730-738.

[23] Ellis, B.J. (1992). Fn. 22, p. 280.

[24] Ellis, B.J. (1992). Fn. 22, p. 278.

[25] Norris, M. and Pawlowski, W. (2010). (Eds). *Business of Consumer Book Publishing 2010*. Stamford, C.T.: Simba Information.

[26] Krentz, J.A. (1992). Trying To Tame The Romance. In J.A. Krentz (ed.) *Dangerous Men and Adventurous Women: Romance Writers on the Appeal of the Romance*. Philadelphia: University of Philadelphia Press. p. 107-114.

[27] Krentz, J.A. (1992). Fn. 26, p. 108-109.

[28] Barlow, L. (1992). The Androgynous Writer: Another Point of View. In J.A. Krentz (ed.) *Dangerous Men and Adventurous Women: Romance Writers on the Appeal of the Romance*. Philadelphia: University of Philadelphia Press. p. 45- 46.

[29] Barlow, L. (1992). Fn. 28, p. 46.

[30] Spencer, B. and Gillen, F.J. (1968 [1899]). *The Native Tribes of Central Australia*. New York: Dover Publications, p. 556-557.

[31] Chagnon, N.A. (1992a). Fn. 14, p.188-189.

[32] Townsend, J.M. and Levy, G.D. (1990). Effects of potential partners' costume and physical attractiveness on sexuality and partner selection. *Journal of Psychology*, **124**, 371-389.

[33] Townsend J.M. and Levy G.D. (1990). Fn. 32, p. 380.

[34] Kaberry, P.M. (1939). *Aboriginal Woman: Sacred and Profane*. London: George Routledge, p. 102.

[35] Shakespeare, W. (1914). Romeo and Juliet. In *The Oxford Shakespeare*. London: Oxford University Press, Act I. Scene I. Lines 175-184.

[36] Crenshaw, T.L. (1984). *Your Guide to Better Sex*. London: Book Club Associates, p. 4.

[37] Chagnon, N.A. (1992b). Fn. 6, p. 149.

[38] *Collins English Dictionary* (1972). London and Glasgow: Collins.

[39] Crenshaw, T.L. (1996). *The Alchemy of Love and Lust: Discovering Our Sex Hormones and How They Determine Who We Love, When We Love, and How Often We Love*. New York: Pocket Books, p. 57.

[40] Kaberry, P.M. (1939). Fn. 34, p. 111-12.

[41] Paagel, M. and Bodmer, W. (2003) A naked ape would have fewer parasites. *Proceedings of the Royal Society of London: Biological Sciences*. (Suppl) **270**: S117-S119.

[42] For a readable summary see, for instance, Jablonksi, N.G. (2006). *Skin: A Natural History*. Berkeley, CA: University of California Press.

[43] Darwin, C. (2004 [1871]). *The Descent of Man, and Selection in Relation to Sex*. London: Penguin, p. 674.

[44] Robins, A.H. (1991). *Biological Perspectives on Human Pigmentation*. Cambridge: Cambridge University Press, p. 207.

[45] Robins, A.H. Fn. 44, p. 206.

[46] Robins, A.H. Fn. 44, p. 206.

[47] Harris, L.J. (1956). Vitamin D and Bone. In G.H. Bourne (ed.), *The Biochemistry and Physiology of Bone* (pp. 581–622). New York: Academic Press, p. 594-597.

[48] Harris, L.J. (1956). Fn. 47, p. 597.

[49] Harris, L.J. (1956). Fn. 47, p. 597.

[50] Brickley, M., Mays, S. and Ives, R. (2008). Evaluation and interpretation of residual rickets deformities in adults. *International Journal of Osteoarchaeology*, **20**, 54-66.

[51] Pinhasi, R., Shaw, P., White, B. and Ogden A.R. (2006). Morbidity, rickets and long-bone growth in post-medieval Britain--a cross-population analysis. *Annals of Human Biology*, **33**, 372-389.

[52] Gibson, C. and Jung, K. (2002). *Historical Census Statistics on Population Totals By Race, 1790 to*

1990, and By Hispanic Origin, 1970 to 1990, For The United States, Regions, Divisions, and States. Population Division, U. S. Census Bureau, Working Paper Series No. 56. < http://www.census.gov/population/www/documentation/twps0056/twps0056.html> (retrieved, 19 December 2012)

[53] Weisberg, P., Scanlon, K.S., Li, R. and Cogswell, M.E. (2004). Nutritional rickets among children in the United States: review of cases reported between 1986 and 2003. *American Journal of Clinical Nutrition*, **80** (6 Suppl), 1697S-1705S.

[54] Craviari, T., Pettifor, J.M., Thacher, T.D., Meisner, C., Arnaud, J. and Fischer, P.R. (2008). Rickets: An Overview and Future Directions, with Special Reference to Bangladesh. *Journal of Health Population Nutrition*. **26**, 112–121. Kabir, M.L., Rahman, M., Talukder, K., Rahman, A., Hossain, O., Mostafa, G., Mannan, M.A., Kumar, S. and Chowdhury, A.T. (2004). Rickets among children of a coastal area of Bangladesh. *Mymensingh Medical Journal*, **13**, 53–58. Another study of 14 rachitic children in Bangladesh found that 70% were not vitamin D deficient: Fischer, P.R., Rahman, A., Cimma, J.P., Kyaw-Myint, T.O., Kabir, A.R., Talukder, K., Hassan, N., Manaster, B.J., Staab, D.B., Duxbury, J.M., Welch, R.M., Meisner, C.A., Haquem S. and Combs G.F. Jr. (1999). Nutritional rickets without vitamin D deficiency in Bangladesh. *Journal of Tropical Pediatrics*, **45**, 291-3.

[55] Craviari, T. and others. (2008). Fn. 54, p. 115.

[56] Roth, D.E., Rashed Shah, R., Black, R.E. and Baqui, A.H. (2010). Vitamin D Status of Infants in Northeastern Rural Bangladesh: Preliminary Observations and a Review of Potential Determinants. *Journal of Health Population Nutrition*. **28**, 458–469.

[57] *Dietary Supplement Fact Sheet: Vitamin D*. Office of Dietary Supplements. National Institutes of Health.

[58] Holick, M.F. (1991) Photosynthesis, metabolism, and biologic actions of vitamin D. In F.H. Glorieux (ed.) *Rickets*. Vol. 2. New York: Raven Press, p. 1–22.

[59] *Dietary Supplement Fact Sheet: Vitamin D*. Office of Dietary Supplements. National Institutes of Health.

[60] Australian Institute of Health and Welfare, Australasian Association of Cancer Registries. (2008). *Cancer in Australia 2010: an overview*. Cancer series no. 42. Cat. no. CAN 56. Canberra: Australian Institute of Health and Welfare.

[61] Australian Bureau of Statistics. (2006). *Causes of Death, Australia, 2006*. <http://www.abs.gov.au/AUSSTATS/abs@.nsf/0/4D3E7129BC863F0ECA25757C00137471?opendocument> (accessed 17-12-2012).

[62] Australian Institute of Health and Welfare. (2006). *Mortality over the twentieth century in Australia: Trends and patterns in major causes of death. Mortality Surveillance Series Number 4*. Canberra: Australian Institute of Health and Welfare, p. 175.

[63] Andersena, S., Laurberga, P., Hvingela, B., Kleinschmidta, K., Heickendorffa, L. and Mosekilde, L. (2012). Vitamin D status in Greenland is influenced by diet and ethnicity: a population-based survey in an Arctic society in transition. *British Journal of Nutrition*, **8**, 1-8.

[64] There has been great interest in the major histocompatibility complex, a group of genes believed to be important for the immune system. One recent study has found that European-Americans from Utah had MHC-disimilar partners while Africans from Nigeria had more MHC-similar partners than would be expected from a random selection. The article does not state whether the Nigerian couples had selected or arranged marriages. If the Nigerian marriages were arranged, then it would be expected that MHC genes would be more similar between partners than amongst the wider population because arranged marriages often occur between cousins. Chaix, R., Cao, C. and Donnelly, P. (2008). Is mate choice in humans MHC-dependent? *PLoS Genetics*. **4 (9)**, e1000184. A similar finding of a lack of genetic diversity between partners was found in a study amongst Amerindians, again, no consideration was made that marriage might be due to arrangement rather than selection. Hedrick, P.W. and Black, F. L. (1997). HLA and mate selection: no evidence in South Amerindians. *American Journal of Human Genetics*. **61**, 505-511. As usual, no consideration has been made for cultural factors, factors that may entirely undermine the intention of the researchers:

how can one investigate selection for diverse genes in marriage amongst people that may not choose their spouse? If the marriages were arranged, greater genetic similarity between partners would be expected compared to a randomly selected group of people from the population. These studies may well suggest that arranged marriages lead to similarity in MHC, that people select extra-maritally for MHC variety, and that people in Western nations are selecting the equivalent of extra-maritally.

The t-shirt experiment, where men wore t-shirts for two nights and women preferred the smell of those males with more dissimilar MHC genes, suggests that there may be a basis for mate selection based on pheromones. Wedekind, C., Seebeck, T., Bettens, F. and Paepke, A. J. (1995). MHC-dependent mate preferences in humans. *Proceedings of the Royal Society of London , B*, **260**, 245-249; Wedekind, C. and Füri, S. (1997). Body odour preferences in men and women: do they aim for specific MHC combinations or simply heterozygosity? *Proceedings of the Royal Society of London, B*, **264**, 1471-1479.

[65] See, for instance, Brennan, P.A. and Kendrick, K.M. (2006). Mammalian social odours: attraction and individual recognition. *Philosophical Transactions of the Royal Society B*. **361**, (1476), 2061-2078.

[66] Swaney, W.T. and Keverne, E.B. (2009) The evolution of pheromonal communication. *Behavioural Brain Research*. **200 (2)**, 239-247.

[67] Doty, R.L. (2010). *The Great Pheromone Myth*. Baltimore: John Hopkins University Press.

[68] Leyden, J.L., McGinley, K.J., Hölzle, E., Labows, J.N. and Kligman, A.M. (1981). The Microbiology of the Human Axilla and Its Relationship to Axillary Odor. *Journal of Investigative Dermatology*, **77**, 413-416.

[69] Zoe, W., Stitt, M.D., Lowell, A. and Goldsmith, M.D. (1995). Scratch and sniff: the dynamic duo. *Archives of Dermatology*, **131**, 997-999. Other odours include: scurvy, putrid (rotten); smallpox, (foul); and yellow fever (a butcher's shop).

[70] Di Natale, C. Macagnano, A., Paolesse, R., Tarizzo, E., Mantini, A. and D'Amico, A. (2000). Human skin odor analysis by means of an electronic nose. *Sensors and Actuators*, **65**, 216-219.

[71] Moshkin, M., Litvinova, N., Litvinova, E.A., Bedareva, A., Lutsyuk, A. and Gerlinskaya, A. (2012). Scent Recognition of Infected Status in Humans. *The Journal of Sexual Medicine*, **9**, 3211-3218.

[72] Franzoi, S.L. and Herzog, M.E. (1987). Judging physical attractiveness: what body aspects do we use? *Personality and Social Psychology Bulletin*, **13**, 19–33. Herz, R.S. and Cahill, E.D. (1997). Differential use of sensory information in sexual behavior as a function of gender. *Human Nature*, **8**, 275–286. Regan, P.C. and Berscheid, E. (1995). Gender differences in beliefs about the causes of male and female sexual desire. *Personal Relations*, **2**, 345–358.

[73] Chagnon, N.A. (1992a). Fn. 14, p. 11.

[74] See, for instance, Spencer, B. (1982). *The Aboriginal Photographs of Baldwin Spencer Selected and Annotated by Geoffrey Walker*. Melbourne: Currey O'Neil. Gardner, R. and Heider, K.G. (1969). *Gardens of War: Life and Death in the New Guinea Stone Age*. London: Deutsch.

[75] Hinsz, V.B. (1989). Facial resemblance in engaged and married couples. *Journal of Social and personal Relationships*, **6**, 223-229. Keller, M.C., Thiessen, D. and Young, R.K. (1996). Mate assortment in dating and married couples. *Personality and Individual Differences*. **21 (2)**, 217-221. Keller and others' data suggests that couples are initially assorted physically and then, for marriage, by personality. Little, A.C., Penton-Voak, I.S. and Burt, D.M. (2003). Investigating an imprinting-like phenomenon in humans: Partners and opposite-sex parents have similar hair and eye colour. *Evolution and Human Behavior*, **24 (1)**, 43-51.

[76] Roberts, S.C., Little, A.C., Gosling, L.M., Jones, B.C., Perrett, D.I., Carter, V. and Petrie, M. (2005) MHC-assortative facial preferences in humans. *Biology Letters*, **1 (4)**: 400-403.

[77] Hinsz, V. B. (1989). Fn. 75.

[78] Penton-Voak, I., Perrett, D. and Pierce, J. (1999) Computer graphic studies of the role of facial similarity in attractiveness judgements. *Current Psychology*, **18**, 104-117.

[79] Chagnon, N.A. (1979b). Is reproductive success equal in egalitarian societies? In N.A. Chagnon and W. Irons (eds.), *Evolutionary Biology and Human Social Organisation: an Anthropological Perspective*.

North Scituate, M.A.: Duxbury Press, 374-401.

[80] Chagnon, N.A. (1980). Kin-selection theory, kinship, marriage and fitness among the Yạnomamö Indians. In G. Barlow and J. Silverberg, *Sociobiology, Beyond Nurture/Nature?* . Boulder, CO: Westview Press, p. 562.

[81] Chagnon, N.A. (1974). *Studying the Yạnomamö*. New York: Holt, Rinehart and Winston, p. 142.

Chapter 7   Male Sexual Selection of Females

[1] Trivers, R.L. (1972). Parental investment and sexual selection. In B. Campbell (ed.) *Sexual Selection and the Descent of Man*. Chicago: Aldine-Atherton, p. 136-179.

[2] Morgan, J. (2002 [1852]). *The Life and Adventures of William Buckley*. Melbourne: The Text Publishing Company, p. 86.

[3] Leblanc, W. (1999). *Naked Motherhood : Shattering Illusions and Sharing Truths*. Milsons Point, N.S.W.: Random House Australia, page 95-97.

[4] Leblanc, W. (1999). Fn 3, p. 94-95.

[5] Leblanc, W. (1999). Fn 3, p. 95-96.

[6] Spencer, B. and Gillen, F.J. (1968 [1899]). *The Native Tribes of Central Australia*. New York: Dover Publications, p. 46.

[7] Chagnon, N.A. (1992a). *Yạnomamö* (4th edn.). Fort Worth: Harcourt Brace Jovanovich, p. 150.

[8] Berndt, R.M. and Berndt, C.H. (1977). *The World of the First Australians* (2nd edn.) Sydney: Ure Smith, p. 192.

[9] Berndt, R.M. and Berndt, C.H. (1977). Fn. 8, p. 192.

[10] Berndt, R.M. (1976). *Love Songs of Arnhem Land*. West Melbourne, Vic.: Thomas Nelson (Australia), p. 57.

[11] Berndt, R.M. (1976). Fn 10, p. 59.

[12] Berndt, R.M. (1976). Fn 10, p. 59.

[13] Berndt, R.M. (1976). Fn 10, p. 60.

[14] Observations of such behaviour include the friendship between the brothers Faben and Figan amongst the chimpanzees of Gombe (see Jane Goodall's *Through a Window*). However, there is also evidence that brothers can be competitors, as Faben and Figan were. One mitochondrial genes study (genes passed through the mother and not the father) found no definite preference between half-brothers over other males for alliances (see Goldberg, T.L. and Wrangham, R.W. (1997). Genetic correlates of social behavior in wild chimpanzees: Evidence from mitochondrial DNA. *Animal Behaviour*, **54**, 559-580.) DNA evidence concerning chimpanzee behaviour and reproductive success is in its infancy, however. For a summary see: Pusey, A.E. (2001). Of genes and apes: chimpanzee social organization and reproduction. In F.B.M. de Waal (ed.) *Tree of Origin: What Primate Behavior Can Tell Us about Human Social Evolution* (pp. 9-37). Cambridge, Mass.: Harvard University Press. Alliance should also depend on age similarity, since males will be more likely to align with a male of their own age group than a very young male, for instance, who is their half-brother. A greater than ten year gap between maternal half-brothers may reduce the likelihood of alliance formation.

[15] Goodall, J. (1999 [1988]). *In the Shadow of Man*. London: Phoenix, p. 77. See also Muller, M.N., Emery Thompson, M. and Wrangham, R.W. (2006). Male chimpanzees prefer mating with old females. *Current Biology*, **16**, 2234-2238.

[16] Goodall, J. (1999 [1988]). Fn. 15, p. 83.

[17] de Waal, F.B.M. (1998 [1982]). *Chimpanzee Politics*. Baltimore, Maryland: John Hopkins University Press, p. 152.

[18] Berndt, R.M. and Berndt, C.H. (1977). Fn. 8, p. 200-01.

[19] Meggitt, M.J. (1974 [1962]) *Desert People: a Study of the Walbiri Aborigines of Central Australia*. Sydney: Angus & Robertson, p. 100.

[20] Warner, W.L. (1969 [1958]). *A Black Civilization: a Study of an Australian Tribe* (rev. edn.). Gloucester,

Mass.: P. Smith, p. 71.

[21] Warner, W.L. (1969 [1958]). Fn. 20, p. 73.

[22] Warner, W.L. (1969 [1958]). Fn. 20, p. 75-76.

[23] Meggitt, M.J. (1974 [1962]). Fn. 19, p. 101.

[24] Morgan. J. (2002 [1852]). Fn. 2, p. 65.

[25] Berndt, R.M. (1976). Fn. 10, p. 34-35.

[26] Valero, H. (1970). *Yanoámo: The Narrative of a White Girl Kidnapped by Amazonian Indians, as told to Ettore Biocca*. Translated by Dennis Rhodes. New York: Dutton, p. 40-41.

[27] Valero, H. (1970). Fn. 26, p. 41-42.

[28] Valero, H. (1970). Fn. 26, p. 197.

[29] Valero, H. (1970). Fn. 26, p. 57-58.

[30] Valero, H. (1970). Fn. 26, p. 58.

[31] Berndt, R.M. (1976). Fn. 10, p. 34-35.

[32] Morgan. J. (2002 [1852]). Fn. 2, p. 115-116.

[33] Morgan. J. (2002 [1852]). Fn. 2, p. 116.

[34] Berndt, R.M. and Berndt, C.H. (1977). Fn. 8, p. 201.

[35] Chagnon, N.A. (2000). Manipulating kinship rules: a form of male Yanomamö reproductive competition. In L. Cronk and N.A. Chagnon (eds), *Adaptation and Human Behavior: an Anthropological Perspective* (pp. 115-132). New York: Aldine de Gruyter, p. 127.

Chapter 8   Sexuality

[1] Mazenod, B., Pugeat, M. and Forest, M.G. (1988) Hormones, sexual function and erotic behavior in women. In J.M.A. Sitsen. (ed.), *Handbook of Sexology. Vol. 6: The Pharmacology and Endocrinology of Sexual Function* (pp. 316-351). New York: Elsevier Science Publishers, p. 339.

[2] For an overview see, for instance, Pillsworth, E.G., Haselton M.G. and Buss D.M. (2004). Ovulatory shifts in female sexual desire. *The Journal of Sexual Desire*, **41**, 55-65.

[3] Crenshaw, T.L. (1984). *Your Guide To Better Sex*. London: Book Club Associates, p. 153.

[4] Mayor, S. (2004). Pfizer will not apply for a licence for sildenafil for women. *British Medical Journal*, **328**, 542.

[5] Day, E. (2004). Sex really is all in the mind for women. *Sunday Telegraph*, February 29. < http://www. telegraph.co.uk/news/1455661/Sex-really-is-all-in-the-mind-for-women.html> (accessed, Nov. 13, 2010).

[6] Kinsey, A.C., Pomeroy, W.B., Martin, C.E. and Gebhard, P.H. (1953). *Sexual Behavior in the Human Female*. Philadelphia: W.B. Saunders Co., p. 163-164.

[7] Chagnon, N.A. (1992b). *Yanomamö: the Last Days of Eden*. San Diego: Harcourt Brace Jovanovich, p. 34-36.

[8] Chagnon, N.A. (1992b). Fn. 7, p. 35.

[9] Meggitt, M.J. (1974 [1962]). *Desert People: a Study of the Walbiri Aborigines of Central Australia*. Sydney: Angus & Robertson, p. 111.

[10] Kinsey, A.C., Pomeroy, W.B., Martin, C.E. and Gebhard, P.H. (1953). Fn. 6, p. 538-539.

[11] Whipple, B., Ogden, G. and Komisaruk, B.R. (1992). Physiological correlates of imagery-induced orgasm in women. *Archives of Sexual Behavior*, **21**, 121-133.

[12] Crenshaw, T.L. (1984) Fn. 3, p. 149.

[13] Valero, H. (1970). *Yanoámo: The Narrative of a White Girl Kidnapped by Amazonian Indians, as told to Ettore Biocca*. Translated by Dennis Rhodes. New York: Dutton, p.133.

[14] Shakespeare, W. (1914). Romeo and Juliet. In *The Oxford Shakespeare*. London: Oxford University Press. Act II Prologue. Lines 10-15.

[15] Buss, D.M. (2000). *The Dangerous Passion: Why Jealousy is as Necessary as Love and Sex*. New York: Free Press, p. 170. Greiling, H. and Buss, D.M. (2000). Women's sexual strategies: the hidden dimension

of extra-pair mating. *Personality and Individual Differences,* **28**, 929-963.

[16] This has been investigated: Meston, C.M. and Frohlich, P.F. (2003). Love at first fright: partner salience moderates roller-coaster-induced excitation transfer. *Archives of Sexual Behavior,* **32: 6**, 537–544.

[17] Masters, W.H., and Johnson, V.E. (1966). *Human Sexual Response.* Boston: Little, Brown and Company, p. 122.

[18] Crenshaw, T.L. (1984). Fn. 3, p. 148.

[19] Masters, W.H., and Johnson, V.E. (1966). Fn. 17, p. 126. Below is the detail of the radiopaque experiment:

> A reasonable facsimile of seminal-fluid content with relation to surface tension, specific gravity, and specific density was developed in radiopaque substance with a liquid base. Since the time interval necessary for normal autolysis of seminal-fluid content is not present during active coition, every attempt was made to have the synthetic material correspond with immediately postejaculatory seminal fluid...
>
> The radiopaque substance was placed in a plastic cap and fitted over the cervix of each of the six participating study subjects. After the cap was placed, a radiographic check plate of the pelvis was taken to assure immersion of the cervix in the experimental material and to rule out the possibility of material spillage. Thereafter, radiographic plates were taken simultaneously with the orgasmic experience and after 10 minutes of the resolution phase. In none of the six individuals was there evidence of the slightest sucking effect on the media in the artificial seminal pool. Nor was there any evidence of the material in the cervical canal or the uterine endometrial cavity. All women were orgasmic during their experimental session. (Pages 122-23.)

[20] Billings, E. and Westmore, A. (2000). *The Billings Method: Using the Body's Natural Signal of Fertility to Achieve or Avoid Pregnancy* (rev. edn.). Richmond, Vic.: A. O'Donovan., 35-36.

[21] Eberhard, W.G. (1996). *Female Control: Sexual Selection by Cryptic Female Choice.* Princeton, NJ: Princeton University Press.

[22] For a readable overview of sperm competition see Birkhead, T.R. (2000). *Promiscuity: an Evolutionary History of Sperm Competition and Sexual Conflict.* London: Faber, pages182-194 for an overview of cryptic female choice.

[23] One recent British survey of 11,161 people aged 16 to 44 found that women were more likely to masturbate if they reported more frequent vaginal sex in the previous four weeks, while men in the same category reported lower levels of masturbation. Gerressu, M., Mercer, C.H., Graham, C.A., Wellings, K., Johnson, A.M. (2008). British National probability data on masturbation prevalence and associated factors. *Archives of Sexual Behavior,* **37**, 266-278. Men and women's response to vaginal intercourse was thus completely opposite, with men less likely to seek extra stimulation, while women were more likely to seek extra stimulation. Considering the number of women who report vaginal intercourse not leading to orgasm, if orgasm relieves discomfort due to arousal without orgasm then such an outcome would be expected.

[24] Masters, W.H., and Johnson V.E. (1966). Fn. 17, p. 119.

[25] Masters, W.H., and Johnson V.E. (1966). Fn. 17, p. 120.

[26] Masters, W.H., and Johnson V.E. (1966). Fn. 17, p. 121.

[27] Masters, W.H., and Johnson V.E. (1966). Fn. 17, p. 121.

[28] See, for instance: Fox, C.A. and Fox, B. (1971). A comparative study of coital physiology with special reference to the sexual climax. *Journal of Reproduction and Fertility,* **24**, 319–336. Fox, C. A., H. S. Wolff, and J. A. Baker. (1970). Measurement of intra-vaginal and intra-uterine pressures during human coitus by radio-telemetry. *Journal of Reproduction and Fertility,* **22**, 243–251.

The criticisms are fairly muted, with Fox and Fox limiting their argument that the 'insuck' method might only apply to men with low semen counts, since orgasm is not essential for fertilisation: 'It has been suggested by Parkes (personal communication) that while female orgasm may facilitate sperm transport, as described in this paper, it is well known that female orgasm is not essential for fertilization, and therefore for effective sperm transport. Our hypothesis must, therefore, be considered to explain only one, albeit the most efficient, method of fertilization. It is possible that the regular uterine contractions, which occur irrespective of orgasm, and in our tracings are abolished by orgasm, may have some significance in sperm transport as they do in some lower mammals. It may be that the achievement of orgasm in the female is necessary where the male's sperm count is low.' (Fox and Fox, p. 258.)

It is worthwhile considering what the 'upsuck' concept requires in reality. It would be the equivalent of a person attempting, without a straw, to 'upsuck' the drops of liquid at the bottom of a glass, but since there is no straw, the effect would need to be on the entire vagina cavity. We are told that this would occur solely through air pressure differences between the uterus and vagina. This air pressure is maintained after the man has ejaculated. Masters and Johnson tested this idea with a cervical cap with a liquid in the cap, and none of the liquid containing radiopaque entered the uterus. This would make sense since sperm enter the cervix and uterus by swimming at a remarkable 30cm/hour (Mortimer, S.T. and Swan, M.A. (1995). Variable kinematics of capacitating human spermatozoa. *Human Reproduction*, **10**, 3178–3182). In experiments, the time taken to reach the fallopian tubes has varied from 10 minutes to 1 hour, a time frame so short that it would seem to undermine any argument for manipulation, except through post-coital masturbation or the storage of sperm. For expulsive orgasm to affect sperm levels would require an impact on the storage of sperm, yet how or where sperm is stored is not clear, let alone whether it could be affected by uterine contractions. For an overview of the area see: Suarez, S.S. and Pacey, A.A. (2006). Sperm transport in the female reproductive tract. *Human Reproduction Update*, **12**, 23-37.

[29] See, for instance, McKee, A., Albury, K. and Lumby, C. (2008). *The Porn Report*. Carlton, Vic.: Melbourne University Publishing, p. 27.

[30] Chagnon, N.A. (1992a). *Yąnomamö* (4th edn.). Fort Worth: Harcourt Brace Jovanovich, p. 111.

[31] Chagnon N.A. (1992a). Fn. 30, p. 112.

[32] See, for instance, Short, R.V. (1979). Sexual selection and its component parts, somatic and genital selection, as illustrated by man and the great apes. *Advances in the Study of Behaviour*, **9**, 131-158. For a summary: Shackelford, T.K., Pound, N. and Goetz, A.T. (2005). Psychological and physiological adaptations to sperm competition in humans. *Review of General Psychology*, **9**, 228–248.

[33] Dixson, A.F. (2009). *Sexual Selection and the Origin of Human Mating Systems*. New York: Oxford, p. 73.

[34] Dixson, A.F. (2009). Fn. 33, p. 73.

[35] Diamond, J.M. (1997). *Why is Sex Fun?: the Evolution of Human Sexuality*. London: Weidenfield and Nicholson, p. 152.

[36] Chagnon, N. A. (1992a). Fn. 30, p. 69.

[37] Crenshaw, T.L. (1984) Fn. 3, p. 48.

[38] Crenshaw, T.L. (1984) Fn. 3, p. 185-186.

[39] Crenshaw, T.L. (1984) Fn. 3, p. 183.

[40] Crenshaw, T.L. (1984) Fn. 3, p. 108.

[41] Joseph, S. (1997). *She's My Wife, He's Just Sex*. Australian Centre for Independent Journalism, University of Technology, Sydney.

[42] Joseph, S. (1997). Fn. 41, p. 57.

[43] Joseph, S. (1997). Fn. 41, p. 61.

[44] Joseph, S. (1997). Fn. 41, p. 64.

[45] Joseph, S. (1997). Fn. 41, p. 70.

[46] Kinsey, A.C., Pomeroy, W.B. and Martin, C.E. (1948). *Sexual Behaviour in the Human Male*. Philadelphia: W.B. Saunders, p. 639.

[47] Joseph, S. (1997). Fn. 41, p. 66.

[48] Joseph, S. (1997). Fn. 41, p. 83.

[49] Joseph, S. (1997). Fn. 41, p. 84.

[50] Joseph, S. (1997). Fn. 41, p. 84.

[51] Joseph, S. (1997). Fn. 41, p. 85.

[52] Joseph, S. (1997). Fn. 41, p. 85.

[53] Joseph, S. (1997). Fn. 41, p. 72-73.

[54] Joseph, S. (1997). Fn. 41, p. 72.

[55] Joseph, S. (1997). Fn. 41, p. 72.

[56] Joseph, S. (1997). Fn. 41, p. 77.

[57] Joseph, S. (1997). Fn. 41, p. 78.

[58] Joseph, S. (1997). Fn. 41, p. 79.

[59] Solomon, S.S., Mehta, S.H, Latimore, A., Srikrishnan, A.K. and Celentano, D.D. (2010). The impact of HIV and high-risk behaviours on the wives of married men who have sex with men and injection drug users: implications for HIV prevention. *Journal of the International AIDS Society*, **13(Suppl 2)**, S7. They cite: Asthana, S. and Oostvogels, R. (1982). The social construction of male 'homosexuality' in India: implications for HIV transmission and prevention. *Social Science and Medicine*, **52**, 707–721.

[60] Solomon, S.S. and others (2010). Fn. 59, S7.

[61] Go, V.F., Srikrishnan, A.K., Sivaram, S., Murugavel, G.K., Galai, N., Johnson, S.C., Sripaipan, T., Solomon, S. and Celentano, D.D. (2004). High HIV prevalence and risk behaviors in men who have sex with men in Chennai, India. *Journal of Acquired Immune Deficiency Syndrome*, **35**, 314–319, p. 314.

[62] Gupta, A., Mehta, S., Godbole, S.V., Sahay, S., Walshe, L., Reynolds, S.J., Ghate, M., Gangakhedkar, R.R., Divekar, A.D., and Risbud, A.R. (2006). Same-sex behavior and high rates of HIV among men attending sexually transmitted infection clinics in Pune, India (1993-2002). *Journal of Acquired Immune Deficiency Syndrome*, **43**, 483–490.

[63] Solomon, S.S. and others (2010). Fn. 59, S7.

[64] Go, V.F., and others. (2004). Fn. 61, p. 316.

[65] Verma, R. and Collumbien, M. (2004). Homosexual activity among rural Indian men: implications for HIV interventions. *AIDS*, **18**, 1845-47.

[66] Brahmam, G,N., Kodavalla, V., Rajkumar, H., Rachakulla, H.K., Kallam, S., Myakala, S.P., Paranjape, R.S., Gupte, M.D., Ramakrishnan, L., Kohli, A. and Ramesh, B.M. (2008). Sexual practices, HIV and sexually transmitted infections among self-identified men who have sex with men in four high HIV prevalence states of India. *AIDS*, **22 Suppl 5**, S45-57.

[67] Kumta, S., Lurie, M., Weitzen, S., Jerajani, H., Gogate, A., Row-kavi, A., Anand, V., Makadon, H. and Mayer, K.H. (2010). Bisexuality, sexual risk taking, and HIV prevalence among men who have sex with men accessing voluntary counselling and testing services in Mumbai, India. *Journal of Acquired Immune Deficiency Syndromes*, **53**, 227-233. Solomon, S.S. and others (2010). Fn. 59, S7.

[68] Beyrer, C., Wirtz, A.L., Walker, D., Johns, B., Sifakis, F. and Baral, S.D. (2011). *The Global HIV Epidemics Among Men Who Have Sex with Men*. Washington: The World Bank.

[69] World Health Organization, Regional Office for South-East Asia. (2010). *HIV/AIDS among men who have sex with men and transgender populations in South-East Asia: The Current Situation and National Responses*. Singapore: WHO Regional Office for South East Asia.

[70] Warner, W.L. (1969 [1958]). *A Black Civilization: a Study of an Australian Tribe* (rev. edn.). Gloucester, Mass.: P. Smith, p. 49.

[71] Warner, W.L. (1969 [1958]). Fn. 70, p. 50-51.

[72] Blanchard, R. (2001). Fraternal birth order and the maternal immune hypothesis of male homosexuality. *Hormones and Behavior*, **40**, 105-114. Cantor, J. M., Blanchard, R., Paterson, A. D. and Bogaert, A. F. (2002). How many gay men owe their sexual orientation to fraternal birth order? *Archives of Sexual Behavior*, **31**, 63-71. The number of older sisters has not been found to be a factor. Birth order has also been found to be a factor for the sexual orientation of pedophiles with men more likely to be attracted to male children than to female children if they have more older brothers: Blanchard, R., Barbaree, H.E., Bogaert, A.F., Dickey, R., Klassen, P., Kuban, M.E. and Zucker, K.J. (2000). Fraternal

birth order and sexual orientation in pedophiles. *Archives of Sexual Behavior*, **29**, 463-478.

[73] Hamer, D. and Copeland, P. (1998). *Living with Our Genes*. London: Pan Books, p. 184.

Chapter 9 Female Alliances

[1] Heider, K. (1979). *Grand Valley Dani: Peaceful Warriors*. New York: Holt, Rinehart and Winston, p. 9-10.

[2] Ekman, P. (1998). Afterword to *The Expression of Emotions in Man and Animals*, by C. Darwin (pp. 363-393). London: HarperCollins, p. 379.

[3] Kaberry, P.M. (1939). *Aboriginal Woman: Sacred and Profane*. London: George Routledge, p. 111-112.

[4] Kaberry, P.M. (1939). Fn. 3, p. 29.

[5] Warner, W.L. (1969 [1958]). *A Black Civilization: a Study of an Australian Tribe* (rev. edn.). Gloucester, Mass.: P. Smith, p. 73.

[6] Warner, L. (1969 [1958]). Fn. 5, p. 75.

[7] Goodale, J.C. (1971 [1980]). *Tiwi Wives: a Study of the Women of Melville Island, North Australia*. Seattle: University of Washington Press, p. 131.

[8] Apter, T. and Josselson, R. (1999). *Best Friends: the Pleasures and Perils of Girls' and Women's Friendships*. New York: Three Rivers Press, p. 147.

[9] Apter, T. and Josselson, R. (1999). Fn. 8, p. 160.

[10] Apter, T. and Josselson, R. (1999). Fn. 8, p. 164.

[11] Apter, T. and Josselson, R. (1999). Fn. 8, p. 174.

[12] Benenson, J.F. and Christakos, A. (2003). The Greater Fragility of Females' Versus Males' Closest Same-Sex Friendships. *Child Development*, **74**, 1123-1129.

[13] Meggitt, M.J. (1974 [1962]). *Desert People: A Study of the Walbiri Aborigines of Central Australia*. Sydney: Angus & Robertson, p. 106.

[14] Berndt, R. M. and Berndt, C.H. (1951). *Sexual Behavior in Western Arnhem Land*. New York: Viking Fund, p. 192.

[15] Chagnon, N.A. (1979a). Mate competition, favoring close kin, and village fissioning among the Yanomamö Indians. In N.A. Chagnon and W. Irons (eds.), *Evolutionary Biology and Human Social Behavior* (pp. 86-131). North Scituate, M.A.: Duxbury Press, p. 122.

[16] Chagnon, N. A. (1992b). *Yanomamö: the Last Days of Eden*. San Diego: Harcourt Brace Jovanovich, p. 184.

[17] Chagnon, N.A. (1979a). Fn. 15, p. 122.

[18] Kanin, E.J. (1957). Male aggression in dating-courtship relations. *The American Journal of Sociology*, **63**, 197-204. The difference was not overwhelming, however, being 35.4 as against 24.5 per cent.

[19] Bourke, J. (2007). *Rape: Sex, Violence, History*. Emeryville, CA : Shoemaker & Hoard, p. 45-46.

[20] Chagnon, N.A. (1981). Terminological kinship, genealogical relatedness and village fissioning among the Yanomamö Indians. In R.D. Alexander and D.W. Tinkle (eds.), *Natural Selection and Social Behaviour: Recent Research and New Theory* (pp. 490-508). New York: Chiron Press, Oxford, p. 500.

[21] Apter, T. and Josselson R., (1999). Fn. 8, p. 296.

[22] Bleske-Rechek, A. and Lighthall, M. (2010). Attractiveness and Rivalry in Women's Friendships with Women. *Human Nature*, **21**, 82-97.

[23] Apter, T. and Josselson, R. (1999). Fn. 8, p. 208.

[24] Apter, T. and Josselson, R. (1999). Fn. 8, p. 208.

[25] Apter, T. and Josselson, R. (1999). Fn. 8, p. 33.

[26] Apter, T. and Josselson, R. (1999). Fn. 8, p. 40.

[27] Apter, T. and Josselson, R. (1999). Fn. 8, p. 31.

[28] Apter, T. and Josselson, R. (1999). Fn. 8, p. 28.

[29] Apter, T. and Josselson, R. (1999). Fn. 8, p. 49.

[30] Tannen, D. (1990). *You Just Don't Understand: Women and Men in Conversation*. London: Virago, p. 98.

[31] Aries, E.J. and Johnson, F.L. (1983). Close Friendship in Adulthood: Conversational Content Between Same-Sex Friends. *Sex Roles*, **12**, 1183-1196. The following study had similar results and discusses previous studies with similar findings: Caldwell, M.A. and Peplau, L.A. (1982) Sex Differences in Same-Sex Friendship. *Sex Roles*, **8**, 721-732.

[32] Buhrmester, D. and Prager, K. (1995). Patterns and functions of self-disclosure during childhood and adolescence. In K.J. Rotenberg (1995). *Disclosure Processes in Children and Adolescents*. New York: Cambridge University Press, 10-56, p. 27.

[33] Maynard, D.B. and Springer-Proverbs, R.F. (2010). An Examination of Gender Differences in Characteristics of Barbadian Adolescents' Same-sex Friendships. *Caribbean Journal of Psychology*, **3**, 40-53.

[34] Kaberry, P.M. (1939). Fn. 3, p. 255.

[35] Meggitt, M.J. (1974 [1962]). Fn. 13, p. 95.

[36] Apter, T. and Josselson, R. (1999). Fn. 8, p. 35.

[37] Campbell, A. (2002). *A Mind of Her Own: the Evolutionary Psychology of Women*. New York: Oxford University Press, p. 155.

[38] Tannen, D. (1990). Fn. 30, p. 104.

[39] Valero, H. (1970). *Yanoáma: The Narrative of a White Girl Kidnapped by Amazonian Indians, as told to Ettore Biocca*. Translated by Dennis Rhodes. New York: Dutton, p. 117.

[40] Apter, T. and Josselson R., (1999). Fn. 8, p. 160.

[41] Warner, W.L. (1969 [1958]). Fn. 5, p. 70.

[42] Burbank, V.K. (1994). *Fighting Women: Anger and Aggression in Aboriginal Australia*. Berkeley: University of California Press, p. 100-101.

[43] Burbank, V.K. (1994). Fn. 42, p. 99. Table 3.

[44] Burbank, V.K. (1994). Fn. 42, p. 98.

[45] Burbank, V.K. (1994). Fn. 42, p. 99. Table 3.

[46] Burbank, V.K. (1994). Fn. 42, p. 104. Table 4.

[47] Burbank, V.K. (1994). Fn. 42, p. 105. Table 5.

[48] Björkvist, K. (1994) Sex Differences in Physical, Verbal, and Indirect Aggression: A Review of Recent Research. *Sex Roles*, **30**, 177-188. Radliff, K.M. and Joseph, L.M. (2011). Girls Just Being Girls? Mediating Relational Aggression and Victimization. *Preventing School Failure*, **55**, 171-179.

[49] Campbell, A. (2004). Female Competition: Causes, Constraints, Content and Contexts. *The Journal of Sex Research*, **41**, 16-26.

[50] Lagerspetz, K.M.J., Björkvist, K. and Peltonen,T. (1988). Is indirect aggression typical of females? Gender differences in aggressiveness in 11- to 12- year-old children. *Aggressive Behavior*, **14**, 403-414. Björkvist, K., Lagerspetz, K.M.J. and Kaukainen, A. (1992) Do Girls Manipulate and Boys Fight? Development Trends in Regard to Direct and Indirect Aggression. *Aggressive Behavior*, **18**, 117-127. One study of 491 3rd to 6th grade children found that 15.6% of boys compared to 0.4% of girls were overtly aggressive, while 17.4% of girls compared to 2.0% of boys were relationally aggressive (or indirectly aggressive) as selected by peers. Overall, 27% of boys compared to 21.7% of girls were deemed to be aggressive: Crick, N.R. and Grotpeter, J.K. (1995). Relational Agression, Gender, and Social-Psychological Adjustment. *Child Development*, **66**, 710-722.

[51] Österman, K., Björkqvist, K., Lagerspetz, K. M. J., Kaukiainen, A., Landau, S. F., Frączek, A. and Caprara, G. V. (1998). Cross-cultural evidence of female indirect aggression. *Aggressive Behavior*, **24**, 1–8.

[52] Crick, N.R., Casas, J.F. and Ku, H. (1999). Relational and physical forms of peer victimization in preschool. *Developmental Psychology*, **35**, 376-385. Burr, J.E., Ostrov, J.M. and Jansen, E.A. (2005) Relational Aggression and Friendship During Early Childhood: "I Won't Be Your Friend!" *Early Education & Development*, **16**, 162-183.

[53] Dijkstra, J.K., Lindenberg, S., Verhulst, F.C., Ormal, J. and Veenstra, R. (2009). The Relation Between Popularity and Aggressive, Destructive, and Norm-Breaking Behaviors: Moderating Effects of

Athletic Abilities, Physical Attractiveness, and Prosociality. *Journal on Research on Adolescence*, **19**, 401-413.

[54] Dijkstra, J.K., Cillessen, A.H.N. and Borch, C. (2012). Popularity and Adolescent Friendship Networks: Selection and Influence Dynamics. *Developmental Psychology*, Advance online publication. doi: 10.1037/a0030098

[55] Apter, T. and Josselson, R. (1999). Fn. 8, p. 71.

[56] Apter, T. and Josselson, R. (1999). Fn. 8, p. 95.

[57] Tannen, D. (1990). Fn. 30, p. 116.

[58] Apter, T. and Josselson, R. (1999). Fn. 8, p. 36.

[59] Apter, T. and Josselson, R. (1999). Fn. 8, p. 171-172.

[60] Apter, T. and Josselson, R. (1999). Fn. 8, p. 171.

[61] Apter, T. and Josselson, R. (1999). Fn. 8, p. 174.

[62] Apter, T. and Josselson, R. (1999). Fn. 8, p. 166.

[63] Apter, T. and Josselson, R. (1999). Fn. 8, p. 167.

[64] Quine, L. (2001). Workplace Bullying in Nurses. *Journal of Health Psychology*, **6**, 73-84.

[65] Einarsen, S. and Skogstad, A. (1996). Bullying at work: epidemiological findings in public and private organizations. *European Journal of Work and Organizational Psychology*, **5**, 185-202.

[66] Einarsen, S. & Raknes. B.I. (1997). Harassment at work and the vicimization of men. *Victims and Violence*, **12**, 247-263.

[67] Niederle, M. and Vesterlund, L. (2006). Do Women Shy Away From Competition? Do Men Compete Too Much? *Quarterly Journal of Economics*, **122**, 1067-1101. In this study women and men undergraduates completed simple mathematical tasks with equal overall ability for each gender. They were paid 50c for each correct answer. They were later given the chance to be paid $2 or nothing in a winner takes all tournament. They were not told of the number of correct answers they gave. Men were twice as likely to apply for the tournament (73% to 35%). Female under-performers were more likely to apply for the tournament than high-performers, while male under-performers were slightly less likely to apply for the tournament than high-performers. (This supports other research that suggests that people are unaware of their level of competence. Kruger, J. and Dunning, D. (1999). Unskilled and Unaware of It: How Difficulties in Recognizing One's Own Incompetence Lead to Inflated Self-Assessments. *Journal of Personality and Social Psychology*, **77**, 1121-34.)

This study has been repeated, with minor modifications, eight times by other researchers with similar outcomes, and once with a finding of no gender difference. In research where the participants were given feedback on their performance, women were more likely to enter the tournament, but still at a lower rate than men. Men were less likely to enter the tournament when they were not given feedback (66% down from 78%), but still entered the tournament more than women (66% to 50%). In research that used language and mathematical tasks, it was found that men are more likely than women to enter the maths test rather than the language test. For an overview, see: Niederle, M. and Vesterlund, L. (2011). Gender and Competition. *Annual Review in Economics*, **3**, 601–30. In an overview of the studies, it was observed that men tend to perform better in tournaments than in basic reward for correct answer scenarios. However, in single-sex tournaments, women also perform better in tournaments. It has thus been suggested that women are less likely to compete in the presence of males. See: Gneezy, U., Niederle, M. and Rustichini, A. (2003). Performance in competitive environments: gender differences. *Quarterly Journal of Economics*, **118**, 1049–74.

[68] Morrison, R.L. (2009). Are Women Tending and Befriending in the Workplace? Gender Differences in the Relationship Between Workplace Friendships and Organizational Outcomes. *Sex Roles*, **60**, 1-13.

[69] Tannen, D. (1990). Fn. 30, p. 99.

[70] Apter, T. and Josselson, R. (1999). Fn. 8, p. 200.

[71] Bearman, P.S. and Moody, J. (2004). Suicide and Friendships Among American Adolescents. *American Journal of Public Health*, **94**, 89-95.

72 Apter, T. and Josselson, R. (1999). Fn. 8, p. 246.

73 Apter, T. and Josselson, R. (1999). Fn. 8, p. 246.

74 Kaberry, P.M. (1939). Fn. 3, p. 128.

75 Kaberry, P.M. (1939). Fn. 3, p. 234-237.

76 Kaberry, P.M. (1939). Fn. 3, p. 237-240.

77 Kaberry, P.M. (1939). Fn. 3, p. 240-245.

78 Kaberry, P.M. (1939). Fn. 3, p. 245-251.

79 Kaberry, P.M. (1939). Fn. 3, p. 253-268.

80 Kaberry, P.M. (1939). Fn. 3, p. 254-255.

81 Kaberry, P.M. (1939). Fn. 3, p. 261.

82 Kaberry, P.M. (1939). Fn. 3, p. 258-259.

83 Kaberry, P.M. (1939). Fn. 3, p. 263-264.

84 Kaberry, P.M. (1939). Fn. 3, p. 264.

85 Kaberry, P.M. (1939). Fn. 3, p. 268.

86 Kaberry, P.M. (1939). Fn. 3, p. 267.

87 Kaberry, P.M. (1939). Fn. 3, p. 267.

88 Apter, T. and Josselson, R. (1999). Fn. 8, p. 248.

89 Tannen, D. (1990). Fn. 30, p. 110.

90 Tannen, D. (1990). Fn. 30, p. 110.

91 Valero, H. (1970). Fn. 39, p. 134.

92 Klitzinger, C. (1996). Toward a politics of lesbian friendship. In J.S. Weinstock and E.D. Rothblum (eds.), *Lesbian Friendships: for Ourselves and Each Other* (pp. 295 – 299). New York University Press: New York.

93 Rose, S.D., Zand, D. and Cini M.A. (1993) Lesbian courtship scripts. In E.D. Rothblum and K.A. Brehony (eds.), *Boston Marriages: Romantic but Asexual Relationships among Contemporary Lesbians* (pp. 70-85). Amherst: University of Mass. Press. Vetere, V.A. (1982) The role of friendship in the development and maintenance of lesbian love relationships. *Journal of Homosexuality*, **8 (2)**, 51-65.

94 Becker, C.S. (1988) *Unbroken Ties: Lesbian Ex-lovers*. Boston: Alyson.

> To varying degrees, lesbian ex-lovers retain their ties to one another after their breakup and use these bonds to rebuild their lives. An ex-lover remains an important part of a woman's evolving identity: as a woman, as a lesbian, and as a participant in intimate relationships. (p. 211)

95 Weinstock, J.S. and Rothblum, E.D. (1996) *Lesbian Friendships: for Ourselves and Each Other*. New York: New York University Press.

96 Apter, T. and Josselson, R. (1999). Fn. 8, p. 263.

97 Apter, T. and Josselson, R. (1999). Fn. 8, p. 264.

98 Goodenow, C., Szalacha, L.A., Robin, L.E. and Westheimer, K. (2008). Dimensions of Sexual Orientation and HIV-Related Risk Among Adolescent Females: Evidence from a Statewide Survey. *American Journal of Public Health*, **98**, 1051-1058.

99 London, S. (2006). Lesbian, Bisexual Women: Misperceptions of Risk Jeopardize Sexual Health. *Perspectives on Sexual and Reproductive Health*, **38**, 53-54.

100 Diamond, L.M. (2003). Was it a phase? Young women's relinquishment of lesbian/bisexual identities over a 5-year period. *Journal of Personality and Social Psychology*, **84**, 352–364.

101 Diamond, L.M. (2000). Sexual identity, attractions, and behavior among young sexual-minority women over a 2-year period. *Developmental Psychology*, **36**, 241–250.

102 Hallpike, C.R. (1977). *Bloodshed and Vengeance in the Papuan mountains: the Generation of Conflict in Tauade Society*. Oxford: Clarendon Press, p. 122; Meggitt, M.J. (1974 [1962]). Fn. 13, p. 57.

103 Warner, W.L. (1969 [1958]). Fn. 5, p. 19.

104 Chagnon, N.A. (1981). Fn. 20, p. 499.

105 Chagnon, N.A. (1988b). Male Yąnomamö manipulations of kinship classifications of female kin for

reproductive advantage. In L. Betzig, M. Borgerhoff Mulder and P. Turke (eds.), *Human Reproductive Behaviour: A Darwinian Perspective* (pp. 23-48). Cambridge: Cambridge University Press, p. 37.

[106] Chagnon, N.A. (1981). Fn. 20, p. 497.

[107] Hawkes, K., O'Connell, J. F. and Blurton Jones, N.G. (1997). Hadza women's time allocation, offspring provisioning, and the evolution of long postmenopausal life spans. *Current Anthropology*, 38, 551-577. Hawkes, K., O'Connell, J.F. and Blurton Jones, N. (1989). Hard-working Hadza grandmothers. In V. Standen and R. Foley (eds.), *Comparative Socioecology: the Behavioural Ecology of Human and Other Mammals* (pp. 341-66). Oxford: Oxford University Press.

[108] Kaplan, H., Hill, K., Lancaster, J. and Magdalena Hurtado, A. (2000). A theory of human life history evolution: diet, intelligence, and longevity. *Evolutionary Anthropology*, 9, 156-85. Hawkes, O'Connell, and Burton-Jones argue that 20% of medium/large mammal carcasses acquired by the Hadza are obtained by scavenging, which must make one wonder about the environment of the Hadza – I have never read of Australian Aborigines ever acquiring meat via scavenging (O'Connell J.F., Hawkes, K. and Blurton Jones, N. (1988). Hadza scavenging: implications for plio/pleistocene hominid. *Current Anthropology*, 29, 356-363).

[109] Kennedy, G.E. (2003). Palaeolithic grandmothers? Life history theory and early Homo. *Journal of the Royal Anthropological Institute*, 9, 549-572.

[110] For instance, 'A son, though he is separated from his mother during periods of initiation, cherishes a strong affection for her during his lifetime. He gives her food and cares for her in old age.' Kaberry, P.M. (1939). Fn. 3, p. 54.

[111] Meggitt, M.J. (1974 [1962]). Fn. 13, p. 236.

[112] Berndt, R.M. and Berndt C.H. (1977). *The World of the First Australians* (2nd edn.) Sydney: Ure Smith, p. 279. Phyllis Kaberry also noted that older women began to be introduced to the sacred male ceremonies. Only men that are initiated or being initiated are allowed to see certain ceremonies: 'He witnesses ceremonies which are dara:gu or sacred and which are fraught with a certain amount of supernatural danger for those who see them for the first time.' 'Only in extreme old age were the Nyigina women permitted to eat some of the dara:gu food...' Kaberry, P.M. (1939). Fn. 3, p. 222.

[113] Warner, W.L. (1969 [1958]). Fn. 5, p. 262.

[114] Price, S. (1984). *Co-Wives and Calabashes*. Ann Arbor: University of Michigan Press, p. 49.

[115] Price, S. (1984). Fn. 114, p. 50.

[116] Price, S. (1984). Fn. 114, p. 50-1

[117] Valero, H. (1970). Fn. 39, p. 133.

[118] Valero, H. (1970). Fn. 39, p. 133.

[119] Sulaiman, T. (2006). Mother-in-law made to pay £35,000 for inflicting four wretched months. *The Times*, July 25. <http://women.timesonline.co.uk/tol/life_and_style/women/article692147.ece> (accessed March 23, 2010).

[120] Wikan, U. (1982). *Behind the Veil in Arabia: Women in Oman*. Baltimore, Maryland: John Hopkins University Press, p. 36.

[121] Meggitt, M.J. (1962 [1974]). Fn. 13, p. 111.

[122] North Gulf Oceanic Society, (2010). *Killer whales of southern Alaska*. http://www.whalesalaska. org/killer_whales_of_southern_alaska.html (Retrieved January 31, 2010). ('NGOS is a non-profit organization that specializes in marine mammal research, conservation and education.')

[123] North Gulf Oceanic Society. (2010). Fn. 122.

[124] Whitehead, H. and Mann, J. (2000). Female reproductive strategies of cetaceans: life histories and calf care. In J. Mann, R.C. Connor, P.L. Tyack and H. Whitehead (eds.), *Cetacean societies: field studies of dolphins and whales* (pp. 219-246). Chicago : University of Chicago Press, p. 231-233.

[125] Conner, R.C., Read, A.J. and Wrangham, R. (2000). Male reproductive strategies and social bonds. In J. Mann, R.C. Connor, P.L. Tyack and H. Whitehead (eds.), *Cetacean Societies: Field Studies of Dolphins and Whales* (pp. 247-269). Chicago: University of Chicago Press, p. 267.

[126] Meggitt, M.J. (1962 [1974]). Fn. 13, p. 136.

[127] Meggitt, M.J. (1962 [1974]). Fn. 13, p. 275.

[128] Meggitt, M.J. (1962 [1974]). Fn. 13, p. 128.

[129] Meggitt, M.J. (1962 [1974]). Fn. 13, p. 150.

[130] Kaberry, P.M. (1939). Fn. 3, p. 106-7. It would be interesting to note if paternity played a part in deciding which children were aborted, and if children created in a sexual selection were less likely to be aborted.

[131] Chagnon, N.A. (1992a). *Yąnomamö* (4th edn.). Fort Worth: Harcourt Brace Jovanovich, p. 199.

[132] United Nations Population Fund (UNFPA) and EngenderHealth (2003). *Obstetric Fistula Needs Assessment Report: Findings from Nine African Countries.* New York: United Nations Population Fund, p. 5.

[133] United Nations Population Fund (UNFPA) and EngenderHealth (2003). Fn. 132, p. 8.

[134] Firstly, it should be emphasised that no one appears to question the heritability of physical prowess, unlike intelligence. Considering the importance that has been placed on physical prowess amongst males in combat, then the importance of the heritability of physical ability should not be underestimated, and is an overlooked area for study. The heritability of intelligence is based on the vast amount of material that has been obtained concerning the heritability of IQ, an indicator of intelligence. Twin studies have pointed to the heritability of IQ to be as high as 0.7, or seventy percent. Bouchard T.J. Jr., Lykken, D.T., McGue, M., Segal, N.L. and Tellegen, A. (1990). Sources of human psychological differences: the Minnesota study of twins reared apart. *Science*, **250**, 223-228. From twin and similar studies, 'The typical conclusion was that about 50% of the observed variance in personality is due to genetic factors.' Bouchard, T.J. Jr. (1994). Genes, environment, and personality. *Science*, **264**, 1700-02. Though with fewer comprehensive studies, it has also been found that for special mental abilities (verbal, spatial, perceptual speed and accuracy, memory) 'Genetic factors strongly influence special mental abilities but less than for general intelligence.' Bouchard T.J. Jr. (1998). Genetic and environmental influences on adult intelligence and special mental abilities. *Human Biology*, **70**, 257-280. For a readable overview see: Wright, L. (1997). *Twins: Genes, environment and the mystery of identity*. London: Phoenix, and Wright, W. (1999). *Born That Way: Genes, behavior, personality*. London: Routledge. See also Jensen, A.R. (1998). *The g Factor: the Science of Mental Ability*. Westport, Conn.: Praeger.

It may also be noted that it would not be surprising if the heritability of the immune system is even more heritable than intelligence. I base this on the workings of the immune system, particularly the specific 'lock-and-key' mechanism that it utilises to attack foreign proteins. The method of our immune system suggests that we are born with our immune response, and that it must be largely inherited by its very nature. Such a conclusion does appear to run against evolutionary arguments (something immediately pointed out by Francis Crick on hearing of the lock-and-key mechanism). Some theorists, such as William Hamilton, have attempted to link the existence of dual sexes to explain this perceived conundrum of two sexes. However, I suspect that any perceived flaw in evolution is due to a misunderstanding of the interaction of the immune system and parasites, rather than a flaw in evolution. In any event, genetics (selection) should play an important role in the immune system, physical and intellectual prowess.

Chapter 10   Communication

[1] Darwin, C. (2004 [1871]). *The Descent of Man, and Selection in Relation to Sex.* London: Penguin, p. 102-103.

[2] Darwin, C. (2004 [1871]). Fn. 1, p. 92.

[3] Darwin, C. (2004 [1871]). Fn. 1, p. 92-93.

[4] Darwin, C. (2004 [1871]). Fn. 1, p. 100.

[5] Butler, R.A. (1960). Acquired drives and the curiosity-investigative motives. In R.H. Waters, D.A. Rethlingshafer and W.E. Caldwell (ed.) *Principles of Comparative Psychology* (pp. 144-176). New

York: McGraw-Hill, p. 164-165. See also Harlow, H.F. (1950). Learning and satiation of response in intrinsically motivated complex puzzle performance by monkeys. *Journal of Comparative and Physiological Psychology*, **43**, 289-294.

6 Butler, R.A. (1960). Fn. 5, p. 168. Butler, R. A. and Harlow, H. F. (1954). Persistence of visual exploration in monkeys. *Journal of Comparative and Physiological Psychology*, **47**, 258-263. Butler, R. A. and Harlow, H. F. (1957). Discrimination learning and learning sets to visual exploration incentives. *The Journal of General Psychology*, **57**, 257-264.

7 Butler, R.A. (1960). Fn. 5, p. 169. Butler, R. A. (1954). Incentive conditions which influence visual exploration. *Journal of Experimental Psychology*, **48**, 19-23.

8 Roes, F, (1998). An interview of Napoleon Chagnon. *Human Ethology Bulletin*, **13**, 6-12, p. 7.

9 Rossen, M. Klima, E.S., Bellugi, U., Bihrle, A. and Jones, W. (1996). Interaction between language and cognition: Evidence from Williams syndrome. In J.J. Beitchman, N.J Cohen, M.M. Konstantareas, and R. Tannock (eds.), *Language, learning and behavior disorders: developmental, biological, and clinical perspectives* (pp. 367-392). New York: Cambridge University Press, p. 367. The children's linguistic capabilities appear to come from rehearsed scripts (WMS is Williams syndrome; DNS is Down syndrome): 'In general on these two subtests, the WMS adolescents display considerable verbal facility, but typically focus on superficial likeness or scripts of personal experience, reflecting undeveloped conceptualization' (p. 371). 'We also found that some of these WMS subjects use the same level of expressivity regardless of how many times or to whom they told the story' (p. 374). Interestingly, the disorder appears to be a pleasant one: 'WMS subjects that come into our lab are often immediately friendly, running up to the examiner or to strangers, requiring little or no "warm up" period. This behavior is unlike the matched DNS subjects. It is also unlike normal children in the same situation' (p. 374).

10 Rossen, M. and others (1996). Fn. 9, p. 375-376. In drawing, the Williams syndrome children tend to be able to replicate details, but not the larger structure. Interestingly, Down syndrome children have the exact opposite problem, drawing the structure but with no details (p. 376). Williams syndrome children have a tremendous capacity for facial recognition, perhaps as a reflection on their focus on details (p. 376).

11 Bellugi, U., Marks, S., Bihrle, A. and Sabo, H. (1988). Dissociation between language and cognitive functions in Williams syndrome. In D. Bishop and K. Mogford (eds) *Language Development in Exceptional Circumstances* (pp. 177-189). London: Churchill Livingstone, p. 189.

12 Bellugi, U. and others. (1988). Fn. 11, p. 179.

13 Bellugi, U. and others. (1988). Fn. 11, p. 188.

14 See, for instance, Rumbaugh, D.M., Savage-Rumbaugh, E.S. and Sevcik R.A. (1996). Biobehavioral roots of language: a comparative perspective of chimpanzee, child and culture. In R.W. Wrangham, W.C. McGrew, F.B.M. de Waal and P.G. Heltne (eds.), *Chimpanzee Cultures* (pp. 319-334). Cambridge: Harvard University Press.

15 CNN, *Koko the gorilla calls for the dentist.* Aug 8, 2004. <http://primates.com/gorillas/koko.html> (accessed Feb 2, 2010):

> About a month ago, Koko, a 300-plus-pound ape who became famous for mastering more than 1,000 signs, began telling her handlers at the Gorilla Foundation in Woodside she was in pain. They quickly constructed a pain chart, offering Koko a scale from one to 10.
> When Koko started pointing to nine or 10 too often, a dental appointment was made. And because anesthesia would be involved, her handlers used the opportunity to give Koko a head-to-toe exam.
> "She's quite articulate," volunteer John Paul Slater said. "She'll tell us how bad she's feeling, how bad the pain is. It looked like it was time to do something."

16 Howitt, A.W. (1996 [1904]). *Native Tribes of South-East Australia.* Canberra: Aboriginal Studies Press, p. 723-725. If women's use of verbal language assisted extra-marital sexual selections, then

the requirement that a widow does not speak after the death of her husband-by-arrangement would reduce her capacity to sexually select during the interim.

[17] Howitt, A.W. (1996 [1904]). Fn. 16, p. 734.

[18] Goldin-Meadow, S. (2007). On Inventing Language. *Daedalus*, **137**, 100-104, p. 103.

[19] Goldin-Meadow, S. (2005). Watching language grow. *Proceedings of the National Academy of Sciences*, **102**, 2271–2272, p. 2271.

[20] Senghas, A. and Coppola, M. (2001). Children creating language: How Nicaraguan sign language acquired a spatial grammar. *Psychological Science*, **12**, 323-328. Sengahs, A., Kita, S. and Özyürek, A. (2004). Children creating core properties of language: Evidence from an emerging sign languague in Nicaragua. *Science*, **305**, 1779-1782.

[21] Sengahs, A. and others (2004). Fn. 20, p. 1781.

[22] Rumbaugh, D.M. and others (1996). Fn. 14, p. 321.

[23] Rumbaugh, D.M. and others (1996). Fn. 14, p. 323.

[24] Rumbaugh, D.M. and others (1996). Fn. 14, p. 321.

[25] Terrace, H.S., Petitto, L.A., Sanders, R.J. and Bever, T.G. (1979). Can an Ape create a sentence? *Science*, **206**, 891-902, p. 895.

[26] Goodall, J. (1986). *The Chimpanzees of Gombe: Patterns of Behaviour*. Cambridge: Cambridge University Press, p. 34.

[27] Petitto, L. A. (2005). How the brain begets language: On the neural tissue underlying human language acquisition. In J. McGilvray (ed.), *The Cambridge Companion to Chomsky* (pp. 84-101). Cambridge, Great Britain: Cambridge University Press, p. 85.

[28] Petitto, L. A. (2005). Fn. 27, p. 86-87.

[29] Savage-Rumbaugh, E.S. and Lewin, R. (1994). *Kanzi: The Ape at the Brink of the Human Mind*. New York: John Wiley and Sons. Savage-Rumbaugh, E.S., Murphey, J., Sevcik, R.A., Brakke, K.E., Williams, S.L. and Rumbaugh, D.M. (1993). Language comprehension in ape and child. *Monographs of the Society for Research in Child Development*. Serial No. 233, **58**, 3-4.

[30] Burling, R. (2005). *The Talking Ape: How Language Evolved*. New York: Oxford University Press, p. 11.

[31] Burling, R. (2005). Fn. 30, p. 11-12.

[32] Burling, R. (2005). Fn. 30, p. 12-13.

[33] For a comparison of chimpanzee and human behaviour, see Kagan, J. (2004). The uniquely human in human nature. *Daedalus*, **133**, 77-88.

[34] Terrace, H.S. and others (1979). Fn. 25, p. 895.

[35] Goldin-Meadow, S. (2007). Fn. 18, p. 100.

[36] Lieberman, P. (1998). *Eve Spoke: Human Language and Human Evolution*. New York: W.W. Norton, p. 5.

[37] Chagnon, N.A. (1992b). *Yanomamö: the Last Days of Eden*. San Diego: Harcourt Brace Jovanovich, p. 49.

[38] Heider, K. (1979). *Grand Valley Dani: Peaceful Warriors*. New York: Holt, Rinehart and Winston, p.75.

[39] Chagnon, N.A. (1988b). Male Yąnomamö manipulations of kinship classifications of female kin for reproductive advantage. In L. Betzig, M. Borgerhoff Mulder and P. Turke (eds.), *Human Reproductive Behaviour: A Darwinian Perspective* (pp. 23-48). Cambridge: Cambridge University Press, p. 45.

[40] Chagnon, N.A. (2000). Manipulating kinship rules: a form of male Yanomamö reproductive competition. In L. Cronk and N.A. Chagnon (eds.), *Adaptation and Human Behavior: an Anthropological Perspective* (pp. 115-132). New York: Aldine de Gruyter, p. 127.

[41] Heider, K. (1979). Fn. 38, p. 50.

[42] Dunbar, R. I. M. (1993). Coevolution of neocortical size, group size and language in humans. *Behavioral and Brain Sciences*, **16**, 681-735. For a readable version see, Dunbar, R.I.M. (1998). *Grooming, Gossip, and the Evolution of Language*. Cambridge, Mass.: Harvard University Press.

[43] Chagnon, N.A. (1992b). Fn. 37, p. 49.

[44] Chagnon, N.A. (1992b). Fn. 37, p. 99.

[45] Halpern, D.F. (2000). *Sex differences in cognitive abilities* (3rd edn.). Mahwah, N.J.: L. Erlbaum Associates, p. 97. The studies quoted are: Martin, D.J., & Hoover, H.D. (1987). Sex differences in educational achievement: A longitudinal study. *Journal of Early Adolescence*, 7, 65-83. Hines, M. (1990). Gonadal hormones and human cognitive development. In J. Balthazart (Ed.), *Brain and behaviour in vertebrates 1: Sexual differentiation, neuroanatomical aspects, neurotransmitters and neuropeptides* (pp. 51-63). Basel, Swtizerland: Karger. Block, R.A., Arnott, D.P., Quigley, B., & Lynch, W.C. (1989). Unilateral nostril breathing influences lateralized cognitive performance. *Brain and Cognition*, 9, 181-190. Bilous, F.R., & Krauss, R.M. (1988). Dominance and accommodation in the conversational behaviours of same- and mixed-gender dyads. *Language and Communication*, 8, 183-194.

Basic test results appear to reflect the same outcomes. In the 2008 national test results in the UK for 14 year old pupils, the percentage of girls compared to boys who reached level 5 or above in English, reading and writing were 80 to 66, 76 to 62, and 83 to 70, respectively. The percentage of girls compared to boys who reached level 6 or above in English, reading and writing were 41 to 26, 41 to 26 and 44 to 30, respectively. By contrast, level 5 results in mathematics and science were similar, 77 to 76 and 71 to 72, as were level 6 results, 56 to 58 and 40 to 42, for girls and boys respectively. (The Guardian. (2008) Sats: Table of Sats results - boys v girls. August, 12. <http://www.guardian.co.uk/education/table/2008/aug/12/sats.results.boysgirls> (accessed Nov. 14, 2010).)

[46] Halpern, D.F. (2000). Fn. 45, p. 96. The studies quoted are: McGuiness, D. (1976). Sex differences in the organization of perception and cognition. In B. Lloyd & J. Archer (eds.), *Exploring sex differences* (pp. 123-156). New York: Academic Press. Smolak, L. (1986) *Infancy*. Englewood Cliffs, NJ: Prentice-Hall. Gazzaniga, M.S., Ivry, R.B. and Mangun, G.R. (1998). *Cognitive neuroscience: The biology of the mind*. New York: Norton. Moore, T. (1967). Language and intellitence: A longitudinal study of the first eight years. *Human Development*, 10, 88-106. Shucard, D.W., Shucard, L.J. and Thomas, D.G. (1987). Sex differences in electrophysiological activity in infancy: Possible implications for langauge development. In S.U. Philips, S. Steele and C. Tanz (eds.), *Language, gender, and sex in comparative perspectives* (pp. 278-295). Cambridge, England: Cambridge University Press. Huttenlocher, J., Haight, W., Bryk, A., Seltzer, M. and Lyons, T. (1991). Early vocabulary growth: Relation to language input and gender. *Development Psychology*, 27, 236-248.

[47] Halpern, D.F. (2000). Fn. 45, p. 174.

[48] Zitzmann, M. and Nieschlag, E. (2001). Testosterone levels in healthy men and the relation to behavioural and physical characteristics: facts and constructs. *European Journal of Endocrinology*, 144, 183-197. Kocoska-Maras, L., Zethraeus, N., Rådestad, A.F., Ellingsen, T., von Schoultz, B., Johannesson, M. and Hirschberg, A.L. (2011). A randomized trial of the effect of testosterone and estrogen on verbal fluency, verbal memory, and spatial ability in healthy postmenopausal women. *Fertility and Sterility*, 95, 152-7. One small study found that hypogonadal men's verbal ability improved with testosterone increases. Alexander, G.M., Swerdloff, R.S., Wang, C., Davidson, T., McDonald, V., Steiner, B. and Hines, M. (1998). Androgen-behavior correlations in hypogonadal men and eugonadal men. II. Cognitive abilities. *Hormones and Behavior*, 33, 85-94. Another small study found that normal men's verbal ability improved with testosterone enhancement: O'Connor, D.B., Archer, J., Hair, W.M. and Wu, F.C. (2001). Activational effects of testosterone on cognitive function in men. *Neuropsychologia*, 39, 1385-94.

[49] Apter, T. and Josselson R., (1999). *Best Friends: the Pleasures and Perils of Girls' and Women's Friendships*. New York: Three Rivers Press, p. 95.

[50] Tannen, D. (1990). *You Just Don't Understand: Women and Men in Conversation*. London: Virago, p. 77.

[51] Tannen, D. (1990). Fn. 50, p. 86.

[52] Tannen, D. (1990). Fn. 50, p. 126.

[53] Cicero, M.T. (1971). *On the Good Life*. Translated by Michael Grant. Harmondsworth: Penguin, p. 241-242

[54] Archer, J. (2009). Does sexual selection explain human sex differences in aggression? *Behavioral and Brain Sciences*, **32**, 249-311. Archer, J. (2004) Sex differences in aggression in real-world settings: A meta-analytic review. *Review of General Psychology*, **8**, 291–322.

[55] Hallpike, C.R. (1977). *Bloodshed and Vengeance in the Papuan Mountains: the Generation of Conflict in Tauade Society*. Oxford: Clarendon Press, p. 131.

[56] Warner, W.L. (1969 [1958]). *A Black Civilization: a Study of an Australian Tribe* (rev. edn.). Gloucester, Mass.: P. Smith, p. 150.

[57] Chagnon, N.A. (1992a). *Yąnomamö* (4th edn.). Harcourt Brace Jovanovich, p. 22.

[58] Hallpike, C.R. (1977). Fn. 55, p. 169-170.

[59] Heider, K. (1979). Fn. 38, p. 70.

[60] Heider, K. (1979). Fn. 38, p. 70.

[61] Heider, K. (1979). Fn. 38, p. 70.

[62] Chagnon, N.A. (1974). *Studying the Yąnomamö*. New York: Hold, Rinehart and Winston, p. 166.

[63] Chagnon, N.A. (1974). Fn. 62, p. 161.

[64] Chagnon, N.A. (1974). Fn. 62, p. 167.

[65] Chagnon, N.A. (1992a). Fn. 57, p. 191.

[66] Chagnon, N.A. (1992a). Fn. 57, p. 191-192.

[67] Chagnon, N.A. (1992a). Fn. 57, p. 193.

[68] Chagnon, N.A. (1992a). Fn. 57, p. 39.

[69] Goodall, J. (1990 [2000]). *Through a Window: Thirty Years with the Chimpanzees of Gombe*. London: Phoenix, p. 47.

[70] Goodall, J. (1990 [2000]). Fn. 69, p. 47.

[71] 'An extended phenotypic character is the product of the interaction of many genes whose influence impinges from both inside and outside the organism'. Dawkins, R. (1999 [1982]). *The Extended Phenotype*. Oxford: Oxford University Press, p. 239.

[72] Flaccus, G. (2009). Jumbo squid invade San Diego shores, spook divers. *The Associated Press*. July 16, 2009. <http://www.breitbart.com/article.php?id=D99FNF0O0> (accessed Nov 14, 2010).

[73] Chagnon, N.A. (1992a). Fn. 57, p. 187-188.

[74] Valero, H. (1970). *Yanoámo: The Narrative of a White Girl Kidnapped by Amazonian Indians, as Told to Ettore Biocca*. Translated by Dennis Rhodes. New York: Dutton, p. 63.

[75] Valero, H. (1970). Fn. 74, p. 63.

[76] Valero, H. (1970). Fn. 74, p. 111.

[77] Berndt, R.M. and Berndt C.H. (1977). *The World of the First Australians* (2nd edn.) Sydney: Ure Smith, p. 170-176. See also chapter 12 of Spencer, B. and Gillen, F.J. (1968 [1899]). *The Native Tribes of Central Australia*. New York: Dover Publications, p. 450-476, which gives examples of the removal of teeth and nose-boring.

[78] Berndt, R.M. and Berndt, C.H. (1977). Fn. 77, p. 180-185.

[79] Berndt, R.M. and Berndt, C.H. (1977). Fn. 77, p. 186-187.

[80] Chagnon, N.A. (1992a). Fn. 57, p. 103.

[81] Chagnon, N.A. (1992a). Fn. 57, p. 102.

[82] Berndt, R.M. and Berndt, C.H. (1977). Fn. 77, p. 389-90.

[83] Hallpike, C.R. (1977) Fn. 55, p. 166.

[84] Hallpike, C.R. (1977) Fn. 55, p. 172.

[85] Chagnon, N.A. (1992a). Fn. 57, p. 164.

[86] Valero, H. (1970). Fn. 74, p. 179.

[87] Chagnon, N.A. (1992a). Fn. 57, p. 196-197.

[88] Heider, K. (1979). Fn. 38, p. 103.

[89] Goodall, J. (1999 [1988]). *In the Shadow of Man*. London: Phoenix, p. 109-111.

[90] Goodall, J. (1999 [1988]). Fn. 89, p. 111.

[91] Pinker, S. and Jackendoff, R. (2005). The faculty of language: what's special about it? *Cognition*, **95**, 201–236, p. 206. It should be noted that this paper was written in response to another paper (Hauser, M.D., Chomsky, N. and Fitch, W.T. (2002). The faculty of language: what is it, who has it, and how did it evolve? *Science*, **298**, 1569–1579.). I support Pinker and Jackendoff's general disagreement with that paper, which suggests that human language was designed for a particular function (navigation is used as an example) and then adapted to other functions (p. 1578). Indeed, this entire chapter is focused on explaining that complex human language developed for very specific purposes – conveying information. Much of this debate is a long-standing debate in linguistics (and elsewhere) as to whether human language (and other unique behaviours and morphologies) evolved for specific purposes or simply happen to have evolved and been adapted without specific evolutionary pressures. The latter argument is implausible: evolution of specific traits requiring specific mutations will not occur by chance.

However, to suggest that the human brain can only process certain concepts with complex language, is, in my opinion, a serious error that elevates complex language to a new state of abstraction. It is suggested by Pinker and Jackendoff that certain concepts can only be understood with language. Thus certain thoughts are impossible without the existence of language. I am yet to be convinced of this. No doubt many people will present complex, abstract ideas as proof of this (none of which exist in the environment in which humans evolved and normally exist), yet I am unsure how many people understand these concepts.

We can work out that a plane was travelling at 400 kilometres per hour, but we really have no understanding of this, except that it is very fast. Further examples might be complex words such as metaphysics, epistemology, pedagogy or paradigm or a number of 'post-modernist' words, but most of these words are rarely used by the majority of the population, are names referring to a large number of concepts (and thus ambiguous), or can be reduced to simpler words. Usually our actual appreciation of complex ideas is quite general and simple. I expect that most people that use complex statistics, for instance, have no understanding of why the statistical application works, only that it does work. This is certainly how statistics, and most complex applications, are taught. Two of the most brilliant ideas in the last two hundred years – natural and sexual selection – can be summarised in a few sentences, and could probably be demonstrated with mime and pictures. The use of the finches of the Galapagos Islands or the fossil record to demonstrate evolution reflects how important visual images are to our understanding of complex ideas. It might be asked how someone could even use a concept that they couldn't imagine.

[92] Pinker, S. and Jackendoff, R. (2005). Fn. 91, p. 206.

[93] Pinker, S. and Jackendoff, R. (2005). Fn. 91, p. 206.

[94] Pinker, S. and Jackendoff, R. (2005). Fn. 91, p. 206.

[95] Einstein, A. (2001 [1920]). *Relativity: the special and general theory*. Translated by Robert W. Lawson. London: Routledge.

Chapter 11   The Evolution of Social Organisation

[1] de Waal, F.B.M. (1998 [1982]). *Chimpanzee Politics*. Baltimore, Maryland: John Hopkins University Press, p. 161.

[2] Chagnon, N.A. (1981). Terminological kinship, genealogical relatedness and village fissioning among the Yanomamö Indians. In R.D. Alexander and D.W. Tinkle (eds.), *Natural Selection and Social Behaviour: Recent Research and New Theory* (pp. 490-508). New York: Chiron Press, Oxford, p. 497

[3] There is evidence of fire occurring 790,000 years ago (Goren-Inbar, N., Alperson, N., Kislev, M.E., Simchoni, O., Melamed, Y., Ben-Nun, A. and Werker, E. (2004). Evidence of hominin control of fire at Gesher Benot Ya'aqov, Israel. *Science*, **304**, 725-727) and 1.6 million years ago (Bellomo, R.V. (1994). Methods of determining early hominid behavioral activities associated with the controlled

use of fire at FxJj 20 Main, Koobi Fora, Kenya. *Journal of Human Evolution*, **27**, 173–195.)

[4] Johanson, D. and Edgar, B. (2006) *From Lucy to Language*. New York: Simon and Schuster, p. 99.

[5] Heider, K. (1979). *Grand Valley Dani: Peaceful Warriors*. New York: Holt, Rinehart and Winston, p. 37

[6] Heider, K. (1979). Fn. 5, p. 77.

[7] Hallpike, C.R. (1977). *Bloodshed and Vengeance in the Papuan Mountains: the Generation of Conflict in Tauade Society*. Oxford: Clarendon Press, p. 65.

[8] Marenky, R.K., Kuroda, S., Vineberg, E.O. and Wrangham, R.W. (1996). The Significance of Terrestrial Herbaceous Foods for Bonobos, Chimpanzees, and Gorillas. In R.W. Wrangham, W.C. McGrew, F.B.M. de Waal and P.G. Heltne (eds.) *Chimpanzee Cultures* (pp. 59-75). Cambridge, Mass.: Harvard University Press, p. 65 and 67.

[9] Tutin, C.E.G. (1996). Reproductive Success Story: Variability Among Chimpanzees and Comparisons with Gorillas. In R.W. Wrangham, W.C. McGrew, F.B.M. de Waal and P.G. Heltne (eds.), *Chimpanzee Cultures* (pp. 181-193). Cambridge, Mass.: Harvard University Press, p. 182.

[10] Lovett, N.C. (1991). An evolutionary framework for assessing illness and injury in nonhuman primates. *Yearbook of Physical Anthropology*, **34**, 117-155, p. 132.

[11] Goodall, J. (1986). *The Chimpanzees of Gombe: Patterns of Behaviour*. Cambridge: Cambridge University Press, p. 98.

[12] Goodall, J. (1986). Fn. 11, p. 98.

[13] van Schaik, C.P. and van Hooff, J.A.R.A.M. (1996). Toward an understanding of the orangutan's social system. In W.C. McGrew, L.F. Marchant, and T. Nishida (eds.) *Great Ape Societies* (pp. 3-15). Cambridge: Cambridge University Press, p. 9.

[14] Marenky, R.K., and others. (1996). Fn. 8, p. 9.

[15] Most of the debate for chimpanzee body structure appears to concern energy restrictions (see, for evidence and an overview, Pontzer, H. and Wrangham, R.W. (2004). Climbing and the daily energy cost of locomotion in wild chimpanzees: implications for hominoid locomotor evolution. *Journal of Human Evolution*, **46**, 315-333.). Considering levels of inter-community competition, chimpanzees appear to live in reasonable population densities, and if energy levels were an issue, then they would exist in lower population densities, as the orang-utan does. One must conclude that energy limitations are not an issue for chimpanzees.

[16] Goodall, J. (1999 [1988]). *In the Shadow of Man*. London: Phoenix, p. 269-270. See also de Waal, F.B.M. (1998 [1982]). *Chimpanzee Politics*. Baltimore, Maryland: John Hopkins University Press.

[17] Goodall, J. (2000 [1990]). *Through a Window: Thirty Years with the Chimpanzees of Gombe*. London: Phoenix, p. 87.

[18] Goodall, J. (1999 [1988]). Fn.16, p. 203.

[19] Goodall, J. (1999 [1988]). Fn. 16, p. 239.

[20] In cricket, bowling the ball at the batsman's head in cricket, even at 150 km/h, is perfectly legitimate, as long as the ball bounces first. Broken bones (jaw, cheek, fingers and other parts of the body) are not uncommon against very fast bowlers. It is a similar matter for the fielders: while the baseball players have a mitt to catch the ball, cricketers use their bare hands, resulting in broken fingers, the splitting of the webbing between their fingers, or bruising from being hit by the ball. The dangers of cricket are reflected in some of the fielding positions, such as 'silly mid-on' and 'silly mid-off'. These positions are dangerously close to the batsman, and fieldsmen can be hospitalised by a well-hit shot that happens to hit the fieldsman's body or head. Most fielders in 'silly' positions now wear protective clothing, such as helmets. The dangers of the cricket ball need to be experienced to be fully appreciated. For instance, in practice, a fieldsman or bowler in cricket may simply use a foot to attempt to stop the ball after it has been firmly hit by the batsman; even if they have time to use their hands, they may refrain from doing so, simply because it is easier to fetch the ball from the nearest fence, than to have a broken finger.

[21] Goodall, J. (1999 [1988]). Fn. 16, p. 86.

[22] Goodall, J. (1999 [1988]). Fn. 16, p. 132.

[23] Stringer, C and Andrews, P. (2005). *The Complete World of Human Evolution*. London: Thames and Hudson, p. 185

[24] Stringer, C and Andrews P. (2005) Fn. 23, p. 18.

[25] Stringer, C and Andrews P. (2005) Fn. 23, p. 120.

Chapter 12 Conclusion

[1] Alexander, R. D. (1990). How Did Humans Evolve? Reflections on the Uniquely Unique Species. *University of Michigan Museum of Zoology Special Publication*, **1**, 1-38.

[2] Sullivan, K. and Tager-Flusberg, H. (1999). Second-Order Belief Attribution in Williams Syndrome: Intact or Impaired? *American Journal on Mental Retardation*, **104**, 523-532. 'One of the most striking features of children with Williams syndrome is their unusual friendliness and strong interest in other people (Beuren, Schulze, Eberle, Harmjanz, & Apitz, 1964; Dilts, Morris, & Leonard, 1990; Udwin & Yule, 1991; Udwin, Yule, & Martin, 1987). Children with Williams syndrome have been observed to be extremely sociable, charming, outgoing, and highly concerned about the well-being of others (Gosch & Pankau, 1994; Sarimski, 1997; Tomc, Williamson, & Pauli, 1990).' 'However, despite their extreme sociability and strong interest in people, children and adolescents with Williams syndrome experience great difficulty with their peers (Greer, Brown, Pal, Choudry, & Klein, 1997). By middle childhood, these children have difficulty interpreting social interactions and social cues, engage in socially inappropriate behavior, and have great difficulty making and sustaining friendships with their peers (Gosch & Pankau, 1994; Sarimski, 1997).' (p. 523).

[3] Goodall, J. (1990 [2000]). *Through a Window: Thirty Years with the Chimpanzees of Gombe*. London: Phoenix, p. 19.

[4] Goodall, J. (1990 [2000]). Fn. 3, p. 19.

[5] Malcolm, L. (2007). The 'thrifty gene' hypothesis and indigenous health. *The Health Report*. http://www.abc.net.au/radionational/programs/healthreport/the-thrifty-gene-hypothesis-and-indigenous-health/3238202> (accessed November 15, 2010).

[6] Goodall, J. (1990 [2000]). Fn. 3, p. 12.

[7] Roes, F. (1998). An interview of Napoleon Chagnon. *Human Ethology Bulletin*, **13 (4)**, p. 6.

[8] Roes, F. (1998). Fn. 7, p. 10.

[9] Stringer, C. and McKie, R. (1996). *African Exodus: the Origins of Modern Humanity*. New York: Henry Holt and Company, p. 28-30.

[10] Stringer, C. and McKie, R. (1996). Fn. 9, p. 30.

[11] Stringer, C. and McKie, R. (1996). Fn. 9, p. 30-31

[12] Berger, D.T. and Trinkaus, E. (1995). Patterns of Trauma among the Neandertals. *Journal of Archaeological Science*, **22**, 841-852. Trinkaus, E. (2012). Neandertals, early modern humans, and rodeo riders. *Journal of Archaeological Science*, **39**, 3691-3693.

[13] Heider, K. (1979). *Grand Valley Dani: Peaceful Warriors*. New York: Holt, Rinehart and Winston, p. 106.

[14] Heider, K. (1979). Fn. 13, p. 107-108.

[15] Medawar, J. and Pyke, D. (2000). *Hitler's Gift: Scientists Who Fled Nazi Germany*. London: Richard Cohen Books., p. 120.

[16] Wilson, E.O. (1998). Consilience among the great branches of learning. *Daedalus*, **127 (1)**, 131-149, p. 132-133.

[17] Wilson, E.O. (1998). Fn. 16, p. 145.

[18] Wilson, E.O. (1998). Fn. 16, p. 145.

[19] See, for instance, Chagnon, N.A. (1992a). *Yąnomamö* (4th edn.). Forth Worth: Harcourt Brace Jovanovich, p. 27.

[20] Watson, J.D. and Crick, F.H.C. (1953). A Structure for Deoxyribose Nucleic Acid. *Nature*, **171**, 737-738.

References

Alexander, G.M., Swerdloff, R.S., Wang, C., Davidson, T., McDonald, V., Steiner, B. and Hines, M. (1998). Androgen-behavior correlations in hypogonadal men and eugonadal men. II. Cognitive abilities. *Hormones and Behavior*, **33**, 85-94.

Alexander, R.D. (1979). *Darwinism and Human Affairs*. Seattle: University of Washington Press.

Alexander, R. D. (1990). How Did Humans Evolve? Reflections on the Uniquely Unique Species. *University of Michigan Museum of Zoology Special Publication*, **1**, 1-38.

Amaro, H., Fried, L.E., Cabral, H. and Zuckerman, B. (1990). Violence during pregnancy and substance abuse. *American Journal of Public Health*, **80**, 575-579.

Andersena, S., Laurberga, P., Hvingela, B., Kleinschmidta, K., Heickendorffa, L. and Mosekilde, L. (2012). Vitamin D status in Greenland is influenced by diet and ethnicity: a population-based survey in an Arctic society in transition. *British Journal of Nutrition*, **8**, 1-8.

Anonymous (1966 [1890]). *My Secret Life*. Secaucus, N.J.: Castle Books.

Apter, T. and Josselson, R. (1999). *Best Friends: the Pleasures and Perils of Girls' and Women's Friendships*. New York: Three Rivers Press.

Archer, J. (2004) Sex differences in aggression in real-world settings: A meta-analytic review. *Review of General Psychology*, **8**, 291–322.

Archer, J. (2009). Does sexual selection explain human sex differences in aggression? *Behavioral and Brain Sciences*, **32**, 249-311.

Aries, E.J. and Johnson, F.L. (1983). Close Friendship in Adulthood: Conversational Content Between Same-Sex Friends. *Sex Roles*, **12**, 1183-1196.

Asthana, S. and Oostvogels, R. (1982). The social construction of male 'homosexuality' in India: implications for HIV transmission and prevention. *Social Science and Medicine*, **52**, 707–721.

Australian Bureau of Statistics. (2006). *Causes of Death, Australia, 2006*. <http://www.abs.gov.au/AUSSTATS/abs@.nsf/0/4D3E7129BC863F0ECA25757C00137471?opendocument> (accessed 17-12-2012).

Australian Institute of Health and Welfare. (2006). *Mortality over the twentieth century in Australia: Trends and patterns in major causes of death. Mortality Surveillance Series Number 4*. Canberra: Australian Institute of Health and Welfare.

Australian Institute of Health and Welfare, Australasian Association of Cancer Registries. (2008). *Cancer in Australia 2010: an overview*. Cancer series no. 42. Cat. no. CAN 56. Canberra: Australian Institute of Health and Welfare.

Axelrod, R. (1984). *The Evolution of Cooperation*. New York: Basic Books.

Axelrod, R. and Hamilton, W.D. (1981). The evolution of cooperation. *Science*, 211, 1390-6.

Ballard, T.J., Saltzman, L.E., Gazmararian, J.A., Spitz, A.M., Lazorick, S. and Marks, J.S. (1998). Violence during Pregnancy: Measurement Issues. *American Journal of Public Health*, **88**, 274-276.

Barlow, L. (1992). The Androgynous Writer: Another Point of View. In J.A. Krentz (ed.) *Dangerous Men and Adventurous Women: Romance Writers on the Appeal of the Romance*. Philadelphia: University of Philadelphia Press.

Batten, M. (1994 [1992]). *Sexual Strategies: How Females Choose Their Mates*. New York: G.P. Putnam's Sons.

Bearman, P.S. and Moody, J. (2004). Suicide and Friendships Among American Adolescents. *American Journal of Public Health*, **94**, 89-95.

Becker, C.S. (1988) *Unbroken Ties: Lesbian Ex-lovers*. Boston: Alyson.

Bellomo, R.V. (1994). Methods of determining early hominid behavioral activities associated with the controlled use of fire at FxJj 20 Main, Koobi Fora, Kenya. *Journal of Human Evolution*, **27**, 173–195.

Bellugi, U., Marks, S., Bihrle, A. and Sabo, H. (1988). Dissociation between language and cognitive functions in Williams syndrome. In D. Bishop and K. Mogford (eds.) *Language Development in Exceptional Circumstances* (pp. 177-189). London: Churchill Livingstone.

Benenson, J.F. and Christakos, A. (2003). The Greater Fragility of Females' Versus Males' Closest Same-Sex Friendships. *Child Development*, **74**, 1123-1129.

Berger, D.T. and Trinkaus, E. (1995). Patterns of Trauma among the Neandertals. *Journal of Archaeological Science*, **22**, 841-852.

Berndt, R.M. and Berndt, C.H. (1951). *Sexual Behavior in Western Arnhem Land*. New York: Viking Fund.

Berndt, R.M. (1976). *Love Songs of Arnhem Land*. West Melbourne, Vic.: Thomas Nelson (Australia).

Berndt, R.M. and Berndt C.H. (1977). *The World of the First Australians* (2nd edn.) Sydney: Ure Smith.

Beyrer, C., Wirtz, A.L., Walker, D., Johns, B., Sifakis, F. and Baral, S.D. (2011). *The Global HIV Epidemics Among Men Who Have Sex with Men*. Washington: The World Bank.

Billings, E. and Westmore, A. (2000). *The Billings Method: Using the Body's Natural Signal of Fertility to Achieve or Avoid Pregnancy* (rev. edn.). Richmond, Vic.: A. O'Donovan.

Birkhead, T.R. (2000). *Promiscuity: an Evolutionary History of Sperm Competition and Sexual Conflict*. London: Faber.

Björkvist, K. (1994) Sex Differences in Physical, Verbal, and Indirect Aggression: A Review of Recent Research. *Sex Roles*, **30**, 177-188.

Björkvist, K., Lagerspetz, K.M.J. and Kaukainen, A. (1992) Do Girls Manipulate and Boys Fight? Development Trends in Regard to Direct and Indirect Aggression. *Aggressive Behavior*, **18**, 117-127.

Blanchard, R. (2001). Fraternal birth order and the maternal immune hypothesis of male homosexuality. *Hormones and Behavior*, **40**, 105-114.

Blanchard, R., Barbaree, H.E., Bogaert, A.F., Dickey, R., Klassen, P., Kuban, M.E. and Zucker, K.J. (2000). Fraternal birth order and sexual orientation in pedophiles. *Archives of Sexual Behavior*, **29**, 463-478.

Bleske-Rechek, A. and Lighthall, M. (2010). Attractiveness and Rivalry in Women's Friendships with Women. *Human Nature,* **21**, 82-97.

Bouchard, T.J. Jr. (1994). Genes, environment, and personality. *Science*, **264**, 1700-02.

Bouchard T.J. Jr. (1998). Genetic and environmental influences on adult intelligence and special mental abilities. *Human Biology*, **70**, 257-280.

Bouchard T.J. Jr., Lykken, D.T., McGue, M., Segal, N.L. and Tellegen, A. (1990). Sources of human psychological differences: the Minnesota study of twins reared apart. *Science*, **250**, 223-228.

Bourke, J. (2007). *Rape: Sex, Violence, History*. Emeryville, CA: Shoemaker & Hoard.

Boyd, R. and Richerson, P.J. (1992). Punishment allows the evolution of cooperation (or anything else) in sizeable groups. *Ethology and Sociobiology*, **13**, 171-195.

Brahmam, G.N., Kodavalla, V., Rajkumar, H., Rachakulla, H.K., Kallam, S., Myakala, S.P., Paranjape, R.S., Gupte, M.D., Ramakrishnan, L., Kohli, A. and Ramesh, B.M. (2008). Sexual practices, HIV and sexually transmitted infections among self-identified men who have sex with men in four high HIV prevalence states of India. *AIDS*, **22 Suppl 5**, S45-57.

Brennan, P.A. and Kendrick, K.M. (2006). Mammalian social odours: attraction and individual recognition. *Philosophical Transactions of the Royal Society B.* **361**, (1476), 2061-2078.

Brickley, M., Mays, S. and Ives, R. (2008). Evaluation and interpretation of residual rickets deformities in adults. *International Journal of Osteoarchaeology*, **20**, 54-66.

Buhrmester, D. and Prager, K. (1995). Patterns and functions of self-disclosure during childhood and adolescence. In K.J. Rotenberg (1995). *Disclosure Processes in Children and Adolescents*. New York: Cambridge University Press, 10-56

Burbank, V.K. (1994). *Fighting Women: Anger and Aggression in Aboriginal Australia*. Berkeley: University of California Press.

Burling, R. (2005). *The Talking Ape: How Language Evolved*. New York: Oxford University Press.

Burr, J.E., Ostrov, J.M. and Jansen, E.A. (2005) Relational Aggression and Friendship During Early Childhood: "I Won't Be Your Friend!" *Early Education & Development*, **16**, 162-183.

Buss, D.M. (2000). The Dangerous Passion: Why Jealousy is as Necessary as Love and Sex. New York: Free Press.

Buss, D.M. (2001). Human nature and culture: an evolutionary psychological perspective. *Journal of Personality*, **69**, 955-978.

Butler, R. A. (1954). Incentive conditions which influence visual exploration. *Journal of Experimental Psychology*, **48**, 19-23.

Butler, R.A. (1960). Acquired drives and the curiosity-investigative motives. In R.H. Waters, D.A. Rethlingshafer and W.E. Caldwell (ed.) *Principles of Comparative Psychology* (pp. 144-176). New York: McGraw-Hill, p. 164-165.

Butler, R. A. and Harlow, H. F. (1954). Persistence of visual exploration in monkeys. *Journal of Comparative and Physiological Psychology*, **47**, 258-263.

Butler, R. A. and Harlow, H. F. (1957). Discrimination learning and learning sets to visual exploration incentives. *The Journal of General Psychology*, **57**, 257-264.

Butterworth, B., Reeve, R., Reynolds, F. and Lloyd D. (2008). Numerical thought with and without words: Evidence from Indigenous Australian children. *Proceedings of the National Academy of Sciences of the USA*, **105**, 13179-13184.

Caldwell, M.A. and Peplau, L.A. (1982) Sex Differences in Same-Sex Friendship. *Sex Roles*, **8**, 721-732.

Calvin, W. H. (1983). A stone's throw and its launch window: timing precision and its implications for language and hominid brains. *Journal of Theoretical Biology*. **104**, 121-35.

Campbell, A. (2002). *A Mind of Her Own: the Evolutionary Psychology of Women*. New York: Oxford University Press.

Campbell, A. (2004). Female Competition: Causes, Constraints, Content and Contexts. *The Journal of Sex Research*, **41**, 16-26.

Cantor, J. M., Blanchard, R., Paterson, A. D. and Bogaert, A. F. (2002). How many gay men owe their sexual orientation to fraternal birth order? *Archives of Sexual Behavior*, **31**, 63-71.

Chagnon, N.A. (1974). *Studying the Yąnomamö*. New York: Hold, Rinehart and Winston.

Chagnon, N.A. (1977). *Yąnomamö: The Fierce People* (2nd edn.). New York: Harcourt Brace Jovanovich.

Chagnon, N.A. (1979a). Mate competition, favoring close kin, and village fissioning among the Yanomamö Indians. In N.A. Chagnon and W. Irons (eds.), *Evolutionary Biology and Human Social Behavior* (pp. 86-131). North Scituate, M.A.: Duxbury Press.

Chagnon, N.A. (1979b). Is reproductive success equal in egalitarian societies? In N.A. Chagnon and W. Irons (eds.), *Evolutionary Biology and Human Social Organisation: an Anthropological Perspective*. North Scituate, M.A.: Duxbury Press, 374-401.

Chagnon, N.A. (1980). Kin-selection theory and the Yąnomamö Indians. In G.W. Barlow and J. Silverberg (eds.) *Sociobiology: Beyond Nature/Nurture? Special Symposium Publication No. 35, American Association for the Advancement of Science* (pp. 545-574). Boulder: Westview Press.

Chagnon, N.A. (1981). Terminological kinship, genealogical relatedness and village fissioning among the Yanomamö Indians. In R.D. Alexander and D.W. Tinkle (eds.), *Natural Selection and Social Behaviour: Recent Research and New Theory* (pp. 490-508). New York: Chiron Press, Oxford.

Chagnon, N.A. (1982). Sociodemographic attributes of nepotism in tribal populations: man the rule-Breaker. In King's College Sociobiology Group (eds.), *Current Problems in Sociobiology* (pp. 291-318). Cambridge: Cambridge University Press.

Chagnon, N.A. (1988a). Life histories, blood revenge, and warfare in a tribal population. *Science*, **239**, 985-992, p. 987.

Chagnon, N.A. (1988b). Male Yąnomamö manipulations of kinship classifications of female kin for reproductive advantage. In L. Betzig, M. Borgerhoff Mulder and P. Turke (eds.), *Human Reproductive Behaviour: A Darwinian Perspective* (pp. 23-48). Cambridge: Cambridge University Press.

Chagnon, N.A. (1992a). *Yąnomamö* (4th edn.). Forth Worth: Harcourt Brace Jovanovich.

Chagnon, N.A. (1992b). *Yanomamö: the Last Days of Eden*. San Diego: Harcourt Brace Jovanovich.

Chagnon, N.A. (1996). Chronic problems in understanding tribal violence and warfare. In G.R. Bock and J.A. Goode (eds.), *Genetics of Criminal and Antisocial Behavior. Ciba Foundation Symposium 194* (pp. 202-232). Chichester, New York: Wiley.

Chagnon, N.A. (2000). Manipulating kinship rules: a form of male Yanomamö reproductive competition. In L. Cronk and N.A. Chagnon (eds.), *Adaptation and Human Behavior: an Anthropological Perspective* (pp. 115-132). New York: Aldine de Gruyter.

Chaix, R., Cao, C. and Donnelly, P. (2008). Is mate choice in humans MHC-dependent? *PLoS Genetics.* **4 (9)**, e1000184.

Chatwin, B. (1987). *Songlines.* London : Jonathon Cape.

Cialdini, R.B. (1993). *Influence: The Psychology of Persuasion* (rev. edn.). New York: Quill.

Cicero (1971). *On the Good Life.* Translated by Grant, M. Harmondsworth: Penguin.

Clutton-Brock, T.H. and Parker, G.A. (1995). Punishment in animal societies. *Nature,* **373**, 209-216.

Clutton-Brock, T.H. and Parker, G.A. (1995). Sexual coercion in animal societies. *Animal Behaviour,* **49**, 1345-65.

CNN, *Koko the gorilla calls for the dentist.* Aug 8, 2004. <http://primates.com/gorillas/koko.html> (accessed Feb 2, 2010).

Collins English Dictionary (1972). London and Glasgow: Collins.

Connolly, B. (2005). *Making 'Black Harvest': Warfare, Film-making and Living Dangerously in the Highlands of Papua New Guinea .* Sydney: ABC Books for the Australian Broadcasting Corporation.

Conner, R.C., Read, A.J. and Wrangham, R. (2000). Male reproductive strategies and social bonds. In J. Mann, R.C. Connor, P.L. Tyack and H. Whitehead (eds.), *Cetacean Societies: Field Studies of Dolphins and Whales* (pp. 247-269). Chicago : University of Chicago Press.

Constable, J., Ashley, M., Goodall, J. and Pusey, A. (2001). Noninvasive paternity assignment in Gombe chimpanzees. *Molecular Ecology,* **10**, 1279–1300.

Coontz, S. (2005). *Marriage, A History: from Obedience to Intimacy or How Love Conquered Marriage.* New York: Viking.

Craig, P., Halavatau, V., Comino, E. and Caterson, I. (2001). Differences in body composition between Tongans and Australians: time to rethink the healthy weight ranges? *International Journal of Obesity,* **25**, 1806-1814.

Craviari, T., Pettifor, J.M., Thacher, T.D., Meisner, C., Arnaud, J. and Fischer, P.R. (2008). Rickets: An Overview and Future Directions, with Special Reference to Bangladesh. *Journal of Health Population Nutrition.* **26**, 112–121.

Crenshaw, T.L. (1984). *Your Guide to Better Sex.* London: Book Club Associates.

Crenshaw, T.L. (1996). *The Alchemy of Love and Lust: Discovering Our Sex Hormones and How They Determine Who We Love, When We Love, and How Often We Love.* New York: Pocket Books.

Crick, N.R., Casas, J.F. and Ku, H. (1999). Relational and physical forms of peer victimization in preschool. *Developmental Psychology,* **35**, 376-385.

Crick, N.R. and Grotpeter, J.K. (1995). Relational Aggression, Gender, and Social-Psychological Adjustment. *Child Development,* **66**, 710-722.

Crosby, A.W. (2002). *Throwing Fire: Projectile Technology through History.* Cambridge: Cambridge University Press.

Daly, M. and Wilson, M. (1982). Homicide and Kinship. *American Anthropologist,* **84**, 372-378.

Daly, M. and Wilson, M. (1998). *The Truth about Cinderella: a Darwinian View of Parental Love.* London: Weidenfelt & Nicolson.

Darlington, P.J. Jr. (1979). Group Selection, altruism, reinforcement, and throwing in human evolution. *Proceedings of the National Academy of Sciences of the United States of America.* **72, No. 9**, 3748-3752.

Darwin, C. (2004 [1871]). *The Descent of Man, and Selection in Relation to Sex.* London: Penguin.

Dawkins, R. (1989). *The Selfish Gene* (rev. edn.). Oxford: Oxford University Press.

Dawkins, R. (1999 [1982]). *The Extended Phenotype.* Oxford: Oxford University Press.

Day, E. (2004). Sex really is all in the mind for women. *Sunday Telegraph,* February 29. < http://www.telegraph.co.uk/news/1455661/Sex-really-is-all-in-the-mind-for-women.html> (accessed, Nov. 13, 2010).

de Waal, F.B.M. (1998 [1982]). *Chimpanzee Politics*. Baltimore, Maryland: John Hopkins University Press.

Dennell, R. (1997). The world's oldest spears. *Nature*, **385**, p. 767-768.

Di Natale, C. Macagnano, A., Paolesse, R., Tarizzo, E., Mantini, A. and D'Amico, A. (2000). Human skin odor analysis by means of an electronic nose. *Sensors and Actuators*, **65**, 216-219.

Diamond, J.M. (1997). *Why is Sex Fun?: the Evolution of Human Sexuality*. London: Weidenfield and Nicholson.

Diamond, L.M. (2000). Sexual identity, attractions, and behavior among young sexual-minority women over a 2-year period. *Developmental Psychology*, **36**, 241–250.

Diamond, L.M. (2003). Was it a phase? Young women's relinquishment of lesbian/bisexual identities over a 5-year period. *Journal of Personality and Social Psychology*, **84**, 352–364.

Dietary Supplement Fact Sheet: Vitamin D. Office of Dietary Supplements. National Institutes of Health.

Dijkstra, J.K., Cillessen, A.H.N. and Borch, C. (2012). Popularity and Adolescent Friendship Networks: Selection and Influence Dynamics. *Developmental Psychology*, Advance online publication. doi: 10.1037/a0030098

Dijkstra, J.K., Lindenberg, S., Verhulst, F.C., Ormal, J. and Veenstra, R. (2009). The Relation Between Popularity and Aggressive, Destructive, and Norm-Breaking Behaviors: Moderating Effects of Athletic Abilities, Physical Attractiveness, and Prosociality. *Journal on Research on Adolescence*, **19**, 401-413.

Dixson, A.F. (2009). *Sexual Selection and the Origin of Human Mating Systems*. New York: Oxford.

Doob, A.N. and Gross, A.E. (1968). Status of frustrator as an inhibitor of horn-honking responses. *Journal of Social Psychology*, **76**, 213-18.

Doty, R.L. (2010). *The Great Pheromone Myth*. Baltimore: John Hopkins University Press.

Duby, G. (1994 [1988]). *Love and Marriage in the Middle Ages*, translated by Jane Dunnett. Cambridge: Polity.

Dunbar, R. I. M. (1993). Coevolution of neocortical size, group size and language in humans. *Behavioral and Brain Sciences*, **16**, 681-735.

Dunbar, R.I.M. (1998). *Grooming, Gossip, and the Evolution of Language*. Cambridge, Mass.: Harvard University Press.

Eberhard, W.G. (1996). *Female Control: Sexual Selection by Cryptic Female Choice*. Princeton, NJ: Princeton University Press.

Einarsen, S. and Raknes. B.I. (1997). Harassment at work and the vicimization of men. *Victims and Violence*, **12**, 247-263.

Einarsen, S. and Skogstad, A. (1996). Bullying at work: epidemiological findings in public and private organizations. *European Journal of Work and Organizational Psychology*, **5**, 185-202.

Einstein, A. (2001 [1920]). *Relativity: the special and general theory*. Translated by Robert W. Lawson. London: Routledge.

Ekman, P. (1998). Afterword to *The Expression of Emotions in Man and Animals*, by C. Darwin (pp. 363-393). London: HarperCollins.

Ellis, B.J. (1992). The evolution of sexual attraction: evaluative mechanisms in women. In J.H. Barkow, L. Cosmides, and J. Tooby (eds.), *The Adapted Mind*. New York: Oxford University Press, p 267-288.

Elman, R.A. (1996). *Sexual subordination and state intervention: comparing Sweden and the United States*. Oxford: Berghahn Books.

Eyre, E. (1964 [1845]). *Journals of expeditions of discovery into Central Australia, and overland from Adelaide to King George's Sound, in the years 1840-1*, volume 2. Adelaide: Libraries Board of South Australia.

Finkelhor, D. and Yllo, K. (1985). *License to Rape: Sexual Abuse of Wives*. New York: Holt, Rinehart & Winston.

Fischer, P.R., Rahman, A., Cimma, J.P., Kyaw-Myint, T.O., Kabir, A.R., Talukder, K., Hassan, N., Manaster, B.J., Staab, D.B., Duxbury, J.M., Welch, R.M., Meisner, C.A., Haquem, S. and Combs G.F. Jr. (1999). Nutritional rickets without vitamin D deficiency in Bangladesh. *Journal of Tropical Pediatrics*, **45**, 291-3.

Flaccus, G. (2009). Jumbo squid invade San Diego shores, spook divers. *The Associated Press*. July 16, 2009. <http://www.breitbart.com/article.php?id=D99FNF0O0> (accessed Nov 14, 2010).

Fox, C.A. and Fox., B. (1971). A comparative study of coital physiology with special reference to the sexual climax. *Journal of Reproduction and Fertility*, **24**, 319–336.

Fox, C. A., Wolff, H. S. and Baker, J. A. (1970). Measurement of intra-vaginal and intra-uterine pressures during human coitus by radio-telemetry. *Journal of Reproduction and Fertility*, **22**, 243–251.

Franzoi, S.L. and Herzog, M.E. (1987). Judging physical attractiveness: what body aspects do we use? *Personality and Social Psychology Bulletin*, **13**, 19–33.

Fraser, A. (1984). *The Weaker Vessel: Woman's Lot in Seventeenth-Century England*. London: Weidenfeld and Nicolson.

Gardner, R. and Heider, K.G. (1969). *Gardens of War: Life and Death in the New Guinea Stone Age*. London: Deutsch.

Gerressu, M., Mercer, C.H., Graham, C.A., Wellings, K., Johnson, A.M. (2008). British National probability data on masturbation prevalence and associated factors. *Archives of Sexual Behavior*, **37**, 266-278.

Gibson, C. and Jung, K. (2002). *Historical Census Statistics on Population Totals By Race, 1790 to 1990, and By Hispanic Origin, 1970 to 1990, For The United States, Regions, Divisions, and States*. Population Division, U. S. Census Bureau, Working Paper Series No. 56. < http://www.census.gov/population/www/documentation/twps0056/twps0056.html> (retrieved, 19 December 2012)

Glass, D. (1995). *All My Fault: Why Women Don't Leave Abusive Men*. London: Virago.

Go, V.F., Srikrishnan, A.K., Sivaram, S., Murugavel, G.K., Galai, N., Johnson, S.C., Sripaipan, T., Solomon, S. and Celentano, D.D. (2004). High HIV prevalence and risk behaviors in men who have sex with men in Chennai, India. *Journal of Acquired Immune Deficiency Syndrome*, **35**, 314–319.

Goetz, A.T., Shackelford T.K. and Schipper, L.D. (2006). Adding Insult to Injury: Development and Initial Validation of the Partner-Directed Insults Scale. *Violence and Victims*, **21 (6)**, 691-706.

Goldberg, T.L. and Wrangham, R.W. (1997). Genetic correlates of social behavior in wild chimpanzees: Evidence from mitochondrial DNA. *Animal Behaviour*, **54**, 559-580.

Goldin-Meadow, S. (2005). Watching language grow. *Proceedings of the National Academy of Sciences*, **102**, 2271–2272.

Goldin-Meadow, S. (2007). On Inventing Language. *Daedalus*, **137**, 100-104.

Gneezy, U., Niederle, M. and Rustichini, A. (2003). Performance in competitive environments: gender differences. *Quarterly Journal of Economics*, **118**, 1049–74.

Goodale, J.C. (1971 [1980]). *Tiwi Wives: a Study of the Women of Melville Island, North Australia*. Seattle: University of Washington Press.

Goodall, J. (1986). *The Chimpanzees of Gombe: Patterns of Behaviour*. Cambridge: Cambridge University Press.

Goodall, J. (1999 [1988]). *In the Shadow of Man*. London: Phoenix.

Goodall, J. (2000 [1990]). *Through a Window: Thirty Years with the Chimpanzees of Gombe*. London: Phoenix.

Goodall, J., Bandora, A., Bergmann, E., Busse, C. Matama, H., Mponga, E., Pierce, A. and Riss, D. (1979). Inter-community interactions in the chimpanzee population of the Gombe National Park. In D.A. Hamburg and E.R. McCown (eds.) *The Great Apes* (pp. 13-54). Menolo Park: Benjamin/Cummings.

Goodenow, C., Szalacha, L.A., Robin, L.E. and Westheimer, K. (2008). Dimensions of Sexual Orientation and HIV-Related Risk Among Adolescent Females: Evidence from a Statewide Survey. *American Journal of Public Health*, **98**, 1051-1058.

Goodman, M. (1997). *The Roman World, 44BC - AD 180*. London: Routledge.

Goren-Inbar, N., Alperson, N., Kislev, M.E., Simchoni, O., Melamed, Y., Ben-Nun, A. and Werker, E. (2004). Evidence of hominin control of fire at Gesher Benot Ya'aqov, Israel. *Science*, **304**, 725-727.

Greiling, H. and Buss, D.M. (2000). Women's sexual strategies: the hidden dimension of extra- pair mating. *Personality and Individual Differences*, **28**, 929-963.

Gupta, A., Mehta, S., Godbole, S.V., Sahay, S., Walshe, L., Reynolds, S.J., Ghate, M., Gangakhedkar, R.R., Divekar, A.D., and Risbud, A.R. (2006). Same-sex behavior and high rates of HIV among men attending sexually transmitted infection clinics in Pune, India (1993-2002). *Journal of Acquired Immune Deficiency Syndrome*, **43**, 483–490.

Hallpike, C.R. (1977). *Bloodshed and Vengeance in the Papuan Mountains: the Generation of Conflict in Tauade Society*. Oxford: Clarendon Press.

Halpern, D.F. (2000). *Sex differences in cognitive abilities* (3rd edn.). Mahwah, N.J.: L. Erlbaum Associates.

Hamer, D. and Copeland, P. (1998). *Living with Our Genes*. London: Pan Books.

Hamilton W.D. (1964). The genetical evolution of social behaviour (I and II). *Journal of Theoretical Biology*. 7, 1-16; 17-52.

Harlow, H.F. (1950). Learning and satiation of response in intrinsically motivated complex puzzle performance by monkeys. *Journal of Comparative and Physiological Psychology*, **43**, 289-294.

Harris, L.J. (1956). Vitamin D and Bone. In G.H. Bourne (ed.), *The Biochemistry and Physiology of Bone* (pp. 581–622). New York: Academic Press.

Hawkes, K., O'Connell, J. F. and Blurton Jones, N.G. (1997). Hadza women's time allocation, offspring provisioning, and the evolution of long postmenopausal life spans. *Current Anthropology*, 38, 551-577.

Hawkes, K., O'Connell, J.F. and Blurton Jones, N. (1989). Hard-working Hadza grandmothers. In V. Standen and R. Foley (eds.), *Comparative Socioecology: the Behavioural Ecology of Human and Other Mammals* (pp. 341-66). Oxford: Oxford University Press.

Hays, H.R. (1964). *The Dangerous Sex: the Myth of Feminine Evil*. New York: G.P. Putnam's Sons.

Hedrick, P.W. and Black, F. L. (1997). HLA and mate selection: no evidence in South Amerindians. *American Journal of Human Genetics*. **61**, 505-511.

Heider, K. (1979). *Grand Valley Dani: Peaceful Warriors*. New York: Holt, Rinehart and Winston.

Hershman, P. (1981). *Punjabi Kinship and Marriage*. Delhi: Hindustan Pub. Corp. (India).

Herz, R.S. and Cahill, E.D. (1997). Differential use of sensory information in sexual behavior as a function of gender. *Human Nature*, **8**, 275–286.

Hesiod (1914). *Works And Days*, translated by H.G. Evelyn-White, lines 69-82. <http://www.sacred-texts.com/cla/hesiod/works.htm> (accessed April 5, 2010).

Hiatt, L.R. (1996). *Arguments about Aborigines: Australia and the Evolution of Social Anthropology*. Cambridge, England: Cambridge University Press.

Highfield, R. (2008). We are natural born mathematicians. Telegraph, August 18. <http://www.telegraph.co.uk/science/science-news/3349815/We-are-natural-born-mathematicians.html> (accessed Nov. 13, 2010.)

Hill, K. and Hurtado, M. (1996). *Ache Life History: the Ecology and Demography of a Foraging People*. New York: Aldine de Gruyter.

Hill, K. and Kaplan, H. (1988). Tradeoffs in male and female reproductive strategies among the Ache. Parts 1 and 2. In L. Betzig, M. Borgerhoff Mulder and P. Turke (eds.), *Human Reproductive Behavior: A Darwinian Perspective*. Cambridge: Cambridge University Press.

Hinsz, V.B. (1989). Facial resemblance in engaged and married couples. *Journal of Social and personal Relationships*, **6**, 223-229.

Holick, MF. (1991) Photosynthesis, metabolism, and biologic actions of vitamin D. In F.H. Glorieux (ed.) *Rickets*. Vol 2. New York: Raven Press, p. 1–22.

Howitt, A.W. (1996 [1904]). *Native Tribes of South-East Australia*. Canberra: Aboriginal Studies Press.

Jablonksi, N.G. (2006). *Skin: A natural history*. Berkeley, CA: University of California Press.

Jeffreys, M.D.W. (1965). The Hand Bolt. *Man*, **65**, 153-154.

Jensen, A.R. (1998). *The g Factor: the Science of Mental Ability*. Westport, Conn.: Praeger.

Joseph, S. (1997). *She's My Wife, He's Just Sex*. Australian Centre for Independent Journalism, University of Technology, Sydney.

Johanson, D. and Edgar, B. (2006) *From Lucy to Language*. New York: Simon and Schuster.

Johnson, A. (2006). My Lai massacre hero buried. *The Australian*, January 12.

Johnson, H. (1995). Risk Factors Associated With Non-Lethal Violence Against Women by Marital Partners. In C. Block and R. Block, *Trends, Risks, and Interventions in Lethal Violence: Proceedings of the Third Annual Spring Symposium of the Homicide Research Working Group*. Georgia, Atlanta: U.S. Department of Justice, 151-168.

Johri, M., Morales, R.E, Boivin, J., Samayoa, B.E., Hoch, J.S., Grazioso, C.F., Barrios Matta, I.J., Sommen, C., Baide Diaz, E.L., Fong, H.R. and Arathoon, E.G. (2011). Increased risk of miscarriage among women experiencing physical or sexual intimate partner violence during pregnancy in Guatemala City, Guatemala:

cross-sectional study. *BMC Pregnancy and Childbirth*, **11**, 49.

Kaberry, P.M. (1939). *Aboriginal Woman: Sacred and Profane*. London: George Routledge.

Kabir, M.L., Rahman, M., Talukder, K., Rahman, A., Hossain, O., Mostafa, G., Mannan, M.A., Kumar, S. and Chowdhury, A.T. (2004). Rickets among children of a coastal area of Bangladesh. *Mymensingh Medical Journal*, **13**, 53–58.

Kagan, J. (2004). The uniquely human in human nature. *Daedalus*, **133**, 77-88.

Kanin, E.J. (1957). Male aggression in dating-courtship relations. *The American Journal of Sociology*, **63**, 197-204.

Kaplan, H., Hill, K., Lancaster, J. and Magdalena Hurtado, A. (2000). A theory of human life history evolution: diet, intelligence, and longevity. *Evolutionary Anthropology*, **9**, 156-85.

Keeley, L.H. (1996). *War Before Civilisation*. New York: Oxford University Press.

Keller, M.C., Thiessen, D. and Young, R.K. (1996). Mate assortment in dating and married couples. *Personality and Individual Differences*. **21 (2)**, 217-221.

Kennedy, G.E. (2003). Palaeolithic grandmothers? Life history theory and early Homo. *Journal of the Royal Anthropological Institute*, **9**, 549-572.

King, M.L. (1991). *Women of the Renaissance*. Chicago: The University of Chicago Press.

Kinsey, A.C., Pomeroy, W.B. and Martin C.E. (1948). *Sexual Behaviour in the Human Male*. Philadelphia: W.B. Saunders.

Kinsey, A.C., Pomeroy, W.B., Martin, C.E. and Gebhard, P.H. (1953). *Sexual Behavior in the Human Female*. Philadelphia: W.B. Saunders Co.

Klitzinger, C. (1996). Toward a politics of lesbian friendship. In J.S. Weinstock and E.D. Rothblum (eds.), *Lesbian Friendships: for Ourselves and Each Other* (pp. 295 – 299). New York University Press: New York.

Kocoska-Maras, L., Zethraeus, N., Rådestad, A.F., Ellingsen, T., von Schoultz, B., Johannesson, M. and Hirschberg, A.L. (2011). A randomized trial of the effect of testosterone and estrogen on verbal fluency, verbal memory, and spatial ability in healthy postmenopausal women. *Fertility and Sterility*, **95**, 152-7.

Krentz, J.A. (1992). Trying To Tame The Romance. In J.A. Krentz (ed.) *Dangerous Men and Adventurous Women: Romance Writers on the Appeal of the Romance*. Philadelphia: University of Philadelphia Press.

Kummer, H. (1995). *In Quest of the Sacred Baboon: A Scientist's Journey*. Translated by M. Ann Biederman-Thorson. Princeton, N.J.: Princeton University Press.

Kumta, S., Lurie, M., Weitzen, S., Jerajani, H., Gogate, A., Row-kavi, A., Anand, V., Makadon, H. and Mayer, K.H. (2010). Bisexuality, sexual risk taking, and HIV prevalence among men who have sex with men accessing voluntary counselling and testing services in Mumbai, India. *Journal of Acquired Immune Deficiency Syndromes*, **53**, 227-233.

Kruger, J. and Dunning, D. (1999). Unskilled and Unaware of It: How Difficulties in Recognizing One's Own Incompetence Lead to Inflated Self-Assessments. *Journal of Personality and Social Psychology*, **77**, 1121-34.

Lagerspetz, K.M.J., Björkvist, K. and Peltonen,T. (1988). Is indirect aggression typical of females? Gender differences in aggressiveness in 11- to 12- year-old children. *Aggressive Behavior*, **14**, 403-414.

Leach, E. (1970). *Lévi-Strauss*. London: Fontana/Collins.

Leblanc, W. (1999). *Naked Motherhood: Shattering Illusions and Sharing Truths*. Milsons Point, N.S.W.: Random House Australia.

Lee, R. (1968). What Hunters Do for a Living, or, How to Make Out on Scarce Resources. In *Man the Hunter*. (eds R. Lee and I. DeVore) Chicago: Aldine.

Lee, R. (1969). Kung Bushmen Subsistence: An Input-Output Analysis. In A. Vayda (ed.), *Environment and Cultural Behaviour*. New York: Natural History Press.

Lévi-Strauss, C. (1969). *The Elementary Structures of Kinship* (rev. edn.). Translated by Bell, J.H. and von Sturmer, J. R. and edited by Needham, R. Boston: Beacon.

Leyden, J.L., McGinley, K.J., Hölzle, E., Labows, J.N. and Kligman, A.M. (1981). The Microbiology of the Human Axilla and Its Relationship to Axillary Odor. *Journal of Investigative Dermatology*, **77**, 413-416.

Lieberman, P. (1998). *Eve Spoke: Human Language and Human Evolution*. New York: W.W. Norton.

Little, A.C., Penton-Voak, I.S. and Burt, D.M. (2003). Investigating an imprinting-like phenomenon in humans: Partners and opposite-sex parents have similar hair and eye colour. *Evolution and Human Behavior*, **24 (1)**, 43-51.

Lloyd, E.A. (2005). *The Case of the Female Orgasm: Bias in the Science of Evolution*. Cambridge, Mass.: Harvard University Press.

London, S. (2006). Lesbian, Bisexual Women: Misperceptions of Risk Jeopardize Sexual Health. *Perspectives on Sexual and Reproductive Health*, **38**, 53-54.

Mackay, J. (2000) *The Penguin Atlas of Sexual Behaviour*. New York: Penguin Reference.

McKee, A., Albury, K. and Lumby, C. (2008). *The Porn Report*. Carlton, Vic.: Melbourne University Publishing.

McKibben, W.F., Goetz, A.T., Shackelford, T.K., Schipper, L.D., Starratt, V.G. and Stewart-Williams, S. (2007). Why do men insult their intimate partners? *Personality and Individual Differences*, **43**, 231-241.

McNamee, C. (2001). Rape. In R.J. Simon (ed.) *A Comparative Perspective on Major Social Problems*. Lanham, Maryland: Lexington Books.

Malinowski, B. (1963 [1913]). *The Family among the Australian Aborigines: a Sociological Study*. New York: Schocken Books.

Marenky, R.K., Kuroda, S., Vineberg, E.O. and Wrangham, R.W. (1996). The Significance of Terrestrial Herbaceous Foods for Bonobos, Chimpanzees, and Gorillas. In R.W. Wrangham, W.C. McGrew, F.B.M. de Waal and P.G. Heltne (eds.) *Chimpanzee Cultures* (pp. 59-75). Cambridge, Mass.: Harvard University Press.

Marlowe, F.W. (2005). Hunter-Gatherers and Human Evolution. *Evolutionary Anthropology*, **14**, 54-67.

Masters, W.H. and Johnson, V.E. (1966). *Human Sexual Response*. Boston: Little, Brown and Company.

Maynard, D.B., and Springer-Proverbs, R.F. (2010). An Examination of Gender Differences in Characteristics of Barbadian Adolescents' Same-sex Friendships. *Caribbean Journal of Psychology*, **3**, 40-53.

Mazenod, B., Pugeat, M. and Forest, M.G. (1988) Hormones, sexual function and erotic behavior in women. In J.M.A. Sitsen. (ed.), *Handbook of Sexology. Vol. 6: The Pharmacology and Endocrinology of Sexual Function* (pp. 316-351). New York: Elsevier Science Publishers.

Medawar, J. and Pyke, D. (2000). *Hitler's Gift: Scientists Who Fled Nazi Germany*. London: Richard Cohen Books.

Meeus, W.H.J. and Raaijmakers, Q.A.W. (1986). Administrative obedience: carrying out orders to use psychological-administrative violence. *European Journal of Social Psychology*, **16**, 311-24.

Meggitt, M.J. (1974 [1962]). *Desert People: A Study of the Walbiri Aborigines of Central Australia*. Sydney: Angus & Robertson.

Meston, C.M. and Frohlich, P.F. (2003). Love at first fright: partner salience moderates roller-coaster-induced excitation transfer. *Archives of Sexual Behavior*, **32: 6**, 537–544.

Milgram, S. (1963). Behavioural study of obedience. *Journal of Abnormal and Social Psychology* **67**, 371-78.

Milgram, S. (1974). *Obedience to Authority: an Experimental View*. London : Tavistock Publications.

Mitani, J.C., Watts, D.P. and Muller, M.N. (2002). Recent developments in the study of wild chimpanzee behavior. *Evolutionary Anthropology*, **11**, 9-25, p. 12-14.

Morgan, J. (2002 [1852]). *The Life and Adventures of William Buckley*. Melbourne: The Text Publishing Company.

Morrison, R.L. (2009). Are Women Tending and Befriending in the Workplace? Gender Differences in the Relationship Between Workplace Friendships and Organizational Outcomes. *Sex Roles*, **60**, 1-13.

Mortimer, S.T. and Swan, M.A. (1995). Variable kinematics of capacitating human spermatozoa. *Human Reproduction*, **10**, 3178–3182.

Moshkin, M., Litvinova, N., Litvinova, E.A., Bedareva, A., Lutsyuk, A. and Gerlinskaya, A. (2012). Scent Recognition of Infected Status in Humans. *The Journal of Sexual Medicine*, **9**, 3211-3218.

Muller, M.N., Emery Thompson, M. and Wrangham, R.W. (2006). Male chimpanzees prefer mating with old females. *Current Biology*, **16**, 2234-2238.

Muller, M.N. and Wrangham, R.W. (2001) The reproductive ecology of male hominoids. In P.T. Ellison (ed.) *Reproductive ecology and human evolution* (pp. 397-427). New York: Aldine.

Myers, J.E.B. (2005). *Myers on Evidence in Child, Domestic and Elder Abuse Cases*, volume 1, fourth edition. New York, Aspen Publishers.

Myles, S., Lea, R.A., Ohashi, J., Chambers, G.K., Weiss, J.G., Hardouin1, E., Engelken, J., Macartney-Coxson, D.P., Eccles, D.A., Naka, I., Kimura, R., Inaoka, T., Matsumura, Y. and Stoneking, M. (2011). Testing the thrifty gene hypothesis: the Gly482Ser variant in PPARGC1A is associated with BMI in Tongans. *BMC Medical Genetics*, **12**: 10.

Neel, J.V. (1962) Diabetes mellitus: A "thrifty" genotype rendered detrimental by "progress"? *American Journal of Human Genetics*, **14**, 353-62.

Neel, J.V. (1977). Health and disease in unacculturated Amerindian populations. *Ciba Foundation Symposium*, **49**, 155-68.

Neel, J.V., Weder, A. B. and Julius, S. (1998). Type II diabetes, essential hypertension, and obesity as 'syndromes of impaired genetic homeostasis': the 'thrifty genotype' hypothesis enters the 21st century. *Perspectives in Biology and Medicine*, **42**, 44-74.

Niederle, M. and Vesterlund, L. (2006). Do Women Shy Away From Competition? Do Men Compete Too Much? *Quarterly Journal of Economics*, **122**, 1067-1101.

Niederle, M. and Vesterlund, L. (2011). Gender and Competition. *Annual Review in Economics*, **3**, 601–30.

Nishida, T. (1979). The social structure of chimpanzees of the Mahale Mountains. In D.A. Hamburg and E.R. McCown (eds.) *The Great Apes* (pp. 73-122). Menolo Park: Benjamin/Cummings.

Norris, M. and Pawlowski, W. (2010). (Eds). *Business of Consumer Book Publishing 2010*. Stamford, C.T.: Simba Information.

North Gulf Oceanic Society, (2010). Killer whales of southern Alaska. http://www.whalesalaska.org/killer_whales_of_southern_alaska.html (Retrieved January 31, 2010).

O'Brien, E. M. (1981). The Projectile capabilities of an Acheulian handaxe from Olorgesailie. *Current Anthropology*, **22**: **1**, 76-79.

O'Connell, J.F., Hawkes, K. and Blurton Jones, N. (1988). Hadza scavenging: implications for plio/pleistocene hominid. *Current Anthropology*, **29**, 356-363.

O'Connor, D.B., Archer, J., Hair, W.M. and Wu, F.C. (2001). Activational effects of testosterone on cognitive function in men. *Neuropsychologia*, **39**, 1385-94.

Österman, K., Björkqvist, K., Lagerspetz, K. M. J., Kaukiainen, A., Landau, S. F., Frączek, A. and Caprara, G. V. (1998). Cross-cultural evidence of female indirect aggression. *Aggressive Behavior*, **24**, 1–8.

Paagel, M. and Bodmer, W. (2003) A naked ape would have fewer parasites. *Proceedings of the Royal Society of London: Biological Sciences*. (Suppl) **270**: S117-S119.

Painter, K. and Farrington, D.P. (1999). Wife Rape in Great Britain. In R. Muraskin (ed.) *Women and Justice: development of international policy*. Amsterdam: Gordon and Breach.

Parkin, R. (1997). *Kinship: an Introduction to Basic Concepts*. Oxford, UK: Blackwell.

Penton-Voak, I., Perrett, D. and Pierce, J. (1999) Computer graphic studies of the role of facial similarity in attractiveness judgements. *Current Psychology*, **18**, 104-117.

Petitto, L. A. (2005). How the brain begets language: On the neural tissue underlying human language acquisition. In J. McGilvray (ed.), *The Cambridge Companion to Chomsky* (pp. 84-101). Cambridge, Great Britain: Cambridge University Press.

Pillsworth, E.G., Haselton M.G. and Buss D.M. (2004). Ovulatory shifts in female sexual desire. *The Journal of Sexual Desire*, **41**, 55-65.

Pinhasi, R., Shaw, P., White, B. and Ogden A.R. (2006). Morbidity, rickets and long-bone growth in post-medieval Britain--a cross-population analysis. *Annals of Human Biology*, **33**, 372-389.

Pinker, S. (2002). *The Blank Slate: The Modern Denial of Human Nature*. London: Penguin

Pinker, S. and Jackendoff, R. (2005). The faculty of language: what's special about it? *Cognition*, **95**, 201–236. Hauser, M.D., Chomsky, N. and Fitch, W.T. (2002). The faculty of language: what is it, who has it, and how did it evolve? *Science*, **298**, 1569–1579.

Pontzer, H. and Wrangham, R.W. (2004). Climbing and the daily energy cost of locomotion in wild chimpanzees: implications for hominoid locomotor evolution. *Journal of Human Evolution*, **46**, 315-333.

Price, S. (1984). *Co-Wives and Calabashes*. Ann Arbor: University of Michigan Press.

Pusey, A.E. (2001). Of genes and apes: chimpanzee social organization and reproduction. In F.B.M. de Waal (ed.) *Tree of Origin: What Primate Behavior Can Tell Us about Human Social Evolution* (pp. 9-37). Cambridge, Mass.: Harvard University Press.

Pusey, A.E. and Packer, C. (1997). The ecology of relationships. In J.R. Krebs and N.B. Davies (eds.), *Behavioural Ecology: An Evolutionary Approach*, 4th edn., (pp. 254-283). Oxford: Blackwell Science.

Quine, L. (2001).Workplace Bullying in Nurses. *Journal of Health Psychology*, **6**, 73-84.

Radcliffe-Brown, A.R., (1975 [1950]). Introduction to A.R. Radcliffe-Brown and D. Forde, (eds.), *African Systems of Kinship and Marriage*. London: Oxford University Press.

Radliff, K.M. and Joseph, L.M. (2011). Girls Just Being Girls? Mediating Relational Aggression and Victimization. *Preventing School Failure*, **55**, 171-179.

Regan, P.C. and Berscheid, E. (1995). Gender differences in beliefs about the causes of male and female sexual desire. *Personal Relations*, **2**, 345–358.

Roach, M. (2008). *Bonk: the Curious Coupling of Sex and Science*. Melbourne: Text Publishing.

Roberts, S.C., Little, A.C., Gosling, L.M., Jones, B.C., Perrett, D.I., Carter, V. and Petrie, M. (2005) MHC-assortative facial preferences in humans. *Biology Letters*, **1 (4)**: 400-403.

Robins, A.H. (1991). *Biological Perspectives on Human Pigmentation*. Cambridge: Cambridge University Press.

Roes, F. (1998). An interview of Napoleon Chagnon. *Human Ethology Bulletin*, **13 (4)**, 6-12.

Rose, S.D., Zand, D. and Cini M.A. (1993) Lesbian courtship scripts. In E.D. Rothblum and K.A. Brehony (eds.), *Boston Marriages: Romantic but Asexual Relationships among Contemporary Lesbians* (pp. 70-85). Amherst: University of Mass. Press.

Rossen, M. Klima, E.S., Bellugi, U., Bihrle, A. and Jones, W. (1996). Interaction between language and cognition: Evidence from Williams syndrome. In J.J. Beitchman, N.J Cohen, M.M. Konstantareas, and R. Tannock (eds.), *Language, learning and behavior disorders: developmental, biological, and clinical perspectives* (pp. 367-392). New York: Cambridge University Press.

Roth, D.E., Rashed Shah, R., Black, R.E. and Baqui, A.H. (2010). Vitamin D Status of Infants in Northeastern Rural Bangladesh: Preliminary Observations and a Review of Potential Determinants. *Journal of Health Population Nutrition*. **28**, 458–469.

Rumbaugh, D.M., Savage-Rumbaugh, E.S. and Sevcik, R.A. (1996). Biobehavioral roots of language: a comparative perspective of chimpanzee, child and culture. In R.W. Wrangham, W.C. McGrew, F.B.M. de Waal, and P.G. Heltne (eds.), *Chimpanzee Cultures* (pp. 319-334). Cambridge: Harvard University Press.

Russell, D.E.H. (1990). *Rape in Marriage*, (rev. edn.). Bloomington: Indiana University Press.

Sadalla, E.K., Kenrick, D.T. and Vershure, B. (1987). Dominance and heterosexual attraction. *Journal of Personality and Social Psychology*, **52**, 730-738.

Sahlins, M. (1972). *Stone-age Economics*. Chicago: Aldine-Atherton.

Saran, A. B. (1974). *Murder and Suicide among the Munda and the Oraon*. Delhi: National Publishing House.

Sarkar, N.N. (2008). The impact of intimate partner violence on women's reproductive health and pregnancy outcome. *Journal of Obstetrics and Gynaecology*, **28**, 266-271.

Savage-Rumbaugh, E.S. and Lewin, R. (1994). *Kanzi: The Ape at the Brink of the Human Mind*. New York: John Wiley and Sons.

Savage-Rumbaugh, E.S., Murphey, J., Sevcik, R.A., Brakke, K.E., Williams, S.L. and Rumbaugh, D.M. (1993). Language comprehension in ape and child. *Monographs of the Society for Research in Child Development*. Serial No. 233, **58**, 3-4.

Senghas, A. and Coppola, M. (2001). Children creating language: How Nicaraguan sign language acquired a spatial grammar. *Psychological Science*, **12**, 323-328.

Sengahs, A., Kita, S. and Özyürek, A. (2004). Children creating core properties of language: Evidence from an emerging sign language in Nicaragua. *Science*, **305**, 1779-1782.

Sereny, G. (2001). *The German Trauma: Experiences and Reflections 1938-2001*. London: Penguin.

Shackelford, T.K., Pound, N. and Goetz, A.T. (2005). Psychological and physiological adaptations to sperm competition in humans. *Review of General Psychology*, **9**, 228-248.

Shakespeare, W. (1914). Romeo and Juliet. In *The Oxford Shakespeare*. London: Oxford University Press.

Shamu, S., Abrahams, N., Temmerman, M., Musekiwa, A. and Zarowsky, C. (2011). Intimate Partner Violence in Pregnancy in Africa. *PLoS ONE*, **6**, e17591.

Short, R.V. (1979). Sexual selection and its component parts, somatic and genital selection, as illustrated by man and the great apes. *Advances in the Study of Behaviour*, **9**, 131-158.

Sinclair, D. and Dangerfield, P. (1998). *Human Growth After Birth* (6th edn.). Oxford: Oxford Medical Publications.

Solomon, S.S., Mehta, S.H, Latimore, A., Srikrishnan, A.K. and Celentano, D.D. (2010). The impact of HIV and high-risk behaviours on the wives of married men who have sex with men and injection drug users: implications for HIV prevention. *Journal of the International AIDS Society*, **13(Suppl 2)**, S7.

Spencer, B. (1914). *Native Tribes of the Northern Territory of Australia*. London: Macmillan and Co.

Spencer, B. and Gillen, F.J. (1968 [1899]). *The Native Tribes of Central Australia*. New York: Dover Publications.

Spencer, B. (1982). *The Aboriginal Photographs of Baldwin Spencer Selected and Annotated by Geoffrey Walker*. Melbourne: Currey O'Neil.

Stanford, J.B., White, G.L. and Hatasaka, H. (2002). Timing intercourse to achieve pregnancy: current evidence. *Obstetrics and Gynecology*, **100**, 1333-41.

Stiles, D. (2001). Hunter-gatherer studies: the importance of context. *African Study Monographs*, Suppl.26: 41-65, p. 55-56.

Stringer, C and Andrews, P. (2005). *The Complete World of Human Evolution*. London: Thames and Hudson.

Stringer, C. and McKie, R. (1996). *African Exodus: the Origins of Modern Humanity*. New York: Henry Holt and Company.

Suarez, S.S. and Pacey, A.A. (2006). Sperm transport in the female reproductive tract. *Human Reproduction Update*, **12**, 23-37.

Sulaiman, T. (2006). Mother-in-law made to pay £35,000 for inflicting four wretched months. *The Times*, July 25. <http://women.timesonline.co.uk/tol/life_and_style/women/article692147.ece> (accessed March 23, 2010).

Sullivan, K. and Tager-Flusberg, H. (1999). Second-Order Belief Attribution in Williams Syndrome: Intact or Impaired? *American Journal on Mental Retardation*, **104**, 523-532.

Swaney, W.T. and Keverne, E.B. (2009) The evolution of pheromonal communication. *Behavioural Brain Research*. **200 (2)**, 239-247.

Tannen, D. (1990). *You Just Don't Understand: Women and Men in Conversation*. London: Virago.

Temkin, J. (2002). *Rape and the Legal Process* (2nd edn.). Oxford: Oxford.

Terrace, H.S., Sanders, R.J. and Bever, T.G. (1979). Can an Ape create a sentence? *Science*, **206**, 891-902.

The Australian Women's Weekly (1980), February 13.

The Electronic Text Corpus of Sumerian Literature. (2006). *A hymn to Inana (Inana C)*. <http://etcsl.orinst. ox.ac.uk/cgi-bin/etcsl.cgi?text=t.4.07.3> (accessed March, 29, 2010).

The Guardian. (2008) Sats: Table of Sats results - boys v girls. August, 12. <http://www.guardian.co.uk/ education/table/2008/aug/12/sats.results.boysgirls> (accessed Nov. 14, 2010).

Thieme, H. (1997). Lower palaeolithic hunting spears from Germany. *Nature*, **385**, 807-810.

Thompson, M.E., Kahlenberg, S.M., Gilby, I.C. and Wrangham, R.W. (2007). Core area quality is associated with variance in reproductive. *Animal Behaviour*, **73**, 510-512.

Townsend, J.M. and Levy, G.D. (1990). Effects of potential partners' costume and physical attractiveness on sexuality and partner selection. *Journal of Psychology*, **124**, 371-389.

Trinkaus, E. (2012). Neandertals, early modern humans, and rodeo riders. *Journal of Archaeological Science*, **39**, 3691-3693.

Trivers, R.L. (1971). The evolution of reciprocal altruism. *Quarterly Review of Biology*, 46, 35-57.

Trivers, R.L. (1972). Parental investment and sexual selection. In B. Campbell (ed.), *Sexual Selection and the Descent of Man*. Chicago: Aldine-Atherton.

Trivers, R. (1985). *Social Evolution*. Menlo Park, Calif.: Benjamin/Cummings Pub. Co.

Tutin, C.E.G. (1996). Reproductive Success Story: Variability Among Chimpanzees and Comparisons with Gorillas. In R.W. Wrangham, W.C. McGrew, F.B.M. de Waal, and P.G. Heltne (eds.), *Chimpanzee cultures* (pp. 181-193). Cambridge, Mass.: Published by Harvard University Press.

Tylor, E. (1880). On the methods of investigating the development of institutions applied to the laws of marriage and descent. *Journal of the Anthropological Institute*, **18**, 245-272.

United Nations Population Fund (UNFPA) and EngenderHealth (2003). *Obstetric Fistula Needs Assessment Report: Findings from Nine African Countries.* New York: United Nations Population Fund.

Valero, H. (1970). *Yanoámo: The Narrative of a White Girl Kidnapped by Amazonian Indians, as Told to Ettore Biocca.* Translated by Dennis Rhodes. New York: Dutton.

van Schaik, C.P. and van Hooff, J.A.R.A.M. (1996). Toward an understanding of the orangutan's social system. In W.C. McGrew, L.F. Marchant, and T. Nishida (eds.) *Great Ape Societies* (pp. 3-15). Cambridge: Cambridge University Press.

Varma, S. C. (1978). *The Bhil Kills.* Delhi: Kunj Publishing House.

Verma, R. and Collumbien, M. (2004). Homosexual activity among rural Indian men: implications for HIV interventions. *AIDS*, **18**, 1845-47.

Vetere, V.A. (1982) The role of friendship in the development and maintenance of lesbian love relationships. *Journal of Homosexuality*, **8 (2)**, 51-65.

Warner, W.L. (1969 [1958]). *A Black Civilization: a Study of an Australian Tribe* (rev. edn.). Gloucester, Mass.: P. Smith.

Watson, J.D. and Crick, F.H.C. (1953). A Structure for Deoxyribose Nucleic Acid. *Nature*, **171**, 737-738.

Watts, D.P., Muller, M., Amsler, S.J., Mbazi, G. and Mitani, J.C. (2006). Lethal intergroup aggression by chimpanzees in Kibale National Park, Uganda. *American Journal of Primatology*, **68**, 161–180.

Wedekind, C. and Füri, S. (1997). Body odour preferences in men and women: do they aim for specific MHC combinations or simply heterozygosity? *Proceedings of the Royal Society of London, B*, **264**, 1471-1479.

Wedekind, C., Seebeck, T., Bettens, F. and Paepke, A. J. (1995). MHC-dependent mate preferences in humans. *Proceedings of the Royal Society of London , B*, **260**, 245-249.

Weinstock, J.S. and Rothblum, E.D. (1996) *Lesbian Friendships: for Ourselves and Each Other.* New York: New York University Press.

Weisberg, P., Scanlon, K.S., Li, R. and Cogswell, M.E. (2004). Nutritional rickets among children in the United States: review of cases reported between 1986 and 2003. *American Journal of Clinical Nutrition*, **80** (6 Suppl), 1697S-1705S.

Whipple, B., Ogden, G. and Komisaruk, B.R. (1992). Physiological correlates of imagery-induced orgasm in women. *Archives of Sexual Behavior*, **21**, 121-133.

Whitehead, H. and Mann, J. (2000). Female reproductive strategies of cetaceans: life histories and calf care. In J. Mann, R.C. Connor, P.L. Tyack and H. Whitehead (eds.), *Cetacean societies: field studies of dolphins and whales* (pp. 219-246). Chicago: University of Chicago Press.

Wikan, U. (1982). *Behind the Veil in Arabia: Women in Oman.* Baltimore, Maryland: John Hopkins University Press.

Wilson, E.O. (1998). Consilience among the great branches of learning. *Daedalus*, **127 (1)**, 131-149.

Wilson, M. and Daly, M. (1985). Competitiveness, Risk-Taking, and Violence: The Young Male Syndrome. *Ethology and Sociobiology*, **6**, 59-73.

Wilson, M. and Daly, M. (1993). Spousal Homicide Risk and Estrangement. *Violence and Victims*, **8**, 3-16.

Wilson, M. and Daly, M. (1995). Uxoricide. In C. Block and R. Block, *Trends, Risks, and Interventions in Lethal Violence: Proceedings of the Third Annual Spring Symposium of the Homicide Research Working Group.* Georgia, Atlanta: U.S. Department of Justice, 169-178.

World Health Organization, Regional Office for South-East Asia. (2010). *HIV/AIDS among men who have sex with men and transgender populations in South-East Asia: The Current Situation and National Responses.* Singapore: WHO Regional Office for South East Asia.

Wrangham, R.W., Wilson, M.L. and Muller M.N. (2006). Comparative rates of violence in chimpanzees and humans. *Primates*, **47**, 14-26.

Wright, L. (1997). *Twins: Genes, environment and the mystery of identity*. London: Phoenix.

Wright, W. (1999). *Born That Way: Genes, behavior, personality*. London: Routledge.

Zitzmann, M. and Nieschlag, E. (2001). Testosterone levels in healthy men and the relation to behavioural and physical characteristics: facts and constructs. *European Journal of Endocrinology*, **144**, 183-197.

Zoe, W., Stitt, M.D., Lowell, A. and Goldsmith, M.D. (1995). Scratch and sniff: the dynamic duo. *Archives of Dermatology*, **131**, 997-999.

Index

www.ingramcontent.com/pod-product-compliance
Lightning Source LLC
Chambersburg PA
CBHW081346280326
41927CB00042B/3078